SCHAUM'S OUTLINE OF

THEORY AND PROBLEMS

OF

MOLECULAR and CELL BIOLOGY

•

WILLIAM D. STANSFIELD, Ph.D.
Emeritus Professor of Biological Sciences
California Polytechnic State University
at San Luis Obispo

JAIME S. COLOMÉ, Ph.D.
Professor of Microbiology
Biological Sciences Department
California Polytechnic State University
at San Luis Obispo

RAÚL J. CANO, Ph.D.
Professor of Microbiology
Biological Sciences Department
California Polytechnic State University
at San Luis Obispo

SCHAUM'S OUTLINE SERIES

McGRAW-HILL

New York   San Francisco   Washington, D.C.   Auckland   Bogotá   Caracas   Lisbon
London   Madrid   Mexico City   Milan   Montreal   New Dehli
San Juan   Singapore   Sydney   Tokyo   Toronto

WILLIAM D. STANSFIELD has degrees in Agriculture (B.S., 1952), Education (M.A., 1960), and Genetics (M.S., 1962; Ph.D., 1963; University of California at Davis). His published research is in immunogenetics, twinning, and mouse genetics. From 1957 to 1959 he was an instructor in high school vocational in agriculture. He was a faculty member of the Biological Sciences Department of California Polytechnic State University from 1963 to 1992 and is now Emeritus Professor. He has written university-level textbooks in evolution andserology/immunology, and has coauthored a dictionary of genetics.

JAIME S. COLOMÉ obtained his Ph.D. in Molcular Biology at the University of California in Santa Barbara. He has taught Microbiology, Genetics, Cell Physiology, and Molecular Biology at California Polytechnic State University in San Luis Obispo for 23 years. Dr. Colomé is a coauthor of *Microbiology, Essentials of Microbiology,* and *Laboratory Exercises in Microbiology.* He is also the author of a number of papers on the history of Science. Presently, Dr. Colomé is researching the molecular biology of viroids and hepatitis D virus.

RAÚL J. CANO is a Professor of Microbiology in the Biological Sciences Department at California Polytechnic State University, San Luis Obispo. A Cuban born, he earned his B.S. in Biology in 1970, an M.S. in 1972, both from Eastern Washington University, Cheney, WA, and his Ph.D. in Microbiology from the University of Montana, Missoula, MT in 1974. His research focuses on molecular aspects of bacterial evolution and microbial diversity. His published research has centered on the evolution of host-parasite relationships utilizing organisms entombed in amber as his experimental model. He has published more than 50 articles in peer-reviewed journals such as *Science, Nature,* and *Applied and Environmental Microbiology,* and has written two widely-used textbooks in Microbiology and a laboratory manual.

Schaum's Outline of Theory and Problems of
MOLECULAR AND CELL BIOLOGY

7 8 9 10 11 12 13 14 15 16 17 18 19 20 VFM VFM 0 9 8 7 6 5 4

ISBN 0-07-060898-9

Sponsoring Editor: Arthur Biderman
Production Supervisor: Pamela A. Pelton
Editing Supervisor: Maureen Walker

Library of Congress Cataloging-in-Publication Data

Stansfield, William D., date
   Schaum's outline of theory and problems of molecular and cell
biology / William D. Stansfield, Jaime S. Colomé, Raúl J. Cano.
      p.   cm. — (Schaum's outline series)
   Includes index.
   ISBN 0-07-060898-9
   1. Molecular biology — Outlines, syllabi, etc.   2. Cytology —
Outlines, syllabi, etc.   I. Colomé, Jaime S.   II. Cano, Raúl J.
III. Title.
QH506.S76   1996
574.87'02'02 — dc20
                                                            96-18244
                                                            CIP

McGraw-Hill

A Division of The McGraw·Hill Companies

Preface

Ever since James Watson and Francis Crick proposed their double-helical structure of DNA in 1953, biology has been in the throes of a revolution in knowledge at the molecular level. The field has now grown so vast that texts over 800 pages are common. These ponderous tomes try to be comprehensive, but in including so much, present major difficulties for the beginner. How is one to know what fundamentals are needed for a preliminary orientation? This *Schaum's Outline of Molecular and Cell Biology* is designed to provide just such an introduction. It rapidly surveys the field to meet the needs of a one-quarter or one-semester college or university course in molecular biology, and provides the preparation necessary for entry into advanced molecular biology courses.

The format used in this Outline presents essential information in the form of didactic questions. Instead of paragraph after paragraph of text, this Outline presents each concept as an answer to a specific question. In biology, this format was first used in the Schaum's Outline Series in *Human Anatomy and Physiology* by Kent M. Van De Graaff and R. Ward Rhees (1987), and then in *Zoology* by Nancy M. Jessop (1988).

In outlining any subject, difficult choices must be made as to what will be included, and this means that virtually no one will be completely satisfied with the result. However, with three coauthors, the choices made have been examined from at least three different viewpoints. We have tried to ask important and engaging questions and to provide some plausible answers. But the answers to these essay-type questions should not be considered definitive for two major reasons. First, in attempting to provide brief answers to complex problems, oversimplification and incompleteness are unavoidable. Second, given the tentativeness of scientific knowledge, particularly that in a fast-growing field such as molecular biology, some answers may be out of date before this book gets into print.

Following the didactic questions are objective questions of several kinds (terms, multiple choice, true-false, etc.). These objective questions can be used as a kind of self-examination to evaluate the extent to which the terms, facts and concepts of the essay questions have been mastered, while also preparing the student for this kind of examination format.

This Outline is aimed primarily at a general student audience in the sophomore through senior college/university years. It is recommended that the beginning student of molecular biology have some basic knowledge of the biological sciences and biochemistry, at least equivalent to that of a year of college-level courses in the biological sciences. Courses in bacteriology and in general genetics would be of particular benefit. However, without all this recommended preparation, a student should find the information in this Outline comprehensible without the necessity of resorting to other references. The chapters should be read in numerical order because much of the information and many of the terms used in early chapters are required to understand topics in later chapters.

WILLIAM D. STANSFIELD   JAIME S. COLOMÉ   RAÚL J. CANO

Contents

Chapter 1

Cells: Their Structure, Metabolism, and Reproduction

INTRODUCTION

A cell is the smallest unit that exhibits all of the qualities we associate with the living state. Cells must obtain energy from some external source to carry on such vital processes as growth, repair, and reproduction. All of the chemical and physical reactions that occur in a cell to support these functions constitute its **metabolism.** Some of these metabolic reactions are **catabolic,** i.e., they break down complex organic molecules to release energy for driving **anabolic** reactions that build (synthesize) the molecules needed by the cell. Both of these types of metabolic reactions are catalyzed by **enzymes.** Enzymes are protein molecules that repeatedly accelerate specific biochemical reactions without being permanently altered or consumed in the process. The structure of each enzyme (or any other protein) is encoded by a segment of a deoxyribonucleic acid (DNA) molecule referred to as a **gene.**

Molecular biology is the science that studies all life processes in cells at the molecular level. In doing so, this science draws upon knowledge and techniques from several scientific disciplines, including biochemistry, cytology, genetics, physiology, microbiology, embryology, and evolution.

Wherever possible throughout this text an attempt has been made to present the similarities and differences between two major kinds of cells (bacteria vs. all other cells). In this chapter, the relationships between structures and biological functions in both of these cell types are presented. In addition, two fundamental properties of life, metabolism and reproduction, are outlined. Details of these and other life processes—such as the composition, function, and synthesis of biomacromolecules, storage and retrieval of genetic information, regulation of gene activity, cellular differentiation, intercellular communication, mutation, selection, and evolution—are presented in subsequent chapters.

CELLULAR ORGANIZATION

1.1. How many basic structural kinds of cells exist? What major structural difference distinguishes these cell types?

Structurally, there are two basic kinds of cells: (1) bacteria and (2) all other cells (including fungi, plants, and animals). Bacterial cells, although far from simple, are generally much smaller and less complex structurally than other cell types. The major difference is that the genetic material (DNA) of bacteria is not sequestered within a double-membrane structure called a **nucleus** (Fig. 1-1) as is that of other cell types. The first kind of cells that evolved on Earth were most likely of this type. Hence, they are called **prokaryotes** (procaryotes is an alternative spelling), which literally means "before there was a nucleus." In all other cell types (including fungi, plants, and animals), the DNA molecules are confined within a nuclear membrane. They are thus referred to as **eukaryotes** (eucaryotes), indicating that they are "truly nucleated."

1.2. Aside from DNA and certain associated proteins in chromosomes (discussed in Question 1.3), what macromolecular aggregates are shared by all prokaryotes and eukaryotes?

Only two kinds of such structures are shared by all kinds of cells. (1) All cells possess a lipid **plasma membrane** that separates the cell from its environment. The chemical nature of membranes, nucleic acids, and other

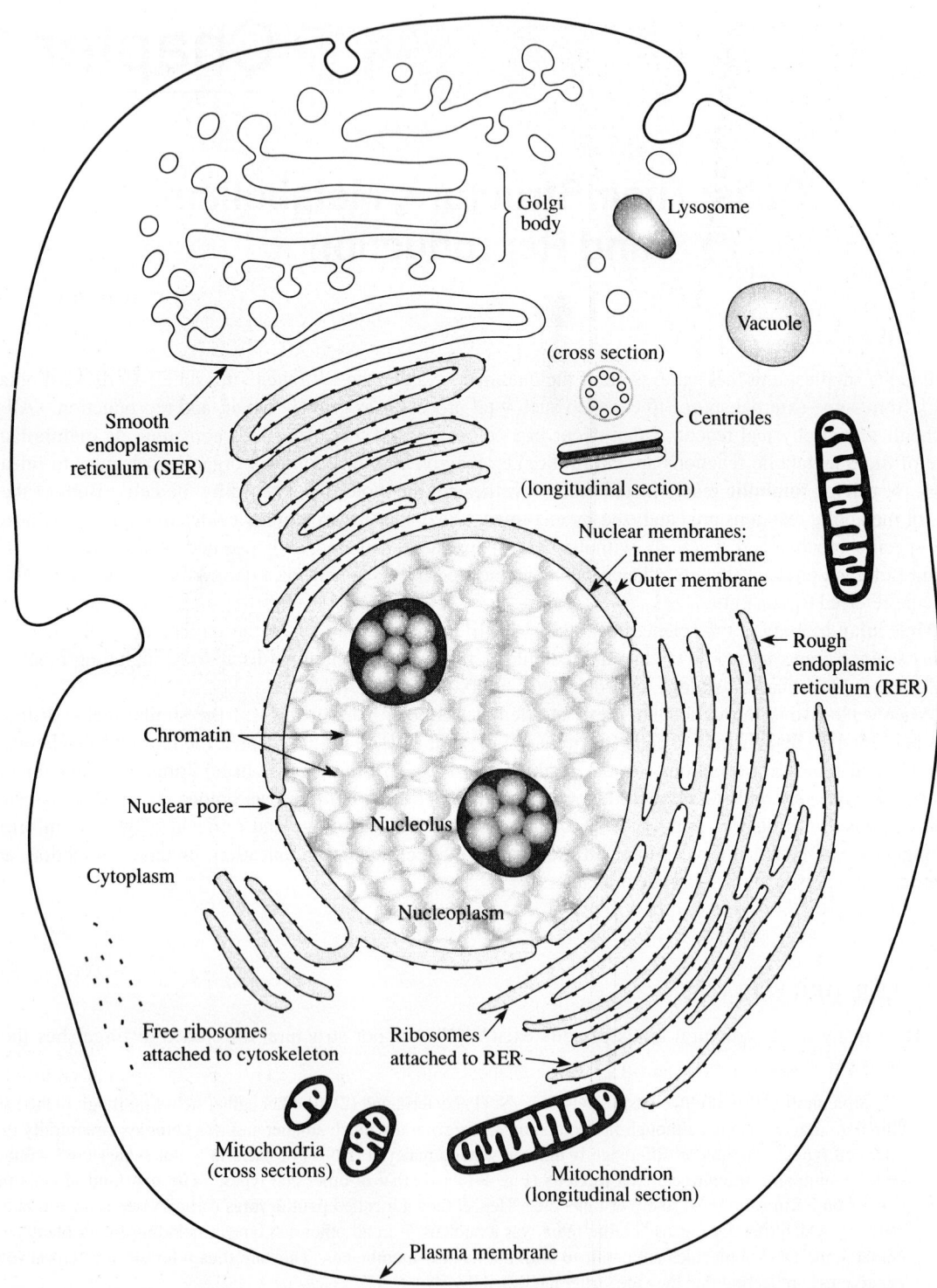

Fig. 1-1. (*a*) Diagram of an animal cell. (*From Schaum's Outline of Theory and Problems of Genetics, Third Edition, W. D. Stansfield, McGraw-Hill, 1991.*) Chromatin is explained in Question 1.3. The nine microtubules in the cross section of the centriole are fused triplets similar to the doublets of Fig. 1-3*a*. Compare this diagram with the illustration of a bacterial cell in Fig. 1-2*a*.

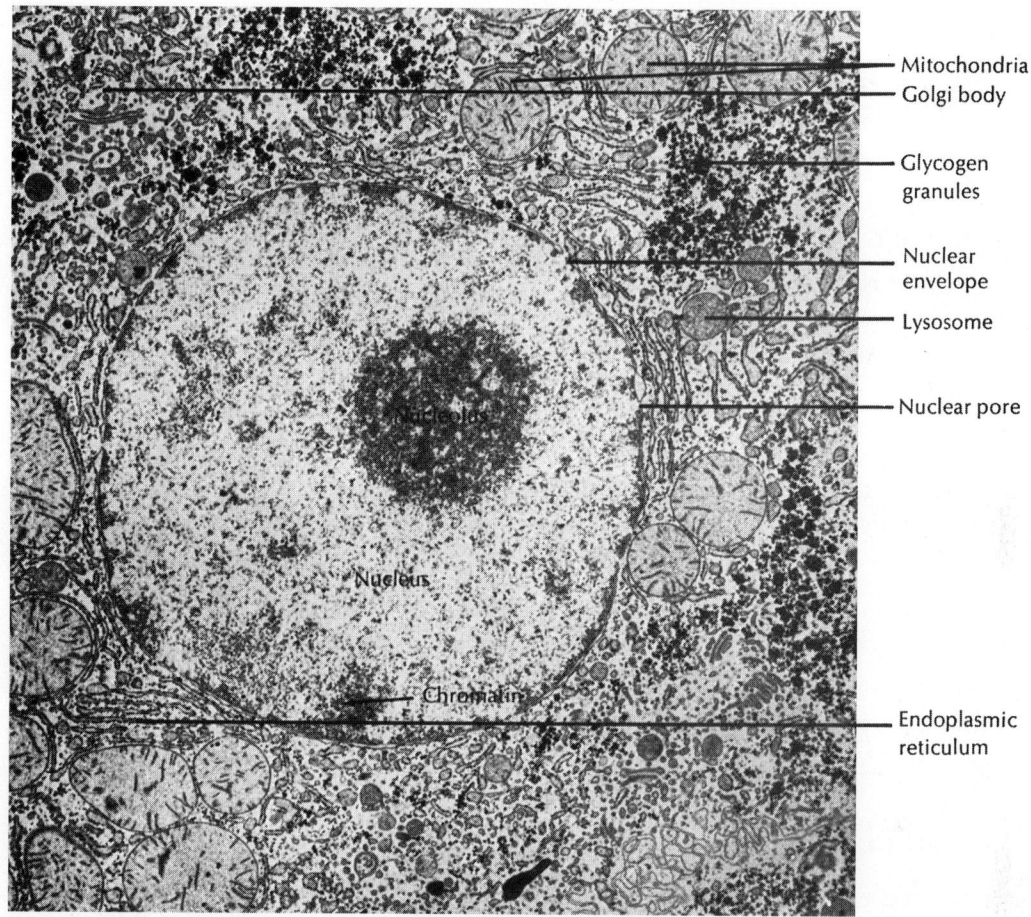

Mitochondria
Golgi body
Glycogen granules
Nuclear envelope
Lysosome
Nuclear pore
Endoplasmic reticulum

Nucleolus
Nucleus
Chromatin

Fig. 1-1. (*Cont.*) (*b*) Electron micrograph of a liver cell from a new-born rat ($\times$ 11,000).

biomacromolecules will be discussed in Chapter 2. (2) All cells have **ribosomes,** made partly of protein and partly of ribonucleic acid (RNA) molecules. Ribosomes function in the synthesis of proteins.

1.3. What structures are characteristic only of eukaryotes?

Among the cellular contents of eukaryotic cells are complex structures of characteristic morphology (appearance) and function called **organelles.** Some textbook authors define an organelle as any subcellular entity that can be isolated by high-speed centrifugation. Others consider only membrane-bound subcellular structures to be organelles.

The nucleus is an organelle that normally contains a complete set of genetic instructions (called a **genome**) in the form of multiple, linear structures called **chromosomes.** Chromosomes consist of the **nucleoprotein** called **chromatin,** i.e., the nucleic acid DNA and various proteins (mainly basic proteins called **histones**). A single, linear DNA molecule runs down the length of each chromosome. The details of chromosome structure, replication, and genetic recombination are presented in Chapter 3.

The intracellular contents between the nucleus and the plasma membrane is called **cytoplasm.** (The fluid portion of the cytoplasm outside the membrane-bound organelles is referred to as the **cytosol** or **hyaloplasm.**) Other organelles besides the nucleus that are surrounded by double membranes, such as mitochondria, and chloroplasts, reside in the cytoplasm. **Mitochondria** (mitochondrion, singular) contain the enzymes for oxidative respiration. **Chloroplasts** contain photosynthetic systems for utilizing the radiant energy of sunlight. With rare exceptions, mitochondria are found in all eukaryotic cells, whereas chloroplasts are found only in multicellular green plants and single-celled algae. Other kinds of nonphotosynthetic **plastids,** such as **chromoplasts** (containing pigments) and **leucoplasts** (starch storage), may be found in some plant cells. Syntheses of most amino acids, the

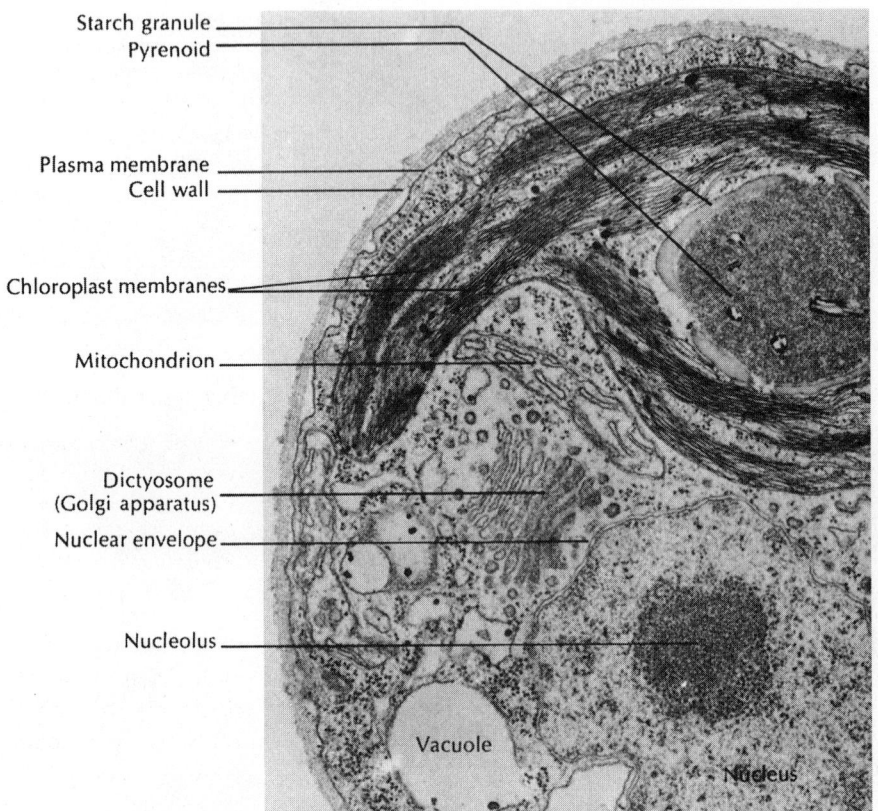

Starch granule
Pyrenoid

Plasma membrane
Cell wall

Chloroplast membranes

Mitochondrion

Dictyosome
(Golgi apparatus)
Nuclear envelope

Nucleolus

Vacuole   Nucleus

Fig. 1-1.   (*Cont.*) (*c*) Electron micrograph of an alga, *Chlamydomonas reinhardi*
(× 30,000). Pyrenoids are proteinaceous areas associated with some algal chloroplasts and may serve as sites for carbohydrate storage-product formation. [*Figs.
1-1b and 1-1c are courtesy of G. E. Palade, University of California, San Diego.*]

purine and pyrimidine components of nucleic acids, and all fatty acids occur in the plastids of plant cells, but in
the cytosol of animal cells.

Some cytoplasmic organelles, such as ribosomes and centrioles, are not surrounded by a membrane. **Centrioles** are associated with the cell's polar regions, toward which chromosomes migrate during cell division (see
Questions 1.9 and 1.41). They are characteristic of animal cells, and are missing from most plant cells. Centrioles lie within a mass of amorphous material called a **centrosome,** which is also devoid of a membrane.

Eukaryotic cells usually contain an extensive cytoplasmic labyrinth of membranes called the **endoplasmic
reticulum** (ER). The ER further compartmentalizes the cell and amplifies the surface area available for specialized biochemical reactions that occur at or across membranes.

Two other organelles are the Golgi bodies and vacuoles. Golgi bodies are discussed in Question 1.16. Vacuoles are the subject of the next question.

1.4.   Membrane-bound vacuoles are found in many eukaryotic cells. What functions do these vacuoles
serve?

Whereas the nucleus, mitochondrion, and chloroplast are each surrounded by a double membrane, the storage
compartments known as **vacuoles** have only a single membrane, called the **tonoplast.** Vacuoles come in different
sizes. The smaller vacuoles are usually referred to as **vesicles** if their main function is intracellular transport of
macromolecules. Vacuoles may arise by the progressive fusion of smaller vesicles emanating from membranes of
the Golgi complex. Several kinds of vacuoles may be found in various eukaryotic cells.

A **food vacuole** forms in some protozoans (single-celled animals such as amoebas) when they ingest bacteria or other small bits of food.   **Lysosomes** occur in the cytoplasm of some animal cells and contain precursors of digestive enzymes but are not considered to be vacuoles.   When a food vacuole coalesces with a lysosome, the enzyme precursors become activated to digest the food internally.

Osmosis is the movement of water (the **solvent**) through a differentially permeable membrane in response to differences in concentrations of dissolved substances (the **solute**) on either side of the membrane.   Many plant cells have one or more **water vacuoles** that absorb and store water because of the relatively high concentration of colloids in these organelles.   The hydrostatic (turgor) pressure created by water moving into these vacuoles pushes the rest of the cell contents and the plasma membrane against the rigid cell wall, thus making the plant tissues turgid.   Loss of such water during drought causes a plant to wilt, and if the loss is extensive the plant will die.   Some protozoans may have specialized water vacuoles, called **contractile vacuoles,** to pump excess water out of the cell and thus protect the cell from bursting like an overfilled balloon as a result of internal osmotic pressure.

1.5.   Aside from the plasma membrane, are prokaryotic cells devoid of other lipid membranes?

Most bacteria lack internal membranes.   However, photosynthetic bacteria usually have extensive internal membranes that arise from invaginations of the plasma membrane.   In some cases, these photosynthetic vesicles seem to be pinched off from the plasma membrane, forming membrane-bound vesicles.   At least some bacteria are thought to have one or more invaginations of the plasma membrane to which the bacterial DNA is attached.   These membranes are called **mesosomes.**   However, no bacteria have organelles surrounded by double membranes comparable to those in eukaryotes.

1.6.   DNA molecules in various kinds of cells may be relatively large or small, linear or circular, normally associated with histone proteins or not, contain mostly essential or nonessential information, and exist in low or high copy number per cell.   On the basis of these criteria, define two major classes of prokaryotic DNA and indicate where they are located in the bacterial cell.

The essential genetic information in bacteria resides in a single, circular DNA molecule called the **genophore** or genome, usually coding for 3,000 to 5,000 genes, depending on the species.   The genophore is loosely referred to as a "chromosome." In cells preparing to divide, a copy of the genophore is replicated; consequently, at least two copies are temporarily present, but normally not more than four, because each copy occupies about 15–25% of the cell's volume.   The region of the bacterial cell occupied by the genophore is referred to as its **nucleoid.** The region between the nucleoid and the plasma membrane is often referred to as the "cytoplasm."

Some bacteria contain one or more (up to 50 per cell) relatively small (containing one to a few hundred genes) circular DNA molecules called **plasmids** (Fig. 1-2).   The genes carried by plasmids are usually nonessential in most environmental conditions.   However, in the presence of an antibiotic such as penicillin, a plasmid carrying a gene that confers resistance against that antibiotic would be essential for life.   Plasmids are usually randomly distributed in the cytoplasm.

Histone proteins are not associated with the bacterial genophore or plasmids, but other types of proteins may take the place of histones in at least some bacterial species.

1.7.   In what respects are the ribosomes of prokaryotes different from those of eukaryotes?

All cells have structures called ribosomes that are not surrounded by, but may be associated with, membranes. Ribosomes function in the manufacture of proteins.   The ribosomes of bacteria are somewhat smaller than those that are free in the cytosol of eukaryotes.   However, the ribosomes found inside mitochondria and chloroplasts are about the same size as those in bacteria.   This ribosomal similarity is one line of evidence supporting the theory that mitochondria and chloroplasts may have evolved from prokaryotes (Chapter 14).

Eukaryotic ribosomes in intimate association with the endoplasmic reticulum give these membranes a granular appearance in electron micrographs, hence the name **rough ER.**   Portions of the ER free of ribosomes are called **smooth ER.**   Mitochondria and chloroplasts have their own small, circular, bacterialike DNA molecules and bacterialike ribosomes for production of some mitochondrial or chloroplast proteins and for synthesis of all ribonucleic acid molecules (rRNAs, mRNAs, and tRNAs) in these organelles.   Proteins used but not encoded by the DNA of these organelles are encoded by nuclear DNA, synthesized on cytoplasmic ribosomes, and transported into the interiors of mitochondria and chloroplasts.   Mitochondria and chloroplasts are thought to reproduce by binary fission, in much the same way that bacteria proliferate.

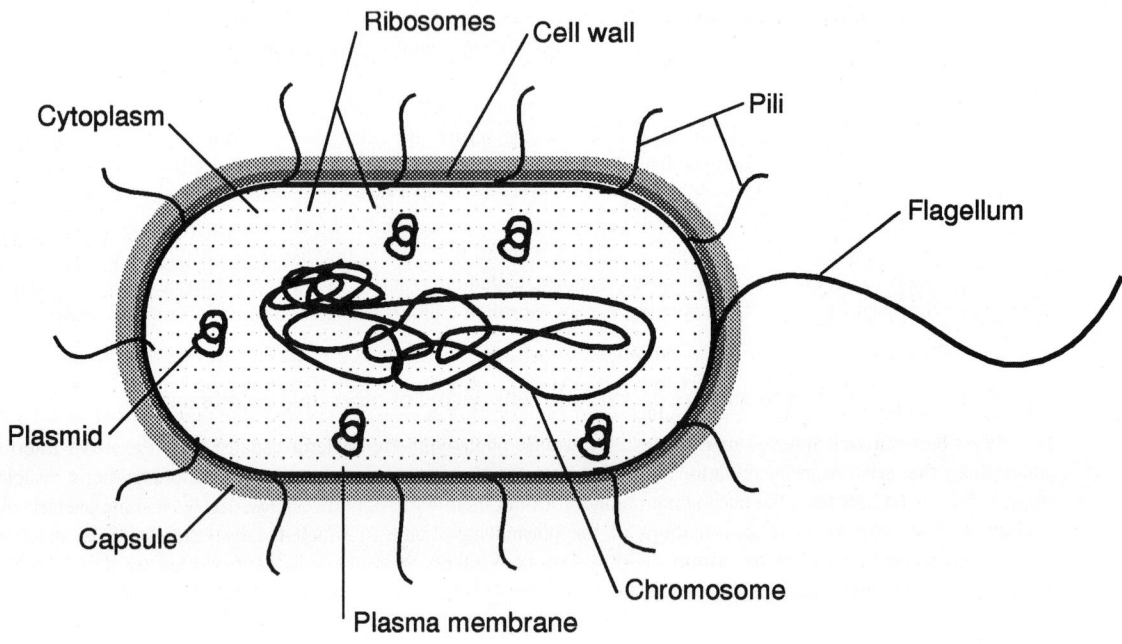

(a)

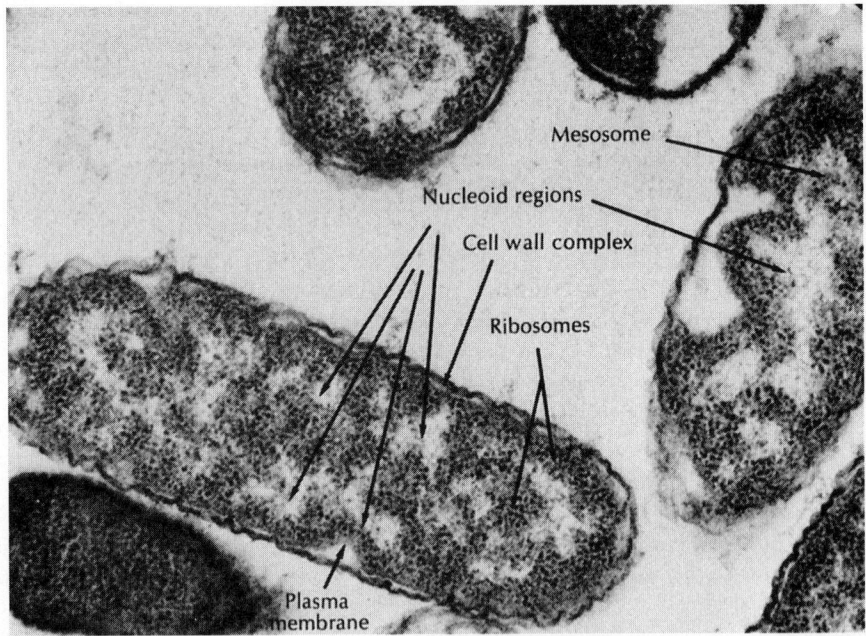

(b)

Fig. 1-2.   (*a*) Diagram of a bacterial cell. Some of the most common structures seen in bacteria with the electron transmission microscope are illustrated. Only some bacteria possess all the structures labeled. (*b*) Electron micrograph of procaryotic cells—*Escherichia coli.* (*Courtesy of J. D. Jamieson, Yale University Medical School.*)

1.8. How does motility in bacteria compare with that in eukaryotes?

Motility can be an important attribute for the dispersal of individual eukaryotic cells or multicellular organisms into new habitats. Motility is also necessary for certain embryonic cells to migrate to specific locations before proliferating and differentiating into specific tissues. Some single-celled eukaryotes—protists such as amoebas—creep along substrates as an undulating mass of constantly changing morphology. Some of the white blood cells in our own bodies that attack bacteria or other foreign cells have this same amoeboid type of locomotion. An extensive network of protein fibers, the **cytoskeleton,** found throughout the cytoplasm of typical eukaryotic cells, is responsible for amoeboid motion. The cytoskeleton has additional functions. Portions of it underlie and strengthen the plasma membrane and confer shape on eukaryotic cells. Actin filaments, microtubules, and intermediate filaments are some of the protein structures of the cytoskeleton. Actin and myosin proteins play key roles in the contraction of muscle cells.

The cells of most bacteria, plants, and fungi are surrounded by rigid cell walls that prevent amoeboid motion. Bacteria have no structures comparable to a cytoskeleton and, consequently, would not be capable of amoeboid motion even if they had no cell walls. In fact, most bacteria are nonmotile. Those that are motile usually are propelled by one or more hairlike appendages called **flagella** (flagellum, singular) that originate in the plasma membrane and rotate like propeller shafts. These filaments are constructed from a protein called **flagellin.** Some eukaryotic cells have long, thin extensions of the plasma membrane that are also called flagella, but they consist of bundles of microtubules made of different proteins, called **tubulins** (alpha and beta). Eukaryotic flagella (such as those in sperm tails) bend back and forth in quasi-sinusoidal waves. Eukaryotic **cilia** (cilium, singular) are structurally similar to flagella, but are much shorter, more numerous, and more rigid on the power stroke. Cilia also beat backward and forward like flagella, and commonly do so in unison as waves.

1.9. Aside from being constructed from different proteins, how do the flagella of eukaryotes differ from those of bacteria?

Unlike a bacterial flagellum, an eukaryotic flagellum does not *originate* in the plasma membrane, but rather develops from a **basal body (kinetosome)** in the cytoplasm. Each eukaryotic flagellum consists of bundles of microtubules arranged in 9 groups of 2 around a central pair (Fig. 1-3). A basal body consists of 9 groups of 3 around a central core devoid of microtubules. Centrioles—which reside in the centrosome and are involved in cell division—have the same general structure as (and may indeed be interconvertible with) basal bodies. The molecular basis for the movement of both cilia and flagella is thought to involve the sliding of microtubules past one another in a manner analogous to the action of actin and myosin filaments during the contraction of muscle cells. A protein complex called **dynein** drives the sliding mechanism of eukaryotic flagella by repetitive cycles of conformational changes in the molecule. Dynein also has the ATPase activity to cleave (by hydrolysis) the high-energy molecule **adenosine triphosphate** (ATP; Fig. 11-2*a*), and thus to extract the energy needed to move eukaryotic flagella and cilia.

Prokaryotic flagella, on the other hand, rotate like propeller shafts and are powered by the movement of hydrogen ions (H^+) or sodium ions (Na^+) across the plasma membrane. Concentrating these ions on one side of a membrane requires the expenditure of energy (e.g., by hydrolysis of ATP) and most likely involves **active transport.** Such an ion concentration gradient creates a voltage energy, which can be utilized (do work) by depolarizing the membrane potential through opening ion channels in the membrane and allowing the ions to diffuse into an area of lower ion concentration, a process known as **passive transport.**

1.10. What other "appendages" do bacteria have besides flagella?

Some bacteria have long, hollow tubes called **pili** (pilus, singular) or **fimbriae** (fimbria, singular) composed of a protein called **pilin.** These structures do not contribute to motility. Instead, they contribute to the adhesiveness of bacteria. One group of pili, the **sex pili,** binds cells of different mating types and facilitates the conjugational transfer of DNA from a donor cell to a recipient cell (see Chapter 7).

1.11. One of the major differences between plants and animals is that plants (and fungi) have cell walls, but animal cells do not. Do bacteria have cell walls or other extracellular structures outside the plasma membrane?

Almost all bacteria (mycoplasmas excepted) have a rigid cell wall surrounding the plasma membrane, but it has a different structure and chemical composition than that in eukaryotes (see Chapter 2). Some bacteria have a

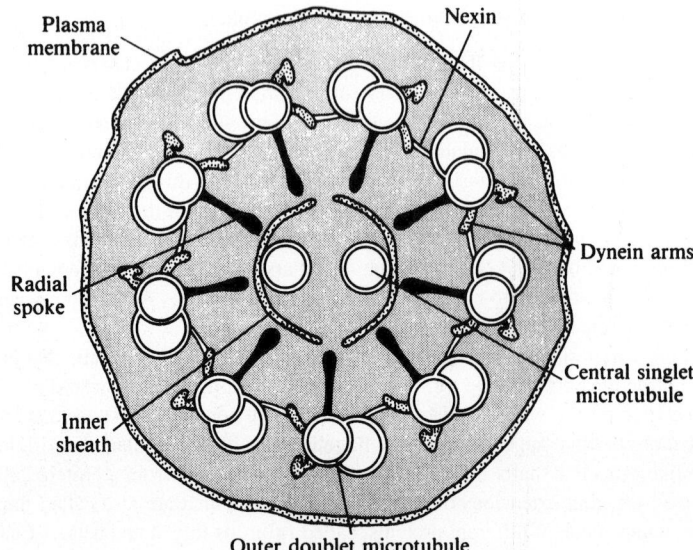

(a)

(b)

Fig. 1-3. (*a*) Diagram of a cross section of a cilium or a flagellum. Each structure contains a ring of nine doublet microtubules surrounding a single pair of singlet microtubules, together with several other proteins (such as nexin and dynein) that contribute to the motion of these structures. (*From Schaum's Outline of Theory and Problems of Biochemistry, P. W. Kuchel, G. B. Ralston, et al., McGraw-Hill, 1988, with permission.*) (*b*) Electron micrograph of a section through cilia and basal bodies in the protozoan *Euplotes patella.* [*Courtesy of L. E. Roth, Ph.D., Experimental Cell Research (Sup. 5), 573–585, 1958.*]

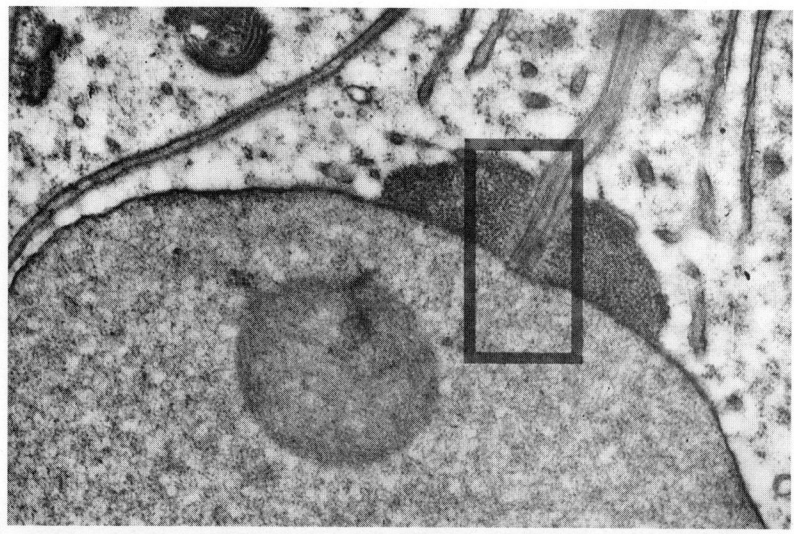

(c)

Fig. 1-3. (*Cont.*) (*c*) Electron micrograph of a maturing spermatid cell of a grasshopper. The tail filament is emerging from a centrosome at the edge of the nucleus (oblong box). (*Courtesy of T. N Tahmisian, Ph.D., and R. L. Devine, M.S., and A Scope Monograph on Cytology, 1965, The Upjohn Company, Kalamazoo, Michigan.*)

capsule of slimy or gummy layers of polysaccharides or polypeptides surrounding the cell wall. These capsules protect the bacteria from predatory cells and promote the attachment of the cell to various objects and to each other. A **glycocalyx** of tangled polysaccharide fibers may extend from the surface of some bacterial cells and serves as the functional equivalent of a capsule. Most eukaryotic cells also have a glycocalyx that covers the surface of the cell and promotes cell adhesions in the formation of specific tissues. A chain or group of bacteria may secrete a stiff polysaccharide **sheath** around the aggregate as an extra layer of protection.

1.12. Describe the structure and function of mitochondria.

   With rare exceptions, all eukaryotic cells contain membrane-bound organelles called mitochondria (Fig. 1-4), where enzymes for electron and proton (hydrogen ion) transport and ATP synthesis reside. Mitochondria are called the "powerhouses" of the cell because ATP is the primary currency of energy exchanges in the cell. Mitochondria have two membranes. The outer membrane contains proteins, called **porins,** that allow passage of molecules with molecular weights as high as 10,000. The inner membrane is much less permeable. Its numerous infoldings, called **cristae,** protrude into the **matrix,** or central space. The cristae increase the surface area for reactions that depend upon the creation of a proton gradient across the cristae and are the sites of the enzymes of respiration and ATP production. Unlike bacterial cells, no respiratory enzymes are associated with the plasma membranes of eukaryotic cells. (The respiratory enzymes of prokaryotes are not sequestered within internal membranes, such as those of the mitochondria of eukaryotes.) Mitochondria have multiple circular DNA molecules that resemble bacterial genophores; they also contain bacterialike ribosomes for production of some of their own proteins from instructions encoded in the mitochondrial DNA.

1.13. What is photosynthesis, and where does it occur in cells?

   Photosynthesis is the process that converts light energy into the chemical bond energy of ATP, which in turn can be used to convert carbon dioxide (CO_2) and water (H_2O) into carbohydrates (e.g., sugars, starch). This occurs primarily in multicellular green plants, in single-celled algae, and in the cyanobacteria. In eukaryotes, this process is compartmentalized into membrane-bound organelles called chloroplasts. In cyanobacteria, photosynthesis occurs in flattened vesicles (thylakoids) formed by invaginations of the plasma membrane.

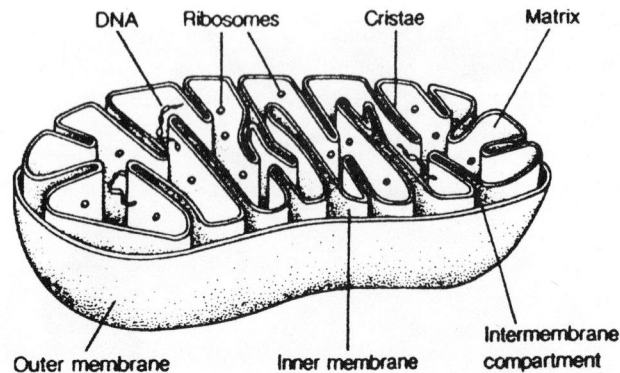

(a)

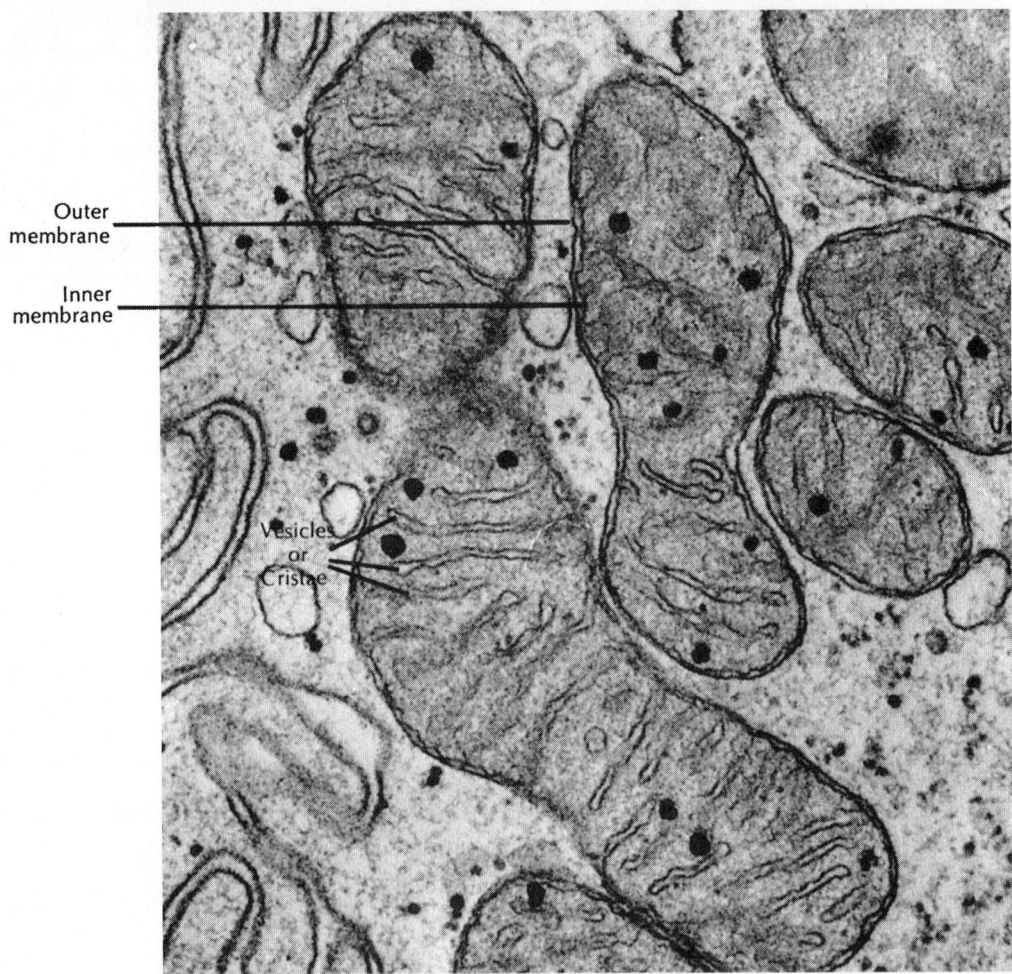

(b)

Fig. 1-4. (*a*) Diagram of a mitchondrion, showing the invaginations of the inner membrane (called cristae), multiple DNA genophores, and ribosomes. (*From Microbiology, R. J. Cano and J. S. Colome, West Publishing Company, 1989, with permission.*) (*b*) Electron micrograph of mitochondria in acinar cells of the pancreas (× 70,000). (*Courtesy of G. E. Palade, University of California, San Diego*).

1.14. Describe the structure of chloroplasts.

Both mitochondria and chloroplasts are surrounded by an inner and an outer membrane, but chloroplasts contain an additional internal system of membranes called **thylakoids** (Fig. 1-5). Chloroplasts contain a bacterialike circular chromosome, and bacterialike ribosomes for synthesizing proteins specified by the chloroplast DNA. Chloroplasts are thought to reproduce by binary fission, but they cannot do so outside the cell because some nuclear genes are essential for this process. As is also true of mitochondria, chloroplasts usually have no fixed location in the cell and can migrate by virtue of cytoskeletal activity.

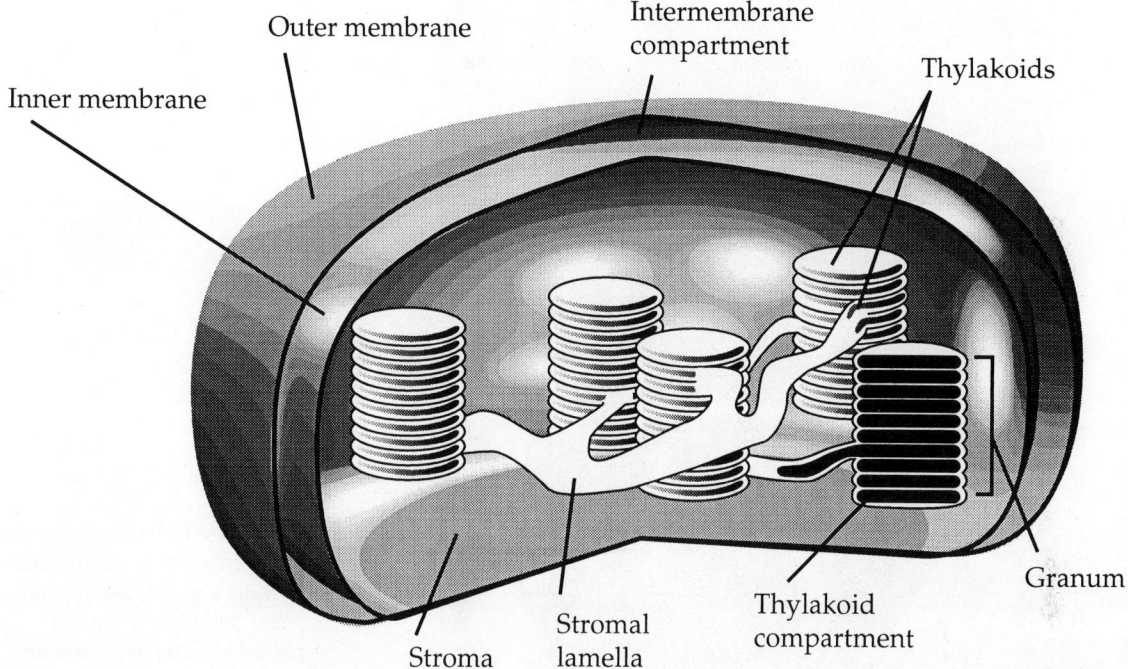

Fig. 1-5. (*a*) Diagram of a chloroplast. The multiple DNA molecules and ribosomes that exist in the stroma are not shown.

The flattened, vesicular thylakoids contain the chlorophylls, the enzymes, and other molecules needed to synthesize ATP by photosynthesis. In some chloroplasts, the thylakoids form discs that are stacked like coins to form **grana** (granum, singular). The poi ion of the chloroplast between the thylakoids and the inner chloroplast membrane is called the **stroma.** In this space, **CO_2 fixation** or **carbon fixation** (conversion of inorganic CO_2 to organic sugars) occurs.

1.15. Contrast mitochondria and chloroplasts with regard to the number of organelles per cell, the number of DNA molecules per organelle, and the size of their respective DNA molecules.

The number of chloroplasts and mitochondria per cell varies with the species and among different tissues of the same species.

Example 1.1. The unicellular alga *Chlamydomonas* has only one large cup-shaped chloroplast, whereas a single leaf cell of a flowering plant may contain over 50 ovoid chloroplasts. The proportion of mitochondria to cell volume is often directly correlated to the rate of oxidative metabolism of the cell. Most eukaryotic cells contain several hundred elongate (usually sausage-shaped) mitochondria.

Except for the noncircular mitochondrial genomes of some algae and protozoans, all mitochondrial genomes consist of a single, circular, double-stranded DNA molecule. All chloroplasts and mitochondria normally contain

Fig. 1-5. (*Cont.*) (*b*) Electron micrograph of a portion of a mesophyll cell showing a chloroplast in profile view. The areas of densely packed lamellae, the grana, are the chlorophyll-containing sites of the plastid. These grana would appear circular in face view.

multiple copies of their genomes. These genomes are located in the stroma of chloroplasts and in the matrix of mitochondria.

> **Example 1.2.** There is not much difference in the number of DNA molecules in these two organelles. Maize has about 20 to 40 DNA molecules per chloroplast. Tetrapod vertebrates generally have about 5 to 10 DNA molecules per mitochondrion.

The number of DNA subunits (called **base pairs,** bp) is commonly used to indicate the size (length) of a DNA molecule. The **chloroplast** DNA (ctDNA) molecules of most plants (2×10^5 bp) are about 10 to 20 times smaller than a typical bacterial genome (e.g., the intestinal bacterium *Escherichia coli* genome has 4×10^6 bp), but ctDNA is about 10 times larger than **mitochondrial** genomes (mtDNA) of many animals (2×10^4 bp).

There is not much variation in the size of chloroplast genomes among plants. Likewise, the size of mitochondrial genomes varies little among different species of animals. The mitochondrial genomes of vascular plants, however, are much larger and more variable, being 10 to 125 times larger than those of most animals. The apparent complexity of plant mtDNA may be due in part to recombination events that produce subgenomic circular molecules that coexist in the same mitochondrion with complete, "master" genomes.

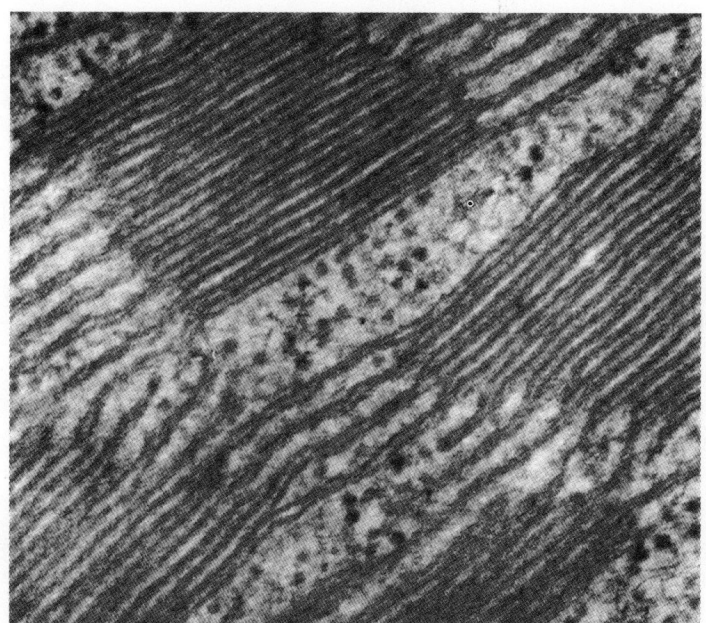

Fig. 1-5. (*Cont.*) (*c*) An electron micrograph of higher magnification than that of Fig. 1-5*b* shows that the grana are distinctly laminate. The variable opacity of the lamellae of the grana is probably the result of macromolecular organization of lipids, protein, and chlorophyll. (*Figs. 1-5b and 1-5c are courtesy of A. E. Vatter, Ph.D., University of Colorado Medical Center, Department of Pathology, Denver, and A Scope Monograph on Cytology, 1965, The Upjohn Company, Kalamazoo, Michigan.*)

1.16. Some of the proteins synthesized within eukaryotic cells are transported to specific locations. How is this protein targeting accomplished?

In eukaryotic cells, a membrane system in the cytoplasm performs this complex task, modifying specific proteins (e.g., by the addition of one or more sugars, a process termed **glycosylation**), tagging them for specific destinations, and packaging them into vesicles for transport within the cell or to the plasma membrane for release from the cell. Some of the carbohydrate markers on proteins may direct them through the **Golgi complex** (also called Golgi apparatus or Golgi body; **dictyosome** in plants; Fig. 1-1) and/or to their target organelles. The Golgi complex consists of closely packed, flat, membranous sacs (**cisternae**) and small, rounded vesicles. It functions to collect, sequester, modify, and target proteins and lipids that have been newly synthesized in the endoplasmic reticulum. The packaging of many proteins into secretory vesicles for secretion from the cell also occurs in the Golgi complex.

Ribosomes in association with the membranes of the endoplasmic reticulum (rough ER) may synthesize proteins directly into the cavities of the ER. Although some of the membranes of the ER are continuous with the nuclear membrane, none of the proteins secreted into the ER channels diffuse to the nucleus. Proteins destined for the plasma membrane are usually packaged into lipid vesicles in the Golgi complex. These vesicles are then transported by fibers of the cytoskeleton to the plasma membrane. There the two membranes fuse. If the proteins are free in the vesicle, they are released to the cell's exterior (**exocytosis**). The reverse of this process (**endocytosis**) can bring materials from outside the cell into an intracellular food vacuole. Vesicles containing digestive enzymes (lysosomes) may be transported to a food vacuole by the cytoskeleton. The fusion of the two vesicles creates a phagolysosome, in which digestion of the exogenous contents occurs.

Bacteria do not have Golgi bodies, and few bacteria have enzymes that glycosylate proteins. Proteins destined for release from bacterial cells are usually synthesized by ribosomes on the plasma membrane in such a way that they are extruded through the membrane.

METABOLISM

1.17. What are the two major carbon sources utilized by cells to synthesize their own organic molecules?

The two major carbon sources utilized by cells are (1) complex organic compounds, such as sugars and amino acids, and (2) single-carbon compounds, such as carbon dioxide (CO_2) or methane (CH_4).

1.18. Define the two most inclusive groups of cells based on the carbon sources they use.

Cells that use CO_2 as their sole source of carbon are known as **autotrophs** ("self-nourished"). In contrast, cells that require complex organic compounds (commonly from other forms of life, living or dead) are referred to as **heterotrophs** ("other-nourished").

1.19. Cells can obtain energy from light or from chemical compounds. Define two classes of cells based upon the energy source used.

Organisms that obtain their energy from light are called **phototrophs,** whereas those that obtain their energy from chemical compounds are designated **chemotrophs.**

1.20. If energy and carbon sources are considered together, most organisms may fit into one of four categories: photoautotrophs, photoheterotrophs, chemoautotrophs, and chemoheterotrophs. Discuss the two types of autotrophic cells.

Chemoautotrophic cells oxidize reduced inorganic molecules such as H_2S, S, NH_3, H_2, and Fe^{2+}. The energy and electrons thereby gained is used to reduce CO_2, thus providing the carbon skeletons for synthesizing the cell's own organic molecules and providing the driving force for energy-requiring (**endergonic**) reactions. Only a few families of living (extant) bacteria make their living by chemoautotrophy.

By far the largest group of autotrophic forms of life today includes **photoautotrophic** plants, algae, and cyanobacteria. These organisms can convert the energy of sunlight into the energy of chemical bonds. They take inorganic CO_2 from the air, water and minerals from the soil or their aqueous environment, and make organic compounds from CO_2 (carbon fixation) by photosynthesis.

1.21. On what basis are the two kinds of heterotrophic cells distinguished?

There are a few photosynthetic organisms (such as purple sulfur bacteria) that require compounds other than CO_2 as a carbon source. Almost all the organic material used by such **photoheterotrophs** is derived from dead organisms.

The vast majority of species (including animals, fungi, and most bacteria) are **chemoheterotrophs** and so require organic compounds as sources of carbon. There are some organisms that can, under special conditions, be classified in more than one of the above categories. For example, in the absence of sunlight, many photosynthetic algae can function as chemoheterotrophs.

1.22. Two types of respiration are recognized among the chemoheterotrophs: anaerobic and aerobic. Define respiration and three classes of chemoheterotrophs based upon their respiratory requirements.

Respiration involves the oxidation of inorganic or organic molecules, the generation of high-energy molecules (such as ATP) by passing pairs of electrons (and hydrogen ions, or protons) through an electron transport system, and the donation of these electrons to an inorganic electron acceptor. **Aerobic respiration** occurs when the terminal electron acceptor is molecular oxygen. **Anaerobic respiration** occurs when the terminal electron acceptor is an inorganic molecule other than oxygen (such as sulfate or nitrate).

Some **strict (obligate) anaerobic** microorganisms use organic molecules, whereas others use inorganic molecules, as terminal electron receptors in their metabolism and cannot survive in the presence of oxygen. *Clostridium botulinum* is an obligate anaerobe that produces the deadly toxin of the "food poisoning" known as botulism.

A **facultative anaerobe** can use either oxygen or organic molecules as terminal electron acceptors. Yeast, for example, metabolize glucose to CO_2 and H_2O in the presence of oxygen, but produce CO_2 and a reduced organic molecule (such as ethanol) in its absence.

Most animals (other than a few of the single-celled forms, called protozoa) are **obligate aerobes,** i.e., organisms that cannot survive in an environment devoid of oxygen.

1.23. Adenosine triphosphate (ATP) is a key molecule in bioenergetics. Describe two ways in which cells may synthesize it.

ATP can be generated from adenosine diphosphate (ADP) by either substrate-level phosphorylation or oxidative phosphorylation. **Substrate-level phosphorylation** involves the synthesis of ATP by the enzyme-mediated transfer of a phosphate group from a phosphorylated organic molecule (such as diphosphoglycerate) to ADP. This process does not require oxygen.

Oxidative phosphorylation occurs when molecules are oxidized and energy is extracted from the electrons by passing them through an electron transport system, where most of the resulting free energy is used to drive the phosphorylation of adenosine diphosphate (ADP), producing **ATP.** A molecule or atom becomes **oxidized** when it loses electrons, and becomes **reduced** when it gains electrons. Coenzymes (Fig. 14-4) such as flavin adenine dinucleotide (**FAD**) and nicotinamide adenine dinucleotide (**NAD**) serve as electron carriers, and together with iron-containing proteins called **cytochromes,** they make up an **electron transport chain.** This coupled series of oxidation-reduction (redox) reactions passes electrons from hydrogen atoms along the membrane-bound carrier molecules. Most of the energy decrease at each of three redox steps of the electron cascade of mitochondria is used to synthesize ATP; the remainder is lost as heat. Oxygen is the final acceptor of hydrogen ions and electrons; water is the waste product of this electron cascade.

Photophosphorylation also synthesizes ATP by oxidative phosphorylation, but uses energy derived from sunlight rather than from the breakdown of organic molecules. This process is the subject of Questions 1.30 and 1.31.

1.24. What is glycolysis?

Glycolysis (also known as the **Embden-Meyerhof pathway;** Fig. 1-6), is a nearly universal process in all cells. During glycolysis, the six-carbon sugar glucose is anaerobically converted, through a series of enzymatically catalyzed steps in the cytosol, to two molecules of the three-carbon compound pyruvate (pyruvic acid). Two molecules of ATP are expended early in glycolysis, but four more are generated later by substrate-level phosphorylation. Thus, there is a net production of two ATP molecules per molecule of glucose during glycolysis. Other pathways for catabolising glucose (e.g. the pertose phosphate shunt) exist, but will not be discussed in this book.

1.25. What is fermentation?

Fermentation is an oxygen-independent process, occurring in the cytosol, that uses organic molecules as terminal electron acceptors; this results in the release of such organic and inorganic molecules as carbon dioxide and molecular hydrogen (gases); lactic, formic, acetic, succinic, butyric, or propionic acids; and ethanol, butanol, or propanol (alcohols). During glycolysis, one molecule of glucose is converted to two molecules of pyruvate. Each fermentation pathway begins with pyruvate and produces one or two of the aforementioned end products, but does not generate or use any ATP molecules. These end products may be "waste products" as far as the cell is concerned, but many of them are valuable commercial products. However, when oxygen is available for aerobic respiration, fermentation essentially ceases, the rate of glucose consumption is decreased, and acid and/or alcohol production is inhibited. This phenomenon is known as the **Pasteur effect.**

1.26. How and where does aerobic respiration occur?

Pyruvate (generated by glycolysis in the cytosol) may enter mitochondria and, if oxygen is available, be enzymatically converted (by oxidative decarboxylation) to acetyl coenzyme A (acetyl CoA; Fig. 14-4) and carbon dioxide. Within the matrix of mitochondria or the cytosol of prokaryotes, the two-carbon acetyl unit of acetyl CoA enters a circular set of enzymatic reactions (Fig. 1-6) known variously as the **Krebs cycle,** the **tricarboxylic acid cycle** (TCA), or the **citric acid cycle.** Two major electron carrier (acceptor) molecules are NAD^+ and FAD (Fig. 14-4). During oxidation of a substrate, NAD^+ accepts two electrons and a hydrogen ion (equivalent to a hydride ion), becoming the reduced form of this carrier, NADH. Likewise, the reduced form of FAD is $FADH_2$. One complete turn of the Krebs cycle produces three molecules of NADH, two molecules of CO_2, and one molecule of $FADH_2$. The electrons and hydrogen ions from NADH and $FADH_2$ are transferred to electron transport chains within the inner membranes (cristae) of mitochondria. The electron carriers of prokaryotes are in the plasma membrane. These chains consist of a series of proteins (such as the iron-containing cytochromes) that first serve as electron acceptors and then as donors to the next complex in the chain. This series of coupled oxidations and reductions results in the terminal transfer of electrons and hydrogen ions to oxygen, forming water as the end product. Along these paths, several molecules of ATP are synthesized from ADP by oxidative phosphorylation.

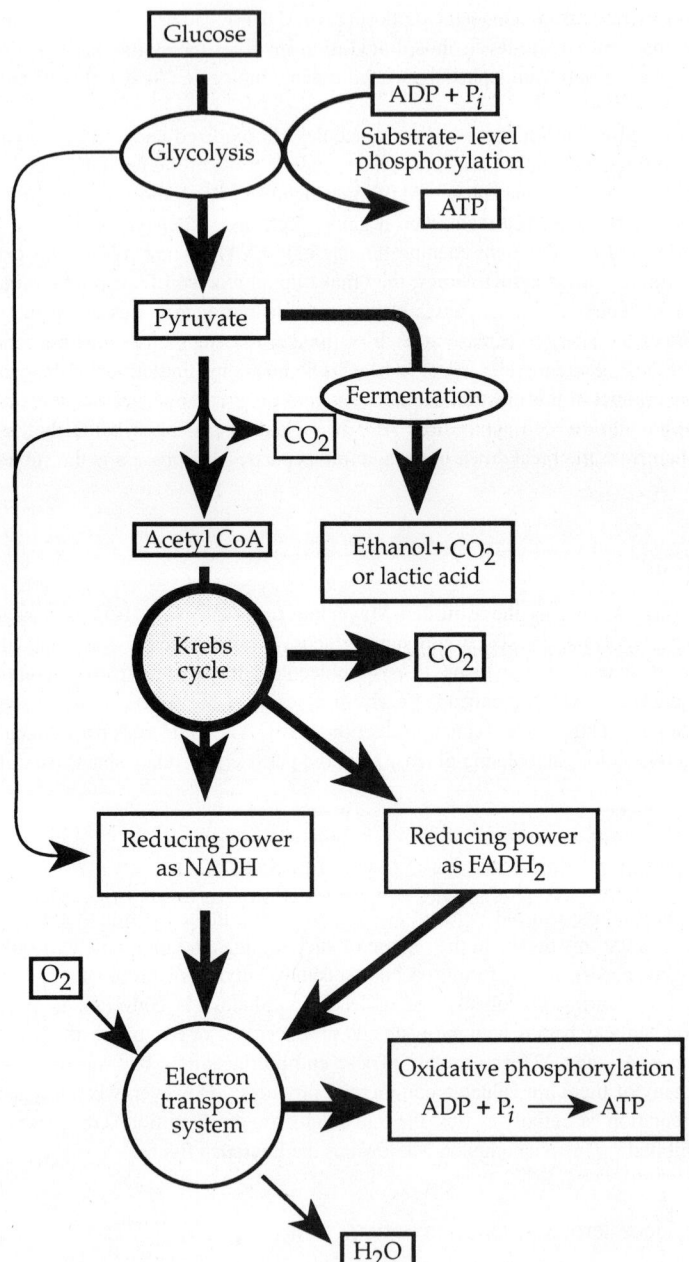

Fig. 1-6. Diagram of chemoheterotrophic metabolism.

The net energy yield from oxidative respiration may be as high as 38 ATP molecules per molecule of glucose (2 each by substrate-level phosphorylation during glycolysis and the Krebs cycle; 34 from oxidative phosphorylation via the electron transport chain). The essence of aerobic respiration of a molecule of glucose is summarized as follows:

$$C_6H_{12}O_6 + 6\ O_2 \rightarrow 6\ CO_2 + 6\ H_2O.$$

The essence of photosynthesis (to be discussed a little later in this section) is the reverse of the above equation.

1.27. What is the chemiosmotic hypothesis?

According to the **chemiosmotic hypothesis,** nutrient or radiant energy is conserved by the cell when electrons and hydrogen ions are separated by membrane-associated electron transport systems. The excess concentration of hydrogen ions on one side of a membrane (such as the mitochondrial inner membrane) creates an electrical potential, or voltage, called a **proton motive force** (PMF) that can be used to drive the synthesis of ATP by enzymes known as **ATP synthetases** (ATPases). The ATP synthetases are channel proteins that span the membrane and conduct ions through the membrane to drive the synthesis of ATP from ADP and inorganic phosphate (P_i).

1.28. What are the major features of chemoautotrophic metabolism?

Chemoautotrophic cells obtain their energy by oxidation of reduced inorganic molecules (e.g., ammonium ions, hydrogen sulfide, molecular hydrogen, ferrous iron, nitrite ions, or sulfur). They pass the resulting electrons and hydrogen ions through an electron transport system to generate membrane potentials. The voltage is used to synthesize ATP by oxidative phosphorylation. For example, bacteria belonging to the genus *Nitrosomonas* are obligate aerobic chemoautotrophs that obtain energy from the oxidation of ammonium ions to nitrite ions. Oxygen is the final acceptor for electrons and hydrogen ions; oxygen is thus reduced to water.

1.29. What are the major features of phototrophic metabolism?

Most phototrophic cells obtain electrons by using light energy to oxidize chlorophyll (a magnesium porphyrin molecule). The excited electrons from chlorophyll are passed through electron transport systems to generate membrane potentials (voltages). ATP is synthesized by phosphorylation of adenosine diphosphate (ADP) on ATP synthetases as hydrogen ions pass through the enzymes and the voltage is consumed. Photophosphorylation can be either cyclic or noncyclic.

1.30. How is cyclic photophosphorylation accomplished?

Anaerobic photosynthetic green sulfur bacteria use **cyclic photophosphorylation** (Fig. 1-7) to phosphorylate ADP to ATP, with light supplying the energy. When a group of pigments (including carotenoids and bacteriochlorophyll) called **photosystem I** (PS-I) absorbs light energy, electrons from PS-I are driven to a higher energy state before being picked up by electron acceptors called **quinones** and then passed on to a transport system that generates membrane voltages to support ATP synthesis. Low-energy electrons return from the electron transport system to reduce the oxidized bacteriochlorophyll, converting it back to its initial state, so that it can be reoxidized by absorbing more photons.

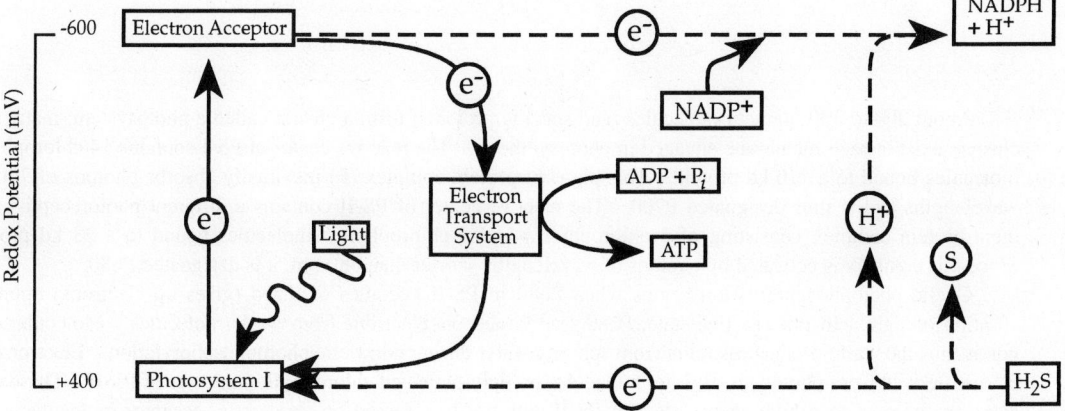

Fig. 1-7. Cyclic photophosphorylation occurs in anaerobic photosynthetic bacteria. The cyclic pathway for the production of ATP is indicated by solid arrows. Broken arrows indicate an alternative pathway for reduction of NADP$^+$. The symbol e$^-$ stands for electrons.

When an anaerobic photosynthetic bacterium has sufficient ATP, the high-energy electrons can bypass the electron transport system and be diverted to oxidized nicotinamide adenine dinucleotide phosphate ($NADP^+$), forming reduced $NADPH + H^+$. Both NADPH and ATP are required for the fixation of carbon dioxide into carbohydrates. ATP supplies the energy, whereas NADPH provides hydrogen atoms and electrons for the synthesis of the carbohydrates. In these anaerobic photosynthetic bacteria, the electrons needed to reduce (add electrons to) oxidized PS-I are obtained by oxidizing (removing electrons from) hydrogen sulfide (H_2S). The **photolysis** ("light-splitting") of H_2S releases sulfur as a by-product. Thus, the cell can make either ATP or NADPH, depending on its needs, by diverting the flow of electrons from the electron transport system to $NADP^+$ instead of cycling them back to PS-I.

1.31. Where does noncyclic photophosphorylation occur, and how does it function?

Noncyclic **photophosphorylation** (Fig. 1-8) occurs in cyanobacteria ("blue-green algae") and green plants, but not in the anaerobic, photosynthetic, green and purple sulfur-oxidizing bacteria. Two membrane-associated, light-absorbing pigment systems are involved: PS-I (thought to have evolved first in ancestors of the green sulfur bacteria) and PS-II (thought to have evolved from ancestors of purple sulfur bacteria).

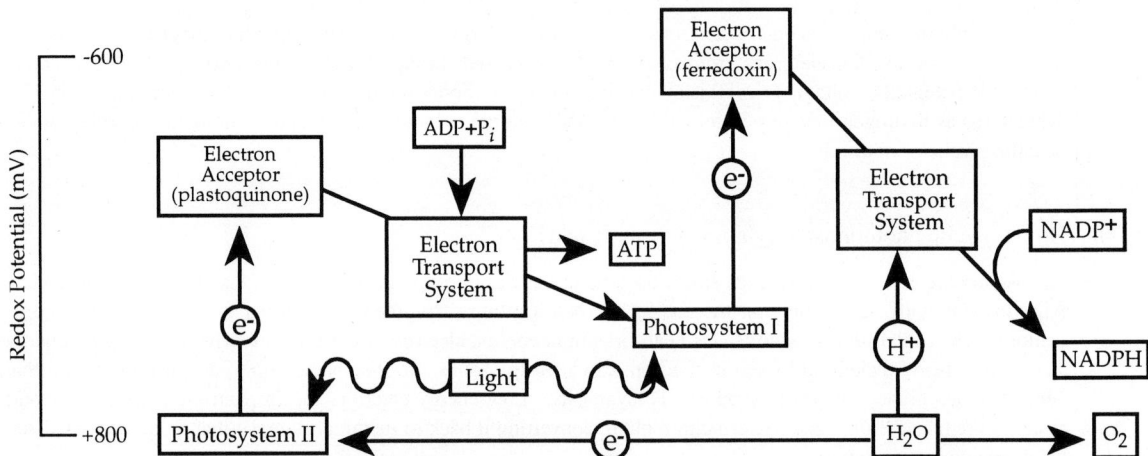

Fig. 1-8.   Two major steps in the noncyclic photosynthesis of plants and cyanobacteria. Photosystem I is similar to that in anaerobic photosynthetic bacteria, but generates only NADPH; photosytem II produces ATP.

About 200 to 300 pigment molecules (and specific proteins) form a cluster called a photosystem; many such clusters exist in each membrane engaged in photosynthesis. The reactive center of PS-I contains 14 chlorophyll *a* molecules bound to a 110 kd protein. This pigment-protein complex (P) maximally absorbs photons of 700-nm wavelengths and is thus designated P700. The reactive center of PS-II contains a different photoreceptive pigment-protein complex consisting of 3 chlorophyll *a* and 3 chlorophyll *b* molecules bound to a 28 kd protein. Since this complex is activated by photons of wavelengths shorter than 680 nm, it is designated P680.

Cyclic photophosphorylation begins when P680 in PS-II becomes oxidized (gives up electrons) upon absorption of light. In this reactive state, P680 readily accepts electrons from water molecules. Molecular oxygen is thus the waste product resulting from splitting water during noncyclic photophosphorylation. Electrons and H^+ are derived from photolysis of water molecules by light-activated chlorophyll molecules of PS-II. These electrons are boosted to a high-energy state in PS-II and are then passed on to electron acceptor molecules called quinones. The quinones transport electrons across thylakoid membranes in chloroplasts of plants (or across the thylakoids in aerobic photosynthetic bacteria), creating a voltage to support ATP formation by ATPases.

The high-energy electrons flow through an electron transport system containing plastoquinone, cytochrome proteins, and the copper-containing protein **plastocyanin.** The copper atoms can alternate between +1 and +2 oxidation states. The proton gradient generated by this electron transport chain is used to synthesize ATP. The electrons are eventually absorbed at an intermediate energy level by a different chlorophyll (P700) in PS-I. Absorption of photons by PS-I boosts the energy of electrons to a higher level than that attained in PS-II. From this energy state, the electrons flow through membrane-bound ferredoxin molecules to soluble **ferredoxin.** Ferredoxin is an iron-containing protein. Reduction of an iron atom (by acceptance of an electron) in each of two ferredoxin molecules is required to form one molecule of NADPH from NADP$^+$.

Alternatively, high-energy electrons produced in PS-I can be recycled through a different cytochrome of the electron transport system back to PS-I, to generate ATP through cyclic photophosphylation. In this case, water is not split, oxygen is not produced, and ATP is generated without the concomitant synthesis of NADPH.

1.32.   In the preceding question, the light-dependent reactions of noncyclic photophosphorylation were discussed. There are, however, light-independent reactions in photosynthesis. What are they?

The reactions that fix CO_2 into organic molecules can occur if there is sufficient ATP and NADPH to drive the chemical reactions. Since light is not required to drive the reactions involved in CO_2 fixation, they are known as light-independent reactions. These reactions can occur in the dark of night, so they are sometimes called "dark reactions." But since they occur faster in the light, most of the light-independent reactions occur during daylight. The light-independent reactions occur in the stroma of chloroplasts (outside thylakoid membranes) as a series of enzymatic steps known as the **carbon-fixation cycle,** or the **Calvin-Benson cycle** (Fig. 1-9).

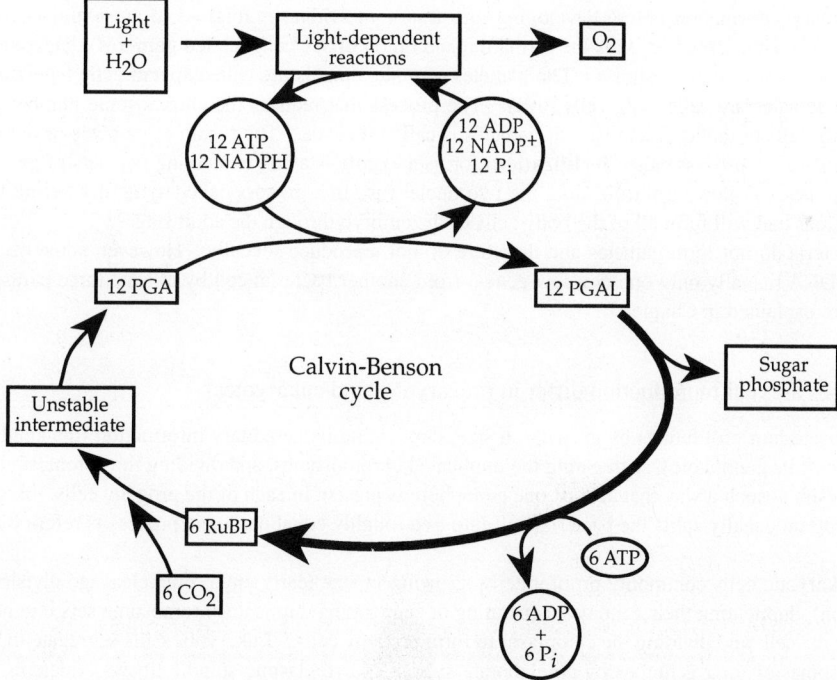

Fig. 1-9.   Diagram of the interactions between the light-dependent and light-independent reactions of photosynthesis. ADP = adenosine diphosphate; ATP = adenosine triphosphate; NADP$^+$ = nicotinamide adenine dinucleotide; NADPH = reduced NADP; P_i = inorganic phosphate; PGA = phosphoglyceric acid (phosphoglycerate); PGAL = phosphoglyceraldehyde; RuBP = ribulose bisphosphate.

In this cycle, the five-carbon sugar, variously known as ribulose 1,5-diphosphate, ribulose bisphosphate, ribulose biphosphate, or RuBP, becomes coupled to carbon dioxide, forming an unstable, six-carbon intermediate (metabolite) that quickly breaks into two molecules of a three-carbon compound called 3-phosphoglyceric acid (PGA), the salt of which is known as 3-phosphoglycerate. The enzyme catalyzing this coupling, ribulose biphosphate carboxylase, is a very sluggish enzyme that sometimes comprises more than 50% of the total protein in chloroplasts. It is claimed to be the most abundant protein in the biosphere. PGA is made more reactive by receiving a phosphate group from the hydrolysis of ATP. The resulting intermediate then obtains electrons and protons (H^+) from NADPH to form a three-carbon compound known as phosphoglyceraldehyde (PGAL). Most of the PGAL (10 molecules out of 12) is converted back into RuBP molecules, with the expenditure of ATP (thus completing the Calvin-Benson cycle); but some of it (2 molecules out of 12) is rearranged into phosphorylated sugar intermediates of the glycolytic pathway, such as glucose-6-phosphate. These phosphorylated intermediates can serve as building blocks for the synthesis of the plant's main carbohydrates (sucrose, starch, cellulose). Six turns of the Calvin-Benson cycle are required to synthesize one molecule of a hexose such as glucose or fructose, because only one carbon atom is reduced in each cycle.

REPRODUCTION

1.33. What is the difference between asexual reproduction and sexual reproduction?

Most cells (exceptions include mature red blood cells and nerve cells of animals and sieve tube elements of plants) are able to reproduce **asexually,** i.e., they are all capable of proliferating or increasing in number without having to exchange or acquire new hereditary information. All bacteria reproduce almost exclusively in this fashion. Most cells that form the bodies of multicellular eukaryotes are also produced asexually.

Some single-celled eukaryotes and most multicellular organisms (plants and animals) predominantly form new individuals from cells that have exchanged or received DNA from other cells. This exchange or acquisition of hereditary information before developing into a new individual is referred to as **sexual reproduction.** The hallmark of sexual reproduction is the formation and union of sex cells (called **gametes**) that contain half the normal complement of chromosomes. The gametes of male animals are called **sperm cells (spermatozoa);** the gametes of females are called **egg cells (ova).** The process that reduces the chromosome number characteristic of most body cells to that characteristic of gametes is called **meiosis.** The fusion of gametes or reduced nuclei from different sexes (a process called **fertilization**) forms a **zygote**—a cell containing two sets of genetically different chromosomes. Following fertilization, the two nuclei fuse in a process called **syngamy** before the zygote starts the divisions that will form all of the body cells of the embryo through the adult stages.

Bacteria do not form gametes and therefore do not reproduce sexually. However, some of them are able to acquire DNA (usually only one or a few genes) from another bacterial cell by at least three parasexual processes, which are explained in Chapter 7.

1.34. How does asexual reproduction differ in procaryotes and eukaryotes?

A bacterium proliferates by growing in size, duplicating its hereditary information (the single, circular DNA molecule of its genophore), segregating the duplicated chromosomes, and dividing the cytoplasm (a process called **cytokinesis**) in such a way that at least one genophore is present in each of the progeny cells. Because division of the cytoplasm usually splits the bacterial cell into two roughly equal parts, the process is referred to as **binary fission.**

Eukaryotic cells commonly proliferate by growing in size (early embryonic cleavage divisions are a notable exception), duplicating their genomes, separating or segregating duplicate chromosome sets into nuclei at opposite poles of the cell, and dividing the cytoplasm to form progeny cells. Eukaryotic cells segregate the multiple, linear chromosomes of their genomes by an elaborate system of cytoplasmic spindle fibers. Bacteria, however, attach duplicate copies of their single genophore ("chromosome") to the plasma membrane; cell growth between these attachment points and cytokinesis by formation of new cell walls eventually segregate the duplicate genophores into progeny cells. Because of these and other distinctive differences in the way that bacteria and eukaryotes undergo asexual cell division, the process in eukaryotes is referred to by a special term, **mitosis.**

1.35. What are the major phases of the asexual reproductive cycle of eukaryotes, commonly called the cell cycle?

The eukaryotic **cell cycle** (Fig. 1-10) contains four major phases. The **S phase** is when DNA synthesis occurs to replicate the chromosomes (details in Chapter 3). The period between the end of the S phase and the beginning of mitosis (**M phase**) is a gap, or growth period, designated **G_2 phase.** Another gap or growth period, called the **G_1 phase,** occurs between the M and S phases to complete the cell cycle. **Interphase,** the period be-

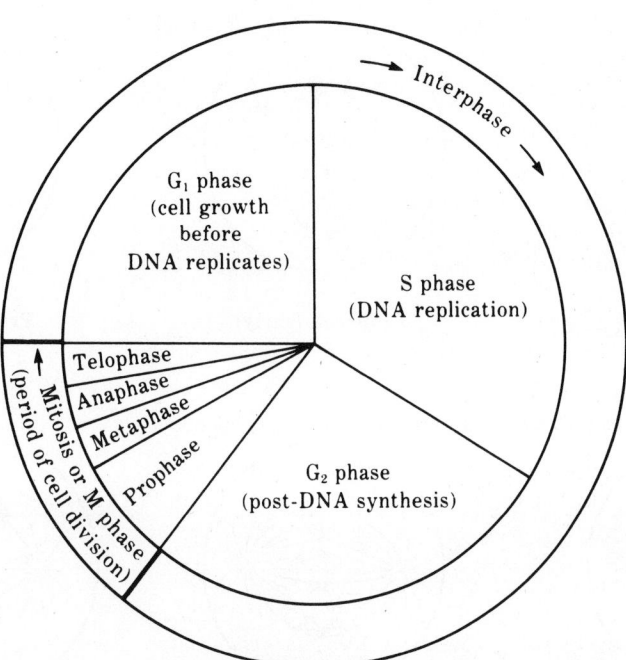

Fig. 1-10.   Diagram of a typical asexual reproductive cycle (cell cycle) in eukaryotic cells.   An additional G_0 phase (not shown) may follow the M phase if growth is inhibited. (*From Schaum's Outline of Theory and Problems of Genetics, Third Edition, W. D. Stansfield, McGraw-Hill, 1991.*)

tween successive mitoses, includes phases G_1, S, and G_2.   The regulatory signals and mechanisms controlling the cell cycle are described in Chapter 11.

1.36.   Mitosis consists of four consecutive phases: prophase, metaphase, anaphase, and telophase.   What major events occur during mitotic prophase?

A generalized diagram of mitosis is presented in Fig. 1-11.   As explained in the previous question, DNA replication occurs during the period between successive mitoses called interphase.   Each replicated chromosome contains two identical DNA molecules (and associated proteins) called **chromatids.**   The long, thin strands of chromosomal DNA become coiled around clusters of histone proteins, forming structures (called **nucleosomes**) resembling beads on a string.   These and other details of chromosome structure are presented in Chapter 3.   As a cell enters mitosis, each extended chromosome begins to shorten and thicken (a process referred to as **condensation**) by supercoiling on itself again and again.   Gradually the chromosomes become visible in the light microscope as individual fibrous structures.   The nuclear membrane eventually dissolves, and a **spindle** of microtubules (assembled from the proteins α- and β-tubulin) forms from one pole of the cell to the other.   All of these events occur during **prophase**—the first of four major mitotic stages.

1.37.   What happens during mitotic metaphase?

During mitotic **metaphase,** the chromosomes line up in the center of the spindle (in a region called the **metaphase plate** or **plane**), which is usually near the middle of the cell.   This movement is thought to develop as a dynamic equilibrium is reached between spindle fibers tugging in opposite directions on a region of each chromosome known as a **centromere** (actually, on protein complexes attached to the centromeres) from each pole of the cell.   The centromere forms the primary constriction on a chromosome that is used to distinguish the shorter arm of the chromosome (designated the **p arm**) on one side of the centromere from the longer arm (designated the

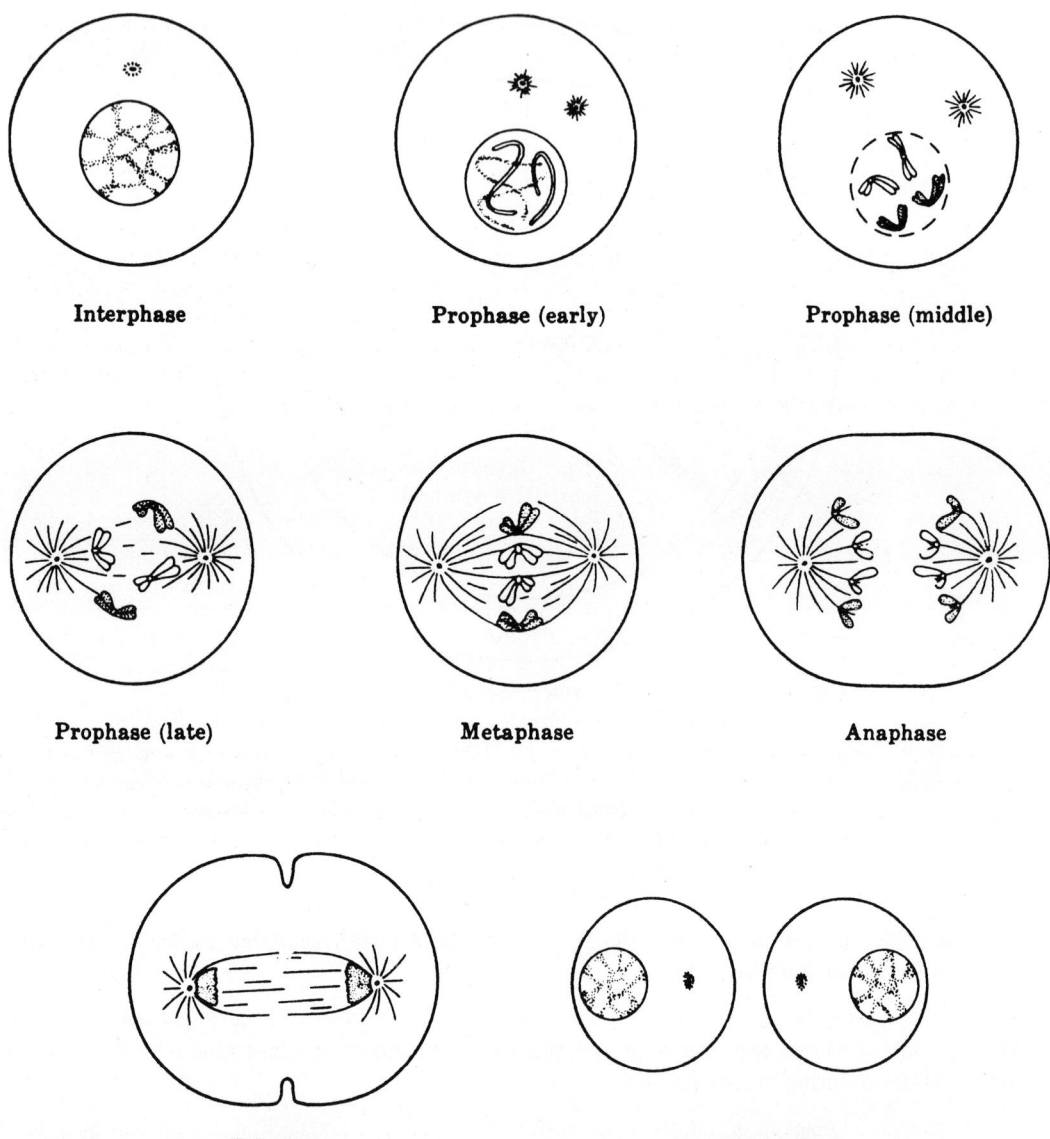

Interphase   **Prophase (early)**   **Prophase (middle)**

Prophase (late)   **Metaphase**   **Anaphase**

Telophase   **Daughter cells**

Fig. 1-11.   Diagram of mitosis in animal cells. Dark chromosomes were derived from one parent (e.g., maternal parent) of the individual now undergoing mitosis; light chromosomes were derived from the
other parent (e.g., paternal parent). One pair of homologues has a centrally located centromere
and so is said to be **metacentric;** the other pair has arms of unequal length since the centromere is
off center and is said to be **submetacentric.** (*From Schaum's Outline of Theory and Problems of
Genetics, Third Edition, McGraw-Hill, 1991.*)

q arm) on the other side of the centromere. Up to this point, the two identical chromatids of each chromosome
remain connected at their centromeric regions. The mechanism that keeps the centromeres of the chromatids together is unknown. Perhaps they are connected by proteins that bind to specific DNA sequences in the centromeric regions. In any event, some signal (perhaps an increase in the cytoplasmic concentration of calcium
ions) causes the centromeres of the two chromatids to dissociate from one another.

1.38.   What events characterize mitotic anaphase?

During mitotic **anaphase,** the two former chromatids of each replicated chromosome are pulled to opposite
poles by depolymerization of those microtubules in the spindle apparatus that are attached to the centromeres.

These former chromatids, being no longer connected at the centromere, are now considered to be new chromosomes, each consisting of a single chromatid.

1.39. What happens during mitotic telophase?

Division of the cytoplasm (cytokinesis) begins in the last phase **(telophase),** as the chromosomes unwind (decondense) and new nuclear membranes form to again enclose the sets of chromosomes at each pole of the cell. Cytokinesis in animal cells involves pinching the progenitor cell into two progeny cells. In plant cells, progeny cells are formed by construction of new membranes and walls in the cytoplasm between the new nuclei. When cytokinesis is completed, the two progeny cells normally (barring mutation) contain identical sets of chromosomes and genetic instructions. Most single-celled eukaryotes replicate new individuals asexually by the process described above or variants thereof. Multicellular fungi, plants, and animals also produce all of their body cells in this general way. All cells of the body that are not involved in sexual reproduction (i.e., all cells incapable of meiosis or the products of meiosis, gametes) are referred to as **somatic cells.**

1.40. What are some of the structural characteristics of the spindle apparatus?

Two kinds of spindle fibers are recognized: (1) **kinetochore microtubules** that extend from one pole to the kinetochore, and (2) **polar microtubules** that extend from one pole to some distance beyond the middle of the cell. Polar microtubules from each pole thus overlap with one another near the middle of the cell.

The **kinetochore**—a darkly staining, multiprotein, fibrous structure approximately 400 nm long—is attached to both a specific DNA sequence in the centromere and to the kinetochore microtubules of the spindle apparatus. During prophase, a spindle of tubulin fibers develops between the polar regions of the cell. At each end of the spindle is a **microtubule-organizing center** (MTOC), or centrosome.

Animal cells ready for mitosis have a pair of organelles called centrioles (Fig. 1-1) within each MTOC; plant cells usually do not have centrioles in their MTOCs. Each centriole is similar in structure to the basal bodies of cilia and flagella. In preparation for cell division, each pair of centrioles is thought to nucleate (or promote the growth of) microtubules into a second pair, but the details of how this occurs have yet to be worked out. As the spindle forms during prophase, one pair of centrioles moves to each pole, taking some of the pericentrosomal material with it. New centrosomes are thus established at each pole.

1.41. The mitotic spindle apparatus accomplishes the separation of the two chromatids of a replicated chromosome by two processes termed anaphase A and anaphase B. What occurs during anaphase A?

In **anaphase A,** the kinetochore microtubules shorten at their kinetochore ends by disassembly of the protein subunits in these microtubules. A similar mechanism may also operate at the polar end of kinetochore fibers. After the centromere replicates, the two former chromatids (now individual chromosomes) are free to move poleward by the shortening of the kinetochore fibers. According to this theory, no energy source (such as ATP) would be necessary for these chromosomal movements. According to an alternative model, the **sliding filament theory,** microtubule-walking proteins (such as dynein or kinesin), which are part of the kinetochore, hydrolyze ATP to provide the energy to pull the chromosome poleward along its kinetochore microtubules, and the microtubules disassemble behind these proteins.

1.42. What occurs during anaphase B?

At the same time that the chromosomes move poleward, the two poles separate in a process called **anaphase B,** accompanied by growth of polar microtubules at their distal ends near the middle of the spindle. The antiparallel polar microtubules in the region of overlap appear to slide away from one another. It is not known whether proteins similar to dynein or kinesin facilitate the sliding motion, but it is known that ATP hydrolysis is required to drive anaphase B. At the same time that polar microtubules are being pushed poleward by sliding past one another, the centrioles may be pulled apart by forces acting on astral microtubules (those radiating away from the centromere toward the cell cortex).

1.43. The predominant form of reproduction in most multicellular eukaryotes is sexual. What are the major features of the sexual life cycle?

The somatic cells of most plants and animals are **diploid** (2n), meaning they have two sets of **homologous** chromosomes. One set (n) is derived from each parent through the gametes that produced the zygote from which the organism developed. Each diploid species has a characteristic chromosome number (e.g., the normal diploid number for human somatic cells is 46). As explained in Question 1.33, meiosis is the process that reduces the chromosome number from the diploid (2n) state (characteristic of body cells) to the **haploid** (n) state characteristic of a gamete (n = 23 for humans). Parents thus contribute essentially an equal number of chromosomes and an equal amount of nuclear DNA to their offspring (see Question 1.49 for sex-linked exceptions).

In multicellular plants (metaphytes) or animals (metazoans), the zygote divides repetitively by mitosis to produce a multicelled embryo. The embryonic cells differentiate (see Chapter 12) to perform specific tasks as the individual matures. All of the somatic cells of an individual are expected to be genetically identical unless mutations have occurred along the way.

The organs of animal reproduction are called **gonads** (ovaries in females and testes in males). At sexual maturity, some diploid germ line cells of the gonads become specialized to undergo meiosis and form haploid gametes. In flowering plants (angiosperms), the male sex organs that produce pollen (containing haploid sperm nuclei) are the stamens (the anthers of which contain some cells that are capable of undergoing meiosis). The female sex organs are the pistils (the ovules of which contain cells capable of undergoing meiosis). Seeds (containing a zygote and a source of nourishment) are produced from fertilized ovules. Meiosis is the process that reduces the diploid number of certain germ line cells in the gonads of sexually mature animals (or their counterparts in plants) to the gametic haploid number. Formation of gametes and their union in fertilization creates a new individual (the zygote), completing the sexual cycle.

1.44. What are the major steps in meiosis?

Meiosis can be visualized as two highly modified cell cycles, back to back (Fig. 1-12). A complete meiotic cycle involves one initial DNA replication and two cytoplasmic divisions, yielding four haploid products, none of which are likely to be genetically identical. The two cell cycles are labeled meiosis I and meiosis II, each of which has its own prophase, metaphase, anaphase, and telophase. The chromosomes replicate in the interphase preceding meiosis I. There is no further DNA replication through both meiosis I and meiosis II.

1.45. What happens during prophase of the first meiotic division?

In the first meiotic prophase, the replicated chromosomes condense, the nuclear membrane dissolves, and a spindle develops as in a mitotic division. But in prophase I, homologous chromosomes appear to pair up with one another, point for point, gene for gene, right down their entire length, in a process known as **synapsis.** The mechanism responsible for synapsis is unknown. For each set of chromosomes that an individual inherits from one parent, there is a comparable set derived from the other parent. These homologous chromosomes share common or similar DNA segments that normally are in some way required for synapsis. Most of these **homologues** look alike; i.e., they are the same length and have the same centromeric position and same banding pattern after staining.

> **Example 1.3.** In human diploid cells, there are 22 pairs of homologues that are not involved in sex determination; these are called **autosomes.** There is also one pair of **sex chromosomes** (labeled X and Y) that do not look alike, but are still considered to be homologous because they synapse to some extent during meiosis and thus have at least some regions of similar DNA content. A diploid female cell has two **X chromosomes;** a diploid male cell has an X and a **Y chromosome.** The diploid chromosome number for the human species is thus 46. Meiosis reduces the number of chromosomes by half. Thus, the haploid chromosome number characteristic of human gametes is 23. Each egg contains 22 different autosomes and an X chromosome. Each sperm also contains 22 different autosomes, but half of the sperm cells are expected to contain an X chromosome and the other half are expected to contain a Y chromosome. Random union of the gametes from opposite sexes produces zygotes, half of which (by the laws of probability) should develop into females (XX + 22 pairs of autosomes) and half of which should develop into males (XY + 22 pairs of autosomes).

A synapsed pair of homologous chromosomes (called a **bivalent**) contains four chromatids (a **tetrad**). Each chromosome usually has one or more regions in which two of the four chromatids break at corresponding sites and reunite with one another, a process called **crossing over** (Chapter 3). A complex, ladderlike, proteinaceous structure called the **synaptonemal complex** develops in the regions where crossing over occurs and is thought to be essential for this process. Crossing over recombines **linked genes** (genes on the same chromosome). Generally, the closer two genes are located on the same chromosome (linked), the less likely it is that they will be recombined

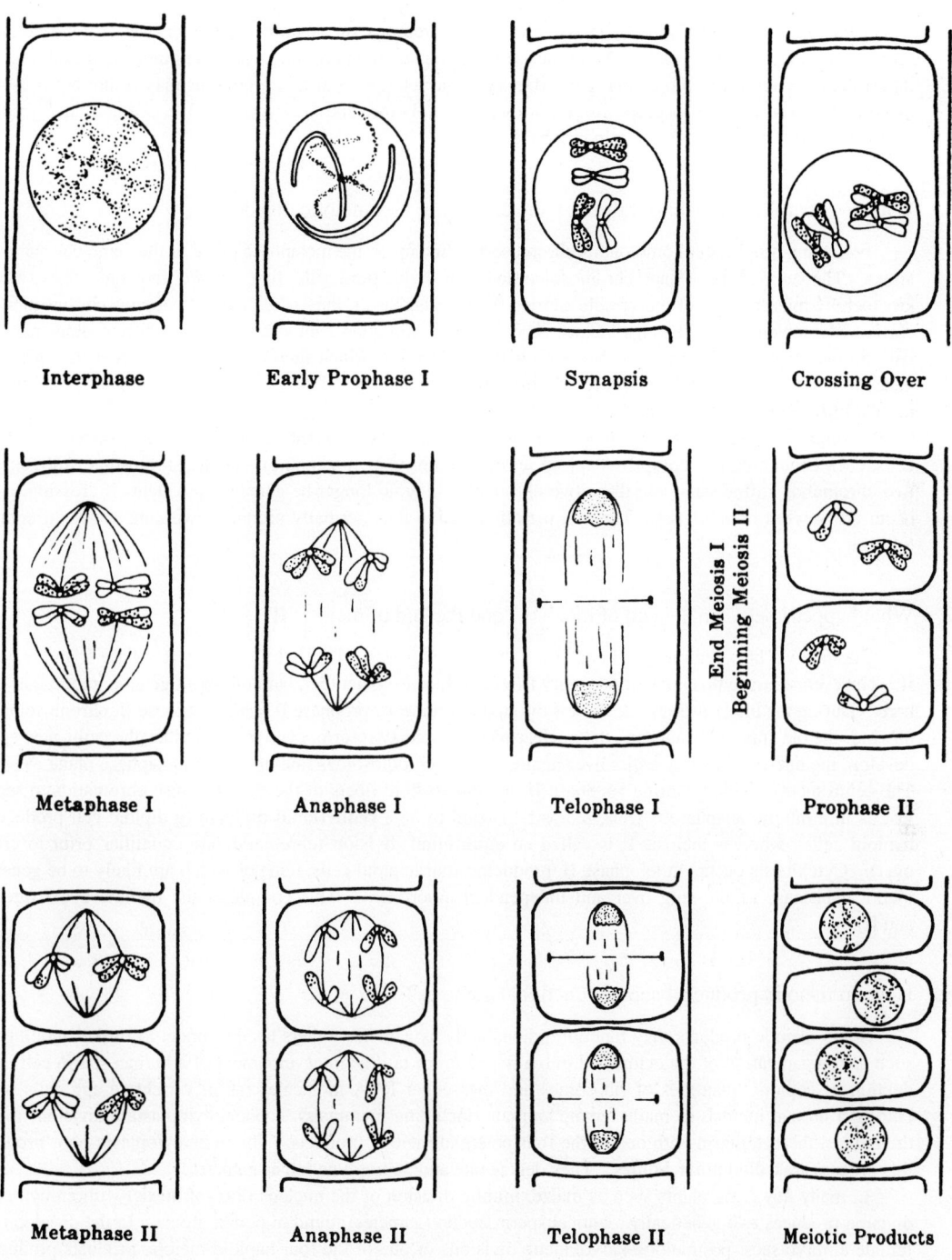

Interphase   Early Prophase I   Synapsis   Crossing Over

Metaphase I   Anaphase I   Telophase I   Prophase II

End Meiosis I
Beginning Meiosis II

Metaphase II   Anaphase II   Telophase II   Meiotic Products

Fig. 1-12.   Diagram of meiosis in plant cells. Dark chromosomes were derived from one parent (e.g., maternal or seed parent) of the individual now undergoing meiosis; light chromosomes were derived from the other parent (e.g., paternal or pollen parent). One pair of homologues is metacentric; the other pair is submetacentric. (*From Schaum's Outline of Theory and Problems of Genetics, Third Edition, McGraw-Hill, 1991.*)

with genes from the homologue by crossing over. For reasons yet unknown, crossing over is not a completely random process. Some regions of a chromosome experience more crossing over than others, and each region has its own characteristic recombination rate. Breakage and reunion of sister chromatids may occur, but it usually is of no genetic consequence because sister chromatids both before and after crossing over are normally genetically identical. All of these events occur in prophase I.

1.46. What events follow prophase I through the end of the first meiotic division?

Following prophase I, homologous chromosomes line up on the metaphase plane of the spindle as pairs (bivalents). The centromere of each chromosome does not divide during the first meiotic division. Thus, when the kinetochore microtubules of the spindle shorten during anaphase I, they segregate homologous chromosomes (not chromatids) toward opposite poles. Since each pair of homologues contains a maternal chromosome and a paternal chromosome and since chance alone normally determines which member of a pair moves to a given pole, genes that are on nonhomologous chromosomes get into gametes independently of one another. This process is termed **independent assortment.**

Cytokinesis in telophase I (shown as cell wall formation in the plant cell of Fig. 1-12) produces two haploid cells, each containing one complete haploid set of chromosomes (a genome). Each chromosome still consists of two chromatids at this stage, but these two chromatids may no longer be genetic equivalents if crossing over has occurred between homologues. The first meiotic division thus normally produces two genetically different haploid cells.

1.47. What happens between the end of meiosis I and the end of meiosis II?

The interval between meiosis I and II is called **interkinesis.** This is followed by prophase II and metaphase II. The events during this interval may vary from one species to another, but nothing of genetic importance occurs here. During the interval from telophase I through interkinesis, prophase II, and metaphase II, chromosomes may unwind, the old spindle disappears, the nuclear membrane may reform, cytokinesis divides the cells, new spindles develop, the nuclear membrane dissolves again, and the chromosomes line up on the metaphase plane as singlets. The centromeres divide. During anaphase II, the kinetochore fibers of the spindle cause chromatids to separate, just as in a mitotic anaphase. Thus, meiosis I is said to be a **reductional division** (a diploid cell produces two haploid cells), whereas meiosis II is called an **equational division** (chromatids are equalities prior to crossing over). Cytokinesis occurs in telophase II, producing four haploid cells, none of which are likely to be genetically identical because of crossing over and independent assortment. This completes the meiotic cycle (meiosis I and II).

1.48. How do meiotic products become functional gametes?

Some meiotic products may function immediately as gametes; others need to undergo further modifications, such as reorganization of the cytoplasm into a flagellum or tail (as in development of a human sperm cell) or production of copious quantities of ribosomes and messenger RNA molecules (as in vertebrate egg cells or ova). During a normal meiosis in many female animals (including mammals), unequal cytoplasmic divisions maintain the bulk of the cytoplasm with one of the four potential meiotic products. The three cytoplasm-poor products of this process are called **polar bodies.** They degenerate and do not function as gametes.

In many flowering plants such as maize, mitotic division of the nucleus (**karyokinesis**) without cytoplasmic division produces two genetically identical sperm nuclei (gametes) found in pollen grains. In the development of female embryo sacs, postmeiotic karyokinetic divisions of one of the four haploid meiotic products produce several identical nuclei, one of which becomes the egg (which, upon fertilization, becomes the zygote); two of which fuse to become a diploid nucleus (which, upon fertilization, develops into a starch-containing nutritive tissue called endosperm); and some of which die a programmed death (apoptosis). In plants with two seed leaves (such as peas and beans), the cotyledons (seed leaves) contain mainly proteins for nourishing the germinating seedling until it can get its leaves into the sunlight and begin photosynthesizing.

1.49. Compare bacteria and eukaryotes with regard to uniparental and biparental heredity.

Bacteria normally undergo binary fission, so that progeny cells derive all their heredity from the lone parent cell (uniparental inheritance). In most multicellular plants and animals, however, offspring receive equal amounts of autosomal genes from each of the two parents (biparental inheritance). Female offspring also get one X chromosome from each parent, but each male offspring receives his X chromosome from his mother and his Y chromosome from his father. The sex-linked (X-linked and Y-linked) genes of males would thus be classified as each having uniparental inheritance.

It is thought that mitochondria and chloroplasts are never produced *de novo,* but instead are produced by growth and division of preexisting organelles, somewhat like bacteria. Furthermore, in many eukaryotic organisms (including humans) with large, cytoplasm-rich egg cells (macrogametes) and tiny, cytoplasm-depleted sperm cells (microgametes), virtually all of the cytoplasm (including all of the mitochondria and chloroplasts) is transmitted to offspring only by the maternal parent. Genes in these organelles would then also be classified as having uniparental inheritance (maternal inheritance, in this case).

Objective Questions

Multiple Choice

Directions: Choose the one best answer.

1. Which of the following is possessed only by some bacteria, not by eukaryotes? (*a*) Centriole (*b*) flagella (*c*) cilia (*d*) cytoskeleton (*e*) capsule

2. Which of the following does not occur within mitochondria? (*a*) Krebs cycle (*b*) glycolysis (*c*) ATP synthesis (*d*) electron transport system (*e*) more than one of the above

3. Which of the following is not surrounded by a double membrane in eukaryotes? (*a*) The cell (*b*) the nucleus (*c*) mitochondria (*d*) chloroplasts (*e*) more than one of the above

4. Which of the following is not characteristic of photosynthesis in plants? (*a*) The oxygen evolved is derived from photolysis of water. (*b*) Light energy is first captured by PS-I and electrons are then passed to PS-II. (*c*) ATP is synthesized by an electron transport chain connecting PS-I and PS-II. (*d*) Light-dependent reactions generate NADPH and ATP. (*e*) Light-independent reactions use energy-rich molecules to reduce CO_2.

5. The phase of the eukaryotic cell cycle during which DNA replicates is designated (*a*) M. (*b*) R. (*c*) S. (*d*) G_1. (*e*) none of the above.

6. The phase of mitosis during which chromatids segregate to opposite poles of the cell is (*a*) telophase. (*b*) metaphase. (*c*) anaphase. (*d*) prophase. (*e*) interphase.

7. The chemiosmotic hypothesis offers an explanation for (*a*) anaphase movement of chromosomes. (*b*) glycolysis. (*c*) ciliary movement. (*d*) activation of ATP synthetases. (*e*) carbohydrate synthesis.

8. The Calvin-Benson cycle (carbon fixation) occurs in the (*a*) nucleus. (*b*) cytosol. (*c*) matrix of mitochondria. (*d*) thylakoids of chloroplasts. (*e*) stroma of chloroplasts.

9. Which of the following occurs in meiosis but not in mitosis? (*a*) Independent assortment (*b*) segregation of homologous chromosomes (*c*) crossing over (*d*) synapsis of homologues (*e*) more than one of the above

10. The meiotic stage during which crossing over occurs is (*a*) prophase I. (*b*) interphase. (*c*) anaphase I. (*d*) metaphase II. (*e*) prophase II.

11. A major function of dictyosomes or Golgi bodies is (*a*) in light-independent photosynthesis. (*b*) in fermentation. (*c*) for isolation of electron transport systems. (*d*) in modification of proteins. (*e*) as foci for spindle fiber formation.

12. The number of turns of the Calvin-Benson cycle required to synthesize a molecule of glucose is (*a*) one. (*b*) two. (*c*) three. (*d*) six. (*e*) twelve.

13. The Embden-Meyerhof pathway begins with glucose and ends with (*a*) pyruvate. (*b*) glucose-6-phosphate. (*c*) ethanol. (*d*) lactic acid. (*e*) citrate.

14. The metabolite that bridges the gap between glycolysis and the Krebs cycle is (*a*) oxaloacetate. (*b*) pyruvate. (*c*) acetyl CoA. (*d*) citrate. (*e*) α-ketogluterate.

15. The five-carbon sugar of the Calvin-Benson cycle that originally captures CO_2 is (*a*) PGAL. (*b*) pyruvate. (*c*) ribose. (*d*) ribulose bisphosphate. (*e*) PGA.

True–False

1. The rotation of prokaryotic flagella is powered by the hydrolysis of ATP.

2. A molecule or atom is oxidized when it loses an electron.

3. Chloroplast DNA of higher plants is about the same size as mitochondrial genomes of most animals.

4. Spindle fibers that attach to chromosomes do so directly to the DNA in the centromeric region.

5. ATP hydrolysis is unnecessary for the shortening of spindle microtubules at their kinetochore ends during anaphase A.

6. The flagella of bacteria are constructed of proteins different from those in eukaryotic flagella.

7. Chromosomes with different lengths are not genetic homologues and do not synapse.

8. Genes that are closely linked are expected to be recombined by meiosis less often than genes that are far apart.

9. All mitochondria and chloroplasts normally contain multiple copies of their single, circular, double-stranded DNA genomes.

10. Complete oxidative phosphorylation of glucose produces at least 15 times more net ATP than fermentation.

11. Cyclic photophosphorylation produces no oxygen.

12. The light-independent phase of photosynthesis occurs primarily at night (in the dark).

13. The end product of the Calvin-Benson cycle is the production of glucose.

14. In meiosis, chromosomal segregation precedes chromatidal segregation.

15. Genes on different chromosomes normally get into gametes independently of one another because homologous chromosome pairs line up randomly on the metaphase plate during meiosis I.

Matching

Directions: Match each item in **Column A** with the one in **Column B** to which it is most closely associated. Each item in **Column B** can be used only once.

Column A	Column B
1. leucoplasts	*A.* microtubule-organizing center
2. ribosome	*B.* protein modification and targeting
3. genome	*C.* protein synthesis
4. mitochondria	*D.* bacterial DNA anchorage
5. centriole	*E.* grana
6. mesosome	*F.* Krebs cycle
7. plasmid	*G.* bacterial DNA apart from the genophore
8. glycocalyx	*H.* starch storage
9. chloroplast	*I.* extracellular polysaccharide cell coating
10. Golgi complex	*J.* complete set of genetic instructions

Terms

Directions: Unless otherwise specified, each correct answer is a single word.

1. The essential complement of genetic instructions of a bacterial cell

2. A kind of bacterial metabolism in which the cell oxidizes inorganic compounds (such as molecular hydrogen, ammonia, or hydrogen sulfide) to produce organic molecules from CO_2 in the absence of light.

3. An adjective describing an organism that can live under more than one specific set of environmental conditions (such as with or without oxygen).

4. Infoldings of the inner mitochondrial membrane to which respiratory enzymes are attached.

5. The energy created by electron transport systems when hydrogen ions become concentrated on one side of a mitochondrial membrane. (3 words)

6. The diffusion of a solvent through a semipermeable membrane separating two solutions of unequal solute concentration.

7. Eukaryotic structures, resembling beads on a string, formed when DNA associates with histones.

8. Nuclear division without cytoplasmic division

9. An anaerobic type of metabolism that oxidizes glucose to ethanol or lactic acid.

10. The end product of glycolysis

11. A fibrous structure that develops at the centromeric region of a chromosome and to which spindle microtubules are directly attached.

12. A synonym for a microtubule-organizing center (MTOC) at a polar region of an animal cell

13. Molecules that accept electrons from photosystem II and transport them across membranes, thereby creating a voltage used for ATP formation.

14. The process of homologous chromosomes forming pairs during meiosis.

15. Vesicles containing photosynthetic enzymes in chloroplasts.

Answers to the Objective Questions

Multiple Choice

1. *e* **2.** *b* **3.** *a* **4.** *b* **5.** *c* **6.** *c* **7.** *d* **8.** *e* **9.** *e* (*a, b, c, d* all occur only in meiosis) **10.** *a* **11.** *d* **12.** *d* **13.** *a* **14.** *c* **15.** *d*

True–False

1. F (proton motive force) **2.** T **3.** F (chloroplast DNA is about ten times larger) **4.** F (attached to kinetochore) **5.** T **6.** T **7.** F (X and Y chromosomes are of different lengths, but they partly synapse in meiosis) **8.** T **9.** T **10.** T **11.** T **12.** F (primarily during day) **13.** F (phosphorylated intermediates of the Embden-Meyerhof pathway) **14.** T **15.** T

Matching

1. *H* **2.** *C* **3.** *J* **4.** *F* **5.** *A* **6.** *D* **7.** *G* **8.** *I* **9.** *E* **10.** *B*

Terms

1. genophore **2.** chemoautotrophy **3.** facultative **4.** cristae **5.** proton motive force (PMF)
6. osmosis **7.** nucleosomes **8.** karyokinesis **9.** fermentation **10.** pyruvate (pyruvic acid)
11. kinetochore **12.** centrosome **13.** quinones **14.** synapsis **15.** thylakoids

Chapter 2

Biomacromolecules

INTRODUCTION

The subject of biochemistry is at the heart of molecular biology. It is a vast discipline in its own right, and there is a *Schaum's Outline of Biochemistry* devoted entirely to it. **Metabolism** is the term used for all the steps by which cells utilize energy and synthesize cellular constituents. The authors assume that the reader has gained at least a rudimentary knowledge of metabolism as an essential background for studying the subjects in this book. Thus, how the small building blocks (**monomers** or **residues**) of large biological molecules (**biomacromolecules**)—such as polysaccharides, membrane lipids, proteins, and nucleic acids—are synthesized will not be discussed in this book. However, the structures and functions of these monomers, as well as the general processes by which they become linked (**polymerized**) into macromolecules called **polymers** will be addressed in this chapter. Details regarding the synthesis of proteins and nucleic acids will be discussed in separate chapters.

CARBOHYDRATES

2.1. What are carbohydrates? How are they classified?

Pure carbohydrates have the empirical formula $(CH_2O)_n$. Remove water, and all that is left is carbon. The smallest carbohydrates are simple sugars, or **monosaccharides.** Glucose (dextrose) is the 6-carbon monosaccharide (hexose) used as a basic source of energy by most heterotrophic cells. Ribose and deoxyribose are the 5-carbon sugars (pentoses) that serve a structural role in the nucleic acids RNA and DNA, respectively. **Oligosaccharides** are small carbohydrate polymers of two to about six monosaccharides. Cane sugar (sucrose) is a disaccharide of the two monosaccharides glucose and fructose (fruit sugar; an isomer of glucose). Sucrose is the major sugar transported between plant cells, whereas glucose is the primary sugar transported between animal cells. Lactose, the major sugar in milk, is a disaccharide of glucose and galactose (an epimer of glucose). Most of the carbohydrate molecules in nature are composed of hundreds of sugar units and are referred to as **polysaccharides.**

2.2. How are polysaccharides constructed? Why are they potentially such a structurally diverse class of biomacromolecules?

The monomers of polysaccharides become covalently connected by a **glycosidic bond** (Fig. 2-1) through the activity of specific enzymes. Polysaccharides can vary widely in composition, construction, and size due to four factors: (1) the kind of monosaccharides that constitute the repeating units, (2) the kinds of linkages between the sugars, (3) the length of individual chains (which is usually indeterminate), and (4) the possibility of branching and the extent thereof. Therefore, any pure sample of a polysaccharide will consist of molecules in various degrees of polymerization and hence various sizes. Two identical monosaccharides can bond in various ways to form 11 different disaccharides. Four different monosaccharides can be linked into 35,560 unique tetrasaccharides. By comparison, four different nucleotides (the monomeric units of nucleic acids) can be arranged into only 24 unique linear tetranucleotides.

Homopolysaccharides consist of a single kind of monosaccharide. Cellulose, starch, and glycogen (animal starch) are homopolysaccharides consisting entirely of glucose subunits. They differ mainly in the amount of branching. **Heteropolysaccharides** contain two or more kinds of monosaccharides. Most heteropolysaccharides contain only two or just a few kinds of monosaccharides in a repeating unit. For example, hyaluronic acid (an important component of connective tissue in animals) consists of a dimeric repeating unit containing D-glucuronic

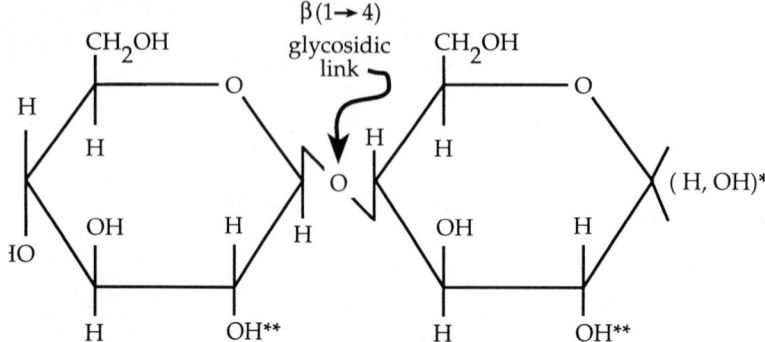

Fig. 2-1. Cellobiose, the basic repeating unit of cellulose, is a disaccharide of
glucose molecules joined by β (1→4) glycosidic linkages. The single asterisk (*) indicates that the H and OH at this position are interchangeable (i.e., free to mutarotate). Replacing the hydroxyl groups
(double asterisks, **) with NHCOCH groups creates the basic repeating unit of chitin known as *N*-acetylglucosamine (NAG).

acid and *N*-acetyl-D-glucosamine. Since the amine group contains nitrogen, the formula for this heteropolysac-
charide deviates from that of a pure carbohydrate. Heteropolysaccharides containing amino sugars, such as
hyaluronic acid, are called **mucopolysaccharides.** Chitin is another important mucopolysaccharide. It consists
entirely of *N*-acetyl-D-glucosamine units and forms the exoskeletons of insects and crustaceans. The cell walls of
fungi and the coats of mold spores may contain cellulose, but chitin is their major constituent.

2.3. What cellular functions do carbohydrates perform?

Carbohydrates serve five major functions in living systems: (1) readily utilizable energy sources in the form
of monosaccharides or oligosaccharides; (2) stored macromolecular energy sources in the form of starch in plants
or glycogen in animals; (3) structural components, such as cellulose in plant cell walls and chitin in the exoskeletons of arthropods; (4) signaling molecules in cellular communication; and (5) specific cell-to-cell attachments.

2.4. The great potential for structural diversity of carbohydrates, discussed in Question 2.2, makes these molecules very useful as cell-recognition markers. How does this recognition phenomenon occur?

Surface carbohydrates are often complexed with proteins as **glycoproteins** or with lipids as **glycolipids** in the
plasma membrane. The kinds and amounts of surface glycoproteins can change in a programmed manner as a
cell develops, differentiates, becomes infected with microbes or viruses, ages, or becomes cancerous. These surface carbohydrates generally do not interact with one another. Rather, each surface carbohydrate "marker" molecule is recognized by a specific surface carbohydrate-binding protein called a **lectin** or **selectin.** The interaction
between a lectin on one cell and its carbohydrate target on another cell is exquisitely specific. This specificity is
explained by the stereochemical structures (the three-dimensional arrangement of atoms in molecules) of the reactants and the properties of weak chemical interactions (hydrogen bonds, ionic bonds, van der Waals forces).
There must be a precise orientation in space and distance between reactants—analogous to a hand in a glove or a
lock and its key—to account for the specificity between a lectin and its cognate carbohydrate, between a protein
enzyme and its substrate, or between a protein antibody and its cognate antigen. Like surface carbohydrates, the
kinds and/or amounts of surface lectins change with a cell's developmental, physiological, and pathological states.

2.5. Give some examples of specific cellular interactions mediated by carbohydrates.

All cells have carbohydrates on their surfaces that may serve as points of attachment between (1) similar cells
(as during embryonic differentiation), (2) dissimilar cells (as when pathogenic bacteria infect specific types of host
cells), (3) hormones and their target cells, and (4) viruses or bacterial toxins and host cells.

Example 2.1. Generally speaking, after all the organs of the embryo have formed, normal cells that were mobile during embryogenesis no longer migrate through tissues. Lymphocytes are a notable exception to this rule. These immunological cells continuously migrate through the body in search of foreign antigenic material. Lymphocytes are known to be specifically selected to migrate through the endothelial lining of small blood vessels and through the tissue spaces of particular organs by attaching to carbohydrate "homing receptors" on the surface of endothelial cells. Also, cancer cells often bear unusual surface carbohydrates that may be essential for the ability of these cells to invade and migrate, or **metastasize,** through not only their tissues of origin, but also throughout the body.

Example 2.2. Bacterial pathogens have not only a specific **host range** (set of one or more vulnerable host species) but often specific cell or tissue types to which they adhere. It is generally believed that bacterial adhesion is mediated by specific lectins on bacteria that bind to complementary carbohydrates on susceptible host cells. This is the first step in the process of infection. For example, the causative organism of gonorrhea, *Neisseria gonorrhoeae,* adheres to genital and oral epithelial cells of humans, but not to cells from other organs or other animal species.

Bacterial lectins may bind to the ends of surface carbohydrates or to more internal parts of these molecules. Different bacteria, having different carbohydrate-binding lectins, may adhere to different parts of the same surface carbohydrate receptor. Sometimes only one face of a surface carbohydrate may be exposed on a host cell. Consequently the cell can be recognized by one bacterial species, but not by others. Thus, the mode of presentation of a surface carbohydrate may also be an important aspect of its ability to act as a recognition molecule.

2.6. The four most prevalent polysaccharides in eukaryotes are cellulose, chitin, starch, and glycogen. Compare the structures of cellulose and chitin. How are these molecules digested?

Cellulose is the most abundant organic compound in the biosphere. Together with pectins, lignins, hemicelluloses, and glycoproteins, cellulose forms the cell walls of plants, providing protection from lysis by internal osmotic pressure and from damage by external abrasion. Cellulose is a homopolysaccharide with repeating subunits of the disaccharide cellobiose, which in turn consists of D-glucose units joined by $\beta(1\rightarrow4)$ glycosidic linkages (Fig. 2-1). Mammals cannot metabolize cellulose because they lack the enzyme **cellulase** that cleaves the $\beta(1\rightarrow4)$ glycosidic linkages. Herbivorous mammals have symbiotic microflora in their gut to do this job.

Chitin is the second most abundant organic compound in nature. It is the major structural component of the exoskeletons of insects and crustaceans. Chitin is composed of repeating subunits of the sugar *N*-acetylglucosamine in $\beta(1\rightarrow4)$ linkage (Fig. 2-1). *N*-acetylglucosamine is a major component of bacterial cell walls (see Question 2.10). The enzyme **chitinase** digests chitin.

2.7. What is the composition of starch? How is it digested?

Starch is the major molecule in which plants store energy. Adenosine triphosphate (ATP) is the primary molecule for supplying energy to cells, but ATP does not accumulate in cells the way that starch can. Like cellulose, starch is a homopolymer of glucose units, but they are joined in $\alpha(1\rightarrow4)$ and $\alpha(1\rightarrow6)$ linkages (Fig. 2-2), whereas cellulose contains $\beta(1\rightarrow4)$ linkages. Furthermore, starch consists of a mixture of two kinds of glucose polymers: an unbranched form called **amylose** and a branched form called **amylopectin.** During digestion by enzymes such as salivary and pancreatic **amylases,** starch is hydrolyzed to maltose and glucose. Maltose is a disaccharide of two glucose units joined by an $\alpha(1\rightarrow4)$ link that can be cleaved by the enzyme maltase to yield two glucose molecules.

2.8. Compare the structures of glycogen and starch.

Glycogen is the most abundant polysaccharide for energy storage in animals. It consists of polymers of glucose units joined by $\alpha(1\rightarrow4)$ linkages, and it also forms branched chains by $\alpha(1\rightarrow6)$ linkages (hence the name "animal starch"). However, starch has fewer $\alpha(1\rightarrow6)$ linkages than glycogen.

2.9. How are polysaccharides synthesized?

The synthesis of polysaccharides such as glycogen, cellulose, and starch or disaccharides such as sucrose requires the glucose monomer to be in an activated form, which in the case of glycogen and sucrose is called uridine

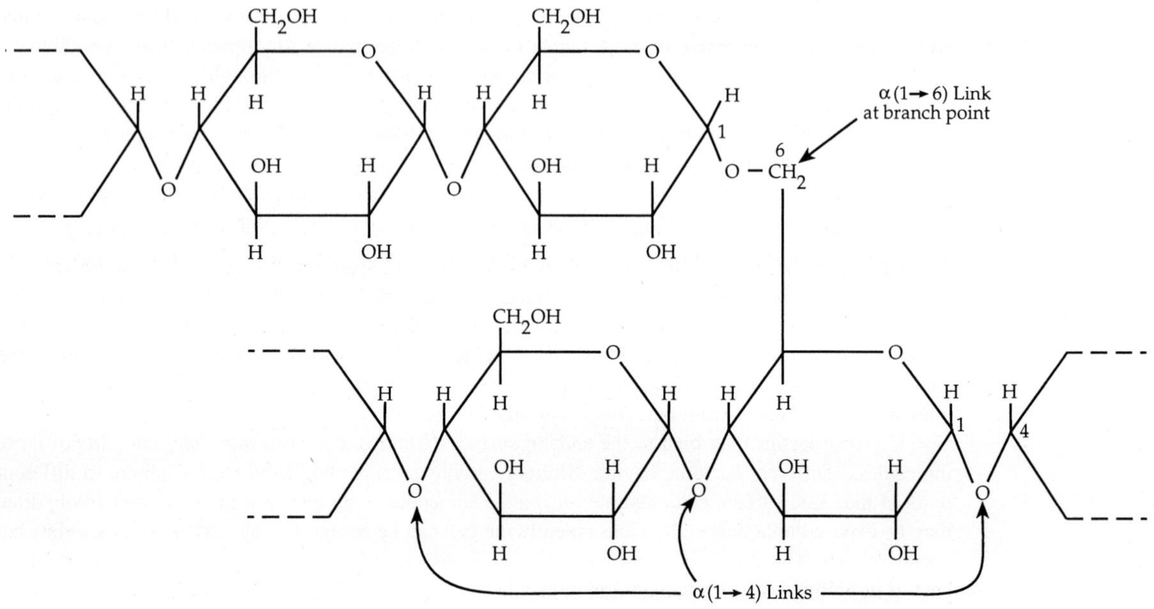

Fig. 2-2. Branching between linear chains of glucose subunits in amylopectin and glycogen. Glycogen is essentially identical to amylopectin except that glycogen is more highly branched. Branching in amylopectin occurs approximately once every 25 glucose units.

diphosphate glucose (**UDP-glucose;** Fig. 2-3). The enzyme glycogen synthetase catalyzes the transfer of glucose from UDP-glucose to the terminal hydroxyl group of a growing glycogen strand.

Biosynthesis of cellulose is thought to involve **GDP-glucose** (the guanine-containing analog of UDP-glucose) or other nucleotide diphosphate sugars (depending on the species of plant), but the details of the process are not well understood.

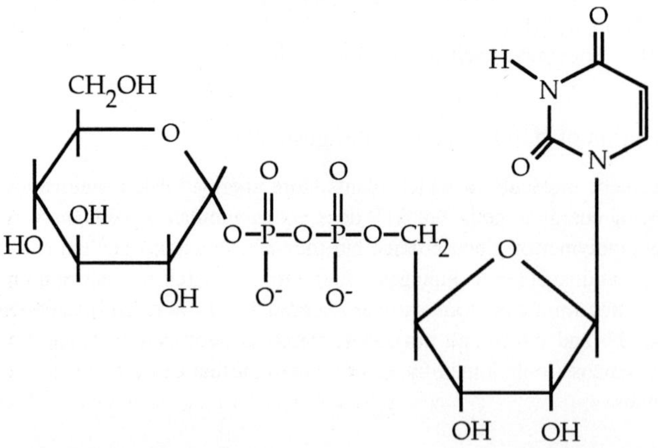

Fig. 2-3. Uridine diphosphate glucose, the activated monomer of glycogen synthesis.

Starch synthesis occurs when glyceraldehyde 3-phosphate (produced in the stroma of chloroplasts by the carbon-fixation cycle) is converted to glucose 1-phosphate, which in turn is used to synthesize the immediate activated precursor of starch, a nucleoside diphosphate sugar called adenosine diphosphate glucose (**ADP-glucose).** Starch is stored as large grains in the stroma during periods of excess photosynthetic activity and is broken down at night to supply the energy needs of the plant. However, much of the glyceraldehyde 3-phosphate produced in chloroplasts is transported to the cytoplasm, where it is converted into glucose 1-phosphate and fructose 6-phosphate by reversal of some steps in glycolysis. Glucose 1-phosphate becomes converted to UDP-glucose, which combines with fructose 6-phosphate to form sucrose phosphate, the immediate precursor of sucrose.

2.10. How is the structure of bacterial cell walls involved in bacterial classification?

The cell walls of most bacteria do not contain cellulose, but rather a different polysaccharide, called **peptido-glycan.** Peptidoglycan consists of two alternating sugars (*N*-acetylglucosamine and *N*-acetylmuramic acid). In addition, a chain of four amino acids (a tetrapeptide) is attached to each *N*-acetylmuramic acid molecule (Fig. 2-4) involved in connecting adjacent peptidoglycans, forming a single giant molecule that functions as the cell wall.

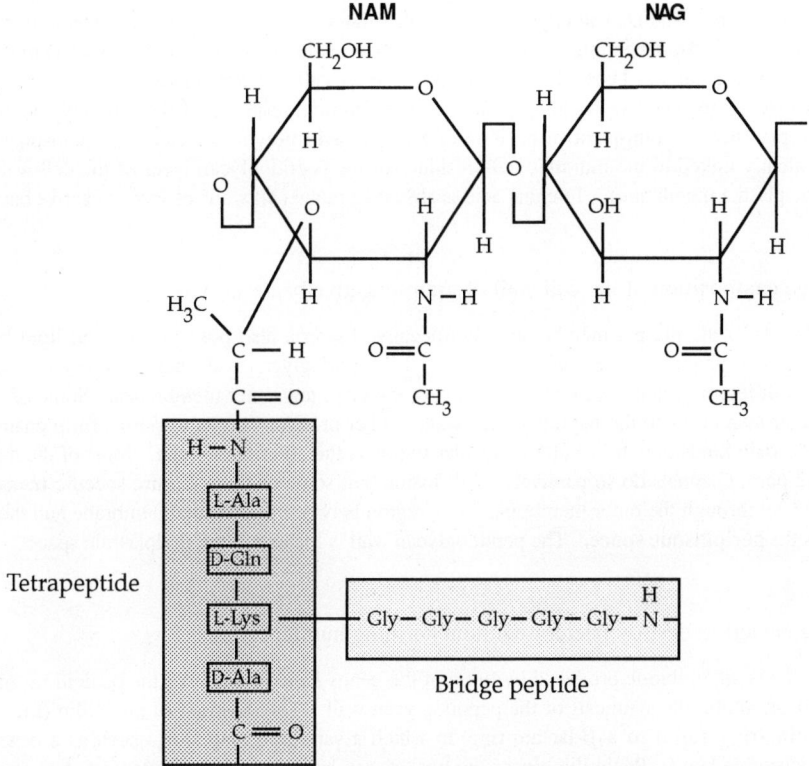

Fig. 2-4. Repeating structural unit of the peptidoglycan in the cell wall of *Staphylococcus aureus*. NAM = *N*-acetylmuramic acid; NAG = *N*-acetylglucosamine; the tetrapeptide consists of L-alanine, D-glutamine, L-lysine, and D-alanine; bridge peptide = peptidoglycine (Gly$_5$). During growth of the macromolecule, a D-alanine is added to the one already in the tetrapeptide. However, this additional D-alanine is subsequently removed when the pentaglycine (bonded at one end to L-lysine) becomes bonded at its other end to a D-alanine in a tetrapeptide of a different polysaccharide chain.

The vast majority of bacteria have a common cell wall component (peptidoglycan), similar protein synthesizing molecules (tRNAs, rRNAs; Question 2.56), and similar RNA synthesizing enzymes (RNA polymerases; Chapter 4). Exceptions exist, however. Mycoplasmas are considered to be the simplest extant (alive) prokaryotes that are capable of a free-living existence; they have no cell wall (see Question 2.15). The archaebacteria are a group of prokaryotes with walls that lack peptidoglycans. Mycoplasmas and the prokaryotes with peptidoglycan in their cell walls are considered to be so structurally similar at the molecular level as to be classified together in a group called **eubacteria.** The **cyanobacteria** (blue-green bacteria, blue-green algae) are prokaryotes with extensive intracellular membranes supporting photosynthesis, which releases molecular oxygen. These major structural and metabolic differences are recognized by many taxonomists who exclude cyanobacteria from the eubacteria. For the purposes of this text, however, the term eubacteria will also include the cyanobacteria. A third major division of prokaryotes is the archaebacteria. These organisms lack peptidoglycan in their cell walls (although some have a similar polysaccharide called pseudopeptide glycan), and their tRNAs, rRNAs, and RNA polymerases are distinctly different from those of eubacteria. The archaebacteria will be more thoroughly discussed in Chapter 14. Thus, the presence or absence of a cell wall or the structure of the cell wall are not the definitive criteria for classifying bacteria into the two major groups—eubacteria and archaebacteria.

2.11. Eubacteria have two kinds of cell walls. How are they distinguished?

Eubacteria having relatively thick peptidoglycan layers retain the dye crystal violet better than those with thin peptidoglycan layers. The Danish physician Hans Christian Gram (1853–1938) discovered that bacteria could be divided into two groups, depending on whether or not they retain crystal violet (blue violet) or safranin (light red) after washing with ethanol. Those that stain deep purple are called **gram-positive** bacteria; those that stain a pink color are called **gram-negative** bacteria. Part of the differential response of these two types of bacteria to Gram stain is the presence in gram-positive bacteria of **teichoic acid** polymers (alternating phosphate and glycerol or ribitol covalently linked to the muramic acid residues of the peptidoglycan layer of the cell wall or to the glycolipids of the plasma membrane). Teichoic acid is absent from the cell walls of gram-negative bacteria.

2.12. What is the composition of the cell wall of gram-negative bacteria?

In addition to the plasma membrane, gram-negative bacteria also possess a second lipid bilayer (called the **outer membrane** or **outer envelope**) surrounding the peptidoglycan wall, but considered to be part of the cell wall. Many different proteins are embedded in both the inner and outer membranes. Some of these proteins anchor the outer membrane to the peptidoglycan wall. Other proteins known as porins form channels or pores that allow only certain kinds of extracellular molecules to transit the outer membrane. Most of the molecules that pass through the porin channels do so passively by diffusion, but some of them require specific **transport proteins** or receptors to get through the outer membrane. The region between the plasma membrane and the outer membrane constitutes the **periplasmic space.** The peptidoglycan wall is located in the periplasmic space.

2.13. How does penicillin destroy bacteria without harming human cells?

Penicillin is an antibiotic produced by fungi of the genus *Penicillium.* All the penicillins, natural and chemically modified, inhibit the synthesis of the peptidoglycan wall. The structure of penicillin (Fig. 2-5) consists of a thiazolidine ring fused to a β-lactam ring, to which a variable group (R)—such as a benzyl group—is attached by a peptide bond. Penicillin blocks the final step in bacterial cell-wall synthesis by irreversibly inhibiting the enzyme **glycopeptide transpeptidase** that cross-links different peptidoglycan strands. Without extensive crosslinking, the cell wall loses its structural integrity. Penicillin is exquisitely specific for the transpeptidase enzyme because of its structural resemblance to the D-Ala-D-Ala terminus of nascent strands of peptidoglycan, which is one of the substrates of this enzyme. Since no human enzyme is known to recognize this sequence, penicillin does not interfere with our own enzymatic processes.

Most bacteria contain enzymes in their cell wall or periplasmic space that, when activated, disrupt the cell wall. During normal growth, these self-destructive enzymes are restrained. But when these cells die, the controls that inhibit these enzymes are released, resulting in cell lysis. These controls are also deactivated when cell-wall construction is inhibited by penicillin. The processes that regulate these control mechanisms, however, remain unknown.

2.14. Under what conditions can penicillin-sensitive bacteria be unaffected by penicillin?

Penicillin-sensitive bacteria can be grown in the presence of penicillin if the external medium is **isotonic** (containing the same amount of dissolved substances or solutes) or slightly **hypertonic** (containing more solutes) to that of the cell. Penicillin has no effect on bacteria in stationary (nongrowing) phase because it does not cause breakdown of bacterial cell walls; it only interferes with cell-wall synthesis.

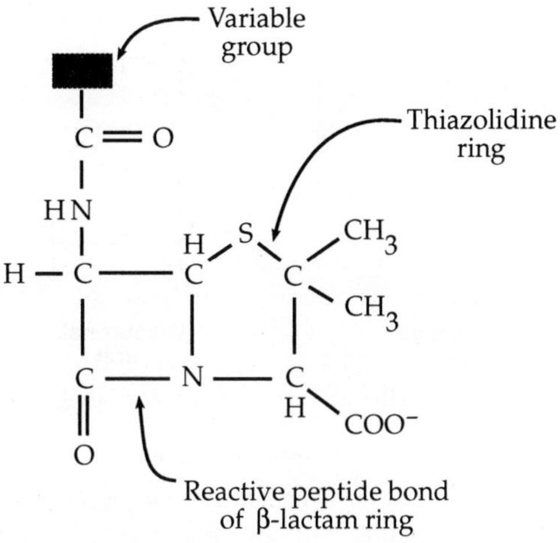

Fig. 2-5. General structure of a penicillin molecule.

2.15. What are protoplasts, and how do they respond to penicillin?

Bacteria devoid of a cell wall are called **protoplasts** or **L** (Lister) **forms,** and they will lyse if transferred to a **hypotonic** medium (containing fewer solutes than those within the cell). Protoplasts exist as mutant (rare) forms in most species of bacteria, and they are usually capable of reverting back to the normal type at very low frequency. However, all bacteria of the genus *Mycoplasma* normally lack cell walls, and they do not mutate to possess cell walls. They are stable in ordinary culture media because of sterols in their membranes. These sterols are not in the membranes of other bacteria or L forms. Penicillin has no effect on protoplasts or on mycoplasmas. **Mycoplasmas** are the group of bacteria considered to be the simplest forms of life capable of a free-living existence. They are highly variable in shape **(pleomorphic)** due to the absence of a cell wall. Penicillin also has no effect on viruses, because they, too, do not have cell walls.

2.16. Why are some bacteria resistant to penicillin?

Some bacteria synthesize and secrete an enzyme (called **penicillinase** or **β-lactamase**) that cuts the amide bond in the β-lactam ring of penicillin to form penicilloic acid, which has no antibacterial properties. The activity of various penicillinases depends mainly on the kind of R group attached to the β-lactam ring. Semisynthetic penicillins have been synthesized to have R groups that cannot be recognized by most penicillinases, making them more medically useful. Another type of penicillin resistance occurs in mutant strains of bacteria that produce only an altered glycopeptide transpeptidase enzyme that penicillin cannot recognize.

2.17. The polysaccharide capsule of the bacterium *Streptococcus* played an important role in the discovery that DNA carries specific genetic information. What is this story?

Streptococcus pneumoniae causes pneumonia in humans, mice, and other susceptible mammals. Normally the cell walls of these bacteria are surrounded by a slimy, glistening, polysaccharide capsule. One type of capsule consists of an alternating sequence of glucose and glucuronate units (Fig. 2-6). This capsule renders the bacteria refractory to destruction by immune cells of the host organism. Since the encapsulated cells form colonies with smooth borders on nutrient agar plates, they are referred to as **S-type** pneumococci. A mutant strain of the pneumococcus lacks the dehydrogenase enzyme that converts uridine diphosphate glucose (UDP-glucose) into UDP-glucuronate in the pathway to capsule formation. These mutant bacteria plate out as colonies with rough borders, and so are termed **R-type** pneumococci.

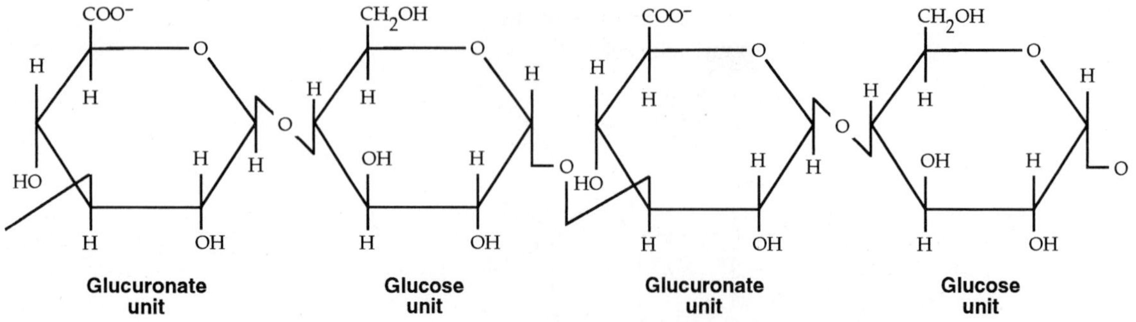

Fig. 2-6.   One type of capsular polysaccharide, consisting of alternating glucose and glucuronate units.

In 1928, Fred Griffith discovered that the **avirulent** (nonpathogenic) R strain could be transformed into the **virulent** (pathogenic) S strain by injecting into mice a mixture of live R strain and heat-killed S strain. Neither R strain alone nor heat-killed S strain alone killed mice.   But the mixture killed mice, and live S strain pneumococci were found in the blood of the dead mice.   The change from R-type to S-type was permanent and hereditary.   This process of **transformation** (assimilation of naked DNA from the environment; Question 7.10) can also occur *in vitro* by adding a cell-free extract of heat-killed S strain to live R strain.   Griffith did not know the nature of the "transforming principle" (as he called it).

Sixteen years later (1944), Oswald Avery, Colin MacLeod, and Maclyn McCarty reported that the transforming principle was deoxyribonucleic acid (DNA).   Their conclusion was based on several facts.   (1) Chemical analysis of purified and highly active transforming principle indicated that its structural components agreed closely with those of DNA.   (2) When subjected to ultracentrifugal, optical, and electrophoretic analyses, transforming principle appeared to have the same characteristics as DNA.   (3) Removal of proteins and lipids had no effect on the transforming ability of the transforming principle.   (4) Treatments with the protein-digesting enzymes (**proteases** or **peptidases**) trypsin and chymotrypsin, as well as treatments with a **ribonuclease** enzyme that digests ribonucleic acids, all had no effect on the transforming principle.   (5) Transforming principle was destroyed when treated with **deoxyribonuclease** (an enzyme that specifically degrades DNA).

Before the report of Avery *et al.,* biologists thought that chromosomal proteins were the genetic material. But the rigorous proofs supplied by these researchers indicated that DNA has genetic specificity.   Although most scientists were reluctant to immediately give up the prevailing theory (**paradigm**) of protein heredity, eventually many other lines of evidence provided by independent workers led to the acceptance that DNA is the hereditary material and that protein contains no genetic information.

LIPIDS

2.18.   What distinguishes lipids from carbohydrates, proteins, and nucleic acids?

Lipids are water-insoluble (nonpolar) molecules that are soluble in weakly polar or nonpolar solvents such as chloroform.   The definition of lipids is based on a physical property, whereas the definitions of carbohydrates, proteins, and nucleic acids are based on their chemical structures.

2.19.   What are the two most important functions that lipids perform in living organisms?

The most important function that lipids perform for all kinds of cells stems from their ability to form sheet-like membranes.   The plasma membranes of both prokaryotic and eukaryotic cells separate the protoplasm within the cell from its external environment, thus allowing the cell to function as a unit of life.   Eukaryotic cells also have internal membranes—such as those of the endoplasmic reticulum, nucleus, mitochondrion, and chloroplast—that further compartmentalize the cell for specific functions.

The other important function of lipids is as energy-storage molecules.   Carbohydrates such as starch (in plants) and glycogen (in animals) also serve this function.   However, the fatty glycerolipids (especially the triglycerides) are by far the most abundant lipids in animals and are much more efficient energy-storage molecules than glycogen or starch.   For example, a gram of anhydrous fat stores more than six times as much energy as a gram of glycogen.

2.20. There are three major kinds of membrane lipids: phospholipids, glycolipids, and sterols. Briefly describe the structure of each of these molecular classes.

The phosphoglyceride type of **phospholipid** (Fig. 2-7a) has a phosphorylated alcohol in the head unit esterified to glycerol, which, in turn, is esterified to two fatty acid chains in their tails. Glycerol, ethanolamine (bacterial membranes), inositol (brain tissue), and choline (eukaryotic cells) are some of the alcohol moieties found in membrane phosphoglycerides. Both of the fatty acyl side chains may be **saturated** (lack double bonds) or **unsaturated** (contain one or more double bonds); or one chain may be saturated and the other unsaturated. The only phospholipid in membranes that is not derived from glycerol is sphingomyelin. It has a phosphoryl choline unit in its polar head. The backbone of sphingomyelin is sphingosine, an amino alcohol linked to an unsaturated hydrocarbon chain by an amide bond.

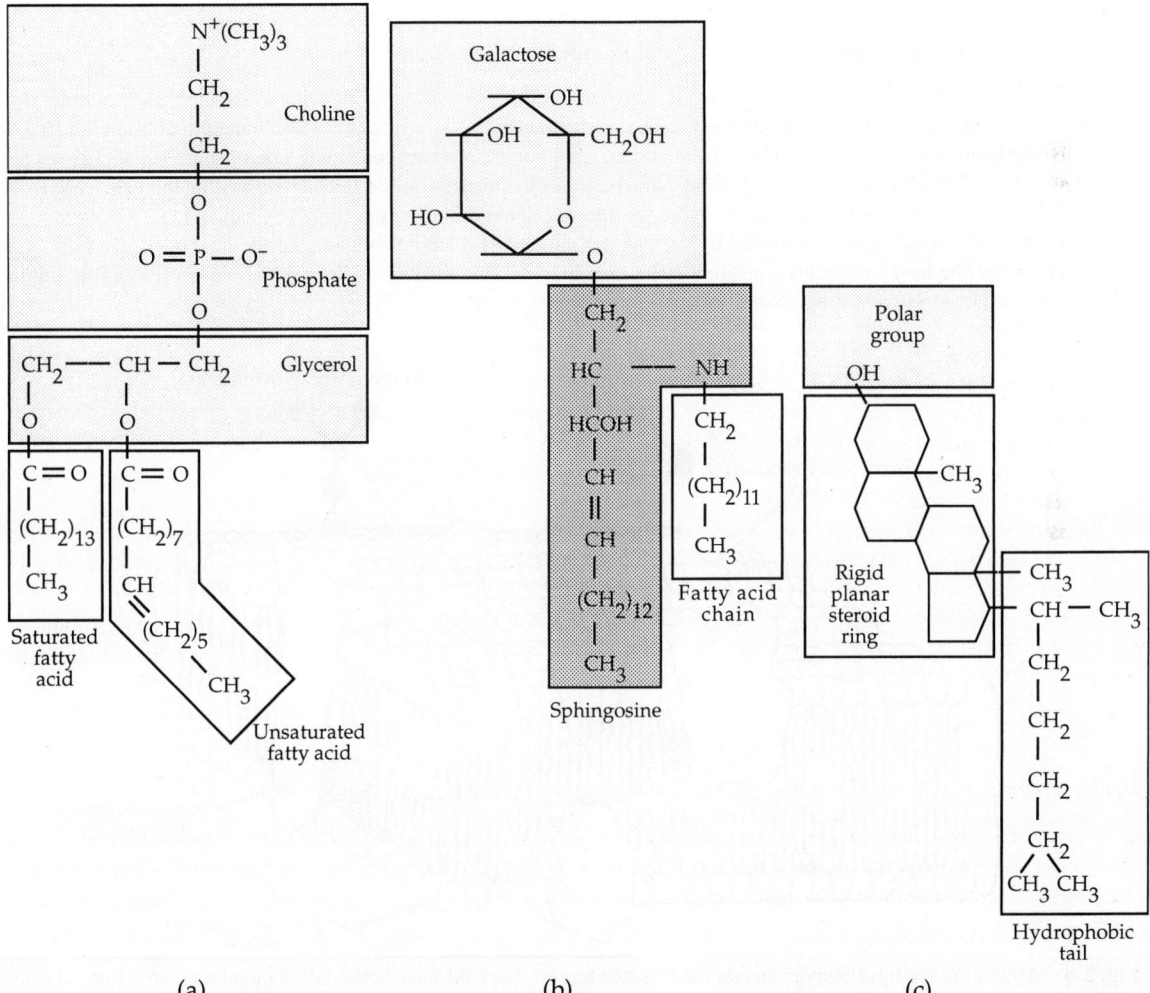

(a)   (b)   (c)

Fig. 2-7. Three major kinds of membrane lipids are (a) phospholipids (e.g., the phosphoglyceride called phosphatidylcholine); (b) glycolipids (e.g., a glycosphingolipid); and (c) sterols (e.g., cholesterol).

Glycolipids in animal cell membranes are glycosphingolipids (Fig. 2-7b) containing one or more polar (water-soluble) sugar units in their heads, with both a hydrocarbon chain of sphingosine and an organic chain or alcohol in their tails. The sugar moieties of glycolipids (or of glycoproteins embedded in the plasma membrane) are invariably located on the external surface of cells. In the glycolipids of bacteria and plants, glycerol forms the connection between the saccharide and the fatty acid chain. Both glycolipids and phospholipids readily associate spontaneously to form a lipid bilayer membrane (see Question 2.23).

Sterols, such as cholesterol, contain a multiring structure similar to the one shown in Fig. 2-7c. Cholesterol has a polar OH^- group in its head; the entire molecule except for the OH^- group is its nonpolar unit. Cholesterol is richly present in the plasma membranes of animal cells, but is relatively meager or absent in the plasma membranes of plants, as well as the membranes of both plant and animal organelles. Sterols other than cholesterol play important roles in plasma membranes of plants. Except in the membranes of micoplasmas, sterols are not present in most prokaryotes. In more complex animals such as mammals, cholesterol is the general biosynthetic precursor for steroids, including sex hormones (such as testosterone, estradiol, progesterone) and hormones of the adrenal cortex. These hormones control gene expression by inhibiting or stimulating protein synthesis. Sterols readily pass through lipid membranes and can react directly with cytoplasmic or nuclear constituents. Other kinds of chemical messengers (such as protein hormones) must be recognized by specific cell-surface proteins called receptors. This kind of protein–protein interaction often leads to a cascade of reactions in the cytosol that eventually influences DNA activities in the nucleus.

2.21. What factors contribute to the fluidity of lipid membranes?

Cellular membranes behave as two-dimensional, semifluid structures, allowing embedded protein molecules to constantly move about rather freely by lateral diffusion (Fig. 2-8). However, membrane proteins are not free to rotate from one side of a membrane to the other. Membrane proteins containing one or more sugars (glycoproteins) and membrane glycolipids always have their sugar moieties on the outside surface of the cell. Many of these saccharide units function as receptors for intercellular communication signals (Chapter 11). The fluidity of prokaryotic membranes is regulated by varying the number of double bonds in, and the lengths of, the fatty acid chains of the lipid molecules constituting the membrane. The quantity of the sterol lipid cholesterol in animal membranes is also a key regulator of membrane fluidity.

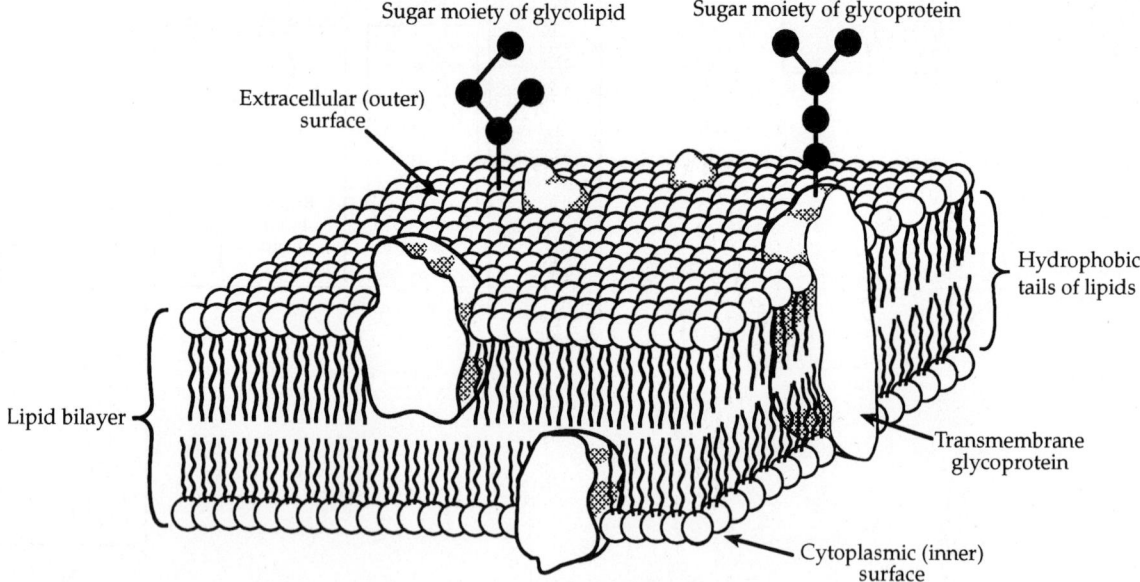

Fig. 2-8. Diagram of a lipid bilayer membrane. According to the fluid mosaic model of membrane structure, various proteins embedded in the membrane may be highly mobile. The mosaic bilayer of polar lipids is approximately 5 nm (50 Å) thick.

2.22. The plasma membrane is a selective filter that controls the entry of nutrients and other molecules needed for cellular processes. Waste products of metabolism pass out of the cell through this membrane. How do these molecules move across the plasma membrane?

The lipids in biological membranes usually have tails 16 to 18 carbon atoms in length. Such membranes have a very low permeability for ions and most polar molecules. Water, however, can easily diffuse passively (without the expenditure of energy) through membranes from areas of high to low concentrations. Most other molecules that contain polar groups cannot pass through lipid membranes, but may enter the cell passively via ion channels formed by specific integral membrane proteins or be transported actively across the membrane by specific membrane carrier proteins with the expenditure of energy. For example, a sodium-potassium transport system exists in plasma membranes to pump Na^+ out of the cell and K^+ into the cell against concentration gradients. Hydrolysis of ATP to ADP provides the energy to run the pump. The ability of cells to control their concentrations of ions and nutrients in various compartments of the cell and to respond to extracellular signals depends almost entirely on the kinds and amounts of specific proteins embedded in various cell membranes.

2.23. Proteins and nucleic acids require a template for their synthesis. Polysaccharides and lipid membranes are not constructed from templates. How are macromolecular lipid membranes synthesized?

The kind of phosphorylated lipids found in membranes are **amphipathic** compounds that have a **hydrophilic** (water-loving) polar "head" region and one or more nonpolar **hydrophobic** (water-fearing) "tails" (usually fatty acid chains alone or with other hydrocarbon compounds). Unlike polysaccharides that require enzymes for their synthesis, membrane lipids tend to spontaneously form a macromolecular **bilayer** aggregate (usually about 50 Å thick) in water when the lipid concentration exceeds a critical level. (One Ångstrom unit is 10^{-10} meter; a nanometer, nm, or millimicron, mμ, is 10^{-9} meter or 10^{-1} Å). Below this concentration, individual lipid molecules remain soluble in the aqueous phase. The charged polar heads of phospholipid molecules are on the outside of the bilayer, where they can interact with water. The water-insoluble hydrocarbon tails of membrane lipids avoid water by associating with one another noncovalently in the center of the bilayer—a phenomenon known as "hydrophobic bonding." This bilayer structure forms what is known as a **unit membrane.** To avoid interaction with water, a lipid bilayer tends to form a self-sealed hollow pancake or sphere called a vesicle. A cell is actually a giant vesicle bounded by a lipid bilayer (a unit membrane) called the plasma membrane.

PROTEINS

2.24. What are proteins and what general features do they share?

Each protein chain is synthesized on ribosomes and consists of an unbranched, linear sequence of amino acid monomers connected by covalent amide linkages called **peptide bonds.** Twenty kinds of amino acids can be used in the construction of proteins. The absolute configuration of amino acids and sugars is referenced to the structure of D-glyceraldehyde. The nonsuperimposable mirror image of the D-form (**enantiomer**) is called the L-configuration (L for levo, or "left handed"), and this is the form found in all biological proteins synthesized on ribosomes. However, some D-amino acids (D for dextro, or "right handed"; the optical isomer of the L-form) are found in certain prokaryotic molecules such as in bacterial cell walls and in some peptide antibiotics such as gramacidin S. All amino acids have the same general structure as indicated in Fig. 2-9. An α-**carbon** is at the center of each amino acid. To its left (as conventionally written) is a basic (when ionized) amino group (NH_3^+). To the right of the alpha carbon is an acidic (when ionized) carboxyl group (COO^-). A hydrogen atom forms a third bond to the α-carbon, and the fourth bond connects to a side-chain group (R). One amino acid differs from another according to the nature of the R group.

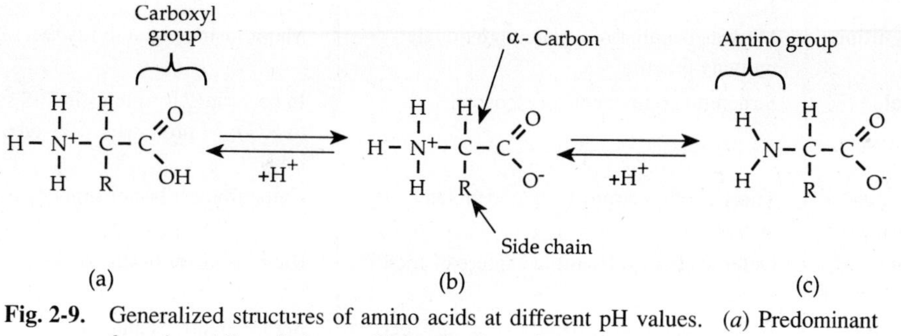

Fig. 2-9. Generalized structures of amino acids at different pH values. (*a*) Predominant form in acidic solutions. (*b*) Predominant form at neutrality (pH 7). (*c*) Predominant form in basic (alkaline) solutions.

2.25. Proteins perform many enzymatic, structural, and other roles in living systems. List some of the major functional classes of proteins and give at least one example for each class.

> See Table 2.1.

2.26. How are amino acids grouped or classified?

> Amino acids can be classified in various ways. Table 2.2 classifies amino acids into eight groups. Aliphatic amino acids have straight or linear carbon side chains. Aromatic amino acids have ring-shaped carbon side chains. Heterocyclic amino acids have at least one ring containing both carbon and nitrogen.

Table 2.1. Some Functions of Proteins and Peptides

Protein Class	Function	Examples
Motility	Allow movement of organelles within the cell and movement of the cell itself (amoeboid motion, endocytosis, exocytosis)	Tubulin: flagellin; actin; intermediate filaments of the cytoskeleton
Anchoring	Attachment of one cell to another or to a substratum	Fibronectin; cadherins
Contractile	Contraction by means of sliding filaments	Actin; myosin
Structural	Provide support and strength	Collagen; keratin; elastin; fibroin; silk
Enzymes	Catalyze biochemical reactions	DNA polymerase; β-galactosidase
Transport	Carry molecules across membranes or from one cell to another	Hemoglobin; galactoside permease; transferrin; haptoglobin; porins
Hormones	Signaling between different cell types; stimulation or inhibition	Insulin; glucagon; follicle stimulating hormone (FSH)
Nutritional	Food storage	Egg albumin; zein
Cell-surface receptors	Label certain cells as targets for hormones, viruses, growth factors; photoreceptors; transmission of nerve impulse	Insulin receptors; phage attachment sites; nerve cell growth factor; rhodopsin; acetylcholine receptor
Nucleo-proteins	Bind to DNA to condense its length; bind to RNA for translation	Histones; translational factors for initiation, elongation, termination
Repressors	Bind to DNA to prevent transcription	*lac* repressor in *E. coli*
Activators	Bind to DNA to stimulate transcription	Catabolite activator protein (CAP)
Neurotrans-mitters	Signaling between neurons	Enkephalins; endorphins
Immuno-globulins	Recognition of antigens; immune functions	Antibodies of classes IgG, IgM, IgA
Complement	Cytolysis; immune functions	C1 through C9
Hemostatic factors	Blood coagulation	Thrombin, fibrin
Self-recognition	Markers peculiar to a given individual; immune function	Major histocompatability (MHC) antigens
Ribosomal	Structural components of ribosomes	In bacteria, 21 proteins in 30S subunit (S1 to S21); 34 proteins in 50S subunit (L1 to L34)
Energy trans-ductions	Transfer of electrons from one molecule to another	Cytochromes; bacteriorhodopsin
Poisons or toxins	Defense chemicals; aid in capture of food	Ricin in castor beans; snake venoms
Homeostasis	Aid in regulation of osmotic pressure, pH, and viscosity of blood	Serum albumins and globulins

Table 2.2. Structural formulas of the 20 amino acids that may exist in proteins made on ribosomes. The amino acids are grouped according to the chemical characteristics of their side groups.

Glycine	Alanine	Valine*	Isoleucine*	Leucine*

ALIPHATIC, MONOAMINO, MONOCARBOXYLIC ACIDS

Lysine*	Arginine	Cysteine　Methionine*	Proline
ALIPHATIC, DIAMINO		**ALIPHATIC, SULFUR-CONTAINING**	

Aspartic Acid　Glutamic Acid		Asparagine　Glutamine	Tryptophan*
ALIPHATIC, DICARBOXYLIC		**ALIPHATIC AMIDES**	

Serine　Threonine*	Phenylalanine*　Tyrosine	Histidine
ALIPHATIC, HYDROXYL-CONTAINING	**AROMATIC**	**HETEROCYCLIC**

*Required in the diet of mammals.

From *A Dictionary of Genetics, Fourth Edition,* 1990, by Robert C. King and William D. Stansfield, Oxford University Press.

Another classification scheme recognizes only three major groups of amino acids, based on their net electric charge under physiological conditions: (1) neutral, (2) acidic, and (3) basic. Among the neutral amino acids, some are nonpolar and hydrophobic (Ala, Val, Ile, Leu, Trp, Pro, Met, Phe); others are polar and hydrophobic (Gly, Ser, Thr, Tyr, Gln, Cys, Asn). The one-letter and three-letter symbols for these amino acids are listed in the legend to Table 5.1. Among the polar and hydrophilic amino acids, two are acidic (aspartic acid or aspartate and glutamic acid or glutamate), and three are basic (lysine, arginine, histidine). Of the two sulfur-bearing amino acids, only cysteines can form a covalent disulfide bond with one another. Oxidation of the thiol side chains of two cysteines joins them together by a disulfide bond, forming the derived amino acid known as cystine. These disulfide bonds aid in stabilizing tertiary and quaternary protein structures. Proline contains a secondary amino group that renders proline incapable of participation in the secondary protein structure known as the α-helix. Proline is often found in the bends of polypeptide chains, where it disrupts α-helix formation. The side chain of proline is bonded to the α-carbon and to its amino group in a cyclic structure; thus, proline is actually an imino acid, but this fact is usually ignored by commonly referring to it as an amino acid.

2.27. What are "essential" amino acids?

Most bacteria can manufacture all of their complex organic molecules from simple carbon sources such as glucose. Most animals are not as metabolically sophisticated as the bacteria in this respect. The kinds of amino acids that are essential may vary from one species to another. For example, humans cannot synthesize eight of the twenty amino acids necessary to make their proteins; these eight, called **essential amino acids,** must be present in the food that we eat. Although tyrosine is not flagged in Table 2.2 as being an essential amino acid, it is synthesized from the essential amino acid phenylalanine; thus, tyrosine might also be considered as an essential amino acid.

2.28. How are amino acids linked together to form proteins by the living cell?

The process whereby amino acids are polymerized into polypeptide chains occurs on ribosomes and is discussed at length in Chapter 5. The peptide bond is enzymatically formed by dehydration synthesis, i.e., by the removal of water (Fig. 2-10). An oxygen atom is removed from the carboxyl group of one amino acid together with two hydrogens from the amino group of a second amino acid. The remaining carbon and nitrogen share an electron as a covalent peptide bond.

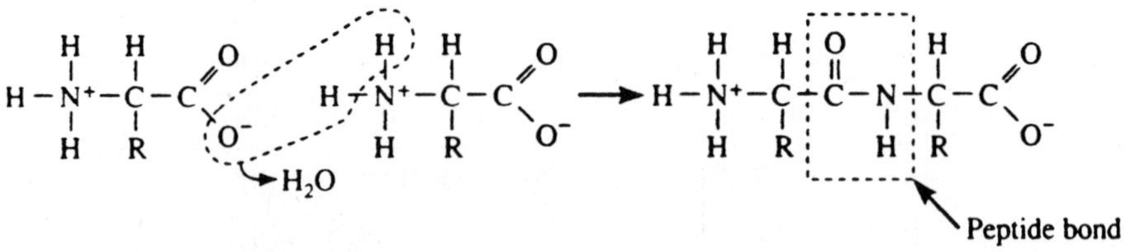

Fig. 2-10. Dehydration synthesis of a dipeptide by the formation of a peptide bond.

2.29. Does a polypeptide chain have polarity?

Yes. At one end of the chain (the left end as conventionally drawn) there is a free amino group. At the other end (the right end according to convention), there is a free carboxyl group. Excluding the side chains and the terminal amino acids of each chain, there are no other free amino or carboxyl groups attached to the α-carbons in a polypeptide chain, because each of those groups present in the free amino acids (from which the chain was made) were destroyed by dehydration synthesis of the peptide bonds that link them all together. Thus there is a N — or amino-terminus and a C — or carboxyl-terminus to each polypeptide chain.

2.30. What is the difference between a polypeptide chain and a protein?

Oligopeptides are relatively small polymers, usually less than ten amino acids in length. **Polypeptides** have molecular weights less than about 5,000 daltons. (A dalton, d, is a unit equal to the mass of a hydrogen atom; 1.67×10^{-24} gram.) Over 5,000 d, amino acid strands are generally called proteins. Many functional proteins consist of a single polypeptide chain of amino acids. In these cases, a complete polypeptide chain would be syn-onymous with a functional protein. In other instances, however, a functional protein may consist of two or more identical chains (**homopolymer**) or different chains (**heteropolymer**). In these cases, a single polypeptide chain would not be equivalent to a functional protein. Unless these distinctions are important, in which case they will be noted, the term protein is synonymous with polypeptide chain in this book.

2.31. How does one polypeptide chain differ from another?

Polypeptides may differ by the number and kinds of individual amino acids (residues) they contain. Some chains contain over a thousand amino acids. Others are shorter than a hundred amino acids. An average polypeptide might contain about 300 residues (monomers). In addition, not only may the composition of amino acids in polypeptide chains differ, but also the sequence in which the amino acids are strung together may differ. In other words, it is conceivably possible for two chains to be of equal length with the same amino acid composi-tion, and yet differ in the sequence of their residues. All completely synthesized polypeptide chains of a specific kind (e.g., the α-chains of hemoglobin) are identical in all respects, thus making it possible to obtain protein crys-tals that can be subjected to x-ray diffraction analysis, which may provide insight into the polypeptide's structure.

The **primary structure** of a functional protein molecule consists of the linear sequence of amino acids in each of its polypeptide chains (Fig. 2-11a).

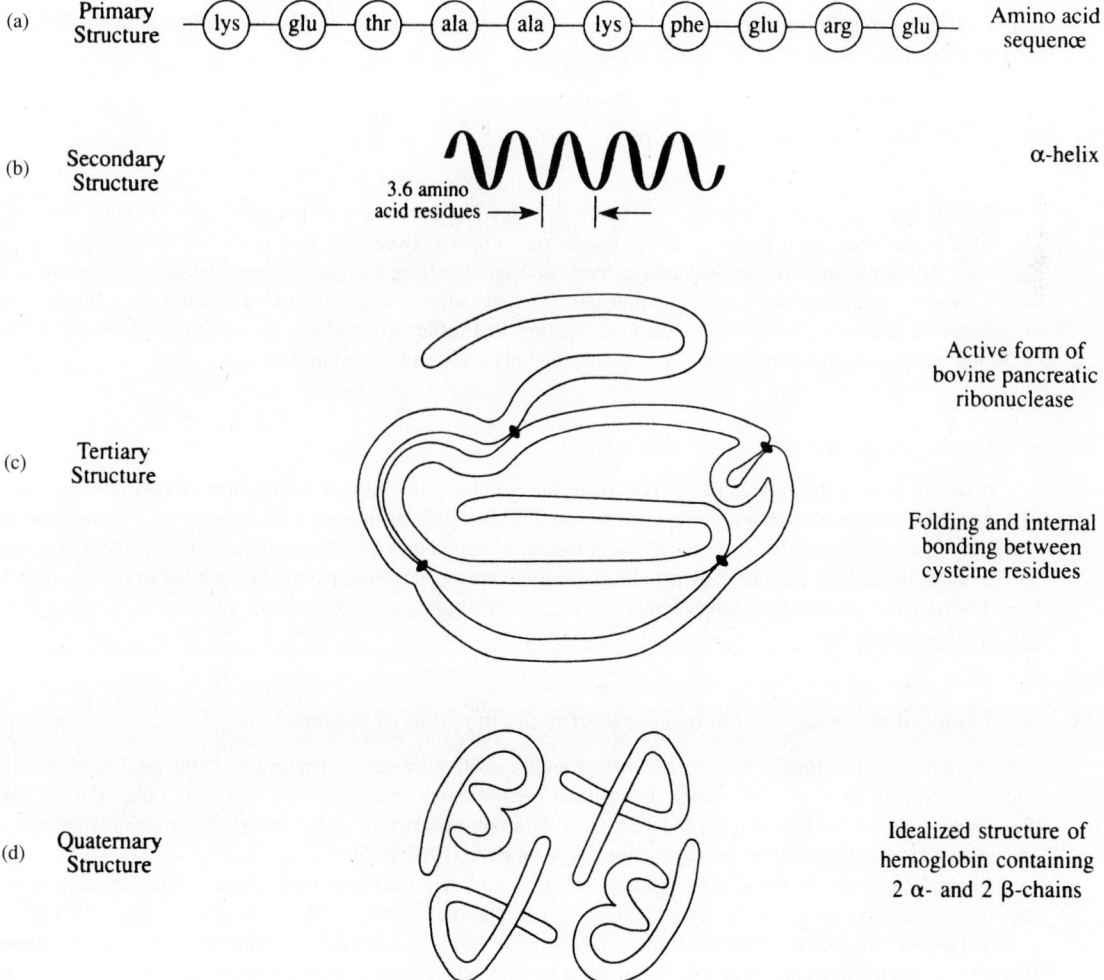

Fig. 2-11. Diagram of four structural levels (a to d) of proteins. (*From Schaum's Outline of Theory and Problems of Genetics, Third Edition, W. D. Stansfield, McGraw-Hill, 1991.*)

2.34. There are two major kinds of *secondary* protein structures: (1) α-helix, and (2) β-pleated sheet. What is an α-helix?

An **alpha helix** (Fig. 2-11*b*) forms when a carbonyl group ($C = O$) adjacent to one peptide bond is linked by a hydrogen bond (by sharing a proton, or hydrogen nucleus) to an imino group (NH) flanking a peptide bond in an amino acid about four residues along the same chain. One turn of the helix occurs each 3.6 amino acid residues. A polypeptide chain may not have an α-helical structure from one end to the other. Often there are **domains** of α-helices interrupted by nonhelical regions. One of the factors that may interrupt the formation of an α-helix is the amino acid proline. Because proline has the nitrogen of its amino group incorporated into a ring, it cannot form a hydrogen bond with a carbonyl group farther along the chain. The rigidity of proline creates a kink or bend in the polypeptide chain.

2.35. What is a *β*-pleated sheet?

Proteins that are insoluble in water tend to have a fibrous structure. Examples of such proteins are the α-keratin in hair, horn, and fingernail; the collagen in ligaments; the stretchy protein elastin; and silk. The polypeptide chains of fibrous proteins may contain relatively high amounts of certain amino acids (e.g., proline, glycine, threonine, asparagine) that tend to destabilize α-helices, thus forming an extended strand. Hydrogen bonds tend to form between identical polypeptide chains of this kind, rather than within a single chain, thus creating a secondary structure called a **β-pleated sheet.** The strands of the sheet may lie side-by-side in parallel (in the same direction) or alternate chains may be oriented in opposite directions (antiparallel). Globular proteins may contain β structures or β sheets in some regions (Fig. 2-12). β-Sheets may also be stacked upon one another and crosslinked by hydrogen bonds (or other weak secondary bonds).

2.36. What constitutes the tertiary structure of proteins?

The polypeptide chain may fold back upon itself, forming weak internal bonds (e.g., hydrogen bonds, ionic bonds) as well as stronger covalent disulfide bonds that stabilize its **tertiary structure** (Fig. 2-11*c*) into a precisely and often intricately folded pattern. These bonds are formed between the side chains of different amino acid residues. **Globular proteins** are soluble in water and tend to form compact shapes characteristic of most enzymes, antibodies, protein hormones, and transport proteins (such as hemoglobin). **Isozymes** are multiple forms of an enzyme that catalyze the same chemical reaction, but differ in net charge or catalytic efficiency, or both. Commonly isozymes differ in amino acid sequence by only one or a few residues.

2.37. What is the quaternary structure of proteins?

If two or more polypeptide chains spontaneously associate, they form a **quaternary structure** (Fig. 2-11*d*). The individual polypeptide chains of a quaternary protein are called **protomers.** Many proteins require more than one polypeptide chain to function normally. For example, some types of collagen (a prominent protein in connective tissue) exist as three identical polypeptide chains (a homotrimer) twisted around each other to form a triple helix. The bacterial enzyme tryptophan synthetase is a heterotetramer consisting of two identical α- and two identical β-polypeptide chains.

2.38. What kinds of processes can diminish or destroy the functions of proteins?

Proteins only function in their native tertiary and/or quaternary state. The primary structural level of a protein is determined by a structural gene. If a normal gene mutates (changes) to another form (allele), it may alter the genetic code so that the protein produced has a different sequence of amino acids. Even a single amino acid change in a protein may alter or destroy that protein's function (Chapter 6).

Proteolytic (protein-digesting) **enzymes,** called peptidases, disrupt peptide bonds of the primary protein structure, creating fragments of various lengths or degrading the protein completely to amino acids. This process is called **protein digestion.** Peptidases that attack peptide bonds of nonterminal amino acids are called **endopeptidases. Exopeptidases** are of two kinds: aminopeptidases cleave amino acids from the amino ends of polypeptide chains; carboxypeptidases cleave amino acids from the carboxyl ends of polypeptide chains.

Protein structure at the secondary or higher levels can be modified in various ways called **denaturation.**

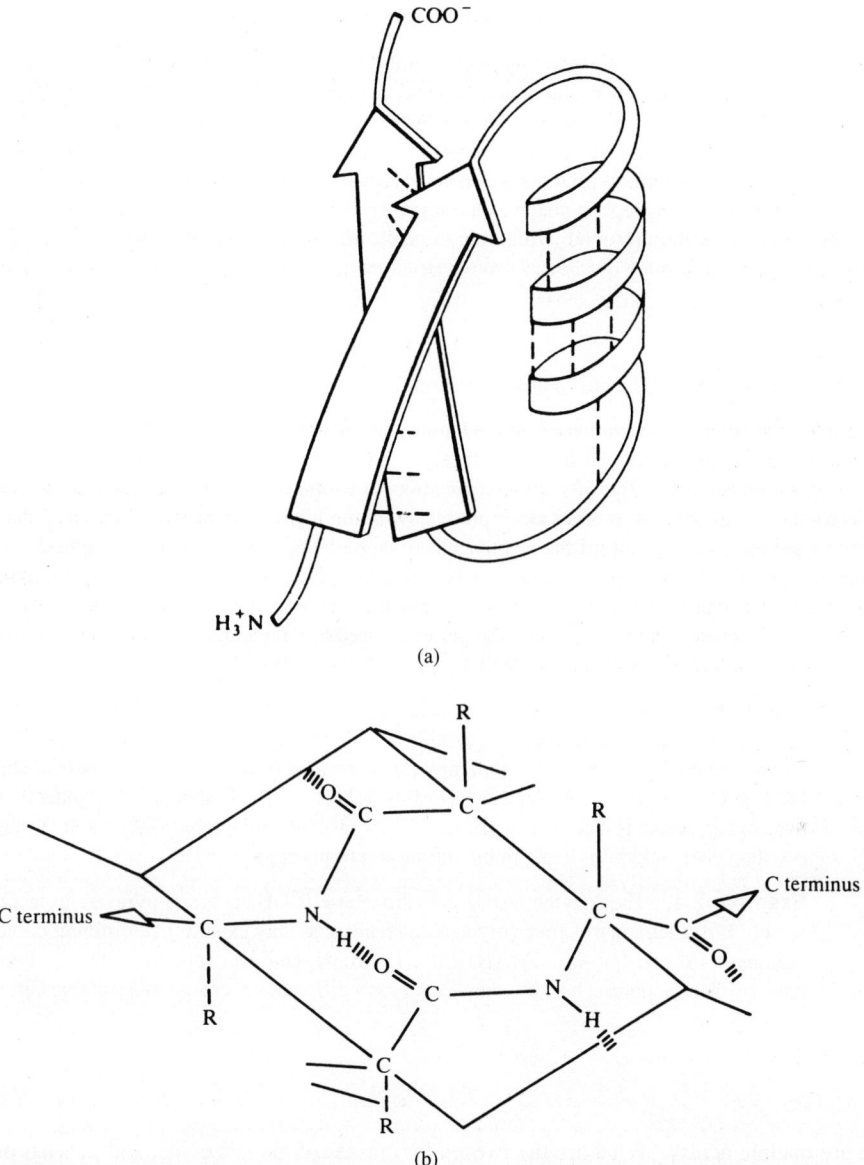

Fig. 2-12. (*a*) Diagram of a supersecondary β-α-β folding unit of a protein. Regions with β-sheet structure are represented by arrows; the α-helical region is represented by the coiled structure. Approximate positions of hydrogen bonds are shown with dashed lines. (*b*) Hydrogen bonding between adjacent polypeptide chains in an antiparallel β sheet. (*From Schaum's Outline of Theory and Problems of Biochemistry, P. W. Kuchel, G. B. Ralston, et al., McGraw-Hill, 1988.*)

Proteins can be denatured by heat (ovalbumin in eggs is denatured by frying), by a large pH change (adding lemon juice to milk denatures and precipitates the milk protein casein), by chemical treatment (exposure to a detergent disrupts the hydrophobic interactions within globular proteins), or by mechanical treatment (whipping egg whites denatures the protein egg albumin). Some proteins become at least partly denatured by only a small change in temperature. These are called **temperature-sensitive** proteins. The gene mutations responsible for these proteins are called temperature-sensitive (*ts*) mutations. This class of mutations has been very important in the study of bacterial and viral genetics.

2.39. Do certain amino acids have a preferential location within a protein?

The ionized side chains of some amino acids readily interact with water (hydrophilic). Hydrophobic amino acids contain nonionized side chains that prefer to avoid contact with water. Thus, when a polypeptide chain folds into a globular tertiary shape, amino acids with hydrophilic groups tend to predominate on the outside of the molecule and hydrophobic segments of the chain tend to predominate in the interior of the molecule. The multiple polypeptide chains of quaternary proteins are usually joined by hydrophobic interactions ("bonds"). Nonpolar groups of the individual polypeptide chains come together as a way of excluding water. Hydrogen bonds and/or ionic bonds, together with intrastrand (within the same chain) and/or interstrand (between chains) disulfide bonds may also participate in forming quaternary protein structures. No other covalent crosslinks are normally found in proteins.

2.40. How can electrophoresis be used to study proteins?

Electrophoresis is the migration of charged molecules in an electrical field. A popular way to electrophoretically separate a mixture of proteins is to introduce the sample into a well in a semisolid slab of agarose or starch gel. The gel is placed in a buffer of suitable ionic strength with electrodes in the buffer at each end. When a direct electric current is applied to the system, proteins with the highest net negative electrical charge will migrate through the gel most rapidly toward the positive electrode (anode). Proteins with the highest net positive charge will migrate most rapidly toward the negative pole (cathode). Proteins that are at their **isoelectric point** in a particular buffer system have no net charge and so will not migrate toward either pole. After a period of time (often an hour or so), the current is turned off and the gel is exposed to a protein stain. Destaining reagents remove the unbound stain, revealing bands of proteins at different locations on the gel.

Example 2.3. Hemoglobin is a heterotetrameric molecule that transports oxygen from our lungs to our tissues. It consists of four polypeptide chains (two identical α-chains and two identical β-chains). Normal hemoglobin contains a glutamic acid at residue 6 from one end of its beta chain. Sickle-cell hemoglobin is a variant protein that has valine at that residue. Being acidic, glutamic acid has a negative charge; valine is electrically neutral. Thus, normal hemoglobin will migrate faster toward the anode than does sickle-cell hemoglobin during electrophoresis.

Example 2.4. The enzyme lactate dehydrogenase (LDH) converts pyruvate to lactic acid or the reverse. It is a tetramer of four polypeptide chains that may exist in five different combinations of two protomers (M and H): 4M, 3M:1H, 2M:2H, 1M:3H, and 4H. The five LDH isozymes can be separated by electrophoresis because they each bear a different net charge in a suitable buffer.

NUCLEIC ACIDS

2.41. What are nucleic acids? What are the two major classes of nucleic acids, and in what order were they discovered? Where in the eukaryotic cell are these two types located?

Nucleic acids, as a distinctive class of macromolecules, were discovered in 1868 by Friederich Miescher. He isolated a substance called "nuclein" from the nuclei of pus cells. Then a similar substance was found in salmon sperm heads. Later, nuclein was shown to be a mixture of basic proteins (mainly histones) and phosphate-containing organic acids polymerized into **deoxyribonucleic acid** (DNA). It is now known that a second kind of nucleic acid polymer called **ribonucleic acid** (RNA) exists both in the nucleus (where it is synthesized) and in the cytoplasm (where it participates in protein synthesis). Both types of nucleic acid (DNA and RNA) are linear, unbranched polymers of subunits (monomers) termed **nucleotides.** The negatively charged phosphate groups of nucleotides give the polymer its acidic property.

2.42. What are the three major components of a nucleotide and how are they connected?

Each nucleotide consists of three major parts: (1) a five-carbon sugar (pentose); (2) a flat, heterocyclic, nitrogen-containing organic base; and (3) a phosphate group (Fig. 2-13). The nitrogenous base in each nucleotide is covalently attached to the sugar by a glycosidic bond. The phosphate group is also covalently linked to the sugar.

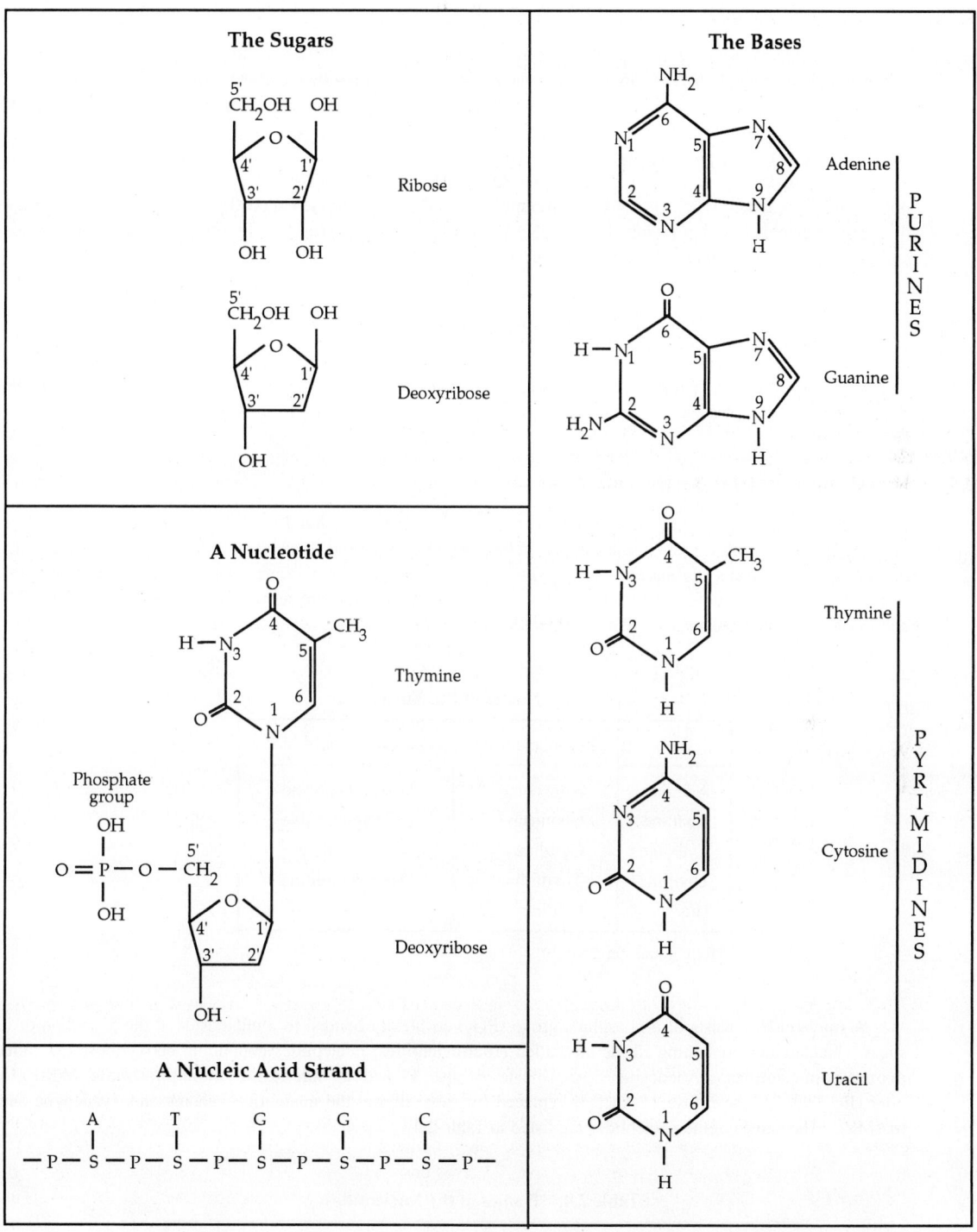

Fig. 2-13. Structural components of nucleic acids. The carbon atoms in the sugar are designated by primed numbers to distinguish them from the carbon atoms in the base. The #1 nitrogen atom of pyrimidines and the #9 nitrogen atom of purines are bonded to the #1′ carbon of ribose or deoxyribose of nucleotides.

2.43. What bases are found in nucleic acids?   How are the bases classified?

The organic bases of the nucleic acids are of two general types: pyrimidines and purines (Fig. 2-13).   **Pyrimidines** are single-ringed; **purines** are double-ringed.   The purines **adenine** (A) and **guanine** (G)—both of which are found in DNA and RNA—differ by their side groups.   Adenine has an amino group (NH_2) attached to the #6 position (atom) of the purine and thus is called 6-aminopurine.   Guanine has an oxygen at the #6 position and an amino group at the #2 position and therefore is 6-oxy-2-aminopurine or 2-amino-6-oxypurine.   Similarly, the pyrimidine **cytosine** (C), found in both DNA and RNA, is 2-oxy-4-aminopyrimidine.   A second pyrimidine, found only in RNA, is **uracil** (U); it is 2,4-dioxypyrimidine.   A third pyrimidine called **thymine** (T), or 5-methyl-2,4-dioxypyrimidine, is found predominately in DNA (a few thymine bases are normally found in transfer RNA molecules, along with other "unusual bases"; see Question 2.59).

2.44. What kinds of carbohydrates are found in nucleotides?

The sugar β-D-ribose (hereinafter to be called simply ribose; Fig. 2-13) is found in the **ribonucleotide** monomers of RNA.   The pentose in the **deoxyribonucleotide** monomers of DNA differs from ribose by the absence of oxygen at the #2 carbon and is thus 2-deoxy-β-D-ribose (hereinafter to be called **deoxyribose**).

2.45. How do nucleosides differ from nucleotides?

A **nucleoside** consists of a base connected by a covalent glycosidic bond from the N_1 of pyrimidines or from the N_9 of purines to the 1' carbon of the sugar.   The plane of each base is perpendicular to that of the sugar, so that the flat surfaces of sequential bases in the polymer are stacked like a pile of coins.   The nucleosides containing ribose are termed **ribonucleosides;** nucleosides containing deoxyribose are termed **deoxyribonucleosides.**   The names of all the nucleosides are listed in Table 2.3.

Table 2.3.   **Names of the Nucleosides**

Base	Ribonucleoside	Deoxyribonucleoside
Adenine	Adenosine	Deoxyadenosine
Guanine	Guanosine	Deoxyguanosine
Cytosine	Cytidine	Deoxycytidine
Thymine	Thymidine*	Deoxythymidine
Uracil	Uridine	—

* Rare in transfer RNA.

A **nucleotide** consists of a phosphate group (PO_4) covalently bonded to a nucleoside at the 5' carbon of its sugar.   Nucleotides containing ribose are called **ribonucleotides;** nucleotides containing deoxyribose are called **deoxyribonucleotides.**   Adenosine 5'-phosphate is also known as **adenosine monophosphate** (AMP) or **adenylic acid.**   If the ribose of AMP is replaced by deoxyribose, the nucleotide becomes deoxyadenylic acid (dAMP).   The names of the nucleotides are listed in Table 2.4.

Table 2.4.   **Names of the Nucleotides**

Base	Ribonucleotide	Deoxyribonucleotide
Adenine	Adenylic acid, AMP	Deoxyadenylic acid, dAMP
Guanine	Guanylic acid, GMP	Deoxyguanylic acid, dGMP
Cytosine	Cytidylic acid, CMP	Deoxycytidylic acid, dCMP
Thymine	Thymidylic acid, TMP	Deoxythymidylic acid, dTMP
Uracil	Uridylic acid, UMP	—

2.46.   What kind of bond connects adjacent nucleotides on the same polynucleotide strand or chain?

In each polynucleotide strand of DNA and RNA, adjacent nucleotides are joined covalently by **phosphodiester bonds** between the 3′ carbon of one nucleotide and the 5′ carbon of the adjacent nucleotide. Polymers of DNA and RNA are synthesized on DNA templates with the aid of specific enzymes and other proteins (Chapters 3 and 4).

2.47.   How do DNA and RNA differ with respect to the number of chains in a molecule?

All cellular RNA molecules are synthesized from DNA templates as single strands, whereas DNA is a two-chain molecule (a double helix). Some regions of RNA chains become joined by hydrogen bonds between complementary base pairs in ribosomes, but they are not paired over their entire length the way DNA molecules are. A few bacterial viruses (such as ϕX174 and M13) have a single strand of DNA for their genome. Most viruses that have single-stranded RNA (ssRNA) genomes (e.g., polio, influenza) become locally double stranded during their replication; reoviruses have double-stranded RNA (dsRNA) genomes; retroviruses (e.g., human immunodeficiency virus, Chapter 10) have ssRNA genomes that, during replication, synthesize a DNA strand from the RNA strand, but the two strands soon separate.

2.48.   What kind of bonds form between the bases of DNA or RNA molecules?

Bases in the nucleotides spontaneously form hydrogen bonds (a weaker or secondary type of bond than the stronger covalent phosphodiester bonds or glycosidic bonds) in a highly specific manner. Adenosine in one strand of DNA normally forms two hydrogen bonds with thymine in a complementary strand of the double helix (A=T). Likewise, it forms two hydrogen bonds with uracil (A=U) in DNA-RNA hybrids, and in RNA-RNA interactions, either between different parts of the same RNA strand or between different RNA strands. Guanine and cytosine normally form three hydrogen bonds (G≡C) in both DNA and RNA.

2.49.   How do the shapes of DNA and RNA molecules differ?

The single chain of an RNA molecule may spontaneously fold back upon itself and form **complementary base pairs** over localized regions in its energetically most stable state. The possible shapes of RNA molecules are thus quite varied. DNA, on the other hand, normally exists in the uniform shape of a double helix, with the complementary chains wound around each another like a spiral staircase. The handrails, or backbones, of this staircase consist of alternating sugar-phosphate groups in each chain; the steps of the staircase are the base pairs. Since a pyrimidine always pairs with a purine, the width of the double helix is uniformly 20 Å.

2.50.   How do DNA and RNA molecules compare with regard to size?

Even the largest RNA molecules are miniscule in comparison to most genomic DNA molecules, some of which contain more than 10^8 nucleotides. Each chromosome contains a single DNA molecule that may contain many thousands of genes in addition to regions that do not code for RNA or protein molecules. By contrast, the largest of the RNA molecules that serve as templates for protein synthesis (mRNA) in bacteria may code for only a dozen or so proteins.

2.51.   What roles do DNA and RNA molecules play?

DNA functions as the molecule of heredity in all cells. It stores genetic information for making RNA molecules and proteins, and each of its strands serves as a template for its own replication (Chapter 3). Cellular RNA molecules do not act as templates for their own replication. All cellular RNA molecules are synthesized on DNA templates. Except in certain viruses (mentioned in Question 2.47), RNA does not serve as genetic information. DNA does not serve as the direct template for protein synthesis. Instead DNA is transcribed into messenger RNA (mRNA) molecules that serve this function. Other kinds of RNA molecules include ribosomal RNAs (rRNAs), which are structural components of ribosomes, and transfer RNAs (tRNAs), which assist in translation of mRNA into proteins.

2.52.   Where are DNA and RNA molecules located in eukaryotic cells?

Almost all the DNA of eukaryotic cells is found in the linear chromosomes of the nucleus. Small amounts of circular DNA are found in cytoplasmic organelles such as mitochondria and chloroplasts. All RNA molecules are synthesized from DNA templates. Most RNA molecules that are synthesized and processed in the nucleus soon find their way into the cytoplasm, where they function in the synthesis of proteins.

2.53.   Compare the stability of DNA and RNA molecules.

Single strands of RNA polymers are relatively quickly digested by enzymes called **ribonucleases** unless they possess much intrastrand base pairing (making them "double stranded" over localized regions) or unless they are extensively complexed with proteins such as those in ribosomes. By contrast, DNA molecules are much more stable because they are double stranded throughout their length and because enzymes are normally present in the nucleus to repair damage to DNA. There are no comparable enzymes either in the nucleus or in the cytoplasm to repair RNA damage. There also are no enzymes to repair damage to extranuclear DNA in mitochondria and chloroplasts. RNA molecules must be continuously replaced as they are being broken down at various rates.

2.54.   What are Chargaff's rules, and what is the chemical basis for them?

Since an A on one strand of DNA forms two hydrogen bonds to a T on the opposite (complementary) strand (and G likewise pairs by three hydrogen bonds with C), the purine/pyrimidine ratio or $(A + G)/(T + C)$ in DNA is unity (1). This ratio is not a constant feature of RNA molecules, however, because of RNA's single-stranded character. Before the structure of DNA was known, Erwin Chargaff discovered that A and T exist in equal molar amounts in DNA; likewise G equals C. These regularities came to be known as **Chargaff's rules.**

2.55.   In 1953, James Watson and Francis Crick proposed a structure for DNA (Fig. 2-14). Explain some of the details of their model.

Watson and Crick proposed that the two strands of a DNA double helix run in opposite directions (are **antiparallel**). That is, if one chain is read from the 5′ phosphate end, the sugar-phosphate linkages would be repetitively 5′ phosphate-3′ sugar, respectively; the opposite chain read *in that same direction* from the 3′ hydroxyl end would be repetitively 3′ sugar-5′ phosphate, respectively. Note in Fig. 2-15 how the sugars point (as it were) in opposite directions on the two chains.

Most DNA molecules are normally right-handed double helices. The stacking of the paired bases by their flat surfaces in the center of the molecule forms a hydrophobic core. Together with the hydrogen bonds between the base pairs, these hydrophobic interactions contribute to the stability of the molecule. At neutral pH, there is no charge on any of the bases, but the phosphates on the surface of the DNA molecule are ionized, giving the molecule a net negative charge.

The double helix makes one complete turn every ten base pairs or 3.4 nm (34 Å). This distance is termed the **pitch** of the helix. Thus, there are 3.4 Å between each level of stacked base pairs; each level of paired bases is rotated 36° around the diad axis relative to the level above and below it. The surface of the double helix has a **major groove** and a **minor groove,** where interactions between specific nucleotide sequences and specific **DNA-binding proteins** may occur. Such specific DNA-protein interactions may either activate ("turn on") or repress ("turn off") genes at or near these binding sites (Chapter 4).

2.56.   There are three major classes of RNA molecules based on their functions: (1) tRNAs, (2) mRNAs, and (3) rRNAs. Describe the structure and function of tRNA molecules.

The smallest RNA molecules (75–80 nucleotides or bases in length) are **transfer RNAs** (tRNAs). These serve as adaptor molecules to position each amino acid in proper alignment on the ribosome for polymerization into polypeptide chains. Transfer RNAs are unique in that they contain a few unusual bases in addition to the standard ones (A, U, G, C; see Question 2.59). Since there can be 20 different kinds of amino acids in biological proteins, there must be at least 20 different species of tRNA molecules, one specific for each amino acid. About 10–20% of total cellular RNA is tRNA.

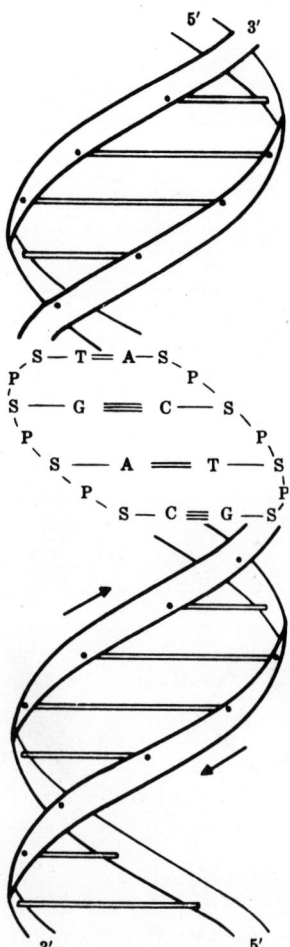

Fig. 2-14. Diagram of the Watson-Crick model of DNA structure. (*From Schaum's Outline of Theory and Problems of Genetics, Third Edition, W. D. Stansfield, McGraw-Hill, 1991.*)

2.57. Describe the structure and function of mRNA molecules.

The genetic code that specifies the amino acid sequences of proteins resides in the deoxyribonucleotide sequences of DNA (Chapter 5), and it becomes transcribed into complementary ribonucleotide sequences of **messenger RNAs** (mRNAs). Ribosomes attach to and move along the mRNA, in ticker-tape fashion. Activated tRNAs, with their attached amino acids, read the mRNA code that specifies the order in which the amino acids are linked into polypeptide strands on ribosomes. The size and base composition of different mRNA molecules can vary greatly. However, barring mutation, the size and composition of all mRNA molecules specified by a single gene are identical.

2.58. Describe the structure and function of rRNA molecules.

Ribosomal RNAs (rRNAs) are structural components of ribosomes. There are three molecular-weight classes of rRNAs in bacteria: 5S, 16S, and 23S. Within each of these three classes, all the rRNA molecules are identical in molecular structure. The base sequences of closely related biological species are identical or very similar; the more distant the genetic relationship between organisms, the more different the base sequences tend to be. The molecular weight symbol S stands for Svedberg units, a measure of the speed with which molecules move during ultracentrifugation. (T. Svedberg invented the ultracentrifuge in 1923). In general, the larger the S value, the higher the molecular weight, although the geometry or shape of the molecule may also be involved. Four molecular weight classes of rRNAs are found in eukaryotes: 5S, 5.8S, 17–18S, and 25–28S. These rRNA molecules interact with tRNAs, mRNAs and approximately 50 ribosomal proteins to accomplish protein synthesis.

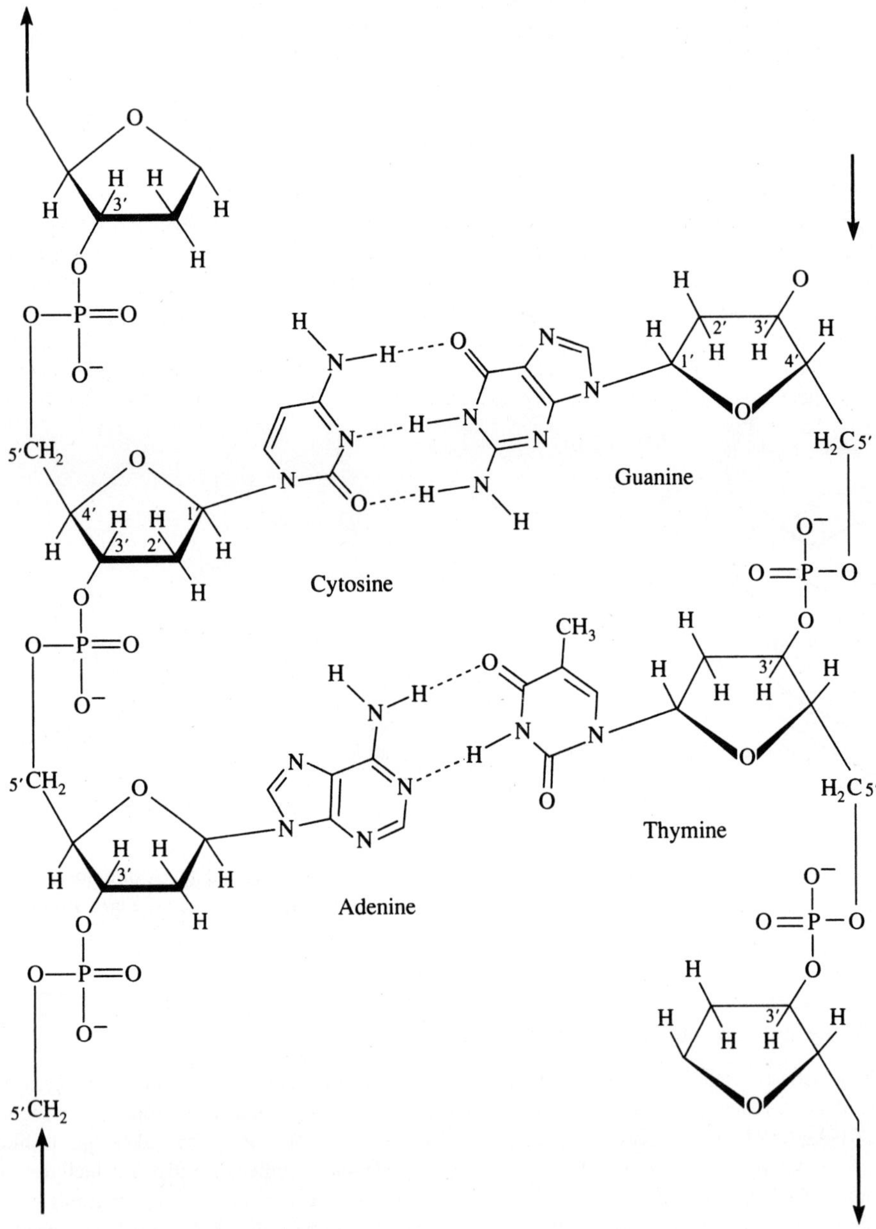

Fig. 2-15. Base pairing in DNA. (*From Schaum's Outline of Theory and Problems of Genetics, Third Edition, W. D. Stansfield, McGraw-Hill, 1991.*)

2.59. What are "unusual bases" in nucleic acids? How do they become part of a nucleic acid? What functions do they perform?

Any base other than adenine, guanine, thymine, or cytosine in DNA is considered to be unusual or rare. The same is true for any base other than adenine, guanine, uracil, or cytosine in RNA. Unusual bases are made by enzymatic modification of nucleotides either before or after polymerization. Two examples follow.

Example 2.5. The DNA of the bacteriophage (bacterial virus) T4 contains 5-hydroxymethylcystosine (HMC) instead of cytosine. A methyl group is added to cytosine bases by phage-encoded enzymes before they become polymerized into the phage genome. The phage encodes another enzyme to digest host DNA that contains cytosine rather than HMC. Phage T4 can use these host degradation products to synthesize its own DNA. By methylation of its own cytosines, phage T4 protects its own DNA from being attacked by the same enzyme that it uses to digest host DNA. After replication of the phage DNA, two phage enzymes glucosylate the HMC bases and thereby protect phage DNA from being digested by a host endonuclease that attacks unmodified HMC bases.

Example 2.6. Normally, neither mRNAs nor rRNAs contain any unusual bases. All tRNAs, however, normally contain a few modified bases. Transfer RNAs become modified by specific enzymes after the normal nucleotides are linked together by $3'$-$5'$ phosphodiester bonds. Adding a methyl group to atom 1 of guanosine forms the unusual ribonucleoside 1-methylguanosine. Some of the unusual bases in tRNAs are presented in Fig. 2-16. Most of the unusual bases seem to be concentrated in the loops, or unpaired regions, of tRNA molecules. These unpaired regions are thus free to base pair with other regions in rRNA or mRNA. The function of most of these unusual bases is not yet known, but some of them have been shown to be involved in regulation of tRNA specificity.

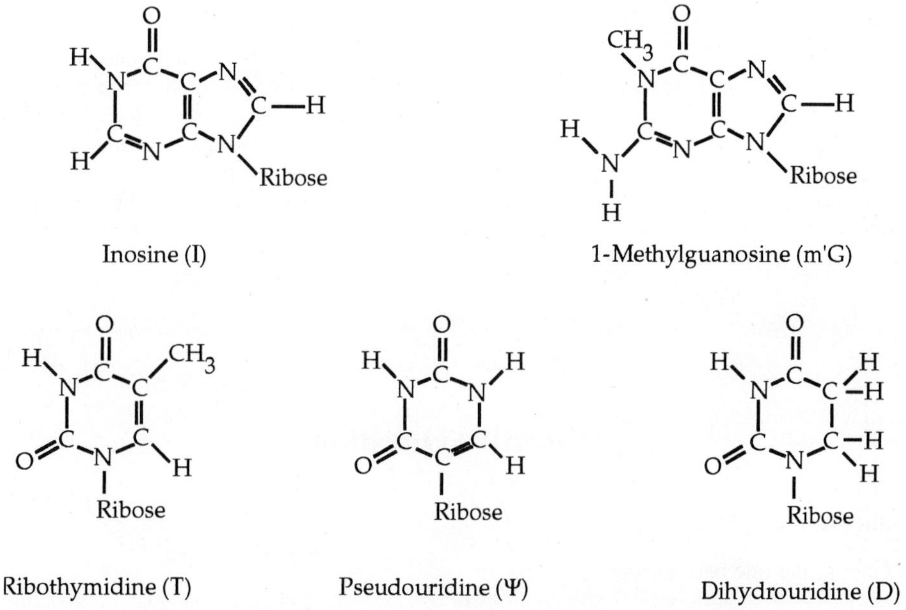

Inosine (I)   1-Methylguanosine (m'G)

Ribothymidine (T)   Pseudouridine (Ψ)   Dihydrouridine (D)

Fig. 2-16. Structures of some modified bases found in tRNA molecules.

2.60. What kind of two-dimensional model can be produced by folding the nucleotide chain back upon itself and maximizing the base pairing of complementary regions in single-stranded tRNA molecules?

A so-called cloverleaf structure is formed as illustrated in Fig. 2-17. The tRNA in this figure is activated, being charged or loaded at its $3'$ terminal adenine with its own species of amino acid (alanine). All tRNAs have this same general cloverleaf structure. The regions where intrastrand base pairing occurs are called **stems,** and the unpaired regions are called **loops.** Transfer RNAs coding for different amino acids vary somewhat in their ribonucleotide sequences, especially in the middle loop that contains a contiguous triplet of ribonucleotides called the **anticodon.** Each kind of amino acid has at least one species of cognate tRNA with a unique anticodon sequence (see Chapter 5).

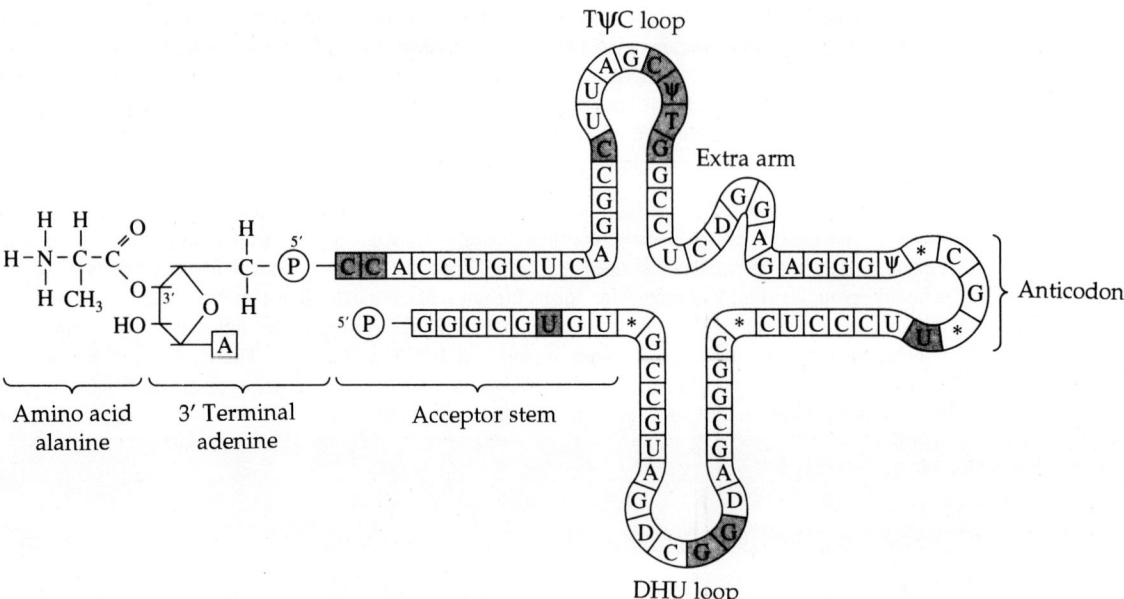

Fig. 2-17. A "cloverleaf model" of alanine transfer RNA from yeast. There are several unusual bases (*) in tRNA molecules (see Fig. 2-16); pseudouridine (ψ) is one of them, and it is found in the loop nearest the 3′ end, called the TψC loop. The loop nearest the 5′ end is called the DHU loop because it contains another unusual base, dihydrouridine (D). All tRNA molecules end with CCA3′. Positions bearing identical bases in almost all tRNA species are indicated by shading. This tRNA molecule has been activated by its cognate aminoacyl tRNA synthetase, which attaches alanine to the 3′ terminal adenine of the tRNA by an aminoacyl bond. (*From Schaum's Outline of Theory and Problems of Genetics, Third Edition, W. D. Stansfield, McGraw-Hill, 1991.*)

Objective Questions

Multiple Choice

Directions: Choose the one best answer.

1. The macromolecules that require a template for their synthesis are (*a*) carbohydrates and lipids. (*b*) carbohydrates and proteins. (*c*) proteins and nucleic acids. (*d*) carbohydrates, proteins, and nucleic acids. (*e*) lipids, proteins, and nucleic acids.

2. The macromolecules whose monomers are bound only by noncovalent forces are (*a*) carbohydrates only. (*b*) lipids only. (*c*) carbohydrates and lipids. (*d*) lipids and nucleic acids. (*e*) none of the above.

3. Penicillin acts as an antibiotic in susceptible bacteria by interfering with (*a*) cell-wall formation. (*b*) protein synthesis. (*c*) the Krebs cycle. (*d*) the electron transport chain. (*e*) DNA synthesis.

4. The substance that is the general biosynthetic precursor of sex hormones and hormones of the adrenal cortex is (*a*) inositol. (*b*) sphingomyelin. (*c*) lecithin. (*d*) phosphatidyl choline. (*e*) cholesterol.

5. A nucleotide base not present in DNA is (*a*) adenine. (*b*) guanine. (*c*) cytosine. (*d*) uracil. (*e*) thymine.

6. RNA molecules differ from DNA molecules in all but which one of the following respects? (*a*) Kinds of purines (*b*) kinds of pyrimidines (*c*) number of strands per molecule (*d*) type of sugar (*e*) shape

7. The kind of nucleic acid that is represented by a cloverleaf model is (*a*) mitochondrial DNA. (*b*) chromosomal DNA. (*c*) rRNA. (*d*) tRNA. (*e*) mRNA.

8. Which of the following is **not** a characteristic of most DNA under physiological conditions? (*a*) A pairs with T by two hydrogen bonds (*b*) antipolarity of complementary chains (*c*) 3.4 Å pitch (*d*) 20 Å diameter (*e*) 10 base pairs per turn of the helix

9. Adenosine is (*a*) a nitrogenous organic base. (*b*) a ribonucleoside. (*c*) a ribonucleotide. (*d*) a deoxyribonucleotide. (*e*) none of the above.

10. What type of function do protein porins perform? (*a*) Cytoskeletal (*b*) gene activation (*c*) structural (*d*) self-recognition (*e*) transport

11. An amino acid that has a secondary amine and disrupts α-helix formation is (*a*) threonine. (*b*) phenylalanine. (*c*) serine. (*d*) glycine. (*e*) proline.

12. The polarity of DNA chains is represented by (*a*) $3'$-$5'$. (*b*) $1'$-$6'$. (*c*) $1'$-$3'$. (*d*) $2'$-$3'$. (*e*) none of the above.

True–False

1. Bacterial cell walls contain cellulose.

2. Most bacterial cell walls contain carbohydrates and amino acids.

3. The only branched biomacromolecules are polysaccharides.

4. Starch, cellulose, and glycogen are all homopolysaccharides.

5. Gram-positive bacteria have both an inner and an outer membrane.

6. R-type pneumococci have no cell walls.

7. Lectins are carbohydrate molecules that serve as cellular recognition markers.

8. Heating digests proteins to amino acids.

9. An α-helix is considered to be one kind of protein tertiary structure.

10. There is no general tendency for any particular kinds of amino acids to be located predominantly on the surface or in the interior of a globular protein.

11. Glucose is the primary molecule for powering biochemical reactions.

12. The purine/pyrimidine ratio in a double-stranded DNA molecule should be 1.

Matching

Directions: Match each item in **Column A** with the one in **Column B** to which it is most closely associated. Each item in **Column B** can be used only once.

	Column A		Column B
1.	actin	A.	catalyst
2.	fibronectin	B.	immunity
3.	collagen	C.	hormone
4.	DNA polymerase	D.	stimulates transcription
5.	galactoside permease	E.	contraction, motility
6.	glucagon	F.	transmembrane transport
7.	rhodopsin	G.	electron transport chain
8.	complement	H.	cell anchorage
9.	cytochrome	I.	photoreceptor
10.	catabolite activator protein	J.	connective tissue

Terms

Directions: Unless otherwise specified, each answer is a single word.

1. Covalent linkages between monomers of a polysaccharide (2 words)

2. The region between the inner and outer membranes of Gram-negative bacteria (2 words)

3. Any bacterium devoid of a cell wall

4. A polysaccharide coat that surrounds the cell wall of streptococci and is associated with virulence.

5. The kind of lipid that is richly present in the plasma membranes of animal cells, but is relatively meager or absent in membranes of organelles and not present in most prokaryotes.

6. The repeating unit in bacterial cell walls, consisting of a pair of sugars and a tetrapeptide.

7. Amide covalent linkages between monomers of a polypeptide chain (2 words)

8. The monomers of RNA chains

9. In double-stranded DNA molecules, adenine and thymine exist in equal molar amounts; the same is true for guanine and cytosine.   These regularities are known as (2 words).

10. The class of RNA molecules to which amino acids are covalently bound.

11. The migration of charged molecules in an electrical field

12. The kind of covalent linkages that join adjacent nucleotides in the same strand of DNA or RNA. (2 words)

Answers to the Objective Questions

Multiple Choice

1. *c*   2. *b*   3. *a*   4. *e*   5. *d*   6. *a*   7. *d*   8. *c*   9. *b*   10. *e*   11. *e*
12. *a*

True–False

1. F (plant cell walls) **2.** T **3.** F (peptidoglycans in bacterial cell walls are also branched) **4.** T
5. F (gram-negative) **6.** F (no capsule) **7.** F (lectins are carbohydrate-binding proteins) **8.** F (denatures 2°, 3°, or 4° structures) **9.** F (secondary structure) **10.** F (hydrophilic outside; hydrophobic inside) **11.** F (adenosine triphosphate, ATP) **12.** T

Matching

1. *E* **2.** *H* **3.** *J* **4.** *A* **5.** *F* **6.** *C* **7.** *I* **8.** *B* **9.** *G* **10.** *D*

Terms

1. glycosidic bonds **2.** paraplasmic space **3.** protoplast or L form **4.** capsule **5.** cholesterols
6. peptidoglycan **7.** peptide bonds **8.** ribonucleotides **9.** Chargaff's rules **10.** transfer (tRNA)
11. electrophoresis **12.** phosphodiester bonds

Chapter 3

Chromosomes: Their Structure, Replication, and Recombination

INTRODUCTION

Cytologists who first studied the nuclei of animal and plant cells found a substance that readily stained with certain dyes. This nuclear material became known as **chromatin.** Cytologists observed that chromatin fragmented and condensed during mitosis or meiosis to form microscopic rod-shaped bodies that came to be called **chromosomes.** Chromatin and the chromosomes that developed from it were shown to consist of large amounts of protein and a nucleic acid now known as **2′-deoxyribonucleic acid,** or **DNA.** Correlations between abnormal chromosomes and aberrant genetic traits provided support for the hypothesis that the hereditary information was contained in the chromosomes.

Experiments in the early 1940s by Oswald Avery and his colleagues strongly implicated DNA as the hereditary information. However, this idea was not generally accepted. Many scientists believed that DNA was too simple a molecule to control the heredity of an organism. They favored protein as the carrier of hereditary information because there is so much of it in chromosomes and because there are so many different types of proteins. In 1953, James Watson and Francis Crick proposed a model for the structure of DNA and an explanation for how it might replicate; soon after that they suggested how DNA alone could specify the characteristics of an organism. By the mid-1960s, numerous scientists had shown conclusively that DNA alone was the hereditary information and that specific nucleotide sequences of DNA, called **genes,** coded for the many proteins and RNAs needed to build and maintain cells.

Originally, the word chromosome had a precise meaning; it referred to any of the microscopic rod-shaped bodies that formed from chromatin during mitosis or meiosis. But over the years, the term also came to be used to describe the hereditary information in viruses, bacteria, and organelles (mitochondria and chloroplasts). This chapter describes the most common kinds of chromosomes encountered in prokaryotic and eukaryotic cells. In addition, it discusses DNA replication and recombination.

CHROMOSOME STRUCTURE

3.1 How are chromosomes constructed in bacteria?

All essential bacterial genes are found in a single, circular, double-stranded DNA (dsDNA) chromosome in the cytoplasm. It is believed that the bacterial chromosome is attached to the plasma membrane. The **bacterial chromosome** specifies between 1000 and 5000 proteins, depending on the species. In addition to the bacterial chromosome, there may be one or more smaller chromosomes called plasmids, which generally specify 20 to 100 proteins. Plasmids are circular, double-stranded DNA molecules that may or may not be attached to the plasma membrane. Most or all of the proteins specified by plasmids are not absolutely essential for the survival of the cell under normal environmental conditions. Many of the plasmids encode proteins that make the transfer of some hereditary information to another cell possible, that promote the metabolism of unusual compounds, or that make the cell resistant to chemicals or heavy metals.

The bacterial chromosome, consisting of DNA, protein, and RNA is highly condensed. Not only is it supercoiled by an enzyme referred to as **topoisomerase,** but it loops in and out from an RNA and protein "core." The many negative charges of the DNA are neutralized by polyamines such as **spermine** and **spermidine,** as well as by basic proteins around which the DNA may be wrapped. DNA from gently lysed bacterial cells looks like beads on a string. Although the bacterial chromosome is highly condensed, it cannot be seen with the light microscope.

The bacterial chromosome under the transmission electron microscope looks very much like the **chromatin** seen in the nuclei of nondividing eukaryotic cells.

3.2.   What are some of the distinctive features of eukaryotic chromosomes?

The genes of eukaryotes are distributed among a number of chromosomes, which commonly vary in size. In contrast to bacterial chromosomes, which consist of circular molecules of DNA, eukaryotic chromosomes contain linear, double-stranded DNA. The DNA and various types of associated proteins constitute a chromosome. RNA does not appear to be a structural component of eukaryotic chromosomes.

> **Example 3.1.**   A monoploid, single-celled yeast used to make alcoholic beverages divides its genes among 4 chromosomes; a haploid, human cell has 23 chromosomes, whereas a haploid carp cell has 52 chromosomes.

> **Example 3.2.**   In human cells from males, the smallest chromosome is called the Y chromosome. It is about 1/5 the size of the largest chromosome, labeled number one.

Eukaryotic chromosomes may be condensed to different degrees. The least condensed structure is the **extended nucleosome form** (Fig. 3-1a). **Nucleosomes** consist of DNA wound twice around nearly spherical bodies composed of an octet of proteins called **histones** (H2a, H2b, H3, and H4). Approximately 200 base pairs of the DNA are wound around the spherical bodies formed by the histones, and about 50 base pairs of DNA connect the nucleosomes. Although the extended nucleosome form of the eukaryotic chromosome resembles the beaded structure seen in bacteria, these structures may not be equivalent. The extended nucleosome form occurs in those regions of the chromosome that are being replicated or are involved in the synthesis of RNA molecules.

A more compact state of eukaryotic DNA is known as the **solenoid form** (Fig. 3-1b). **Histone H1** binding to nucleosomes induces them to associate into a ring of 6 nucleosomes and the rings to associate into a cylinder called a **solenoid**. During **interphase,** most of the eukaryotic chromosome exists in the solenoid form. Those portions that are undergoing replication or being expressed (transcribed into RNA molecules) **decondense** to the extended nucleosome form. Remember, DNA replication occurs during interphase in a period known as the S stage. Gene expression occurs during all stages of interphase (G_1, S, and G_2). Eukaryotic cells that are not proliferating (progressing through the cell cycle) are said to be in stage G_0 and look like interphase cells. Chromatin, seen in nuclei during interphase, consists of DNA mostly in the solenoid form.

The most compact state of the DNA is referred to as the **looped solenoid form** (Fig. 3-1c). DNA-binding proteins stimulate the solenoid to loop back and forth from a central core of proteins called the **scaffold**. In some eukaryotes, 18 loops of the solenoid form a disklike structure. The chromosome condenses as hundreds of disks stack together. The looped solenoid form is observed during mitosis and meiosis. Compact chromosomes are essential when many long chromosomes have to be moved through the cell and might become entangled during movement.

3.3.   Does the degree of chromosomal condensation play a role in controlling DNA replication and gene expression (transcription)?

DNA replication and gene expression in eukaryotes are affected by the degree of chromosomal condensation. Replication and gene expression are inhibited by highly condensed chromosomes, apparently because the enzymes that synthesize DNA and RNA as well as accessory proteins essential for replication and gene expression cannot bind to condensed DNA. On the other hand, decondensed DNA in the extended nucleosome form can associate correctly with the enzymes that synthesize new DNA or express genes.

3.4.   What are heterochromatin and euchromatin? What is their importance in DNA replication and gene expression?

Heterochromatin is highly condensed DNA that remains in the condensed solenoid form throughout the cell cycle except during DNA replication, when it decondenses. Because of its condensed state, heterochromatin is as thick as a mitotic chromosome and can be seen with the light microscope during interphase. Most of the genes associated with heterochromatin are not expressed because of the DNA's condensed state. In contrast, **euchromatin** is decondensed DNA that exists in the solenoid form or in the extended nucleosome form. Euchromatin is

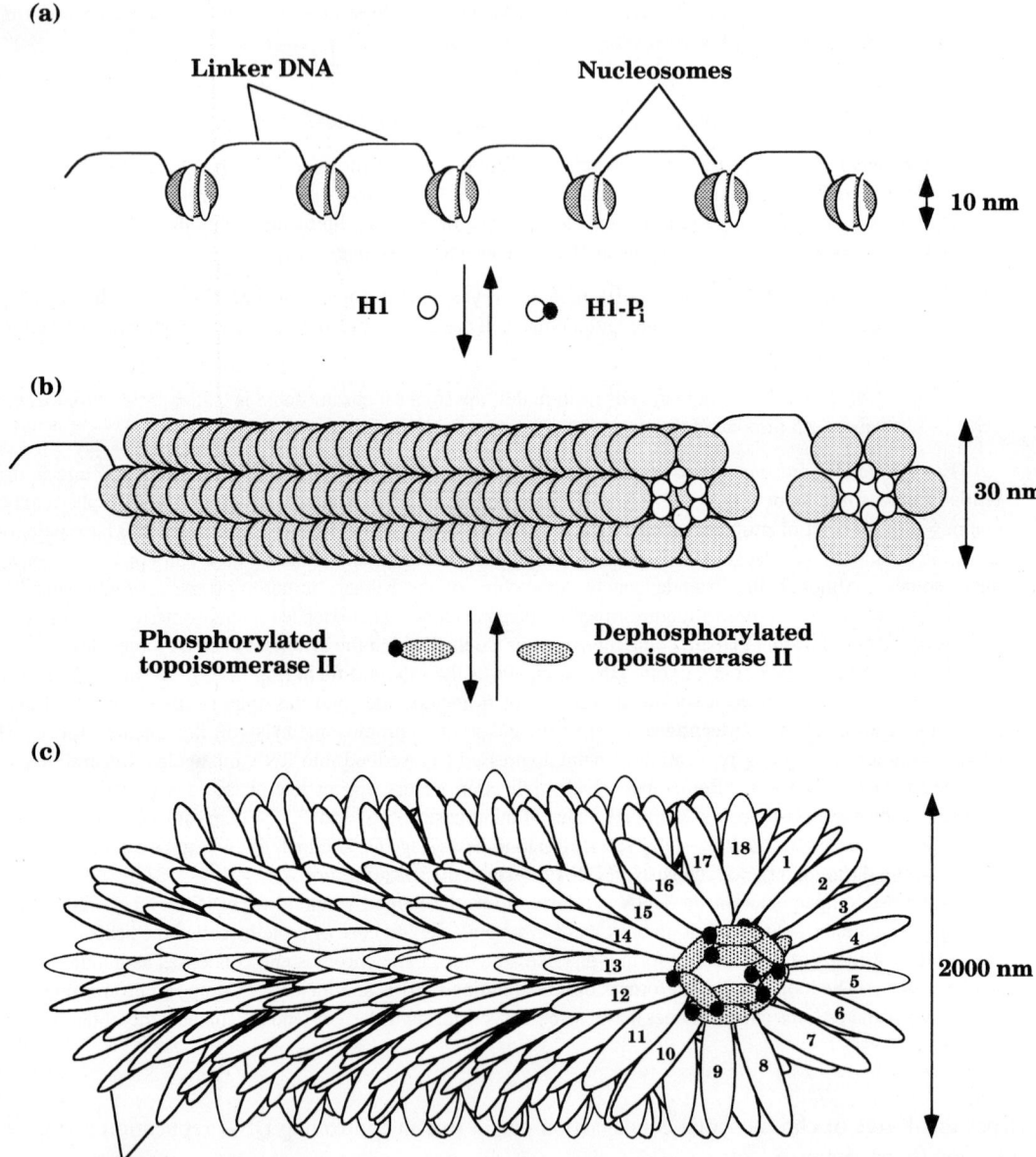

Fig 3-1.   Different forms of a eukaryotic chromosome.   (*a*) Extended nucleosome form: Histones H2a, H2b, H3, and H4 form an octamer around which dsDNA winds.   The histone core and the attached DNA is referred to as a nucleosome.   Six nucleosomes are shown connected by dsDNA strands called "linkers." (*b*) Solenoid form: Histone H1 binding to nucleosomes induces six of them to coalesce into a ring.   The rings associate, forming a solenoid.   Phosphorylation of histone H1 results in the dissociation of the solenoid into an extended nucleosome form.   (*c*) Looped solenoid form: Topoisomerase II (and other proteins not shown) form a scaffold to which the solenoid binds.   The solenoid binding to the scaffold forms a disc of approximately 18 loops.   The discs stack, forming a chromosome that is visible under the light microscope.   Dephosphorylation of topoisomerase II (and other proteins) causes the dissocation of the scaffold and results in the decondensation of the chromosome to the solenoid form.

"true" chromatin, appearing somewhat granular and dispersed under the light microscope. Euchromatin in the extended nucleosome form can be replicated or expressed, whereas that in the solenoid form cannot.

In mammals, cells from females have two X chromosomes (XX), whereas those from males have a single X chromosome (XY). One of the X chromosomes in female cells generally exists as heterochromatin. The highly condensed X chromosome is often referred to as a **Barr body,** named after the scientist who discovered it. Very few genes on the Barr body are expressed because of the high degree of condensation along most of the chromosome. One possible explanation for cells having one highly condensed X chromosome and one decondensed X chromosome is to maintain the same balance of gene products in female cells as in male cells.

3.5.   What is a centromere and how does it differ from a kinetochore?

A centromere is a highly constricted region of a mitotic or meiotic chromosome where spindle fibers attach (Fig. 3-2*a*). Complex sequences of DNA constitute centromeres (Fig. 3-2*b*). Centromeres from budding baker's yeast *Saccharomyces cerevisiae* are about 220 base pairs long and are protected from endonuclease digestion by various proteins distinct from the histones that associate with the remainder of the DNA. Although the centromeric region is protected from endonuclease attack by special proteins, it is free of nucleosomes and decondensed. This apparently accounts for the highly constricted appearance of the centromeric regions in condensed chromosomes during mitosis and meiosis. On either side of the 220–base-pair sequence of the centromere are endonuclease-sensitive sites whose function may be to promote the cutting of DNA so that chromatids can separate from each other during anaphase (Fig. 3-2*c*). Endonucleases are enzymes that cut within nucleic acids at specific sites.

The centromeres from *S. cerevisiae* contain a core area about 120 base pairs long consisting of three regions: one that is required for the separation of chromatids at anaphase (8 base pairs long); a second that consists of more than 90% A==T base pairs and is required for microtubule attachment and chromatid separation (about 85 base pairs); and a third that is required for the attachment of microtubules to kinetochores.

Analysis of the structure of centromeres from different organisms and their interaction with microtubules suggests that centromeres vary considerably.

> **Example 3.3.**   The centromeric regions from the fission yeast *Schizosaccharomyces pombe* are very large, containing between 40,000 and 80,000 base pairs. These centromeres consist of numerous sequences repeated and organized in various combinations. At least four different sequences have been characterized in the regions surrounding a central core sequence.

> **Example 3.4.**   The centromeric regions from the budding yeast *Saccharomyces cerevisiae* are unable to interact with the microtubules of multicellular eukaryotes.

Protein complexes associated with the centromeric regions of mitotic and meiotic chromosomes are called kinetochores (Fig. 3-2*a*). Kinetochores bind microtubules of the spindle bundle; they can be seen only with the aid of the electron microscope. Kinetochores vary considerably. In many animal cells, kinetochores are disclike structures about 2 μm in diameter that form on one side of each chromatid. These kinetochores have three distinct layers: a 40–60-nm inner layer contacting the proteins associated with the centromere, a 25–30-nm central layer whose function is unknown, and a 40–60-nm outer layer that attaches to the ends of microtubules. Microtubules pass through the three layers of the kinetochore discs. Kinetochores can be detached from the chromosome by treating interphase (G_1, S, G_2) cells with caffeine. In most higher eukaryotes, 15 to 35 microtubules attach to the outer layer of a kinetochore.

> **Example 3.5.**   Human kinetochores bind approximately 15 microtubules, whereas kinetochores from *Saccharomyces* bind only one microtubule.

Kinetochores in higher plants are ball-like structures that form around the centromere of each chromatid. No clear-cut layers have been detected in these kinetochores. In many eukaryotes (protozoa, some fungi, and insects), no kinetochores can be seen because the proteins are too sparse or disintegrate during preparation.

3.6.   What are the functions of centromeres and kinetochores?

Centromeres and associated kinetochores are required for the distribution and maintenance of eukaryotic chromosomes as cells proliferate. Kinetochores not only bind microtubules in prophase but actively depolymerize microtubules during anaphase. This activity draws the chromatids (at this point referred to as daughter chromosomes) toward the poles.

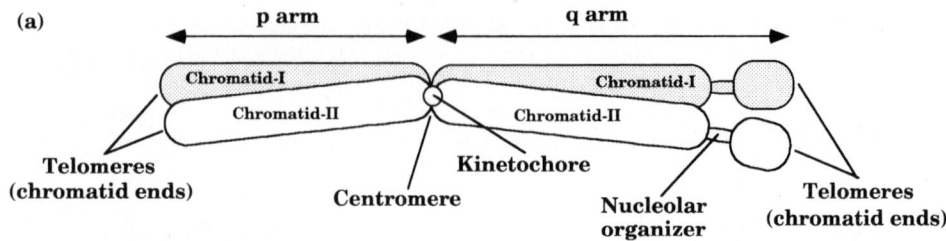

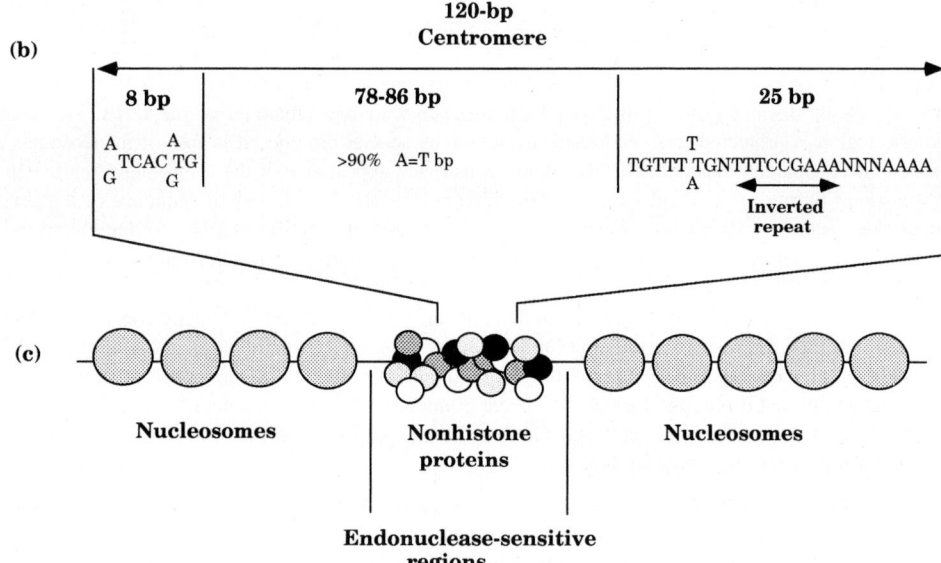

Fig 3-2. Structures associated with the eukaryotic chromosome. (*a*) Fully condensed metaphase chromosome: The constricted regions where kinetochores form are called centromeres. Spindle fibers attach to the kinetochores. The long, constricted regions (far right on each chromatid) where the genes for rRNA are located are known as nucleolar organizers. The ends of the chromatids (and chromosomes) are referred to as telomeres. The short and long arms of the chromatids are known as the p and q arms, respectively. (*b*) Centromere from *Saccharomyces*: The centromere is about 120 base pairs (bp) long, consisting mainly of A═T bp. Mutations in the 8-bp region have no effect on attachment of microtubules, but do disrupt the separation of chromatids. Changes in the nucleotide sequence of the 25-bp region disrupt microtubule attachment. (*c*) Proteins associated with the centromere: The centromere is complexed with nonhistone proteins that protect it from endonuclease digestion. Stretches of DNA on either side of the centromere are sensitive to endonuclease digestion, a finding suggesting that they are not protected by proteins. DNA on either side of the endonuclease-sensitive regions is associated with histones in nucleosomes.

3.7. What special DNA sequences are necessary to propagate, maintain, and correctly distribute pieces of DNA in proliferating eukaryotic cells?

Propagation and maintenance of any piece of DNA require the presence of one or more eukaryotic **origin of replication sites** (*oriR*) and special ends called **telomeres.** Origin of replication sites are special DNA sequences where DNA replication initiates. Telomeres protect the ends of linear chromosomes from cellular enzymes that degrade nucleic acids from their ends. These enzymes are known as exonucleases. DNA molecules can be distributed to progeny cells during the eukaryotic cell cycle only if they have centromeres (Fig. 3-2).

Artificial chromosomes can be constructed from any piece of DNA by adding *oriR* sites, telomeres, and a centromere. A number of yeast artificial chromosomes (YACs) have been constructed this way. The minimum size for these artificial chromosomes seems to be about 100,000 base pairs.

3.8. How are individual eukaryotic chromosomes recognized?

During mitosis (and meiosis), chromosomes in the cells of multicellular eukaryotes condense maximally and can be seen under the light microscope (Fig. 3-1c). Each kind of chromosome has its own characteristic length, centromere location, and staining (banding) pattern. If the centromere is in or very near the middle, the chromosome is said to be **metacentric.** A chromosome having a centromere between the middle and the end is referred to as a **submetacentric** or **acrocentric** chromosome. If the centromere is at or very near the tip, a chromosome is known as a **telocentric** chromosome. The short and long arms of a chromosome with respect to the centromere are designated as **p** and **q**, respectively.

Special staining techniques reveal that each chromosome has a specific pattern of dark and light regions called **bands.** Homologous chromosomes have the same banding pattern, whereas nonhomologous chromosomes have very different patterns. Thus, nonhomologous chromosomes that are of approximately the same length and have similarly located centromeres can be distinguished by their different banding patterns.

3.9. Compare the amount of DNA found in prokaryotes with that found in eukaryotes.

Although many of the multicellular eukaryotes have 100 to 1000 times more DNA per cell than the single-celled eukaryotes and bacteria, they probably do not have that many more genes to account for all of the extra DNA. Ninety to 95% of the DNA in the "higher" eukaryotes is not expressed as RNA or protein. Much of this noncoding DNA consists of short stretches of random base pairs repeated hundreds to thousands of times, thus the name **repetitive DNA.** A small fraction of the noncoding, repetitive DNA comprises the centromeric and telomeric regions of the chromosomes. The remaining noncoding DNA consists of **spacer DNA** between genes. Ninety to 95% of the DNA that specifies messenger RNA consists of **introns** within genes. Introns separate a gene's coding regions known as **exons** from each other. Introns in **primary transcripts (pre-RNA)** are cut out of the pre-RNA and exons are spliced together to form mRNA. The noncoding DNA and the intron sequences in DNA are thought to promote recombination between genes and within genes, respectively.

Example 3.6. Bacterial genomes range in size from 10^6 to 3×10^7 base pairs, whereas the monoploid baker's yeast *Saccharomyces cerevisiae* divides its 7×10^7 base pairs among 8 chromosomes. In comparison, a diploid human cell divides its 5.6×10^9 base pairs among 46 chromosomes, whereas the frog *Rana pipiens* divides its 46×10^9 base pairs among 26 chromosomes. These frog cells have nearly 10 times more DNA than human cells.

3.10. What are lampbrush chromosomes, and what is their function?

Lampbrush chromosomes are found in immature animal oocytes (developing germ cells) during a portion of prophase I of meiosis called diplonema. Lampbrush chromosomes, first studied in amphibian immature oocytes, were so named because they resemble the brushes used to clean lamp chimneys in the late nineteenth century. Prophase I is divided into 5 stages: **leptonema,** when the homologous chromosomes are still well-dispersed and very thin; **zygonema,** when the homologous chromosomes have paired (synapsed) with each other to form the bivalent; **pachynema,** when all four chromatids of a tetrad become thick enough to be seen under the light microscope and where the synaptonemal complex (visible with the transmission electron microscope) promotes crossing over between one chromatid of each chromosome; **diplonema,** when the homologous chromosomes pull away from each other and crossing over (chiasmata) between chromatids becomes visible under the light microscope; and **diakinesis,** when the homologous chromosomes have separated so that all remaining chiasmata are near the chromosomal ends. When used as adjectives, the first four stages are spelled differently. For example, leptonema is interchangeable with leptotene stage.

Extensive transcription of RNA molecules from lampbrush chromosomes during diplonema causes the DNA to loop out in an extended nucleosome form. The DNA loops can be seen with the light microscope because they are thickened by the great number of nascent RNA transcripts. Each chromosome of a synapsed pair (bivalent) consists of two chromatids, and so there are two loops associated with each site being transcribed. The high concentration of RNA transcripts produced and stored during diplotene stage provides the mRNAs, tRNAs, and 5S rRNAs the zygote needs to rapidly divide after the oocyte is fertilized. As many as 20,000 different messenger RNAs have been detected in the oocytes of some amphibians during prophase I. In addition, some proteins (like those specifying histones, ribosomal proteins, and yolk proteins) that will be required in large amounts are produced during diplotene stage.

The DNA (rDNA) that specifies ribosomal RNAs (28S, 5.8S, 18S rRNA) is not looped out from the chromosomes of the bivalent nor is the bivalent rDNA expressed during diplotene stage. Nevertheless, rRNA molecules are produced during diplotene stage in the many nucleoli found in the nucleoplasm. **Nucleoli** of amphibian oocytes (and even human oocytes) are dark-staining bodies in the nucleus; they are produced during pachytene stage when the bivalent's rDNA is amplified and released into the nucleoplasm as circular, double-stranded DNA (Fig. 3-3a). The circular, dsDNA nucleotide sequences specifying rRNA have been observed replicating themselves by a **rolling circle mechanism** (Fig. 3-3b). Transcription of the amplified rDNA in nucleoli is initiated during diplotene stage, and the large number of active genes can be visualized in electron micrographs (Fig. 3-3c).

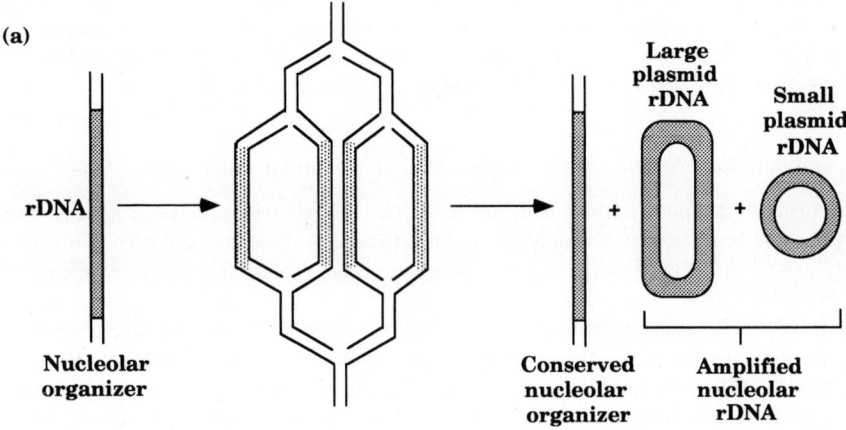

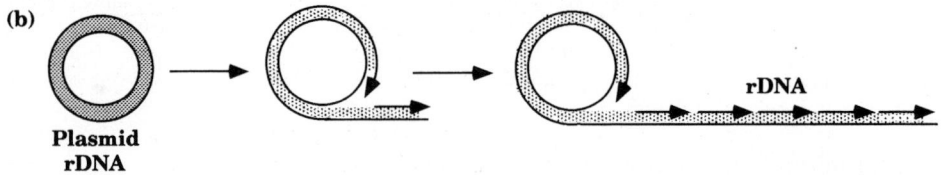

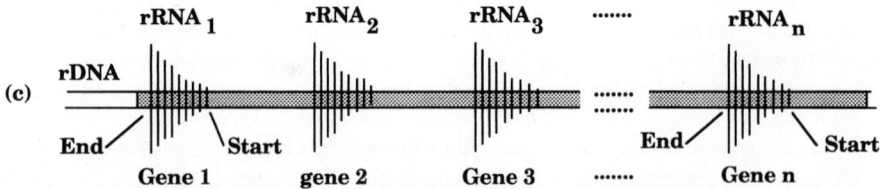

Fig 3-3. Nucleolar organizer and ribosomal DNA. (a) Nucleolar organizer: A long sequence of DNA that specifies rRNA called the nucleolar organizer is amplified through localized DNA replication. Some of the newly replicated DNA forms plasmids of different sizes that specify rRNA. The plasmids become part of a nucleolus that develops around the conserved nucleolar organizer on the chromosome. (b) A plasmid: Plasmids encoding rRNA are replicated by the rolling circle mechanism. This results in linear molecules that have many genes for rRNA. The circular molecules with linear "tails" are part of nucleoli. (c) Expression of rDNA: A linear molecule of rDNA that specifies rRNA is being expressed. The vertical lines represent rRNA molecules that are specified by numerous genes (1, 2, 3, ..., n). The short vertical lines represent rRNA that is just beginning to be synthesized, whereas the long vertical lines are rRNA molecules that are near completion.

3.11. What are polytene chromosomes and what might their function be?

Polytene chromosomes are extremely thick, long, and highly banded chromosomes that can be seen under the light microscope during all stages of the cell cycle (Fig. 3-4). They are usually observed in salivary gland cells in the larvae of certain insects, such as the fruit fly *Drosophila*. Adult flies, however, do not have polytene chromosomes, except in a few cells in their footpads. Although polytene chromosomes have been observed in the macronuclei of ciliated protozoans and in certain embryonic plant tissues, they are rare in most animal or plant species.

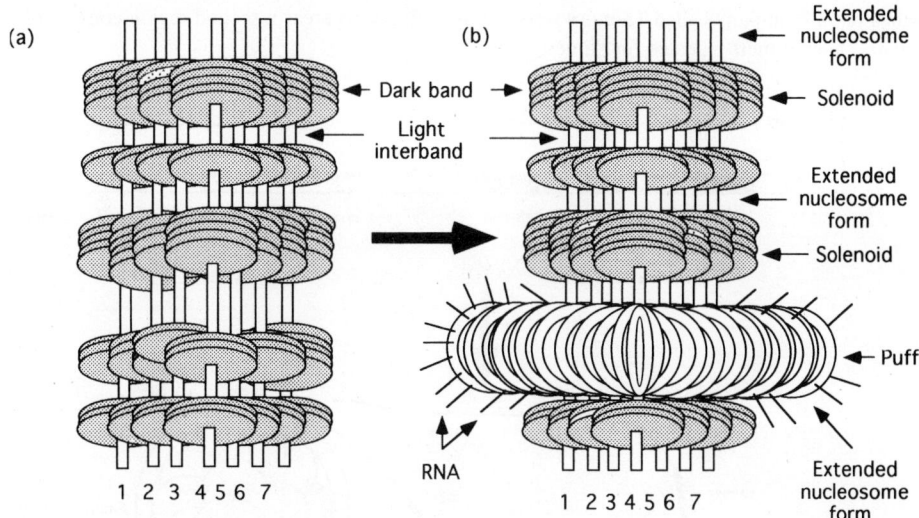

Fig 3-4. Polytene chromosome. (*a*) A polytene chromosome showing seven of a possible 1000 chromatids. The dark bands represent chromatin in the solenoid form, whereas the light interbands show chromatin in the extended nucleosome form. The interband material may represent spacer DNA that does not specify RNAs or proteins. (*b*) A polytene chromosome in which a dark band has decondensed from the solenoid to the extended nucleosome form. This decondensed dark band is known as a puff and is the site of extensive RNA synthesis.

During *Drosophila* larval development, chromosomes in salivary gland cells are stimulated to replicate. DNA replication is initiated from most, but not all, of the *ori*R sites. Ten synchronized replications in the absence of chromatid separation lead to a polytene chromosome consisting of over 1000 chromatids ($2^{10} = 1024$). The chromatin in the dark bands is believed to be in the solenoid form, whereas that between the dark bands is in the extended nucleosome form (Fig. 3-4*a*). The more than 5000 dark bands on the four chromosome pairs of *D. melanogaster* are each thought to contain four to six genes to account for the 20 to 30 thousand genes the fly is estimated to have. The regions between the bands may also contain genes, but they probably represent mostly noncoding spacer DNA. Some polytene chromosomes are over 20 μm in diameter and 200 μm long.

At different times during *Drosophila* larval development, one or more dark bands disappear and the DNA appears to form puffs in those regions (Fig. 3-4*b*). These DNA **puffs** contain genes in the process of being transcribed. The puffs consist not only of DNA but also of newly synthesized RNA, as shown by the rapid incorporation of radioactive uridine. RNA polymerase II and the various enzymes involved in processing new RNA transcripts are also found associated with puffs. The puffing pattern changes during the development of larvae. Puffs appear and then disappear, to be replaced with other puffs. Certain puffs can be artificially induced by hormones. Puffing can be induced by injecting the hormone **ecdysone,** which induces molting. Genes that usually puff during molting can be made to puff earlier by injecting the ecdysone.

During larval development, some gene products are required in such large amounts that they cannot be produced in sufficient quantity by a single gene. **DNA amplification** and **gene amplification** occur as the polytene chromosome forms. The gene amplification allows a cell to produce over 1000 RNA transcripts and a 1,000-fold

more of the products in the time it would take to make one from a normal chromosome. The synthesis of polytene chromosomes is one way to handle the problem of producing large amounts of RNA and proteins rapidly.

DNA REPLICATION

3.12. What conditions are required for the initiation of DNA replication in prokaryotes?

DNA replication of the *Escherichia coli* chromosome begins at a single site called the **origin of replication**
(oriC) and proceeds bidirectionally to a termination site located approximately halfway around the circular chromosome. The *oriC* site is approximately 245 base pairs long (Fig. 3-5a). One part of the *oriC* site contains
stretches of 13 base pairs called **13-mers** repeated three times, whereas a second region consists of stretches of 9
base pairs called 9-mers repeated four times.

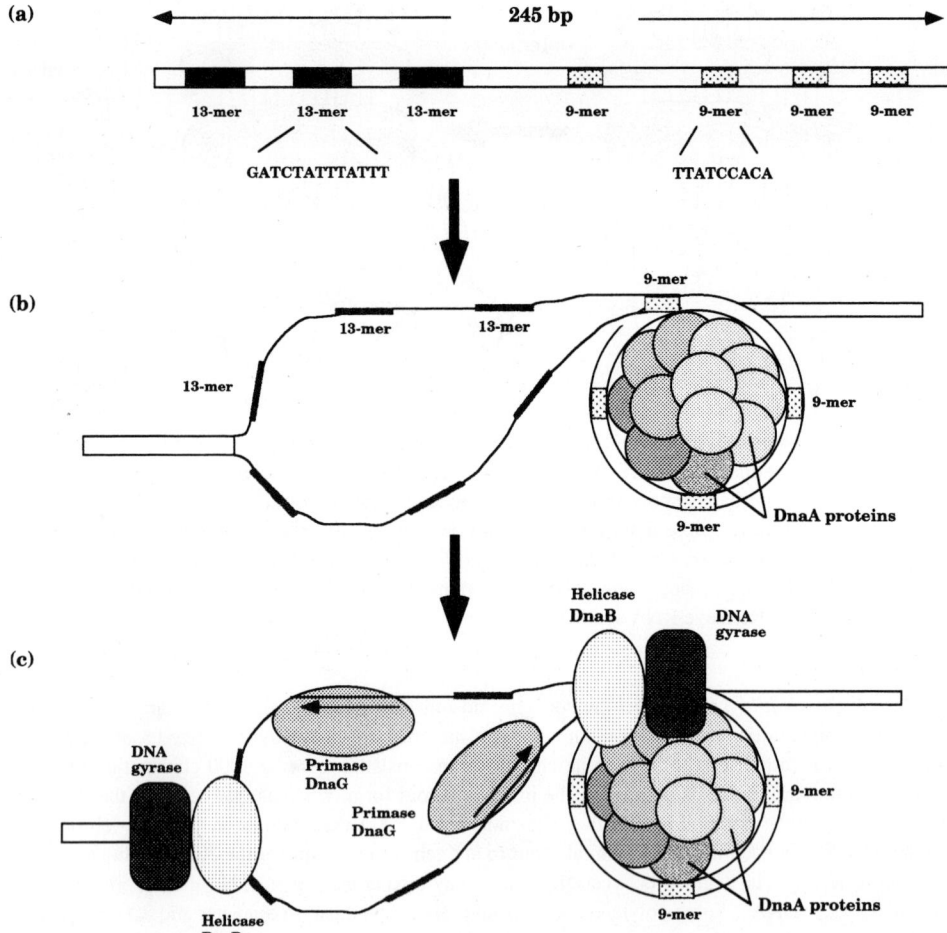

Fig 3-5. Origin of replication in bacteria. (*a*) The origin of replication site (*oriC*) in *E. coli* is approximately 245 bp long. One portion of the *ori* site contains repeated sequences of 13 bp
called 13-mers, whereas another region consists of repeated sequences of 9 bp referred to
as 9-mers. (*b*) The initiation of DNA replication requires the binding of DnaA proteins
(specified by gene *dnaA*) to the 9-mers. The formation of a DnaA complex causes the
DNA to loop around the complex. This looping of the DNA stimulates the unwinding of
the two strands in the portion of the *ori* site containing 13-mers. (*c*) The initiation of DNA
replication continues with the binding of DNA gyrases, helicases, and primases to the
DNA. The DNA gyrases remove positive supercoiling while the helicases unwind the two
strands of the DNA. The primases synthesize RNA primers. The RNA primers are required for deoxyribonucleotide polymerization (DNA synthesis).

DNA replication is initiated when a protein encoded by the gene *dnaA* binds to the 9-mers and forms a protein core around which the DNA coils (Fig. 3-5*b*). This coiling of the DNA apparently stimulates the region containing the three 13-mers to unwind. A number of enzymes bind to the unwound DNA before DNA replication can begin.

The unwinding of DNA introduces **positive superhelical turns** in the rest of the circular DNA, making it energetically unfavorable to continue unwinding the strands. To unwind DNA, positive superhelical turns have to be removed by cutting the DNA and allowing it to relax or by introducing negative superhelical turns to compensate for the positive superhelical turns. The introduction of negative superhelical turns requires energy and an enzyme called **topoisomerase** (DNA gyrase). **DNA gyrase** is an enzyme that can both remove positive supercoils or introduce negative supercoils into the DNA and thereby make unwinding energetically more favorable. DNA gyrase is absolutely required for all stages of DNA replication. Presumably, the DNA gyrase binds ahead of the unwound DNA. **Helicases** specified by *dnaB* and inhibitory proteins encoded by *dnaC* bind in the region containing the 13-mers. Dissociation of protein **DnaC** from the helicase allows the helicase to unwind the DNA.

DNA replication does not begin with the synthesis of new DNA because DNA polymerases are unable to initiate DNA replication. They can only elongate DNA from a nucleic acid with an available OH-group on its 3′ end. The nucleic acid required for the initiation of DNA replication is an RNA molecule known as the **primer.** In *E. coli*, primers are synthesized by RNA polymerase called the **primase,** specified by *dnaG* (Table 3.1). Primases binding to *oriC* synthesize RNA primers about 30 nucleotides long using both DNA strands as templates (Fig. 3-5*c*). The helicase and the primase subsequently form a complex enzyme system known as the **primosome,** which synthesizes primers on the leading strand after DNA synthesis begins.

Table 3.1.   Enzymes Involved in the Replication of Bacterial Chromosomes

Enzymes	Genes	Activities
DNA gyrase (topoisomerase)	*nalA, novR**	Eliminates positive superhelical turns or introduces negative superhelical turns. In bacteria, cuts intertwined daughter chromosomes at the end of DNA replication.
Helicase[†]	*dnaB*	Unwinds the two strands of DNA.
Primase[†]	*dnaG*	Synthesizes RNA primers (RNA polymerase III).
DNA polymerase III	*polC*	Synthesizes DNA
DNA repair enzyme	*polA*	Replaces the RNA primers with DNA.
DNA ligase	*dnaL*	Connects DNA, removing nicks left by the repair enzyme.

* Resistance to naladixic acid (*nalA*) and novobiocin (*novR*).

[†] A primosome consists of both helicase and primase subunits.

3.13.   What happens during DNA replication in prokaryotes?

For simplicity, only one of the two replication forks is considered (Fig. 3-6). A replication fork is the site where both strands of double-stranded DNA are being unwound and are being used as templates for the synthesis of new complementary DNA. Topoisomerase II (also known as DNA gyrase in bacteria) is required for the initiation of DNA replication. It introduces negative supercoiling around the origin of replication site. In addition, topoisomerase II is required for the synthesis of RNA primers and DNA from primers. It removes positive supercoils ahead of the replication fork. After the synthesis of RNA primers, the "free" primases detach from the DNA template and RNA primer, whereas the primases associated with the helicases dissociate only from the template. The two catalytic subunits of **DNA polymerase III** (PolC) associate with the templates and the 3′ ends of the primers and begin to polymerize deoxyribonucleotides into DNA. DNA gyrase (topoisomerase) continues to remove positive superhelical turns and/or introduces negative superhelical turns ahead of the primosome (helicase and primase) that is uncoiling the two strands of DNA. At various intervals, the template signals the primase (RNA polymerase III) portion of the primosome to polymerize primer RNAs about 30 nucleotides long on only one template at the replication fork (Fig. 3-6*a*). DNA polymerase III, encoded by polC, polymerizes DNA 5′ to 3′ from each of the primers at the replication fork. One strand of DNA is polymerized toward the replication fork

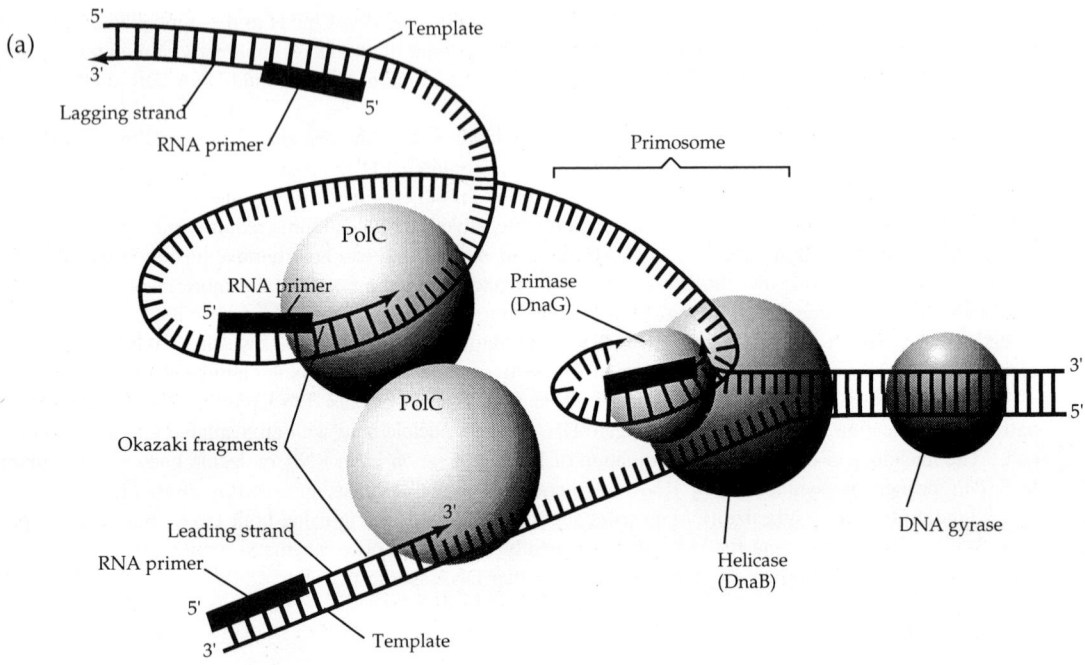

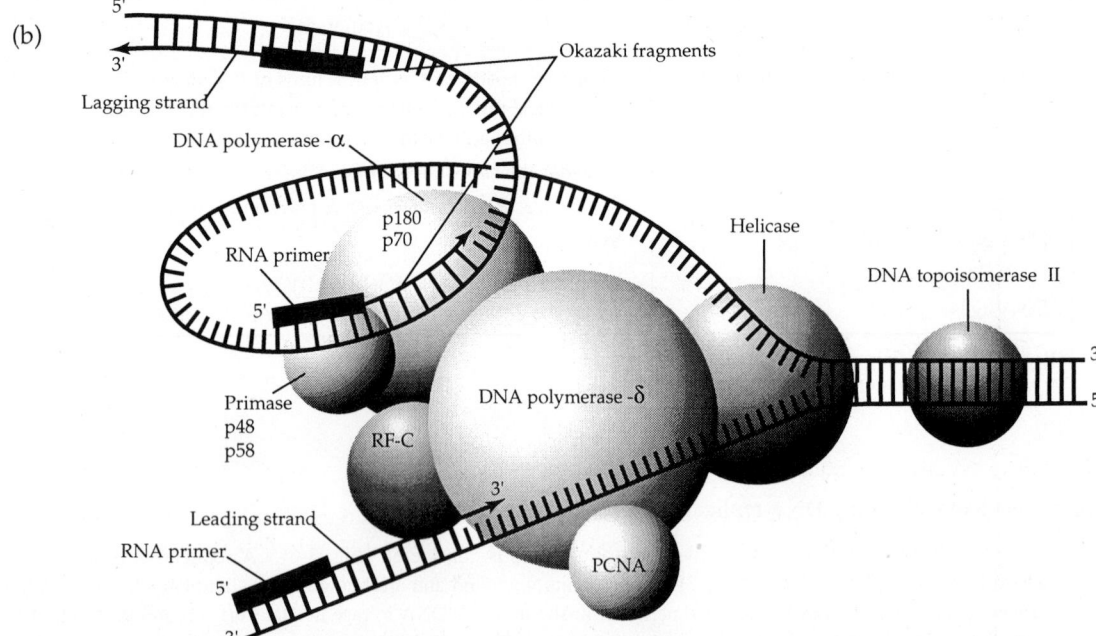

Fig 3-6. DNA replication at a replication fork. (*a*) Prokaryotic DNA replication: A DNA gyrase (far right) is inserting negative superhelical coils (or removing positive superhelical coils). A primosome consists of a helicase (that is unwinding the strands of the DNA) and a primase (that is synthesizing RNA primers on the template that is used to synthesize DNA discontinuously). Different DNA polymerases (PolC) synthesize the continuous and discontinuous strands from the RNA primers. (*b*) Eukaryotic DNA replication: A DNA topoisomerase II (far right) is removing positive superhelical coils. A helicase follows behind, unwinding the two strands of DNA. A complex of DNA primase and DNA polymerase-α synthesizes RNA primers and DNA, respectively, and accounts for the synthesis of the discontinuous DNA. Primase consists of two proteins (p48 and p58). The continuous strand of DNA is synthesized by DNA polymerase-δ with the help of additional peptides like PCNA and RF-C; PCNA = proliferating cell nuclear antigen, RF-C = replication factor-C. RF-C may connect DNA polymerase-δ to DNA polymerase-α.

and continues to be elongated as the DNA unwinds further. The second strand of DNA is polymerized away from the replication fork. As the DNA unwinds further, a new primer is synthesized away from the replication fork and the DNA polymerase synthesizes DNA from the last primer toward the previous RNA primer.

The DNA synthesized toward the replication fork is synthesized in a continuous manner and is called the **leading strand.** The opposite DNA strand is synthesized in a discontinuous manner away from the replication fork and is referred to as the **lagging strand.** The leading and lagging strands are synthesized halfway around the bacterial chromosome until they encounter the lagging and leading strands synthesized at the other replication fork. The RNA-DNA fragments that initially constitute the lagging strand are known as **Okazaki fragments,** named after the scientist who discovered them and showed that lagging and leading strand DNA are synthesized differently.

The RNA primers are removed by a DNA **repair enzyme** called DNA polymerase I specified by *polA* (Table 3.1). It uses neighboring DNA as a primer and polymerizes DNA from it, displacing the RNA primer. A **DNA ligase** removes the nicks in the DNA by connecting the DNA fragments together. Topoisomerase IV is required to separate daughter chromosomes. Topoisomerase IV cuts both strands of one daughter chromosome and passes the other daughter chromosome through the breach. The enzyme subsequently repairs the cut.

3.14. What conditions are required for the initiation of DNA replication in eukaryotes?

DNA replication in eukaryotic chromosomes generally is initiated from many origin of replication *(oriR)* sites. The *oriR* sites studied in eukaryotes are presently called **autonomously replicating sequences** (ARSs). They resemble the bacterial *oriC* sites in that they consist of two regions that bind a distinct set of proteins which destabilize the double helix (Fig. 3-7). The eukaryote's A region, with repeating 11-mers, binds **transcriptional factors** such as **E2F,** as well as topoisomerase, helicase, and an RNA polymerase. The transcriptional factors are produced continuously in cells, but are kept inactive by complexing with tumor-suppressor proteins such as the retinoblastoma protein (Rb). E2F is released from its E2F-Rb complex by stimulation of signal transduction pathways that lead to the phosphorylation and inactivation of Rb.

The B region binds **transcriptional activators** such as **Oct-1** and **AP-1, Fos** and **Jun,** or **CTF** and **NF-I,** depending upon the ARS. Most of the transcriptional activators, however, are not present in sufficient quantities to initiate DNA replication or the synthesis of RNA. Initiation of DNA replication can occur only after genes for transcriptional activators are expressed. Generally, the expression of transcriptional activators is induced by peptide or steroid hormones called growth factors (discussed in Chapter 11). Growth factors stimulate signal transduction pathways, which induce the expression of genes that specify transcriptional activators.

Transcriptional activators that bind to the B region interact with transcriptional factors in the A region, causing the DNA to bend. This distortion of the DNA and/or the interactions between the transcriptional activators and transcriptional factors promote the separation (melting) of paired DNA strands in the *oriR* region and the initiation of RNA synthesis. The RNA that is synthesized initially may serve as the primer for synthesis of the leading strand.

3.15. How does DNA replication take place in eukaryotes and how does it compare with that in prokaryotes?

Eukaryotes have numerous, linear, double-stranded DNA chromosomes, each with many sites where DNA replication begins. These origin of replication sites, symbolized as *oriR,* may be located as close as every 10,000 kilobase pairs. Thus, chromosomes that are 100,000 kilobase pairs long (specifying 10,000 genes) might have 10 *oriR* sites and be replicated as rapidly as shorter bacterial chromosomes. In fact, it turns out that eukaryotic chromosomes may have many more *oriR* sites than would be predicted from the number per genome in bacteria.

Example 3.7. Yeast chromosomes have as many as 400 *oriR* sites. However, not all the *oriR* sites are active at the same time. Some function early, while others function late in S phase of the cell cycle. Radioactive labeling experiments with Chinese hamster cells (in tissue culture) stimulated to replicate their chromosomes suggest that the longer ones may have more than 1000 *oriR* sites.

Example 3.8. The large number of *oriR* sites in eukaryotic chromosomes in relation to the number of prokaryotic chromosomes means that eukaryotes may at times replicate their chromosomes faster than bacteria. In the fruit fly *Drosophila,* nuclei are duplicated every 10 minutes during early embryonic development, whereas in the bacterium *E. coli,* replication may occur every 20 minutes. Later in development, however, the fly's nuclei may take 10 hours to duplicate. During this time, fewer *oriR* sites are active.

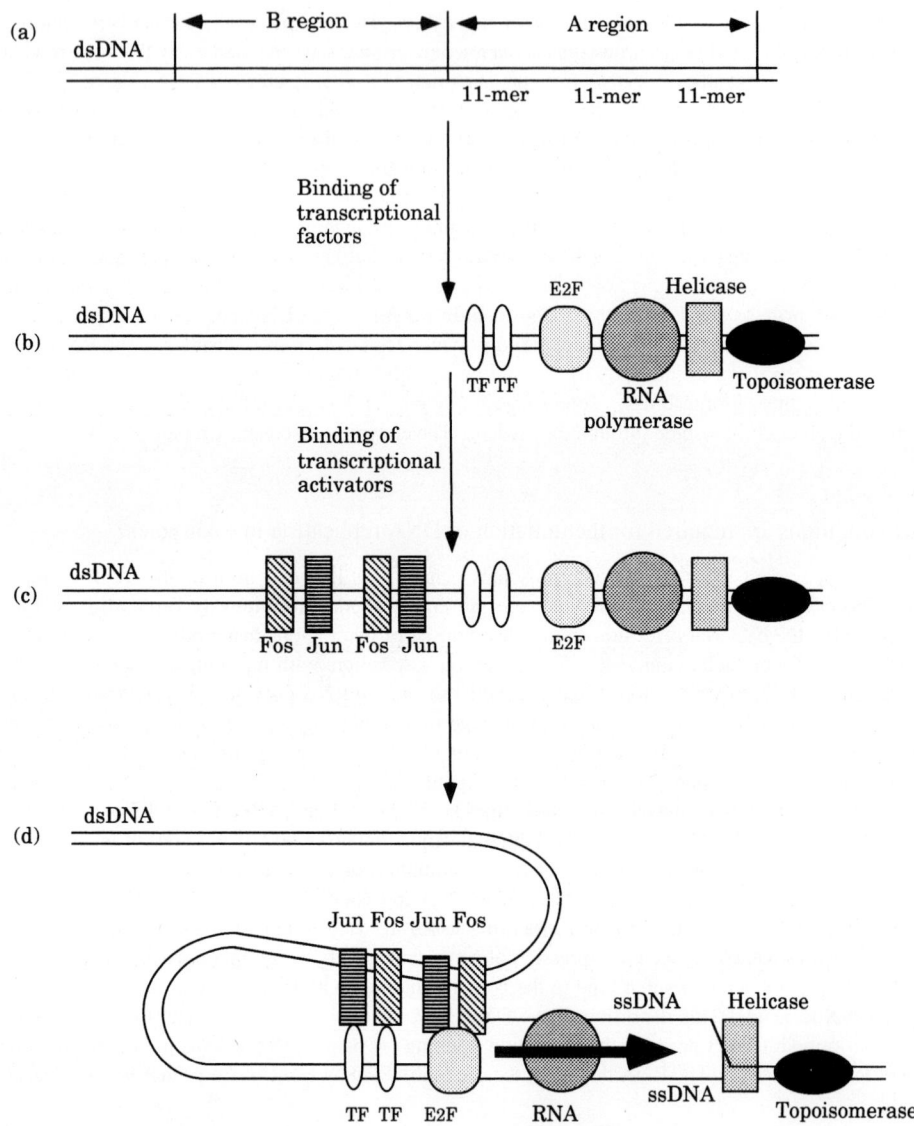

Fig 3-7. Origin of replication (autonomously replicating sequence) in eukaryotes. (*a*) The origin of replication site consists of two regions, A and B. The A region contains repeated stretches of DNA with 11 base pairs referred to as 11-mers. The B region has
not been characterized. (*b*) The A region is bound by transcription factors (such as
E2F) and by RNA polymerase (primase), helicase, and topoisomerase. TF represents
other transcriptional factors (in addition to E2F) that may be required for the initiation
of DNA replication. (*c*) Primer synthesis can occur only after protein transcriptional
activators (such as Fos and Jun) associate with the B region. (*d*) Interactions between
the transcriptional activators (Fos and Jun) and the transcriptional factors (TFs and
E2F) cause the DNA to loop. This looped state of the DNA is required for the unwinding of the two strands of DNA and the synthesis of a primer.

Enzymes similar to those involved in bacterial DNA replication are found in eukaryotes. Numerous topoisomerases, helicases, and RNA polymerases have been found in eukaryotes. **DNA topoisomerase II** is involved in relieving positive supercoils in the DNA, whereas a **helicase** activity unwinds the two strands (Fig. 3-6*b*).

At least five different DNA polymerases (α, β, δ, γ, and ϵ) have been characterized in eukaryotic cells (Table 3.2). The primase is associated with **DNA polymerase-α,** which synthesizes lagging-strand DNA. **DNA polymerase-δ,** however, catalyzes leading-strand synthesis. The two smallest subunits (p48, p58; proteins of molecular weights 48,000 and 58,000 daltons, respectively) associated with DNA polymerase-α have primase (RNA polymerase) activity. The p180 subunit of polymerase-α catalyzes nucleotide polymerization, whereas p70 is a helper factor. **DNA polymerase-ϵ** and **DNA polymerase-β** are responsible for replacing the nucleotide gaps created when RNA primers are removed by endonucleases. A **DNA ligase** repairs single-stranded nicks (unconnected adjacent nucleotides) left in the DNA after DNA repair.

Table 3.2.   DNA Polymerases in Prokaryotes and Eukaryotes

Enzyme	Synthesis	Exonuclease		Function
PROKARYOTIC				
Pol I	$5' \rightarrow 3'$	$5' \rightarrow 3'$,	$3' \rightarrow 5'$	Gap filling, DNA repair
Pol II	$5' \rightarrow 3'$	——	$3' \rightarrow 5'$	Gap filling, DNA repair
Pol III	$5' \rightarrow 3'$	$5' \rightarrow 3'$,	$3' \rightarrow 5'$	Replication
EUKARYOTIC				
Pol-α	$5' \rightarrow 3'$	——	——	Replication of lagging strand in nucleus; works with pol-δ
Pol-β	$5' \rightarrow 3'$	——	——	Gap filling & DNA repair in nucleus
Pol-δ	$5' \rightarrow 3'$	——	$3' \rightarrow 5'$	Replication of leading strand in nucleus; works with pol-α
Pol-γ	$5' \rightarrow 3'$	——	$3' \rightarrow 5'$	Replication in mitochondria
Pol-ϵ	$5' \rightarrow 3'$	——	$3' \rightarrow 5'$	Gap filling & DNA repair in nucleus

3.16. What are telomeres? How are RNA primers at the ends of linear eukaryotic chromosomes replaced by DNA?

The ends of eukaryotic chromosomes are called **telomeres.** They are unusual in that they consist of short nucleotide sequences repeated a great number of times. In some cases, the ends of each strand are covalently linked. This protects the ends from exonuclease degradation.

> **Example 3.9.** The protozoan *Tetrahymena* has chromosomes that end with repeats of the sequence
> 5′TTGGGG3′. There may be hundreds of these repeats at the telomeres. The 3′ end of one strand is
> elongated by a special enzyme. This elongated end is able to form hydrogen bonds with itself, so that
> the end has a hairpin structure (Fig. 3-8).

To complete the replication of a chromosome, RNA primers at each end of the chromosome have to be removed and replaced by DNA. Although RNA primers can be removed by exonucleases, none of the usual DNA polymerases (Table 3.2) are able to replace the RNA with DNA without a primer. An unusual type of DNA polymerase known as a **telomerase** consists of protein and an RNA template that the protein portion copies into DNA. This extends one strand of the telomere (Fig. 3-8).

One model proposes that the RNA primer at the 5′ end of a strand is digested by an exonuclease (Fig. 3-8*a*). The telomerase attaches to the 5′TTGGGG3′ of the DNA through its RNA subunit template that is complementary (3′AAAACCCCAACUUA5′). The protein portion of the telomerase reads its subunit template, catalyzing the addition of another repeat of 5′TTGGGG3′. In some cases, this newly synthesized DNA may form a **hairpin structure** that can function as a new primer. The stretches of G are able to engage in **non-Watson-Crick base pairing** (Fig. 3-9). Watson-Crick base pairing is the base pairing generally found in dsDNA: adenines hydrogen bonded to thymines and guanines hydrogen bonded to cytosines. Non-Watson-Crick base pairing would include any pairings other than A$=$T and G$\equiv$C pairings. The gap-filling DNA polymerase-β or ϵ uses the free 3′-OH group at the end of the looped hairpin structure as a primer.

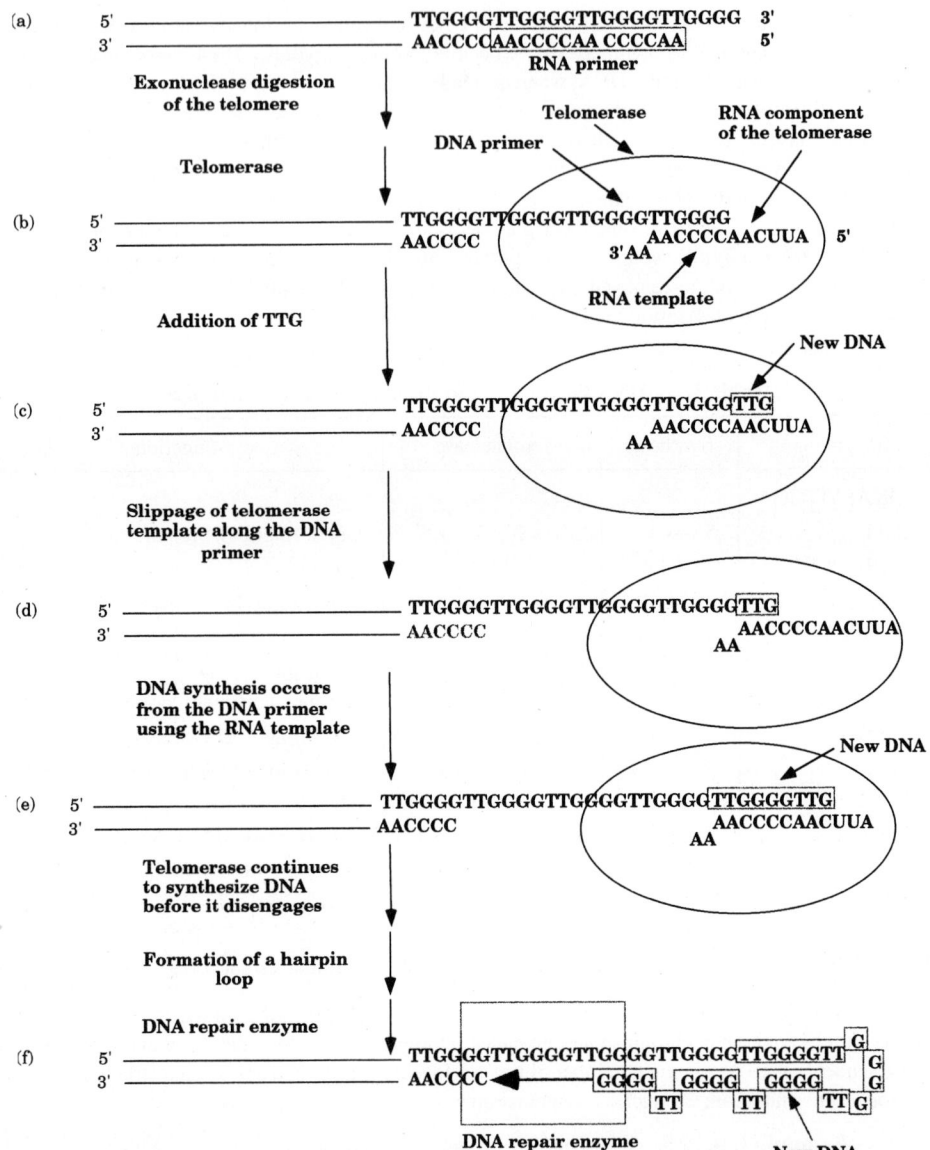

Fig 3-8. Telomeres and telomerases. (*a*) An exonuclease digests the RNA primer in the telomere at the 5′ end, leaving single-stranded DNA that will function as a primer. (*b*) A telomerase binds to the single-stranded remains of the telomere (primer) through its RNA component that will function as a template. (*c*) The protein component of the telomerase polymerizes deoxyribonucleotides from the 3′ end of the DNA primer using its RNA as a template. Only two thymine nucleotides are added before there is slippage of the telomerase template. (*d*) The telomerase slips along the DNA, exposing its RNA template once more to its polymerase activity. (*e*) The protein component of the telomerase once again polymerizes deoxyribonucleotides from the 3′ end of the DNA primer. Again, the telomerase uses its RNA as a template. (*f*) The telomerase disengages from the elongated single-stranded DNA. The elongated ssDNA forms a hairpin-like structure by establishing non-Watson-Crick base pairs between guanines. A DNA repair enzyme fills in the gap and a ligase seals the remaining nick.

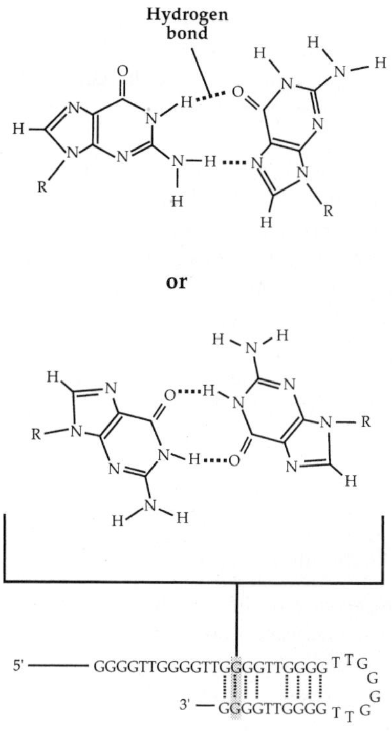

5' ——— GGGGTTGGGGTTGGGGTTGGGG $_{T}^{T}$ G G G

3' — GGGGTTGGGG $_{T T}$ G

Fig 3-9. Non-Watson-Crick base pairing. The possible non-Watson-Crick base pairings between guanines that are involved in the formation of hairpinlike structures during telomere synthesis are illustrated.

3.17. Characterize the DNA found in mitochondria and chloroplasts.

Virtually all mitochondria and chloroplasts have circular, double-stranded, DNA chromosomes that specify all tRNAs, all rRNAs, and a few of the proteins required in the organelles. These organelles may each have as many as 50 to 100 chromosomes. In plant mitochondria, chromosomes often recombine with each other, forming molecules that differ in size. The protozoans *Tetrahymena* and *Paramecium* are exceptions to the "rule" that mitochondrial chromosomes are circular; their mitochondria have linear chromosomes. The green alga *Acetabularia* has linear chromosomes in its chloroplasts. The mitochondrial DNA in animal cells specifies three rRNAs (5S, 16S, and 23S rRNAs) in the mitochondrial ribosomes, 22–30 tRNAs needed in protein synthesis, and 13–15 proteins that function as structural components and enzymes. Depending on the organism, one polypeptide is **cytochrome b,** four proteins constitute **cytochrome c oxidase,** three to four are subunits of the **ATPase** complexes, and six to seven proteins make up the **NADH reductase.** One ribosomal protein is reportedly specified by some mitochondria. All other proteins are encoded by the nuclear genome, synthesized on cytoplasmic ribosomes, and transported into the mitochondria.

The mitochondrial chromosomes in animals are among the smallest, ranging in size from 16 to 20 kilobase pairs. Compare this with the *E. coli* chromosome that contains about 5000 kilobase pairs. The chromosomes in plant mitochondria range in size from 75 to 2500 kilobase pairs, the larger ones containing mostly "filler" DNA (sometimes called "junk" or "selfish" DNA) that does not encode anything.

The chloroplast chromosomes range in size from 80 to 600 kilobase pairs. The size differences are mainly due to repeated nonfunctional sequences. Chloroplast DNAs may specify between 75 and 120 proteins, many more than mitochondrial chromosomes. Many of the genes specify ribosomal proteins and many have become **pseudogenes.** A pseudogene is a DNA sequence bearing a close resemblance to a known gene at a different locus, but rendered nonfunctional by additions or deletions in its structure or nontranscribable because of a missing promoter region. The chloroplast chromosomes also encode the rRNAs for their ribosomes and the tRNAs needed for reading the genetic code.

3.18. How does DNA replicate in mitochondria and how does this replication compare with that in prokaryotic and eukaryotic nuclei?

The two strands of the mitochondrial chromosome can be distinguished on the basis of density. The light strand (L) functions as the template for the continuous synthesis of a new heavy strand (H) at an origin of replication site that corresponds to a transcriptional promoter on the L chain. It is believed that transcriptional factors are required along with a topoisomerase and a helicase for the initiation of primer synthesis by an RNA polymerase. However, none of these enzymes have been characterized. DNA replication carried out by **DNA polymerase-γ** begins at RNA primers. Initially, DNA replication is unidirectional because the promoter-origin site *(ori-2)* on the H strand cannot be used as a template until the DNA is melted by the unidirectional DNA synthesis initiated from *ori-1*. About 2/3 of the way around the chromosome, the unwinding opens a second origin of replication site *(ori-2)* for DNA replication using the H strand as a template. The synthesis of DNA on the H strand is continuous and in the opposite direction to the synthesis on the L strand. When DNA replication is completed using the L strand template, the nascent chromosome separates from the incomplete chromosome. A topoisomerase is required for this separation. In addition, a nuclease and a repair enzyme are required to eliminate the RNA primers at the beginning of each new strand since DNA polymerase-γ has no $5' \rightarrow 3'$ exonuclease activity (Table 3.2). These enzymes have not been characterized, however. An enzyme called ligase repairs any single-stranded nicks in the DNA. Both nascent strands are synthesized continuously (like leading strands), and no Okazaki fragments are produced. Unlike nuclear DNA replication that occurs only during the S phase, replication of mitochondrial DNA is not restricted to the S phase of the cell cycle, but can occur during all stages of interphase $(G_1 \rightarrow S \rightarrow G_2)$.

3.19. How does DNA replication in chloroplasts compare with that in mitochondria?

The replication of chloroplast chromosomes proceeds much like the replication of mitochondrial chromosomes, except that their replication from two origin of replication sites *(ori-1* and *ori-2)* appears to be initiated simultaneously. Each strand is initiated by an RNA primer and each is synthesized continuously around the circular template. There is no discontinuous synthesis of DNA. Once the new strands pass each other, the synthesis superficially resembles bidirectional replication.

3.20. DNA-binding proteins fall into at least four groups based on the protein domains that bind to DNA. What are these groups?

The largest group of DNA-binding proteins characterized to date has sequences of amino acids that fold into a **helix-turn-helix domain** (HTH domain) (Table 3.3). A HTH domain is also known as a **homeodomain,** whereas the portion of a gene that codes for a HTH domain is known as a **homeobox.** Genes with homeoboxes are referred to as **homeotic genes** (homeogenes) because the first of these genes were found to differentiate homologous body segments (such as the abdominal segments of insects) in most animals. The products of homeotic genes are called **homeotic proteins.**

A second group of DNA-binding proteins has one or more helices and amino acids bound to zinc ions. These amino acid sequences allow proteins to bind DNA and are referred to as **helix–zinc-finger domains.** Zinc ions establish and maintain the protein conformation needed for DNA binding. Hunchback (hb) and Kruppel (Kr), *Drosophila* proteins important in the development of segments, have been reported to have zinc fingers. Most of the steroid hormone receptors have one or more helices and zinc fingers. The receptor for *Drosophila's* steroid hormone ecdysone is a helix–zinc-finger protein. Ecdysone stimulates pupation and metamorphosis.

Another group of DNA-binding proteins has a single **bent-helix domain** that binds to the major grooves of DNA. Generally, these proteins function as dimers. The dimers are held together by a second protein domain in each subunit that contains multiple leucines. This second domain that holds subunits of a regulatory protein together is known as the **leucine zipper.** Transcriptional activators—such as Fos and Jun—that initiate DNA replication and activate the synthesis of mRNA in eukaryotes fall into this category of DNA-binding proteins.

A fourth group of DNA-binding proteins has a **β-ribbon structure** that allows them to bind DNA. Runt *(rnt)*, a *Drosophila* gene that controls segment formation and nerve development, falls into the β-ribbon group. The bacterial protein used by the bacteriophage λ to insert its hereditary information into the bacterial chromosome also has the β-ribbon structure.

DNA RECOMBINATION

3.21. Chromosomes are capable of exchanging DNA with each other or fusing with each other through recombination. What is recombination and what is its value?

Table 3.3.   Families of DNA-Binding Proteins

Domains	Example	Function
Helix-turn-helix*	Bicoid (Bcd)	Development of fly's head region
	Bithorax (Bx)	Differentiation of fly's segments
	Antennapedia (Antp)	Differentiation of fly's segments
	Trp repressor (TrpR)	Repressor of bacterial tryptophan genes
	Lac repressor (LacI)	Repressor of bacterial lactose genes
	Lambda repressor (CI)	Repressor of viral genes
	Receptor for cAMP (Crp)	Receptor for cAMP
Helix–zinc finger	Hunchback (Hb)	Development of fly's anterior segments
	Kruppel (Kr)	Development of fly's middle segments
	Activator (Gal4)	Transcriptional activator of Gal-1 & Gal-10
	Ecdysone receptor (Ecd)	Needed for pupation & metamorphosis
	Testosterone receptor (Dev)	Steroid hormone needed for development of primary and secondary male characteristics in mammals
	Transcriptional factor (ATFIIIA)	Eukaryotic transcriptional factor
	Transcriptional activator (Sp1)	Eukaryotic transcriptional activator
Bent helix plus leucine zipper	Activator (Myc)	Transcriptional activator in mammals
	Activator (Fos)	Transcriptional activator in mammals
	Activator (Jun)	Transcriptional activator in mammals
	Activator (Gcn4)	Transcriptional activator in yeast
β-Ribbon	Runt (Rnt)	Development of fly's segments & nerves
	Recombination factor (lhf)	Integration host factor for λ-phage
	Met repressor (MetJ)	Repressor of methionine biosynthetic genes

* Identical to helix-loop-helix homeodomains.

In bacteria, pieces of DNA that enter a cell (through various processes to be discussed in Chapter 7) may become part of the main chromosome or one of the plasmids. The integration process, known as **genetic recombination,** generally occurs at points where the two DNAs are identical or nearly so. Recombination can be a beneficial process for a species because it may create new genetic information. Totally new genes and alleles of genes are created in a population not only by mutation but also by some kinds of recombination events. New genes and new alleles may allow a population to survive in adverse or changing environmental conditions. In addition, the uptake of DNA and the recombination of genes provides organisms with a mechanism by which they can replace genes that have been severely damaged or even deleted.

In eukaryotes, genetic recombination leads to genetic diversity among progeny produced by sexual reproduction. During meiosis, the process of crossing over produces linkage arrangements in gametes different from those that exist in the parent. Favorable gene combinations tend to be perpetuated by natural selection. The recombination that takes place during meiosis is also a means of repairing DNA or replacing lost DNA. Repairs can be made by copying foreign DNA that is similar to the damaged or lost DNA. Because there are a number of genetic systems that may catalyze recombinations, the process varies considerably.

3.22. There are two major types of genetic recombination: site-specific recombination and general recombination. What are the hallmark characteristics that distinguish them?

Site-specific recombination requires only short sequences of homology between recombining molecules of DNA, and it usually changes the relative positions of chromosomal segments. **General recombination** occurs only between homologous DNA molecules, and it does not normally alter the order in which gene loci occur in their respective chromosomes.

3.23. What are some of the different types of site-specific recombination?

Recombination between different chromosomes can occur if they both carry identical, double-stranded regions called recombination sites. There are at least four types of recombination that depend on specific recombination sites (Table 3.4). These recombinations depend upon restriction endonuclease sites (RESs), insertion sequences (ISs), long terminal repeats (LTRs), and signal sequences (SSs). Recombination between bacterial and viral genomes, which depends upon restriction endonuclease sites, and recombination between prokaryotic plasmids and their bacterial hosts, which depends upon ISs, will be discussed in Chapter 7.

Table 3.4.   Types of Recombination

SITE-SPECIFIC RECOMBINATION		
(*a*)	Restriction Endonuclease Site (RES) Dependent	
	Example:	Bacteriophage λ & the host's λ *att*B site
	Enzymes required:	Integrase, excisionase, & integration host factor
	Size of RES:	15 bp* in λ (*att*P) and in the host (*att*B) are identical; 240 bp around *att*P are required also
(*b*)	Insertion Sequence (IS) Dependent	
	Example:	Insertion sequences, plasmids, & complex transposons
	Enzymes required:	Transposase & possibly a resolvase
	Size of IS:	800–1500 bp
(*c*)	Long Terminal Repeat (LTR) or Delta Repeat (DR) Dependent	
	Example:	Retrotransposons & retroviruses
	Enzymes required:	Reverse transcriptase & an integrase
	Size of LTR:	276 bp (Copia in *Drosophila*), 340 bp (Ty in yeast), & 479 bp (Gypsy in *Drosophila*)
(*d*)	Signal Sequence (SS) Dependent	
	Example:	Antibody & T-cell receptor genes
	Enzymes required:	?
	Size of SS:	29 and 39 bp
GENERAL RECOMBINATION		
(*e*)	Homologous Region (HR) Dependent	
	Example:	Homologous chromosomes
	Enzymes required:	RecBCD and RecA
	Size of homology:	Many hundreds of bp

* bp = base pairs.

3.24. How does general recombination (site-independent recombination) take place in prokaryotes?

A commonly encountered type of recombination in bacteria and other organisms is the recombination between two homologous chromosomes or a DNA fragment and a homologous region in a chromosome. Because this type of recombination does not require a "special" site, it is referred to as **general recombination** or **site-independent recombination** (Table 3.4). It does require extensive homology between the DNAs and involves DNA synthesis. Site-independent recombination in many bacteria is dependent upon the products of the *rec* genes *(recA, recB, recC, recD)* (Fig. 3-10).

If a DNA fragment is single-stranded, the **RecA protein** binds to it and guides it to a homologous region on the cell's chromosome (Fig. 3-10*a*). The RecA protein melts the cell's chromosome where it is homologous to the fragment and promotes hydrogen bonding between the fragment and one strand of the chromosomal DNA. The minimal amount of nucleotide base homology required for successful general recombination is about 60 base pairs but usually involves hundreds of bases. Eventually, the **UvrABC endonuclease** excises the melted extra strand from the chromosome. A ligase repairs the nicks. If there is a mismatch, UvrABC cuts on either side of a mismatched strand and a repair enzyme (DNA polymerase I) pushes away the mismatched strand and replaces it

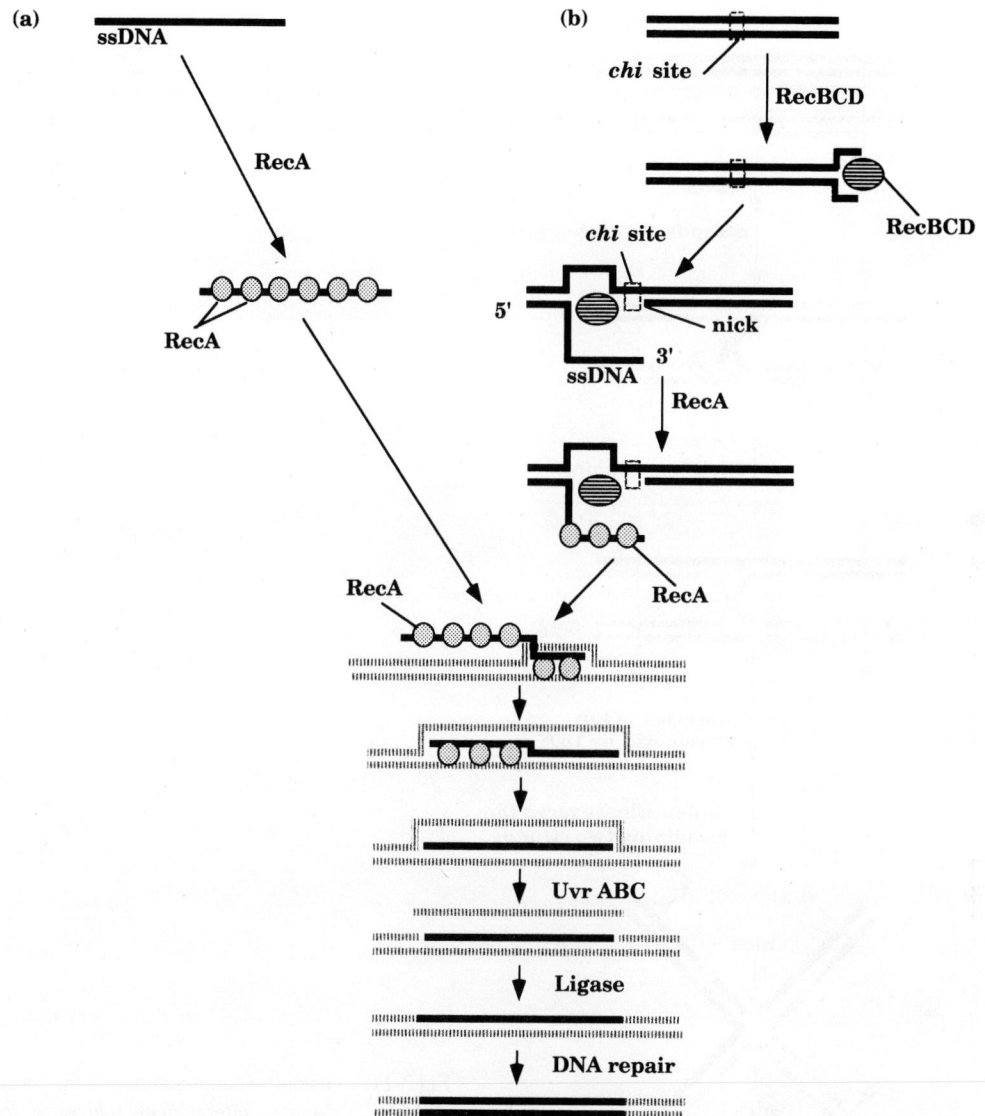

Fig 3-10. General recombination (site-independent recombination). (*a*) In bacteria, transforming DNA enters the cell as a single-stranded molecule. The RecA protein binds to the single-stranded DNA. Subsequently, the RecA protein binds to the bacterial chromosome and melts the double helix while it searches for a region of homology. Hydrogen bonding occurs between the transforming DNA and a complementary region of the cell's DNA. An enzyme such as UvrABC cuts away the unpaired portion of the melted DNA. A ligase repairs the nicks. If the two new strands are not completely homologous, there are small distortions that are corrected by a DNA repair system. The repair process eliminates one of the nonhomologous strands. In this example, the original DNA strand is lost. (*b*) In bacteria, conjugation generally results in a double-stranded molecule within the recipient cell. The RecBCD protein searches the DNA for a *chi* (χ) site, where it cuts one strand of the DNA (creating a "nick"). The single-stranded DNA that results is coated with RecA protein. RecA protein binds to the bacterial chromosome and melts the double helix while it searches for a region of homology. The steps that lead to the incorporation of the new DNA are believed to be identical to those discussed in part (*a*).

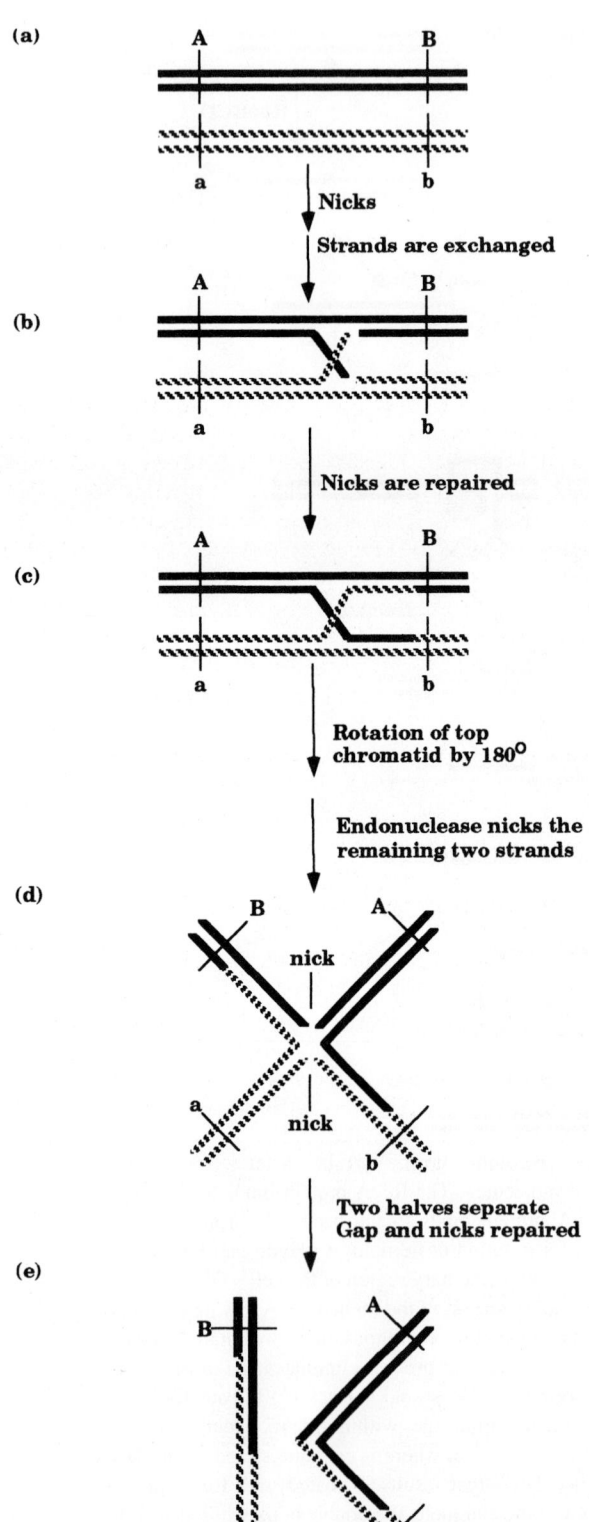

Fig 3-11. General recombination and the formation of a Holliday intermediate (*chi* or χ form). (*a*) Two homologous chromatids synapse. (*b*) A strand of each homologous chromatid is nicked and exchanged. (*c*) DNA synthesis may extend the exchanged strands subsequent to DNA nick repair. A Holliday intermediate can be created by rotation of the top chromatids by 180°. (*d*) The enzyme complex RuvAB may promote branch migration so that the uncut strands can be cut by the endonuclease RuvC at tetranucleotide sequences, 5'-(A/T)TT(G/C)-3'. This is the Holliday intermediate. (*e*) The nicking of the uncut strands of the chromatids creates separate recombined chromatids. Nicks in the recombined chromatids are repaired by a ligase.

with new DNA that matches the complementary strand. Ligase repairs the nicks in the DNA. Half the time a new allele replaces the original because of strand exchange. This may result in an altered cellular characteristic known as a **transformation.**

When a DNA fragment is double-stranded, it must be converted to single-stranded DNA before recombination can occur (Fig. 3-10*b*). Beginning at one end of the fragment, the **RecBCD protein** works its way to a *chi* (χ) **site,** where the RecD subunit of RecBCD nicks one strand of the DNA. Then, the RecA protein binds to the single-stranded DNA and to a homologous region on the cell's chromosome. The single-stranded DNA forms hydrogen bonds with homologous chromosomal DNA, which becomes part of the chromosome, as described in the previous paragraph.

3.25. How is general recombination mediated in eukaryotes?

General recombination normally occurs during prophase I of meiosis. Five stages are recognized during prophase I; they are (in order from earliest to latest) leptonema, zygonema, pachynema, diplonema, and diakinesis. During leptotene stage, the chromosomes begin to condense into the looped solenoid form, but are not yet visible under the light microscope. The pairing of homologous chromosomes occurs during zygotene stage, when a ladderlike protein structure begins to develop between numerous regions of homologous chromosome pairs. This structure is about 100 nm wide and is known as the **synaptonemal complex.** Very large protein complexes called **recombination modules** (about 90 nm in diameter) are found at intervals along the synaptonemal complex, and are thought to function as a multienzyme "recombination machine." At each recombination module, two of the four chromatids break and rejoin with one another in an event termed **crossing over.**

During the **pachytene stage** of prophase I, the chromosomes thicken and the synaptonemal complex begins to breakdown. **Desynapsis** marks the beginning of the **diplotene stage.** Chiasmata first become apparent during diplonema. **Chiasmata** are the regions where homologous chromosomes appear to be stuck together in cross-shaped structures. The intertwined recombinant chromatids at each chiasma function to bind the maternal and paternal homologues together on the spindle until anaphase I. Little is known about the protein molecules of the synaptonemal complex, the recombination modules, or the mechanisms that cause synapsis of homologous chromosomal segments. It is believed that endonucleases in the recombination modules nick a single strand of each chromatid capable of recombination and that helicases unwind the DNA, creating single-stranded regions (Fig. 3-11). A protein similar to RecA is proposed to catalyze the pairing of the single-stranded DNAs to the complementary strands on the homologous chromatids. A DNA polymerase may extend the exchanged strands, and a DNA ligase is thought to eliminate the nicks in the strands (Fig. 3-11*c*). DNA at this stage has been isolated from *E. coli* and is referred to as the **chi** (χ) **form,** the **Holliday intermediate,** or the **Holliday junction,** after the geneticist who proposed the model in 1964 (Fig. 3-11*d*). The *chi* form is converted into two linear DNA molecules by an endonuclease (specified by *ruvC*) that cleaves the previously uncut strand of each chromatid. The resulting DNA molecules are recombinants of each other (Fig. 3-11*e*). The RuvC endonuclease cuts at specific tetranucleotide sequences, 5'-(A/T)TT(G/C)-3', between the T and (G/C). When there is no appropriate tetranucleotide sequence at a Holliday junction, a RuvAB protein complex promotes a migration of the Holliday junction to a region of the chromosomes that contains tetranucleotide sequences that can be cut by RuvC.

Objective Questions

Multiple Choice

Directions: Choose the one best answer.

1. The filamentous DNA and protein that can be stained in interphase nuclei is called *(a)* solenoid. *(b)* nucleosome. *(c)* chromatin. *(d)* polytene. *(e)* leptotene.

2. What kind of structure does replicating eukaryotic DNA have? *(a)* Extended nucleosome form *(b)* solenoid form *(c)* looped solenoid form *(d)* naked DNA form *(e)* positively supercoiled form.

3. Which histone protein is *not* part of a nucleosome? *(a)* H1 *(b)* H2a *(c)* H2b *(d)* H3 *(e)* H4

4. Which histone protein is involved in the transition between the solenoid form and the extended nucleosome form? *(a)* H1 *(b)* H2a *(c)* H2b *(d)* H3 *(e)* H4

5. A DNA sequence required for the partition of eukaryotic chromatids during mitosis is a *(a)* telomere. *(b)* kinetochore. *(c)* centrosome. *(d)* centromere. *(e)* centriole.

6. A protein structure on eukaryotic chromosomes to which spindle fibers (microtubules) bind is a *(a)* telomere. *(b)* kinetochore. *(c)* centrosome. *(d)* centromere. *(e)* centriole.

7. Approximately how large are the centromeres in the yeast *Saccharomyces cerevisiae?* *(a)* 12 bp *(b)* 120 bp *(c)* 1200 bp *(d)* 12,000 bp *(e)* 120,000 bp

8. About how large are the centromeres in the yeast *Schizosaccharomyces pombe?* *(a)* 40 bp *(b)* 400 bp *(c)* 4000 bp *(d)* 40,000 bp *(e)* 400,000 bp

9. Each kinetochore on human chromosomes binds approximately how many microtubules? *(a)* 1 *(b)* 15 *(c)* 50 *(d)* 68 *(e)* 102

10. One strand of telomeric DNA consists of repeated units of *(a)* 5'TTGGGG3'. *(b)* 5'AAAAAAA3'. *(c)* 5'TTTTTTTTT3'. *(d)* 5'TATATATA3'. *(e)* none of the above.

11. Mitochondrial DNA is replicated from *(a)* a single *ori* site bidirectionally. *(b)* two different *ori* sites in the same direction. *(c)* two different *ori* sites at different times in opposite directions. *(d)* many sites bidirectionally, like nuclear chromosomes. *(e)* none of the above.

12. Chloroplast DNA is replicated from *(a)* a single *ori* site bidirectionally. *(b)* two different *ori* sites in the same direction. *(c)* two different *ori* sites simultaneously and in opposite directions. *(d)* many sites bidirectionally, like nuclear chromosomes. *(e)* none of the above.

13. Heterochromatin is DNA in which form? *(a)* Naked DNA form *(b)* extended nucleosome form *(c)* solenoid form *(d)* looped solenoid form *(e)* none of the above

14. Approximately how many genes do the higher eukaryotes have? *(a)* 10,000 *(b)* 100,000 *(c)* 1 million *(d)* 10 million *(e)* 100 million

15. In eukaryotes, the lagging strand DNA is synthesized by DNA polymerase- *(a)* α. *(b)* β. *(c)* γ. *(d)* δ. *(e)* ϵ.

16. What is found in a nucleolus? *(a)* rDNA *(b)* 45S rRNA *(c)* 5.8S, 18S, and 28S rRNA *(d)* rRNA-protein complexes *(e)* all of the above

17. In bacteria, which enzyme unwinds double-stranded DNA until it meets a χ site and then cuts one strand of the DNA? *(a)* RecA *(b)* RecBCD *(c)* UvrABC *(d)* UvrD *(e)* DnaB

18. In bacteria, which enzyme binds single-stranded DNA, denatures double-stranded DNA and matches the single-stranded DNA with complementary denatured DNA? *(a)* RecA *(b)* RecBCD *(c)* UvrABC *(d)* UvrD *(e)* DnaB

19. The Holliday intermediate is also known as the *(a)* α-form. *(b)* β-form. *(c)* γ-form. *(d)* δ-form. *(e)* χ-form.

True–False

1. Positively supercoiled DNA can be converted to nonsupercoiled DNA by enzymes called topoisomerases.

2. The more positively supercoiled DNA is, the easier it is for a helicase to unwind the double helix.

3. Topoisomerase II is the enzyme that separates newly synthesized bacterial chromosomes from each other.

4. Metaphase chromosomes that are telocentric have either very short or no visible p arms.

5. The shortest "arms" of metaphase chromosomes are the p arms.

6. It is possible to see bands on metaphase chromosomes in the light microscope without staining.

7. The frog *Rana pipiens* has more DNA than the human *Homo sapiens*.

8. The origin of replication site in the *E. coli* chromosome is about 245 bp long.

9. The origin of replication is also known as the replication fork.

10. In bacteria, the same type of DNA polymerase synthesizes the leading strand DNA and the lagging strand DNA.

11. In bacteria, RNA polymerase III specified by *dnaG* is known as a primase.

12. The nucleolar organizer is located in the nucleolus.

13. The primase in eukaryotes is associated with DNA polymerase-δ.

14. The initiation of DNA replication in eukaryotes requires transcription factors such as E2F.

15. The initiation of DNA replication in eukaryotes requires transcription activators such as Fos and Jun.

16. Mitochondrial chromosomes specify more products than chloroplast chromosomes.

17. Polytene chromosomes consist of as many as 1,000 chromosomes in the solenoid form.

18. Puffs are associated with lampbrush chromosomes.

19. The bands of polytene chromosomes can be seen only if the chromosomes are stained.

20. The protein known as RecA associates with single-stranded DNA and matches this DNA with complementary DNA during the first stages of general recombination.

21. General recombination between two chromosomes requires that they each have short identical sites in which crossing over occurs.

Matching

Directions: In each of the following matching sets, every term in **Column B** may be matchable to the numbered descriptions in **Column A.** Also, a term may apply to more than one description.

Column A	Column B
1. The structure a chromosome has during mitosis when the chromosome can be seen with the light microscope	A. lampbrush chromosome
2. A chromosome found during meiosis that is actively synthesizing many mRNAs	B. extended nucleosome form
3. A chromosome found in the salivary glands of certain insect larvae	C. solenoid form
4. The structure a chromosome has when it is replicated or actively transcribed	D. looped solenoid form
5. The structure a chromosome has during interphase	E. polytene chromosome

Column A	Column B
6. The DNA found at the ends of linear eukaryotic chromosomes	*A.* nucleolar organizer
7. The protein structure on eukaryotic chromosomes where spindle fibers attach	*B.* centromere
8. A DNA site where kinetochores are found	*C.* telomere
9. In eukaryotes, a DNA site where transcriptional factors and transcriptional activators bind	*D.* kinetochore
10. A DNA sequence that serves as the template for the synthesis and amplification of ribosomal DNA (rDNA)	*E.* origin of replication

Column A	Column B
11. This enzyme synthesizes the leading strand of eukaryotic DNA.	*A.* telomerase
12. This enzyme removes positive superhelical turns from DNA.	*B.* DNA polymerase-δ
13. This enzyme maintains the ends of eukaryotic chromosomes.	*C.* DNA polymerase-α
14. This enzyme unwinds the two strands of the double helix.	*D.* topoisomerase
15. This enzyme synthesizes the lagging strand of eukaryotic DNA.	*E.* helicase

Column A	Column B
16. The stage of meiosis in which chiasmata become visible	*A.* leptonema
17. The stage of meiosis in which homologous chromosomes synapse	*B.* zygonema
18. The stage of meiosis in which lampbrush chromosomes are most active	*C.* pachynema
19. The stage of meiosis in which crossing over actually occurs	*D.* diplonema
20. The stage of meiosis in which large amounts of RNA are synthesized	*E.* diakinesis

Column A	Column B
21. Steroid hormone receptors	*A.* helix-turn-helix
22. Homeotic proteins (homeoproteins)	*B.* zinc finger
23. Transcriptional activator proteins such as Fos and Jun	*C.* β-ribbon
24. Many regulatory proteins (*Lac* repressor, *Trp* repressor, λ repressor, cAMP-receptor protein) found in bacteria	*D.* leucine zipper

Terms

Directions: Unless otherwise specified, each answer is a single word.

1. Enzymes that introduce negative supercoils in DNA. (two words)

2. DNA that does not decondense during interphase.

3. The central core of proteins, including topoisomerase II, in fully condensed eukaryotic chromosomes

4. X chromosomes that do not decondense during interphase. (two words)

5. The ends of linear eukaryotic chromosomes

6. Sites on eukaryotic chromosomes required for their correct partitioning during cell division.

7. The enzyme required to maintain and complete the ends of eukaryotic chromosomes.

8. The origin of replication sites in eukaryotic chromosomes (three words)

9. The molecule that serves as a template for telomerase.

10. An adjective used to describe chromosomes with their centromeres at one end.

11. The protein-coding regions of eukaryotic pre-RNA molecules that have to be spliced together to make mRNA.

12. The protein that binds to the 9-mers in the *oriC* site of *E. coli*.

13. The region where the chromosome strands are unwound and where DNA (and in some cases primer RNA) is being continuously synthesized. (two words)

14. During DNA replication, short segments of the lagging strand that are joined by DNA ligase. (two words)

15. The enzyme that unwinds the two strands of DNA.

16. The enzyme that replicates DNA in mitochondria. (two or three words)

17. The extended regions where RNA synthesis (transcription) occurs on polytene chromosomes.

18. DNA-binding proteins that use a helix-turn-helix domain to bind to DNA and that determine the characteristics of an animal's segments. (one or two words)

19. The DNA that specifies the helix-turn-helix domain of certain DNA-binding proteins.

20. The intermediate formed after two strands from homologous chromatids have recombined with each other. (two words)

Answers to the Objective Questions

Multiple Choice

1. *c*   2. *a*   3. *a*   4. *a*   5. *d*   6. *b*   7. *b*   8. *d*   9. *b*   10. *a*   11. *c*
12. *c*   13. *d*   14. *b*   15. *a*   16. *e*   17. *b*   18. *a*   19. *e*

True–False

1. T   2. F (negatively supercoiled)   3. T   4. T   5. T   6. F (must be stained to see bands)
7. T   8. T   9. F (replication fork is where DNA is being polymerized)   10. T   11. T
12. F (nucleolar organizer is a region of DNA on the chromosome that specifies rRNA; nucleolar organizer is not found in nucleoli that detach from chromosomes)   13. F (primase is associated with DNA polymerase-α)
14. T   15. T   16. F (chloroplast chromosomes specify between 70 and 120 proteins, whereas mitochondrial chromosomes specify only about 13 proteins)   17. T   18. F (puffs are associated with polytene chromosomes found in the larvae of certain insects)   19. F (bands of polytene chromosomes can be seen with the light microscope in the absence of staining)   20. T   21. F (long stretches of homology are required)

Matching

1. *D*   **2.** *A*   **3.** *E*   **4.** *B*   **5.** *C*   **6.** *C*   **7.** *D*   **8.** *B*   **9.** *E*   **10.** *A*   **11.** *B*
12. *D*   **13.** *A*   **14.** *E*   **15.** *C*   **16.** *D*   **17.** *B*   **18.** *D*   **19.** *C*   **20.** *D*   **21.** *B*
22. *A*   **23.** *D*   **24.** *A*

Terms

1. DNA gyrases   **2.** heterochromatin   **3.** scaffold   **4.** Barr bodies (also sex chromatin)   **5.** telomeres
6. centromeres   **7.** telomerase   **8.** autonomously replicating sequences   **9.** RNA   **10.** telocentric
11. exons   **12.** DnaA   **13.** replication fork   **14.** Okazaki fragments   **15.** helicase   **16.** DNA
polymerase-γ **17.** puffs   **18.** homeotic proteins (homeoproteins)   **19.** homeobox
20. Holliday intermediate or χ form

Chapter 4

Transcription and Gene Regulation

INTRODUCTION

An organism's characteristics are determined by its hereditary information and the hereditary information's response to the environment. The **central dogma** of molecular genetics, first suggested in the mid-1950s by Francis Crick, proposes that the information in DNA is used to make RNA molecules, through a process known as **transcription,** and that the information in some of the RNA is used to make proteins, by a process called **translation. Gene expression** always involves transcription, and most gene expression also requires translation. Transcription of DNA templates is carried out by RNA polymerases, whereas translation of RNA is catalyzed by enzymes associated with ribosomes. The RNA molecules and proteins synthesized during the development and/or maintenance of an organism, as well as those synthesized in response to the environment, are responsible for an organism's characteristics.

The information for synthesizing a particular RNA is located in only one of the two strands of DNA. The strand that contains the information for making an RNA molecule and that is "read" by an RNA polymerase is called the **template strand,** or just **template.** Some authors refer to the template strand as the **sense strand,** since it contains the information for polymerizing nucleotides into RNA. The strand of DNA complementary to the template is sometimes referred to as the **nonsense strand,** since it provides no information for the making of RNA or protein. Not all templates coding for RNAs occur on the same strand of DNA, however. Some templates specifying RNAs are found on one strand, whereas others are on the opposite strand. Messenger RNA (mRNA) that specifies the synthesis of a protein is called **sense RNA,** whereas RNA complementary to sense RNA is known as **antisense RNA.**

The genes that code for rRNAs, tRNAs, and the many different mRNAs are regulated, so that they are expressed at the appropriate time and at the level needed to maintain the cell or to promote its growth and proliferation. Under some adverse conditions, such as nutrient starvation, many genes may be turned off and only those genes needed to keep the cell alive continue to be expressed. In addition, many studies have shown that chemicals in the environment or in the cell regulate the expression of genes.

> **Example 4.1.** If the environment does not contain a particular sugar (such as arabinose), bacteria do not express the genes that might be involved in its transport into the cell, nor do they produce the enzymes needed to catabolize that sugar. On the other hand, if the sugar is available in the environment, bacteria may produce the enzymes necessary for transporting it into the cell, concentrating it, and initiating its catabolism.

This chapter will be concerned with the mechanisms by which genes are transcribed, how the primary products of gene expression are processed into their functional forms, and how these functions are regulated. Transcription and gene regulation will be discussed first in prokaryotes, then in eukaryotes.

TRANSCRIPTION AND GENE REGULATION IN PROKARYOTES

Structural Genes, Controlling Sites, and Operons

4.1. What is the difference, if any, between structural genes and cistrons?

> **Structural genes** are nucleotide sequences of DNA that serve as templates for the synthesis of RNAs, such as transfer RNA (tRNA), ribosomal RNA (rRNA), and messenger RNA (mRNA). The average length of structural genes specifying proteins in prokaryotes is about 1,000 base pairs. In contrast, structural genes in eukaryotes av-

erage 10,000 base pairs, largely due to copious amounts of noncoding sequences (introns). **Cistrons** are structural genes that have been defined by a ***cis/trans* test** (Fig. 6-6). The *cis/trans* test involves *cis* and *trans* mutations. *Cis* mutations are linked (on the same DNA molecule), whereas *trans* mutations are unlinked and generally exist on different chromosomes. Two mutations define a structural gene, or cistron, when the two mutations in *trans* position result in a mutant phenotype, but in *cis* position produce a wild-type (normal) phenotype in a diploid organism. In addition, cistrons are structural genes that encode RNAs or proteins that affect gene expression of both *cis* and *trans* genes. This distinguishes them from controlling sites that only affect *cis* genes (see Question 4.3).

4.2. What are regulatory proteins?

Regulatory proteins (encoded by **regulatory genes**) are proteins that affect the expression of structural genes by binding to controlling sites (see Question 4.3) near the structural genes and either activating transcription or repressing it. A regulatory protein that stimulates gene transcription falls into one of two categories: a **transcriptional factor** or **transcriptional activator.** Sometimes these proteins are referred to simply as **activators. Repressors** are proteins that inhibit the initiation of transcription when bound to controlling sequences called **operators.** Proteins that terminate transcription are referred to as **terminators.** In general, activators stimulate RNA polymerase binding to promoter sites on DNA at the beginning of structural genes, whereas repressors inhibit RNA polymerase binding.

4.3. What are controlling sites?

Controlling sites are short nucleotide sequences of DNA, usually 15 to 30 base pairs in length, that control the expression of structural genes *cis,* but not *trans,* to them. Controlling sites are categorized as promoter sites (promoters), operator sites (operators), initiator sites (initiators), attenuator sites (attenuators), and transcriptional terminator sites (terminators). In most cases, controlling sites are bound by regulatory proteins or RNA polymerases.

Controlling sites are differentiated from regulator genes (cistrons) by the different effects mutations have. Mutations in controlling sites only affect the expression of genes *cis* to them, whereas mutations in regulator genes may affect the expression of genes both *cis* and *trans* to them. See Question 4.6 for more information about controlling sites.

4.4. What is an operon?

Structural genes and the controlling sites that regulate the rate of transcription of these genes are called **operons** (Fig. 4-1). The simplest operons consist of one structural gene and one **promoter** to serve as a binding site for RNA polymerase. These operons are **constitutive;** that is, they are expressed at all times, and the level of expression is determined entirely by how efficiently RNA polymerase binds to the promoter. The regulatory operon shown in Fig. 4-1 is an example of a simple operon. Some simple operons may be regulated by an **attenuator** site, which specifies an RNA that causes the RNA polymerase to prematurely cease transcribing. Attenuation is discussed in more detail in Questions 4.6, 4.10, and 4.35.

Most operons in bacteria, however, consist of numerous structural genes and controlling sites. Because the first operons studied all had operators and were turned off (inactivated) by repressors (i.e., were under **negative control**), it was thought that all operons had operators. Consequently, operons were defined initially as one or more structural genes under the control of an operator. This definition, however, does not allow for structural genes that might be regulated only by activators or by the efficiency of RNA polymerase binding to a promoter site. The histidine operon, concerned with the biosynthesis of the amino acid histidine, is not regulated by either a repressor or an operator; it is regulated exclusively by an attenuator site. It is clear, then, that operators should not be part of the definition of an operon. It is less restrictive to think of an operon as one or more structural genes (coding for RNAs) that are coordinately expressed because they are under the control of a single promoter site.

In bacteria, many operons contain more than one structural gene, or cistron. These **polycistronic operons** are transcribed into a single mRNA referred to as **polycistronic mRNA.** Each protein-coding region in the polycistronic mRNA is defined by a start codon (a contiguous group of three nucleotides), where protein synthesis is initiated, and a nonsense codon, where protein synthesis is terminated. Operons in eukaryotes are generally monocistronic, that is, they contain a single gene.

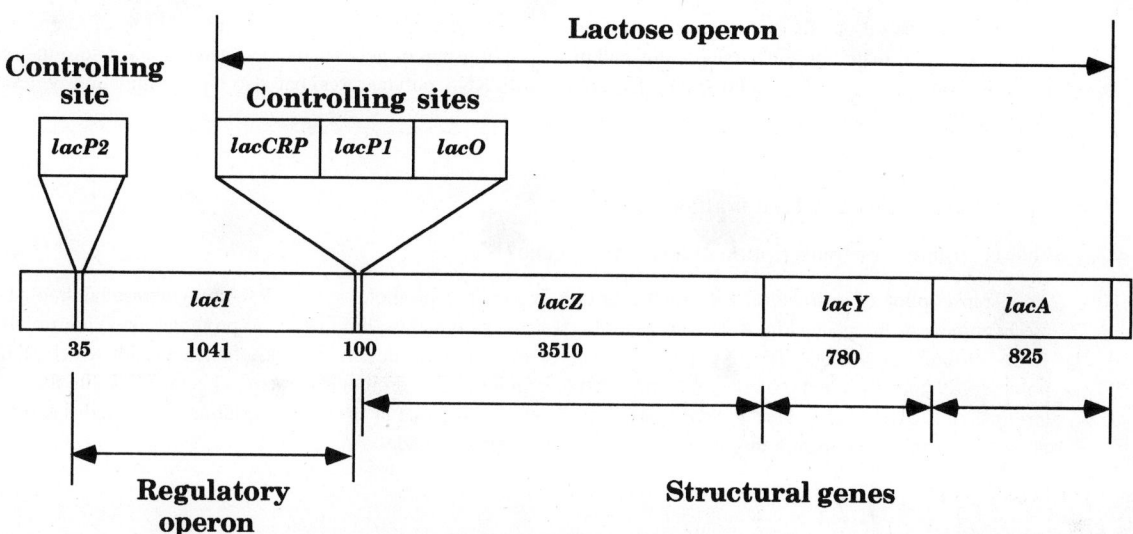

Fig. 4-1.　The small regulatory operon that controls the lactose operon, and the large lactose operon that is involved in the catabolism of the milk sugar lactose. The regulatory operon consists of a single controlling site (*lacP2*) and a single structural gene (*lacI*). The lactose operon consists of three controlling sites (*lacCRP*, *lacP1*, and *lacO*) and three structural genes (*lacZ*, *lacY*, and *lacA*). The controlling sites are quite small, each about 35 base pairs (bp) long. The structural genes vary in size from 780 bp to 3,510 bp. RNA polymerases bind to the promoter sites (*lacP2* and *lacP1*) and a cyclic adenosine monophosphate (cAMP) receptor protein attaches to the activator binding site (*lacCRP*). The *lacI* gene specifies a subunit of a repressor protein that binds to the operator site (*lacO*) and represses transcription from *lacP1*.

4.5.　What is a regulon?

A **regulon** is a group of operons that are under the control of a regulatory protein. The operons of a regulon are generally not contiguous. An important regulon referred to as the **SOS regulon,** because it responds to cellular distress, is induced when there is DNA damage. The SOS regulon is under the control of a repressor protein encoded by the *lexA* gene. The **LexA protein** represses four operons: one specifying the RecA protein, another encoding the UvrA protein, a third producing the UvrB protein, and a fourth coding for the UvrC protein. In addition, the LexA protein also represses its own operon (an example of **autoregulation**). The RecA protein is involved in recombination and DNA repair. The UvrABC nuclease is important in the repair of DNA that has been damaged by UV light.

DNA damage that creates regions of single-stranded DNA stimulates the protease activity of the RecA protein. The activated RecA protein cleaves the LexA protein and consequently **derepresses** (activates) all the operons blocked by the LexA protein.

4.6.　Do all controlling sites regulate gene expression in the same way?

There are at least five categories of controlling sites. Some controlling sites are simply binding sites for regulatory proteins. Promoters, operators, and initiators are examples of controlling sites where regulatory proteins bind. RNA polymerases, enzymes that synthesize RNA molecules from DNA templates, bind to promoters and initiate transcription near these sites. Repressors bind to operators and block the attachment of RNA polymerase and therefore transcription. Activator proteins bind to **initiator** sites and promote the binding of RNA polymerase to a promoter and thus enhance the initiation of transcription.

Other controlling sites, called **attenuators** and **terminators,** code for regions of RNA that interact with RNA polymerase and generally induce the RNA polymerase to stop transcription. When present, attenuator sites are generally found near the 5′ end (beginning) of newly synthesized RNA molecules. In their active form, attenuators block further transcription. Attenuators may be found in the middle of a cluster of genes (operon) that is coordinately turned on or off. Attenuators allow genes that are transcribed into a single mRNA to be differentially expressed. Terminator sites are found at the end of a gene or group of genes. As their name implies, terminators end transcription. Some terminators not only interact with RNA polymerases, but also bind termination proteins that stimulate the process.

Transcription: Initiation and Termination

4.7.   What is required for transcription of a structural gene?

Transcription of a structural gene can occur only if there is a promoter site for RNA polymerase binding near the beginning of the gene (Fig. 4-2). In bacteria, the RNA polymerase binding sites (promoter sites) are usually 15 to 30 base pairs long. The base-pair sequence of the promoter determines the efficiency with which RNA polymerase binds to it and thus the efficiency of transcription. The promoter is a crucial controlling site that determines not only whether there will be transcription, but also the rate at which transcription is initiated (i.e., how often transcription occurs, not the speed with which a transcript is made).

4.8.   What proteins normally bind to prokaryotic promoter regions?

In many cases, RNA polymerase binding to bacterial promoter sites is dependent upon a family of proteins called **sigma (σ) factors** (Fig. 4-2). Sigma factors can be thought of as transcriptional factors. It is common for a bacterial cell to have a number of different sigma factors. Each type determines the kind of promoter that RNA polymerase will recognize. Both a sigma factor and the RNA polymerase are required for efficient binding to most promoters. After initiation of transcription, the sigma factor dissociates from the RNA polymerase. Since sigma factors function as activators, it is clear that activator binding sites may overlap the promoter site.

Although prokaryotic promoter sites vary considerably, two short regions are shared in common by most promoters. A region on the nonsense strand (complementary to the template strand) about 10 base pairs before transcription initiates (-10 sequence) has a consensus sequence 5′TATAAT3′ and another region about 35 base pairs before transcription initiates (-35 sequence) has a consensus sequence 5′TTGACA3′. A **consensus sequence** contains the nucleotide sequence most commonly encountered in a wide variety of prokaryotic cells and their viruses. The -10 sequence is known as the **Pribnow box** and superficially looks like the TATA box found in eukaryotic promoters (discussed in Question 4.39), whereas the -35 sequence is called the **recognition sequence.** The promoters of bacteria and their viruses generally consist of a central sequence high in AT base pairs (10/12) flanked by sequences high in GC base pairs (10/12 and 8/12).

4.9.   What is the function of each of the controlling sites where regulatory proteins bind?

Often, transcription of genes is blocked by repressor proteins that bind to operator sites near the beginning of genes and interfere with either the binding of RNA polymerase to promoters or with its progress (Fig. 4-3). Because of this, repressors are referred to as regulatory proteins. The binding sites for repressors in bacteria are usually 15 to 30 base pairs long. Operator sites may be distinct from promoter sites, or they may overlap one another. Operator sites are important controlling sites because repressors binding to them can decrease transcription more than 100-fold. The level of transcription depends on the structure of repressors and their controlling sites.

The expression of a gene may require the binding of one or more activators near the beginning of the gene. Activators are also considered regulatory proteins since their binding to DNA can increase transcription 100- to 1,000-fold. The activator binding sites have been given many different names. In bacteria activator binding sites are also known as initiator sites (initiators) or may be considered as part of the promoter site.

In bacteria, activator binding sites are usually 15 to 30 base pairs long. Some of them are distinct from the promoter site, whereas others may be part of the promoter. The fact that activators and RNA polymerases may bind to the same stretch of DNA has led to much confusion as to exactly what a promoter site is. In some cases, an activator binding site next to an RNA polymerase binding sites is considered part of the promoter site. For simplicity, we will define the promoter site as the RNA polymerase binding site and the **promoter region** as any sequence of nucleotides adjoining the promoter site that are necessary for the efficient initiation of transcription.

Activators appear to have different functions. Some, like bacterial sigma factors and the eukaryotic transcriptional factors (discussed in Question 4.37), are required for RNA polymerase binding to appropriate promoter sites. The preliminary binding of RNA polymerase is to double-stranded DNA. Other activators, like bacterial

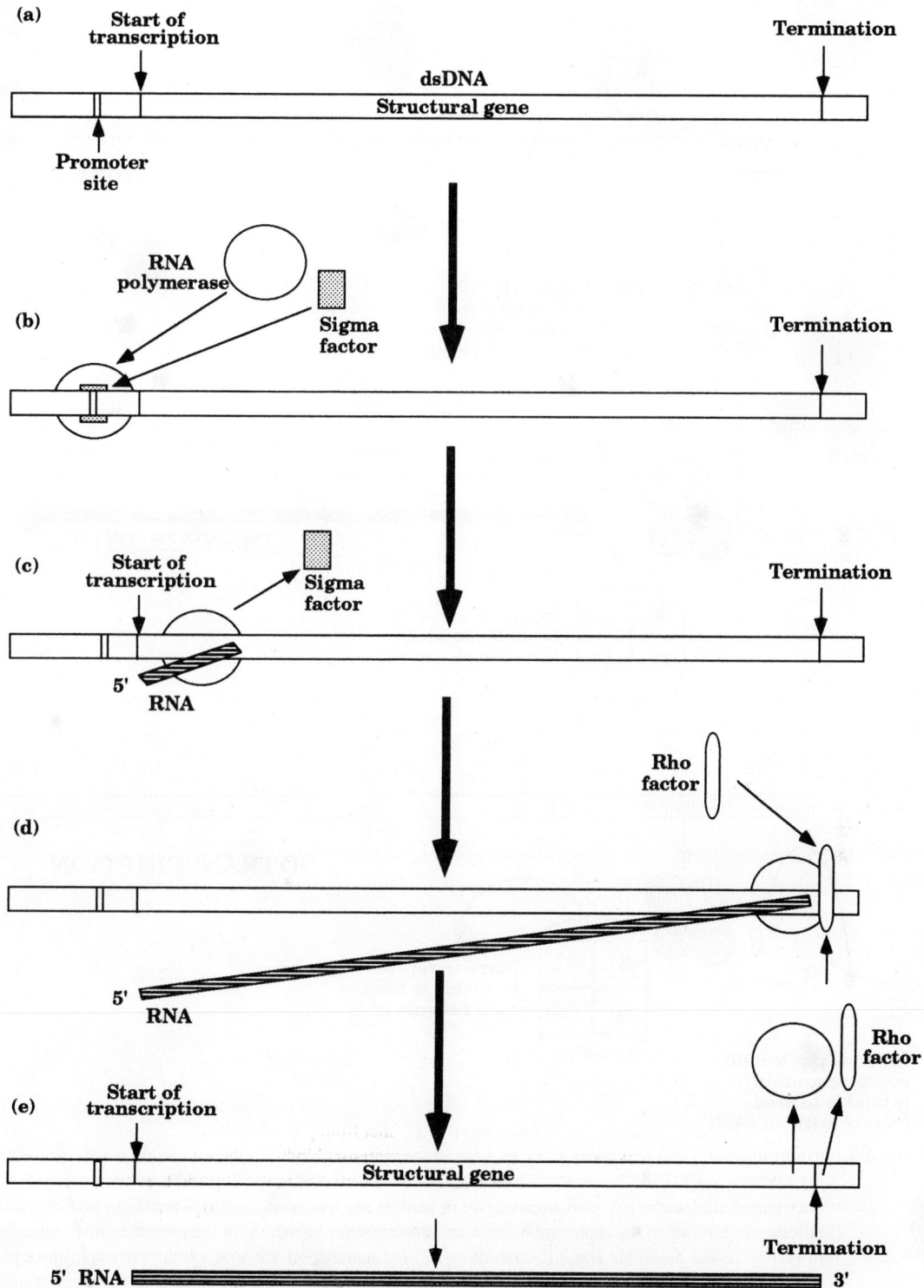

Fig. 4-2. Roles of Sigma and Rho. (*a*) An operon consisting of a promoter site, a structural gene, and a termination site is illustrated. The start site and the termination site affect transcription. (*b*) Sigma factor (a protein) attaches to the RNA polymerase and stimulates polymerase binding to the promoter site. RNA polymerase cannot bind efficiently to the promoter site without sigma factor. (*c*) Soon after RNA polymerase begins transcription, sigma factor dissociates from the enzyme. (*d*) When RNA polymerase reaches the termination site, it stalls and Rho factor (a protein) attaches to the RNA polymerase, promoting its dissociation from the DNA and the release of RNA. (*e*) RNA polymerase, Rho factor, and the nascent RNA dissociate from the DNA.

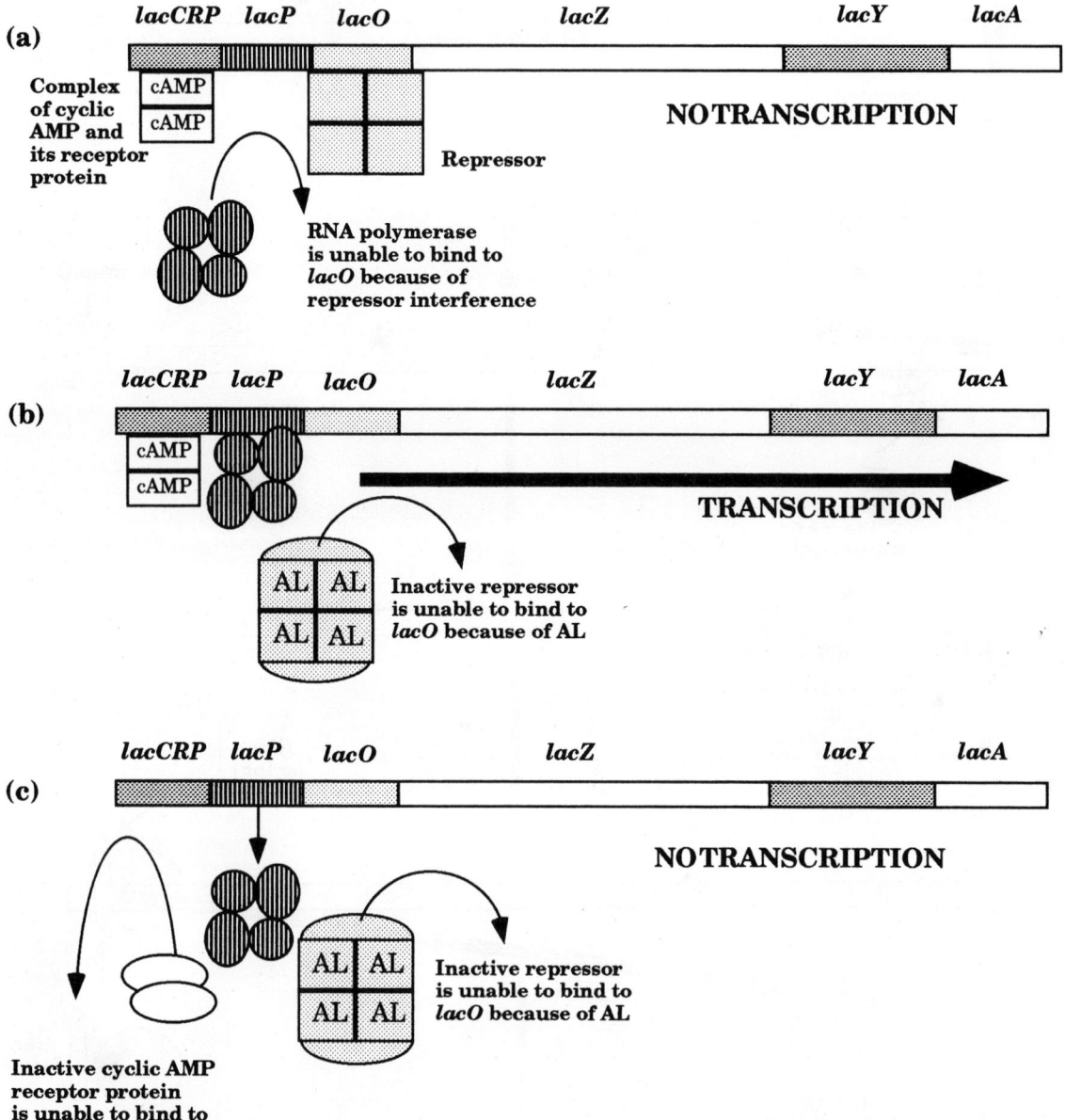

Fig. 4-3. Controlling sites and regulatory proteins. (*a*) In the absense of lactose, the repressor protein is able to bind to the operator (*lacO*) and block transcription by interfering with the binding of RNA polymerase. (*b*) In the presence of allolactose (AL), an altered form of lactose, the repressor protein is unable to bind to the operator. Allolactose binding to the repressor induces an allosteric transformation in the protein to a form that is unable to recognize the promoter site. If anabolism and sugar transport are low, cyclic adenosine monophosphate (cAMP) levels are high. This effector binds to the cAMP receptor protein (CRP) and converts it into a form that is able to bind to *lacCRP*. The attachment of the CRP-cAMP complex to *lacCRP* stimulates the binding of RNA polymerase and the initiation of transcription. (*c*) A high anabolic rate depletes the cell of cAMP. Without cAMP, CRP is unable to stimulate RNA polymerase binding. Thus, even in the presence of allolactose there is no transcription.

CRP (discussed in Questions 4.11 and 4.21) and eukaryotic transcriptional activators (discussed in Question 4.37) are necessary for the initial unwinding of the promoter region. Some activators may help the RNA polymerase bind to the single-stranded template and/or prevent transcription of the nonsense strand of DNA.

4.10. How do attenuators and terminators affect transcription?

Attenuators and terminators are regions of RNA molecules that establish secondary and tertiary structures that interact with the transcribing RNA polymerase. This interaction with the RNA polymerase causes its dissociation from the DNA and the release of the nacent (newly synthesized) RNA. Leader and attenuator regions of mRNA molecules will be discussed in more detail when the tryptophan operon is considered in Question 4.35.

Most terminators at the end of a mRNA molecule code for a single-stranded region that folds on itself due to hydrogen bonding between complementary base pairs and codes for a final region containing numerous uracils (poly-U). The hairpin structure that forms interacts with the RNA polymerase and stimulates its detachment from the DNA. Some terminators require **terminator proteins** such as **Rho** (ρ) to be active (Fig. 4-2). These Rho-dependent terminators lack a poly-U region in the $3'$ end of the terminator. Rho, consisting of six identical subunits, attaches to the RNA. It "walks" along the RNA until it encounters an RNA polymerase stalled at a terminator site; it then catalyzes the release of the RNA polymerase and itself. Sometimes an RNA polymerase may stall because a hairpin structure formed near the end of the mRNA inhibits the polymerase's activity. The **NusA** protein is another termination factor that apparently binds directly to the RNA polymerase and catalyzes its release when it comes to termination sites.

4.11. How can chemicals regulate operons?

When some chemicals bind to regulatory proteins they induce a conformational change in the proteins that alters the protein's ability to bind to DNA.

> **Example 4.2.** When the sugar galactose binds to the repressor of the galactose operon, it causes the repressor to undergo a conformational change to a form that is unable to bind DNA. Since this inactivated repressor cannot block the binding of RNA polymerase to the promoter site, the galactose operon is expressed.

A number of operons involved in the breakdown of sugars require not only the inactivation of their repressors, but also the activation of a common regulatory protein. This common activator is a **cyclic AMP receptor protein,** called CRP (Figs. 4-3 and 4-4). This protein has also been referred to as the **catabolite activator protein,** or CAP. When CRP binds the nucleotide **cyclic adenosine monophosphate** (cAMP), CRP undergoes a conformational change (an **allosteric transformation**) that promotes its binding to an activator binding site. The CRP-cAMP complex bound to its activator site is necessary for RNA polymerase to be able to bind to DNA and initiate transcription. Without cAMP, CRP is unable to bind to DNA and consequently is unable to activate a number of operons.

4.12. What is an effector?

An **effector** is a small molecule like an amino acid or sugar that binds to a regulator protein and changes the protein's activity. For example, galactose binding to the repressor of the galactose operon inactivates the repressor.

4.13. What is an inducer?

An **inducer** is anything physical or chemical (light, heat, amino acids, sugars, etc.) that turns on an operon.

> **Example 4.3.** The sugar lactose induces, or turns on, the expression of the lactose operon. In other words, lactose causes transcription of structural genes in the lactose operon. Structural genes in the lactose operon are involved in the transport of lactose into a cell and its catabolism to galactose and glucose. Lactose is not only cleaved by β-galactosidase, but it may be converted to another disaccharide called allolactose (Fig. 4-4c,d). **Allolactose** is an effector of the lactose operon since it binds to the repressor and inactivates the repressor. This relief from repression (derepression) results in ex-

pression of the operon. Lactose is an inducer of the lactose operon, but it is not an effector because it does not bind to the repressor. Lactose must be converted to allolactose before it has any effect on the lactose operon. Consequently, cells that lack β-galactosidase cannot be induced by lactose. Some readers may see a paradox in getting the lactose operon started, but it will be resolved by the phenomenon of "sneak synthesis" discussed in Question 4.17.

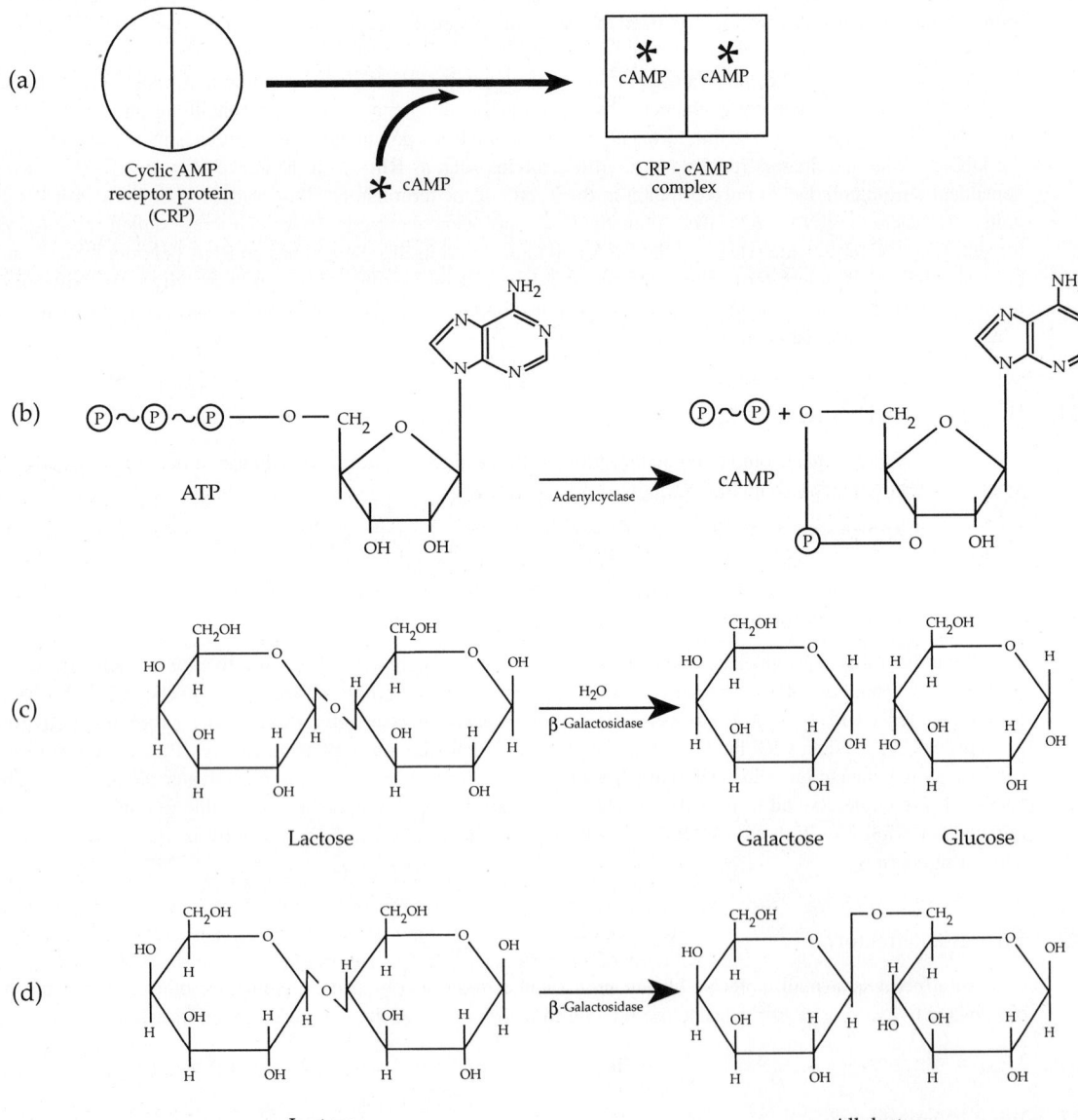

Fig. 4-4. Effectors that affect the lactose operon. (*a*) The cyclic AMP receptor protein (CRP) may bind two molecules of cAMP. CRP is unable to bind to the lactose operon when it is free of cAMP. The CRP-cAMP complex binds to the lactose operon and stimulates transcription. (*b*) ATP is converted to cAMP by the enzyme adenylcyclase. (*c*) Generally, the enzyme β-galactosidase (lactase) catalyzes the hydrolysis of lactose to galactose and glucose. (*d*) Very small amounts of allolactose are produced by β-galactosidase. Allolactose is both an inducer and an effector since its binding to the tetrameric repressor (R) causes the repressor to undergo a conformational change and detach from the promoter site.

Some inducers are not substrates for the enzymes induced.

Example 4.4. The chemical **isopropyl-β-D-thiogalactoside (IPTG)** is both an inducer and an effector of the lactose operon. Although IPTG binds to the repressor and inactivates it, IPTG is not cleaved by β-galactosidase because of its sulfur content.

4.14. Why have cells evolved such complex mechanisms to regulate their genes?

Regulatory proteins and controlling sites allow the cell to conserve energy by blocking unneeded gene expression or allowing only necessary transcription. For instance, if sugars are not present in the environment or within the cell, enzymes for their catabolism are not needed. Repressors keep the cell from wasting energy in making mRNA and enzymes that have no use. If sugars should enter the environment, however, the repressors are quickly inactivated when sugars bind to them. This allows mRNA and enzymes for the metabolism of the sugar to be made.

When the cell is metabolizing excessively, regulatory proteins and controlling sites provide a mechanism by which the cell can control its metabolism. It would be detrimental to any cell if it should devote all its resources and energy to one metabolic pathway. The maintenance, as well as the growth and proliferation of a cell, requires that many metabolic pathways operate at an appropriate level. Chemical effectors binding to regulatory proteins and regulatory proteins selectively binding to controlling sites tie metabolic pathways together, helping to maintain balance within the cell and permitting it to respond adaptively to the environment.

The Lactose System

4.15. What is the lactose operon, and what is its function?

The **lactose operon** consists of three structural genes (*lacZ*, *lacY*, and *lacA*) as well as three controlling sites (*lacCRP*, *lacP*, and *lacO;* Fig. 4-1, Fig. 4-3). The structural gene *lacZ* codes for a subunit of the homotetrameric (4 identical polypeptide chains) enzyme β-galactosidase (LacZ), whereas gene *lacY* specifies the protein **lactose permease** (LacY) and gene *lacA* encodes the enzyme **lactose transacetylase** (LacA). The controlling sites *lacCRP*, *lacP*, and *lacO* are binding sites for the cAMP receptor protein, RNA polymerase, and the lactose repressor, respectively.

Catabolism of lactose in bacteria is dependent upon the lactose operon. The lactose permease transports lactose into the cell and concentrates it, while β-galactosidase cleaves lactose into galactose and glucose. Galactose is converted to glucose, and then the glucose is metabolized by various metabolic pathways such as the Embden-Meyerhof pathway. The significance of transacetylase, which adds acetyl groups to lactose, in the metabolism of lactose is not clear.

4.16. What is the lactose regulatory operon and what are its functions?

The **lactose regulatory operon** consists of one structural gene *(lacI)* and one controlling site (*lacP2;* Fig. 4-1). The structural gene codes for a subunit of the tetrameric lactose repressor, whereas the controlling site is an RNA polymerase binding site. In the absence of an inducer, the repressor binds to *lacO* and blocks the binding of RNA polymerase to *lacP1* and the initiation of transcription of the lactose operon. The repressor may also block transcription of the lactose regulatory operon and, consequently, its own production (autorepression).

4.17. How is the lactose operon regulated in the absence of the sugar lactose or any other inducer?

In the absence of an inducer, expression of the lactose operon is inhibited by the binding of the lactose repressor at *lacO* (Fig. 4-3a). The repressor sterically hinders the binding of RNA polymerase to *lacP* and the initiation of transcription. As the concentration of repressor decreases because of protein turnover (degradation) and dilution by cell growth and division, *lacO* is occasionally free of repressor and an RNA polymerase is able to bind to *lacP* and initiate transcription. This occasional synthesis of mRNA may be thought of as **sneak synthesis.** A single mRNA containing the information for synthesizing β-galactosidase, permease, and transacetylase is made. This type of mRNA is referred to as a polycistronic messenger. Translation of one polycistronic messenger results in very low levels of the three enzymes. These low levels of enzymes are referred to as **basal enzyme levels.** Thus, the cell is never completely devoid of the lactose catabolic enzymes. Although there are about 15 molecules of the repressor in each *E. coli* cell, they manage to find the single operator site out of the more than 3.5 million base pairs that make up the cell's genome. Sneak synthesis is responsible for the basal enzyme levels of many different enzymes and structural proteins.

4.18. How do inducers turn on the lactose operon?

When an inducer like isopropyl-β-D-thiogalactoside (IPTG; see Example 4.4) is added to a population of bacteria that is capable of catabolizing lactose, the lactose operon is turned on. Within minutes of adding inducer, the rate of β-galactosidase, permease, and transacetylase production increases to a maximum and remains steady as long as environmental conditions do not change. IPTG is transported through the membrane and concentrated within cells by permeases in the plasma membrane. Remember, cells have basal levels of the enzymes encoded by the lactose operon. Within a few seconds, all the repressor molecules in the cell undergo conformational changes to their inactive forms because of IPTG binding. Because these inactivated repressors no longer bind to operators, RNA polymerases are free to bind to promoters and initiate transcription (Fig. 4-3*b*). As soon as transcription is initiated by the first RNA polymerase, another RNA polymerase binds to the promoter and initiates transcription. The lactose operon is soon being transcribed by many RNA polymerases. Translation can begin before the first mRNA is completed, and each mRNA may be translated 40 or more times. The rapid production of many copies of mRNA and the repeated translation of each mRNA result in a rapidly increasing rate of enzyme production. Enzyme levels may go from 0.1 unit/mg protein/minute to 50 units within 10 or 15 minutes.

Units of enzyme activity are frequently measured in light absorbance units. For example, the activity of β-galactosidase is determined by adding a colorless chemical called **ortho-nitrophenol galactoside (ONPG)** to a cell extract and then measuring the amount of yellow color that develops over a period of time. The light absorbance by the yellow color might be measured in a Klett colorimeter as **Klett units.** ONPG is cleaved by β-galactosidase into galactose and a yellow compound called ortho-nitrophenol. Unlike lactose or IPTG, ONPG is not an inducer of the lactose operon.

4.19. What regulatory problem does a cell encounter when an operon is induced?

If biosynthesis were unregulated, a cell would rapidly fill with induced mRNA molecules and enzymes. This would be wasteful and counterproductive since the synthesis of these molecules requires a lot of energy—energy that would be better used in making repairs or for cell reproduction. In addition, excessive catabolism is very wasteful because large pools of catabolites tend to leak from the cell. If a mixture of carbohydrates is present in the environment, it is wasteful to make transport proteins and catabolic enzymes for all the carbohydrates when the metabolism of one is sufficient. Since the glucose transport proteins and catabolic enzymes are constitutively synthesized at high levels, it would be especially wasteful to induce operons for the utilization of various compounds when glucose is present. To avoid the excessive production of mRNA, enzymes, and catabolites, a system of controls that limits the synthesis of these molecules has evolved.

Catabolite Repression

4.20. Once an operon has been turned on, what keeps the cell from filling with mRNA and protein within a few minutes?

One reason that a cell does not rapidly fill with mRNA and protein when an operon is induced is that much of the mRNA has a **half-life** of about 2.5 minutes. This means that 2.5 minutes after mRNA is synthesized, half of it is degraded by cellular endonucleases. After 5 minutes, only 25% of the mRNA survives. In about 15 minutes, 98% of the original mRNA is degraded. In those 15 minutes, however, more mRNA is synthesized. Thus, mRNA accumulates, but not as fast as it would if it were stable. Proteins, on the other hand, accumulate more rapidly because they are usually more stable than mRNA. Another reason that a cell does not rapidly fill with mRNA and protein is that the excessive synthesis of mRNA and proteins depletes the cell of energy. The depletion of certain molecules, such as phosphoenolpyruvate (PEP), causes many operons involved in catabolism to be quenched. This control system is superimposed upon that exerted by repressors and is called **catabolite repression.** When a cell is rapidly metabolizing, catabolite repression shuts down the expression of a number of catabolic operons. Catabolite repression involves a small effector molecule called cyclic adenosine monophosphate (cAMP; Fig. 4-4*b*) and the protein to which it binds, the cAMP receptor protein (CRP).

4.21. How do cAMP (cyclic adenosine monophosphate) and CRP (cAMP receptor protein) affect the regulation of catabolic operons such as the lactose operon?

The regulation of the lactose operon is slightly more complicated than previously indicated. RNA polymerase is able to bind efficiently to *lacP1* and initiate transcription only if a complex of CRP and cAMP is bound to the activator binding site *lacCRP* (Fig. 4-3*b*). In most writings, the *lacCRP* site is shown as part of *lacP1,* but in this book the sites are separated (Fig. 4-1). The higher the level of cAMP in the cell, the more cAMP binds to CRP and the more CRP-cAMP complex is present in the cell. A high concentration of the CRP-cAMP complex means that it will be attached to *lacCRP* most of the time. Consequently, RNA polymerases are able to bind efficiently and initiate one round of transcription after another. The maximum rate of expression of the lactose operon is determined not only by the efficiency of the RNA polymerase binding to *lacP1,* but also by the efficiency of CRP-cAMP complex binding to *lacCRP.*

When the rate of metabolism is low, the levels of cAMP are high. Consequently, there will be high levels of CRP-cAMP complex to stimulate transcription. On the other hand, when the rate of metabolism is high, the levels of cAMP are low. This means that there will be little CRP-cAMP complex to stimulate transcription (Fig. 4-3*c*).

4.22. What is the connection between a high metabolic rate and decreased levels of cAMP and transcription?

A protein known as **enzyme-III** (also known as enzyme-IIa) may be in a phosphorylated (E-IIIP) or unphosphorylated (E-III) state. The major source of phosphates is **phosphoenolpyruvate (PEP)** and **acetylphosphate (AP).** E-IIIP (enzyme-III in its phosphorylated state) is required for the activity of **adenylcyclase** (adenylate cyclase), the membrane-bound enzyme that catalyzes the synthesis of cAMP from ATP (Fig. 4-5). When E-III predominates, adenylcyclase is inactive and no cAMP is made. E-III directly affects cellular components. However, it stimulates a **phosphatase,** an enzyme that catalyzes the breakdown of cAMP; it opens membrane pores for cAMP, thereby promoting the rapid loss of cAMP from cells; and it inhibits the lactose permease, thereby slowing inducer uptake.

As the anabolism (synthesis) of both lactose mRNA and enzymes increases, the availability of phosphoenolpyruvate (PEP) and acetylphosphate (AP) for phosphorylating enzyme-III decreases because these compounds are metabolized to provide energy for making nucleotides and amino acids. Depletion of PEP and AP results in mostly unphosphorylated enzyme-III. Unphosphorylated enzyme-III causes a rapid decrease in the concentration of cAMP. When cAMP levels are low, much of the CRP is free of cAMP and therefore unable to bind to *lacCRP.* When CRP-cAMP is not bound to the activator binding site, RNA polymerase is unable to bind efficiently to *lacP1* or initiate transcription. As a result, no new lactose mRNA is synthesized, and the enzymes involved in lactose catabolism do not increase significantly. Previously synthesized mRNA continues to be translated, but because it is rapidly digested, the amount of new protein synthesis rapidly decreases.

4.23. How do the growth and division of cells affect the expression of catabolic operons such as the lactose operon?

As cells grow and divide, the concentration of enzymes per cell is reduced. For example, if cells grow and divide after transcription of the lactose operon has been blocked by high rates of metabolism, the accumulated lactose mRNA can be reduced to less than 1% of its maximum level in 20 minutes. The more stable enzymes are reduced in number more than 50% because of dilution by growth and cell division. Eventually, the expression of the lactose operon decreases to a point at which the pools of PEP and AP increase. As PEP and AP increase, more and more enzyme-III becomes phosphorylated and the concentration of cAMP rises once again. E-IIIP inhibits the phosphatase that breaks down cAMP, blocks cAMP pores, and stimulates adenylcyclase to synthesize cAMP. This change in the cell's physiology leads to new rounds of transcription. If there are no drastic changes in the concentration of nutrients in the environment, a steady-state equilibrium is reached that depends upon the concentration of catabolites, the concentration of cAMP, and the initiation of transcription.

4.24. What is diauxic growth?

The growth kinetics that result from completely catabolizing one sugar before using another is referred to as **diauxic growth.** When cells are grown in a medium containing a mixture of two sugars, such as glucose and lactose, the sugars are not used simultaneously. Glucose is used first, and little or no lactose is catabolized until the glucose is gone. Glucose catabolism involves an initial lag phase, an exponential phase, and a stationary phase.

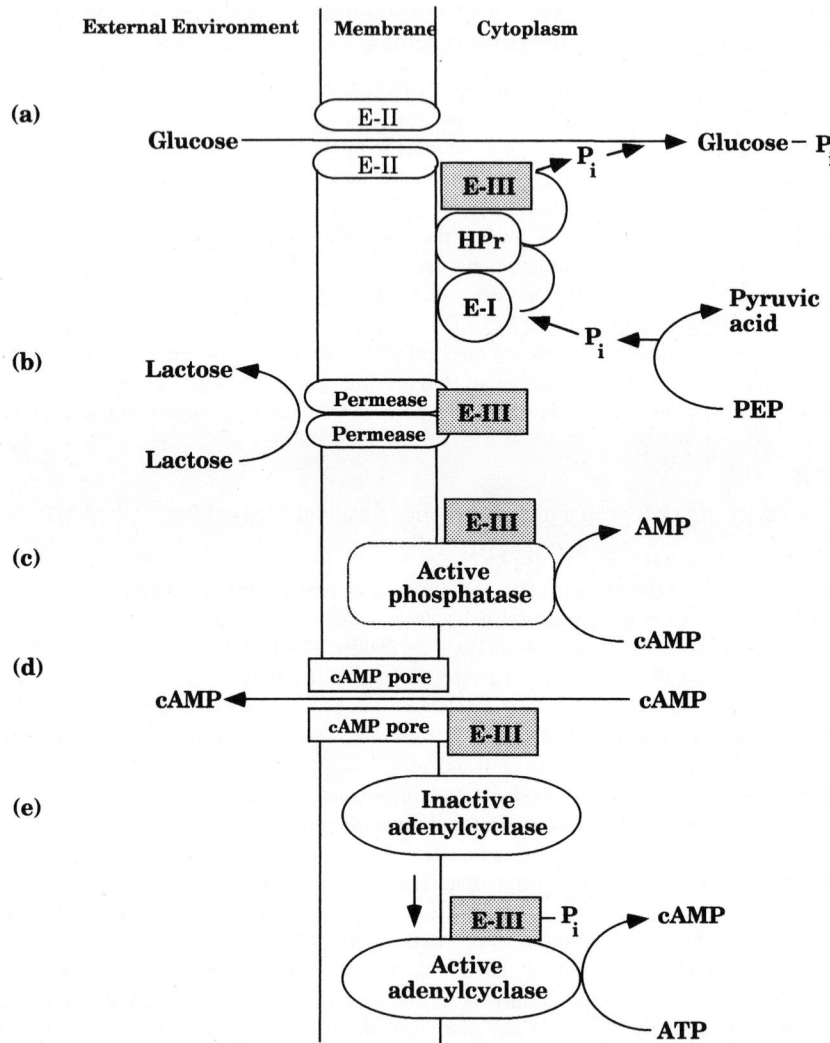

Fig. 4-5. Catabolite repression by glucose. (*a*) Glucose transport diverts inorganic phosphate (P_i) from enzyme-III (E-III). The unphosphorylated form of E-III regulates a number of transport systems and enzymes. (*b*) E-III blocks the lactose permease. (*c*) E-III stimulates phosphatase. (*d*) E-III opens cAMP pores. (*e*) E-III has no effect on adenylcyclase; however, the phosphorylated form of E-III (E-IIIP_i) stimulates adenylcyclase.

Lactose catabolism, which follows the stationary phase after glucose catabolism, promotes renewed growth. The population enters another lag phase, a second exponential phase, and a final stationary phase when all the lactose has been digested.

4.25. What is catabolite repression? How is it the cause of diauxic growth?

Catabolite repression is the inhibition of operons by glucose and other sugars (mannose, fructose, *N*-acetyl-glucosamine, and *β*-glucosides) that are transported into the cell by the phosphotransferase system. The **phosphotransferase system** is an important transport system used by many bacteria to transport and concentrate sugars within the cell. The hallmark characteristics of the phosphotransferase system is that it is powered by the hydrolysis of phosphoenolpyruvate (PEP) and the sugar transported becomes phosphorylated. At one time, it was believed that metabolic products from glucose metabolism (catabolites) inhibited sensitive operons, hence the name **catabolite repression.** It is now known that catabolites do not function as effectors turning off operons. Operons are turned off by inhibiting permeases that transport inducers and/or by deactivating the cyclic adenosine monophosphate receptor (CRP). The deactivation of CRP is achieved when the concentration of cAMP is decreased by increasing its efflux. Catabolite repression is the cause of diauxic growth in a broth with glucose and lactose. Rapid glucose transport and anabolism indirectly inhibit the lactose permease and indirectly reduce the concentration of cAMP in the cell. Thus, little or no lactose is able to enter the cell, and any that might enter is unable to induce the operon because CRP is deactivated. Consequently, operons such as the lactose operon cannot turn on while glucose is present.

4.26. How does glucose indirectly inhibit the lactose permease and reduce the concentration of cAMP within the cell?

In many organisms, glucose is transported by the phosphotransferase system (Fig. 4-5a). In this system **enzyme-II** binds glucose and facilitates its movement through the membrane. The transport is powered by the hydrolysis of phosphoenolpyruvate (PEP), catalyzed by **enzyme-I.** The hydrolysis of PEP results in a phosphate group being attached to enzyme-I. Enzyme-I donates the phosphate to a **heat-stable protein** called **HPr.** Subsequently, HPr donates the phosphate to enzyme-III, and then enzyme-III passes the phosphate on to glucose. When glucose is present in the environment and rapidly transported into the cell, most of the enzyme-III is unphosphorylated because it quickly gives up its phosphate to the incoming glucose (Fig. 4-5a).

In its unphosphorylated form, enzyme-III inhibits (Fig. 4-5b) the lactose permease, preventing lactose uptake, and stimulates the activity of phosphatase (Fig. 4-5c) that converts cAMP to AMP. Thus, inducer uptake is blocked and CRP is deactivated. In addition, enzyme-III stimulates the opening of pores, allowing cAMP to rapidly leave the cell (Fig. 4-5d). This also contributes to CRP deactivation. Finally, unphosphorylated enzyme-III does not stimulate adenylcyclase (Fig. 4-5e). The failure to make new cAMP and keep cAMP at high levels also results in CRP deactivation. Clearly, rapid glucose transport and anabolism that depletes PEP shut the lactose operon off through a number of mechanisms.

After glucose is consumed, enzyme-III remains phosphorylated. Phosphorylated enzyme-III is unable to inhibit lactose uptake, stimulate cAMP departure from the cell, or promote cAMP hydrolysis. Thus, lactose can be transported into the cell and the cAMP level can increase. In the absence of glucose transport and anabolism, cAMP pores close, adenylcyclases are stimulated to synthesize cAMP, and cAMP phosphorylases are no longer stimulated.

Although glucose induces the most severe catabolite repression, other molecules such as lactose, ribose, acetate, succinate, glycerol, and lactate also promote catabolite repression to varying degrees. This catabolite repression is dependent upon how rapidly the organism is growing. If the organism is growing slowly (slack anabolism), there is minimal catabolite repression; if it is growing rapidly, there is maximal catabolite repression. This can be explained by assuming that a slowly growing organism is not rapidly using all its PEP and acetylphosphate (AP) to build cellular material, but is using much of it to phosphorylate enzyme-III (through E-I and HPr). Phosphorylated enzyme-III allows sugar transport and stimulates the synthesis of cAMP, which, in turn, stimulates operons fully. Rapid catabolism and growth of cells, in contrast, use up PEP and AP, so that enzyme-III remains in its unphosphorylated state. This shuts down many catabolic operons that rely on cAMP for their activity.

Some operons not involved in the catabolism of nutrients have been reported to be sensitive to catabolite repression by glucose. For example, many of the operons that encode the proteins for flagella and for locomotion by flagella are affected by catabolite repression. Presently, it is not clear whether the mechanisms for the catabolite repression of these operons is similar to those seen for catabolic operons.

Mutations in the Lactose System and How They Affect Gene Expression

4.27. What is a negatively controlled inducible operon?

The lactose operon is repressed by a specific regulatory protein encoded by *lacI*. This regulatory protein does not influence any other operon, except possibly its own. When a regulatory protein is specific for an operon and represses the operon, the operon is said to be **negatively controlled.** When an operon is turned on by the addition of an inducer or effector, the operon is said to be **inducible.** The galactose and lactose operons and the SOS regulon (see Question 4.5) are examples of negatively controlled inducible systems.

4.28. How do mutations in the controlling sites affect the lactose operon?

In the lactose operon, numerous point mutations in *lacO* have been discovered that alter repressor binding. The consequence of these operator mutations is a constitutive operon, that is, the genes are expressed all the time, even in the absence of an inducer. Because operator mutations make the operon constitutive, they are symbolized *lacO^c*. Generally, enzyme levels in cells that carry operator constitutive mutations are 100% of those found in induced wild-type cells.

Mutations in the *lacCRP* site have been isolated that reduce the expression of the operon down to only 2 to 8% of that in the induced wild type. These mutations eliminate CRP binding.

Mutations in the *lacP1* site have been discovered that reduce the expression of the operon down to 7% of that in the induced wild type. These mutations drastically reduce RNA polymerase binding to *lacP1*. The *lacCRP* and *lacP1* mutations can be distinguished from each other by measuring enzyme levels when these mutations are placed in bacteria that are missing CRP *(crp⁻)*, the cAMP receptor protein, and adenylcyclase *(cyc⁻)*. It was found that the *(lacCRP⁻) crp⁻ cyc⁻* mutants continue to express the lactose operon at levels of 2 to 8% of the induced wild type. On the other hand, the *(lacP1⁻) crp⁻ cyc⁻* mutants express the lactose operon at levels of 0.1% of the induced wild type. In these mutants, inefficient RNA polymerase binding, due to the *lacP1⁻* mutation, is made more inefficient because CRP-cAMP complexes are not available for transcription initiation.

Another group of *lacP1* mutations have been discovered that allow the operon to be expressed at nearly the induced wild-type level. One revertant, UV5, produces almost normal levels of β-galactosidase and is completely independent of CAP and cAMP.

4.29. How do mutations in the structural genes affect the lactose operon?

Mutations in the *lacI* gene that delete or alter the gene so that the protein produced is unable to bind to *lacO* are indicated by the symbol *lacI⁻*. Mutants that are *lacI⁻* have constitutive lactose operons; that is, the lactose operon is always turned on whether or not an inducer is present. Thus, *lacI⁻* mutants and *lacO^c* mutants have the same phenotype, a constitutive lactose operon.

Normally, *lacI⁻* alleles are recessive to the *lacI⁺* (wild-type) allele. For example, in partially diploid cells that are *lacI⁻/lacI⁺*, both of the lactose operons are turned off by the repressor produced by *lacI⁺* and are only expressed when an inducer is present. Certain *lacI* mutations, symbolized *lacI^{-d}*, are dominant to the *lacI⁺* allele. In partially diploid cells that are *lacI^{-d}/lacI⁺*, both lactose operons are expressed, even in the absence of an inducer. The defective subunits produced by *lacI^{-d}* mix with the normal subunits encoded by *lacI⁺*, resulting in repressors that are unable to bind to the operator site efficiently. Remember, the lactose repressor consists of 4 identical subunits. One *lacI^{-d}* subunit will disrupt the normal conformation of a repressor. If the defective and normal subunits are produced in equal amounts, only one out of every 16 repressors will be completely normal. If d = frequency of the defective subunit, r = frequency of the wild-type subunit, $d + r = 1$, and $d = r$, the expansion $(r + d)^4 = r^4 + 4r^3d + 6r^2d^2 + 4rd^3 + d^4$ demonstrates that only 1/16 repressor molecules (r^4) will be fully active. The 1/16 is determined from the coefficient. There is a one in front of r^4 out of a total of 16 possibilities $(1 + 4 + 6 + 4 + 1)$.

Another mutant allele of the *lacI* gene is one that modifies the repressor subunits so that they are unable to bind inducer. Consequently, the repressor cannot be removed from the operator by an effector. Repressors that are unaffected by effectors are referred to as **super-repressors.** The allele is symbolized *lacI^s*.

Although *lacI^s* is dominant to *lacI⁺* and *lacI⁻*, it is recessive to *lacO^c*. Cells that are *lacI⁻/lacI^s* do not express the lactose operon even if an inducer is present. On the other hand, cells that are *lacO^c/lacI^s* are expressed constitutively since the super-repressor is unable to bind to *lacO^c*.

Mutations that destroy the activity of β-galactosidase, permease, and transacetylase are symbolized, respectively, *lacZ⁻*, *lacY⁻*, and *lacA⁻*. Some *lacZ* mutations, symbolized *lacZ^{-ns}*, eliminate the activity of all three enzymes. These mutations are nonsense mutations near the beginning of the *lacZ* gene. During transcription, the mRNA sequence following nonsense mutations is unprotected by ribosomes. Exonucleases degrade the unprotected mRNA and reach the start codons for permease and transacetylase before protein synthesis can begin there.

The Arabinose System

4.30. Is there such a thing as a positively controlled inducible operon?

When a regulatory protein is specific for an operon and activates the operon, the operon is said to be **positively controlled.** When an operon is turned on by the addition of an inducer or effector, the operon is said to be

inducible. Many operons involved in the catabolism of sugars have turned out to be positively controlled inducible operons; examples include the operons involved in the catabolism of the sugars L-arabinose, maltose, and L-rhamnose.

4.31. How does a positively controlled inducible operon work?

Transport of L-arabinose into the cell and its catabolism requires four operons: *araE*, *araFGH*, *araC*, and *araBAD*. Each of the capital letters that follow *ara* represents a gene coding for a different polypeptide chain. The operons *araE* and *araFGH* are involved in the binding of L-arabinose and its transport into the cell. The *araE* gene encodes a protein involved in the low-affinity transport of L-arabinose, while the *araFGH* genes encode proteins that function in the high-affinity transport of L-arabinose. The *araBAD* and *araC* operons specify enzymes involved in the catabolism of L-arabinose and a required regulatory protein, respectively. The operons *araE*, *araFGH*, and *araBAD* constitute a regulon under the control of the regulatory operon *araC*. In addition, the *araC* operon autoregulates itself.

The L-arabinose operon *(araBAD)* was the first operon discovered to be positively controlled by a regulatory protein (Fig. 4-6). Because this operon's regulation has been worked out in such detail, studying the L-arabinose operon can provide further insight into gene regulation in prokaryotes. The L-arabinose operon *(araBAD)* consists of 3 structural genes involved in the conversion of the pentose sugar L-arabinose to D-ribulose-5-phosphate. An enzyme called kinase is encoded by *araB*, an enzyme referred to as isomerase is encoded by *araA*, and an enzyme referred to as epimerase is encoded by *araD*. The isomerase converts L-arabinose to L-ribulose. The kinase phosphorylates L-ribulose, converting it to L-ribulose-5-phosphate. This reaction is followed by the conversion of L-ribulose-5-phosphate to D-xylulose-5-phosphate by the epimerase. The order in which the genes are transcribed is BAD.

The operon that regulates the *araBAD* operon consists of a single structural gene, *araC*, and is located near the *araBAD* operon. The regulatory protein encoded by *arcC* binds to a number of controlling sites ($araI_2$, $araI_1$, $araO_1$, and $araO_2$) as a dimer (Fig. 4-6). The regulatory protein often exists as a tetramer, however, because dimers binding to multiple sites interact with each other. A very important characteristic of the regulatory protein is its ability to function as a repressor or as an activator. The regulatory protein functions as a repressor in the absence of L-arabinose, but is converted into an activator of the *araBAD* operon when L-arabinose binds to it. On the other hand, the activator can block transcription of the *araC* operon when high concentrations of the activator are present (Fig. 4-6d).

Most of the controlling sites for both the *araC* and *araBAD* genes are between *araC* and *araBAD* (Fig. 4-6a). For the *araBAD* genes, there are the following: a promoter site, $araP_1$; an initiator site, $araI_2$; a second initiator site, $araI_1$; a CRP-cAMP binding site, *araCRP*; and an operator site, $araO_1$. Overlapping $araO_1$ is $araP_2$, the promoter site for the *araC* structural gene. A second operator site, $araO_2$, is located 100 base pairs into the *araC* gene.

In the absence of L-arabinose, repressor proteins binding to $araO_2$ and to $araI_1$ associate with each other (Fig. 4-6b). Expression of the *araC* gene is inhibited because RNA polymerase is unable to pass $araO_2$. Expression of *araBAD* requires the binding of a CRP-cAMP complex at *araCRP* and the conversion of the repressor into an activator and its binding at $araI_1$ and $araI_2$. In the absence of activator, there is no transcription from P_1. Similarly, in the absence of the *araC* gene product, there is no transcription from P_1 even when inducer is present. This supports the idea that the repressor and activator are alternate forms of the same protein. The *araBAD* operon is negatively regulated by the protein encoded by *araC* (repressor), but the *araBAD* operon is positively regulated by the same protein (activator) if L-arabinose is present to induce a conformational change in the protein.

4.32. How is the *araC* operon regulated?

In the absence of L-arbinose, the *araC* protein (repressor) binds to both $araO_2$ and $araI_1$ and blocks transcription of *araC* at $araO_2$ (Fig. 4-6b). Approximately 100 nucleotides of *araC* are transcribed before blockage occurs. When L-arbinose is added (Fig. 4-6c), the *araC* protein becomes an activator that disengages from $araO_2$ and associates with $araI_2$. This allows complete transcription of *araC* from $araP_2$, as well as the initiation of transcription from $araP_1$.

Transient derepression of *araC* ceases when the activator binds to both $araO_2$ and $araO_1$ (Fig. 4-6d). This repression by the *araC* gene product-arabinose complex demonstrates that the "activator" of the *araBAD* operon also functions as a "repressor" of the *araC* operon.

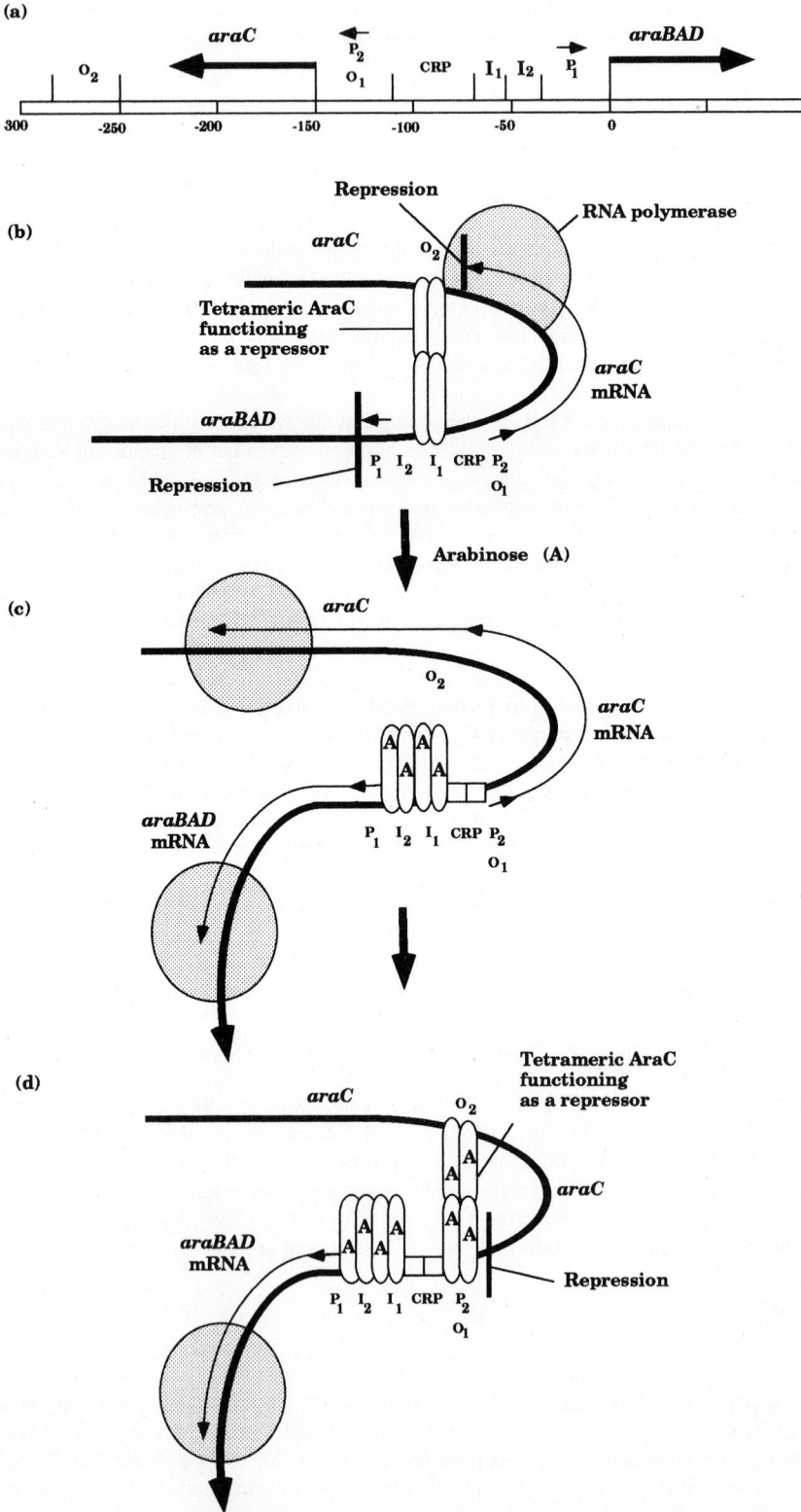

Fig. 4-6.   (Caption on page 103.)

Fig. 4-6. The L-arabinose operon and its regulatory operon. (*a*) The controlling sites in the arabinose operon and in its regulatory operon are illustrated. P_1 is the promoter for the arabinose operon, whereas P_2 is the promoter site for the regulatory operon. I_1 and I_2 are initiator sites where the regulatory protein binds and functions as an activator. Similarly, O_1 and O_2 are operator sites where the regulatory protein binds and functions as a repressor. CRP is the binding site for the cAMP receptor protein. (*b*) In the absense of arabinose, the regulatory protein binds to I_1 and O_2, blocking transcription of the regulatory operon. The arabinose operon cannot be transcribed because there is no activator helping the RNA polymerase to bind to P_1. (*c*) The addition of arabinose (A) converts the regulatory protein into an activator that binds to I_2 and I_1. This stimulates transcription of the arabinose operon. (*d*) The activator also functions as a repressor when it binds to O_1 and O_2. This complex blocks transcription of the regulatory operon. The regulatory protein in its "activator state" also binds to O_1 and O_2, but when binding to these controlling sites the regulatory protein functions as a repressor. This illustrates the fact that the same regulatory protein may function as an activator or as a repressor, depending on the type of site to which it binds.

4.33. How do mutations in the regulatory proteins and regulatory sites affect the expression of the L-arabinose operon?

Most mutations that delete or damage the *araC* gene product shut down the *araBAD* operon. The operon is still inducible, however, if a functional *araC* gene is present (such as in the partial diploid *araC⁻ araBAD⁺/araC⁺ araBAD⁻*). This indicates that the *araC* gene product is necessary for expression of the *araBAD* operon and acts not only in *cis* position (in the same DNA molecule), but also in *trans* position (between different DNA molecules).

Some mutations in *araC* convert the regulatory protein directly into an activator. These mutations, symbolized *araCᶜ*, make the *araBAD* operon constitutive. The wild-type (normal) *araC⁺* allele is dominant to *araCᶜ*. For example, in the absence of L-arabinose, the expression of the *araBAD* operon is reduced 15-fold in a partial diploid like *araCᶜ araBAD⁺/araC⁺ araBAD⁻*. This dominance over *araCᶜ* can be explained by assuming that wild-type subunits induce conformational changes in mutant subunits, rendering them incapable of binding to *araI₂* in the absence of L-arabinose.

Mutations in *araP*, designated *araPᶜ*, make *cis* but not *trans araBAD⁺* operons constitutive in the absence of *araC*. The *araBAD⁺* operon with a *cis araPᶜ* is also expressed in cells lacking both CRP and adenylcyclase; however, the *araBAD⁺* operon is not expressed fully in cells with these deficiencies.

The Tryptophan System

4.34. Describe the regulation of a negatively controlled operon involved in the synthesis of the amino acid tryptophan.

The tryptophan biosynthetic operon, required for the synthesis of the amino acid tryptophan, has been extensively studied because it is an example of a negatively controlled repressible operon. Remember, a negatively controlled operon is one that is inhibited by a regulatory protein such as a repressor. A **repressible operon** is one that is inhibited when an effector molecule is present.

The **tryptophan biosynthetic operon** consists of five structural genes: *trpE, trpD, trpC, trpB,* and *trpA* that code for enzymes involved in the conversion of chorismic acid to tryptophan (Fig. 4-7). The controlling sites lie next to *trpE* and consist of a promoter site, *trpP*, an overlapping operator site, *trpO*, and a leader region, *trpL*, that codes for a leader peptide and for an RNA attenuator. The leader peptide is involved in the attenuation mechanism discussed in Question 4.35. The tryptophan operon can be symbolized by *trpPOLEDCBA*.

The **tryptophan regulatory operon** consists of a single structural gene *trpR* that codes for an inactive regulatory protein called the **aporepressor.** The promoter region of the regulatory operon has not been characterized. Whether the regulatory operon is constitutive or regulated has not been determined. It is known, however, that the regulatory operon is located far from the tryptophan operon. Consequently, the tryptophan biosynthetic operon and its regulatory operon are said to be unlinked even though all bacterial genes are part of the same chromosome.

In the absence of tryptophan, the aporepressor (inactive regulatory protein) has no effect on the tryptophan operon. Consequently, the enzymes for synthesizing the amino acid are produced and the cell makes tryptophan. Tryptophan is referred to as the **corepressor** since it acts as an effector to convert the aporepressor into the active repressor, called the **holorepressor.** When tryptophan binds to the aporepressor, it induces a conformational change that converts the regulatory protein into the tryptophan holorepressor. The holorepressor binds to *trpO* and blocks the attachment of RNA polymerase to *trpP*. This shuts off the transcription of the *trpPOLEDCBA* operon.

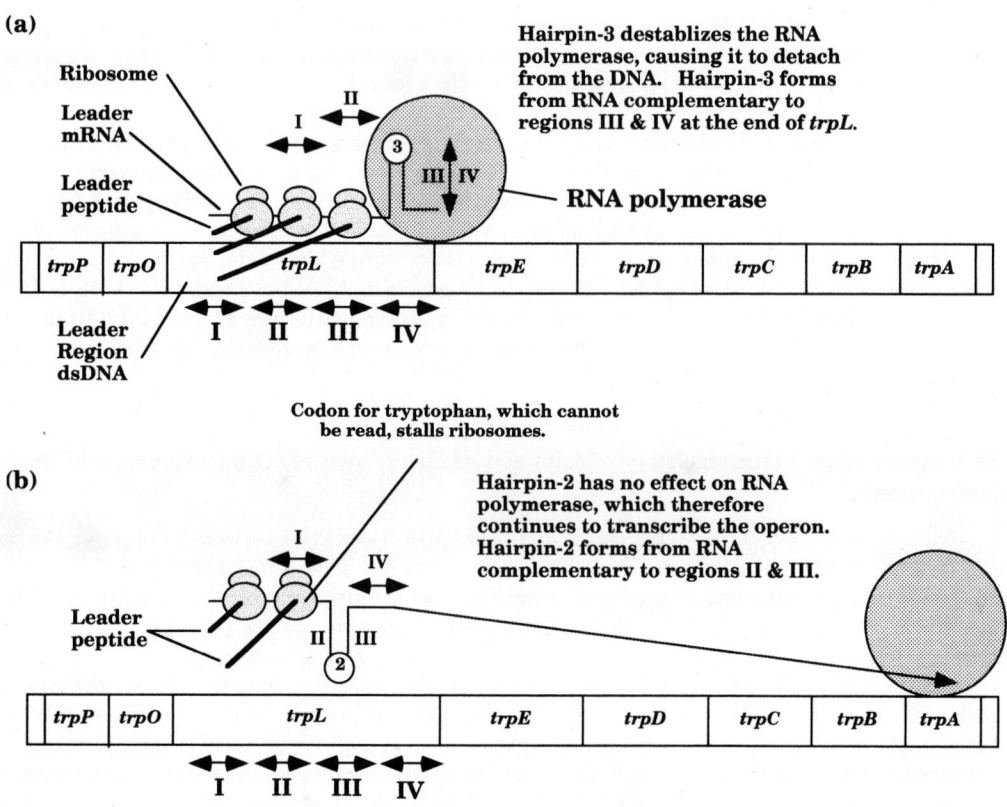

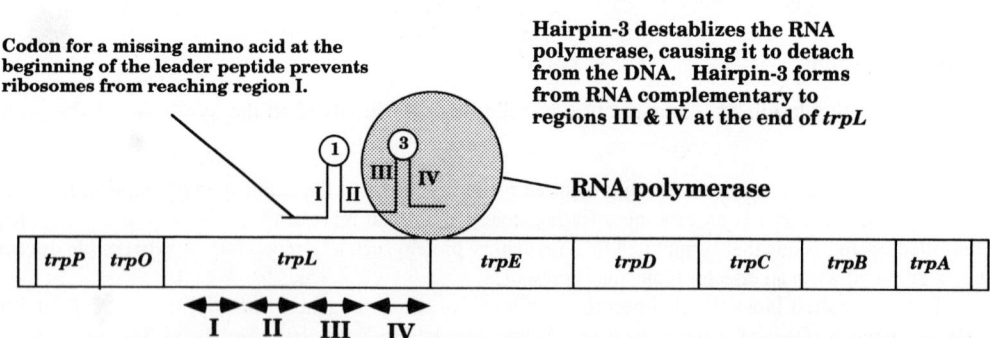

Fig. 4-7. The tryptophan operon and its regulation. (*a*) In the presence of excess tryptophan (and other amino acids), the leader region of mRNA is translated. Translation of the 5′ proximal mRNA prevents region I of the leader from hydrogen bonding with a leader sequence further downstream. The overlapping ribosomes also block region II from hydrogen bonding with region III. Thus, the only regions that can hydrogen bond with each other are regions III and IV. This results in the formation of the hairpin-3 structure that interacts with RNA polymerase as a transcription termination signal, causing detachment of the polymerase from the DNA. (*b*) When a cell is starved for tryptophan (but still has plenty of other amino acids), region I of the leader is not translated beyond the codons specifying tryptophan. Nevertheless, the ribosome, stuck at the codons specifying tryptophan, covers region I and prevents hydrogen bonding between regions I and II. Consequently, region II can hydrogen bond with region III, creating hairpin-2. Since hairpin-2 cannot inhibit the movement of RNA polymerase, transcription proceeds past the leader region. (*c*) When a cell is starved for the first few amino acids of the leader peptide the tryptophan codons (such as methionine, lysine, alanine, and isoleucine), both hairpin-1 and hairpin-3 form. Hairpin-3 blocks progression of the RNA polymerase and terminates transcription.

Because the *trpPOLEDCBA* operon is shut off by a repressor, it is negatively regulated. Since the addition of tryptophan to the medium causes the *trpPOLEDCBA* to be repressed, the operon is repressible.

4.35. Transcription of the tryptophan operon is regulated not only by a repressor but also by an attenuator. How does the attenuator prevent transcription?

An attenuator is a hairpin structure near the end of the **leader region** *(trpL)* of mRNA that causes RNA polymerase to dissociate from the DNA it is transcribing. Attenuation in prokaryotic cells differs from that in nuclear genes of eukaryotic cells, because the bacterial attenuation relies on concomitant translation to regulate the folding of the attenuator into a hairpin structure that terminates transcription. However, an attenuation process involving additional proteins does occur in eukaryotes (Question 4.37).

The leader region codes for RNA that, under certain circumstances, folds on itself and forms **hairpin-3,** which inhibits RNA polymerase (Fig. 4-7*a*). The formation of the inhibitory hairpin occurs when the leader region is translated. When tryptophan and the amino acids in the leader peptide are present in excess, the leader peptide is synthesized and completed. Since the RNA specifying the leader peptide contains a nonsense codon just after region I, the leader peptide terminates near the middle of the leader region. Ribosomes that are translating region I, however, overlap region II. Consequently, region II is unable to prevent regions III and IV from forming hairpin-3. Hairpin-3 destabilizes the RNA polymerase, causing it to detach from the DNA. This prevents transcription of the rest of the operon.

When all amino acids except tryptophan are present, translation of the leader region is inhibited at codons for tryptophan (Fig. 4-7*b*). Because of the pattern of ribosomes on the leader RNA, it folds into **hairpin-2** rather than hairpin-3. In other words, the formation of hairpin-2 prevents the formation of hairpin-3. Since hairpin-2 has no effect on the movement of RNA polymerase, transcription of the *trpPOLEDCBA* operon into mRNA is completed. Other ribosomes can attach to ribosome-binding sites preceding each of the coding regions on the polycistronic mRNA of the tryptophan operon and thereby initiate translation of all five proteins of the operon.

In the situation where amino acids other than tryptophan are in short supply, translation of the leader region may be blocked almost immediately (Fig. 4-7*c*). This results in the formation of **hairpin-1,** which does not prevent the formation of hairpin-3. Hairpin-3 interacts with RNA polymerase and causes termination of transcription. Starvation for any one of 4 amino acids (methionine, lysine, alanine, and isoleucine) found at the beginning of the leader peptide also causes attenuation of the tryptophan operon, even when tryptophan is limiting. Attenuation is advantageous to the cell, because in the absence of many amino acids, very few of the cell's proteins can be made. Therefore, by turning off the tryptophan operon, the cell conserves its energy and nutrient resources by not trying to make enzymes that can't be completed.

4.36. Why are there two regulatory mechanisms controlling the *trpPOLEDCBA* operon?

Repression is a mechanism for turning off the *trpPOLEDCBA* operon when tryptophan is present at acceptable levels. Repression does not allow for fine tuning of the tryptophan operon when tryptophan is at acceptable levels. In addition, repression does not respond to the depletion of other amino acids that must be synthesized or obtained from the environment. Repression is like the off switch for a light. The operon either is turned off completely or is on full blast.

Even when there is sufficient tryptophan in the cell, the repressor occasionally detaches momentarily from *trpO,* and RNA polymerase attaches to *trpP* and initiates transcription (a sneak synthesis). If transcription were completed and translation ensued, much energy would be wasted making enzymes to synthesize tryptophan. This waste does not occur, however, because of the attenuator. Attenuation is a mechanism for finely regulating the rate of tryptophan synthesis in response to the concentration of tryptophan as well as other amino acids. Attenuation becomes less severe as tryptophan levels fall and more severe as tryptophan levels increase. Thus, attenuation is like a dimmer switch for a light.

TRANSCRIPTION AND GENE REGULATION IN EUKARYOTES

In eukaryotes, just as in bacteria, genes are regulated so that they are expressed at the right times and at the correct levels to maintain the cell or to promote its growth and proliferation. Cells in multicellular organisms must not only respond to the chemicals and conditions of their environment but must also work together. Thus, the cells of multicellular organisms carry out complex signaling that often affects gene activity (Chapter

11). Most cells produce protein growth factors (called cytokines), steroid hormones, and/or neurotransmitters that signal cells to turn genes on or off. The messages are sent through signal transduction pathways that often begin at the surface of the cell and end at the chromosomes in the nucleus. This section of Chapter 4 will be concerned with the mechanisms by which genes are expressed in eukaryotes.

Eukaryotic Controlling Sites and Transcription

4.37. Do eukaryotic controlling sites differ from those in bacteria?

The controlling sites regulating transcription in eukaryotes are similar to those found in bacteria; however, there are many more controlling sites and regulatory proteins affecting each eukaryotic gene. Some of the controlling sites are many hundreds or even many thousands of base pairs from the promoter site (Fig. 4-8a). Because of these spacial differences, at least three classes of activators have been characterized: transcriptional factors (TFs), transcriptional activators (TAs), and coactivators. **Transcriptional factors** (TFs) attach to **activator binding sites** in promoter regions and stimulate RNA polymerase binding to the promoter site (Fig. 4-8b). Transcriptional factors are produced constitutively since great numbers of genes depend on these factors for their expression. Promoters, operators, and initiators are relatively close to each other, being separated by only a few hundred base pairs. Controlling sites located many hundreds or even thousands of base pairs from the promoter are known as enhancers. **Enhancers** are bound by **transcriptional activators** (TAs), proteins that are **induced** (i.e., synthesized only in response to specific signals). Some induced activators such as Fos and Jun bind very near promoter sites, which demonstrates that not all enhancer sites are distant from promoter sites. Other induced transcriptional activators such as Gal4, however, bind many thousands of base pairs from the promoter. Transcriptional activators cause the DNA to loop back on itself when they interact with the TFs near the promoter. This interaction between enhancer sites and initiator sites is usually necessary for transcription above a basal level (Fig. 4-8c). **Coactivators** are activator proteins that often connect TFs and TAs and may be essential for expression of genes (Fig. 4-8d).

Still other controlling sites code for regions at the beginning of the mRNA (near the 5′ terminus) that interact with RNA polymerase, causing it to terminate transcription when certain regulatory proteins are present. These regions of mRNA function in much the same manner as those found in bacteria and consequently are called attenuators. RNA regions in the middle or at the end of a gene that interact with the RNA polymerase and terminate transcription are known as terminators.

The original definition given for a promoter site has been obscured by recent discoveries in eukaryotes and their viruses. The terms "promoter" or "promoter region" are used to describe all the clustered regulatory sites involved in eukaryotic transcription; they do not include the enhancer sites, which are many hundreds or thousands of base pairs upstream or downstream from the RNA polymerase binding site.

4.38. How do eukaryotes make use of different RNA polymerases to regulate transcription?

Three different RNA polymerases are involved in eukaryotic nuclear transcription, whereas only one functions in bacterial cells. The three eukaryotic RNA polymerases initiate transcription only with specific combinations of transcriptional factors and transcriptional activators. **RNA polymerase I** (Pol I) transcribes the gene that specifies 5.8S, 28S, and 18S rRNA (Table 4.1). This polymerase is often found associated with chromosomes at the nucleolar organizing regions or in the nucleoli (singular, nucleolus). Nucleolar organizing regions are large stretches of DNA with thousands of gene repeats for rRNA molecules. Nucleolar organizing regions are restricted to a few chromosomes and may give rise to nucleoli in certain cells. Nucleoli are structures in the nucleus consisting of rDNA, rRNA, and ribosomal proteins. Ribosomal RNA is processed, and ribosomal 40S and 60S subunits begin to form within nucleoli. The promoters bound by RNA polymerase I are distinct from those that are recognized by other RNA polymerases. RNA polymerase I does not, however, transcribe 5S rRNA genes. These genes are transcribed by RNA polymerase III (see below).

RNA polymerase II (Pol II) transcribes from promoters that control the synthesis of **pre-mRNA,** which consists of coding and noncoding regions. The coding regions of pre-mRNA are called **exons,** whereas the noncoding regions are referred to as **introns.** Pre-mRNA is processed in the nucleus into mature mRNA, which is translated into protein in the cytoplasm. **Pre-mRNA processing** involves the elimination of all introns and the splicing together of exons (discussed in Questions 4.46 to 4.50).

RNA polymerase III (Pol III) recognizes promoters that control the synthesis of relatively short RNAs such as 5S rRNA, transfer RNAs (tRNAs), **small nuclear RNAs (snRNAs),** and **signal recognition particle RNAs (srpRNAs).** The 5S rRNA is a structural component of the 60S ribosomal subunit. The tRNAs read the informa-

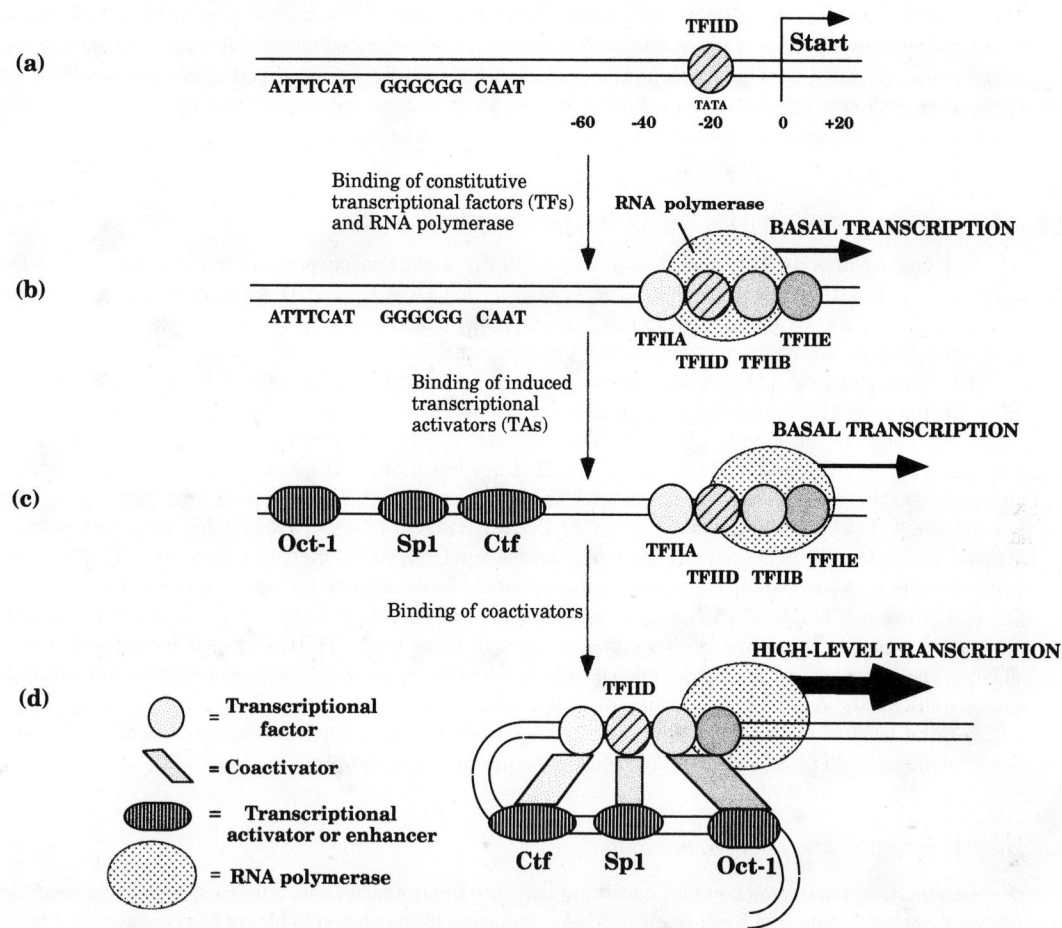

Fig 4-8. Transcriptional activation in eukaryotes. This figure is explained in Questions 4.39 and 4.40. (*a*) An activator called a transcriptional factor (TF) binds to the TATA box in the promoter region. This stimulates the formation of an activator complex and the binding of RNA polymerase II. (*b*) RNA polymerase II is able to initiate low levels of transcription that result in basal levels of RNA and protein. (*c*) The binding of activators, called transcriptional activators (TAs) or sometimes enhancers, is the next step that leads to increased levels of transcription.   (*d*) Full induction of the operon generally requires coactivators to bring the DNA into the appropriate conformation for maximal transcription.

Table 4.1.   Eukaryotic RNA Polymerases

Polymerase	Symbol	Binding Site	Function*
RNA polymerase I	Pol I	Varies	Transcribes rRNA (5.8S, 28S, 18S)
RNA polymerase II	Pol II	Generally TATA boxes	Transcribes all mRNAs
RNA polymerase III	Pol III	Frequently TATA boxes	Transcribes 5S rRNA, tRNAs, snRNAs, and srpRNAs

* rRNA = ribosomal RNA, mRNA = messenger RNA, snRNA = small nuclear RNA, tRNA = transfer RNA, srpRNA = signal recognition particle RNA.

tion in mRNA, whereas the snRNAs are part of enzymes involved in pre-mRNA processing. These enzymes, called **spliceosomes,** consist of **small nuclear ribonucleoproteins (snRNPs)** and splice exons in pre-mRNAs together to produce mRNA. **Signal recognition particle RNAs (srpRNAs)** are structural components of the **signal recognition particles** that guide certain nascent proteins to the endoplasmic reticulum. The small nuclear ribonucleoproteins and the signal recognition particles are discussed in Chapter 10.

4.39. What proteins normally bind to eukaryotic promoter sites?

The binding of RNA polymerase to promoter sites is dependent on proteins called transcriptional factors (TFs), such as **TFIID,** which binds to promoter regions having TATA boxes. This transcriptional factor is comparable to the sigma factors found in bacteria. Transcriptional factors are activators that function in a promoter region and that often directly interact with the RNA polymerase.

The binding of RNA pol-II to its promoter site requires a number of TFs. TFIID is the first transcriptional factor to bind close to the promoter at an initiator site about -20 to -10 base pairs before the transcriptional start site. TFIID binds to TATA boxes (Fig. 4-8a) and consequently has been referred to as the TATA-box binding protein (TBP). The TATA box is also known as the **Hogness box.** The TATA box is a distinctive sequence of nucleotides found in eukaryotic promoter regions. TATA boxes are about -25 base pairs from the RNA polymerase binding site with a consensus sequence $5'T(C/G)TATA(T/A)AAA(T/C)A3'$. Remember, a consensus sequence contains the nucleotide sequence most commonly encountered in a wide variety of eukaryotes. TFIID sits on the TATA box like a saddle and organizes other transcriptional factors required for initiation of RNA synthesis. One part of TFIID binds TFIIA, whereas another part binds a tissue-specificity factor. A **tissue-specificity factor** is a protein that is produced only in cells belonging to a specific tissue type. TFIID may also interact with TFIIB and TFIIE. The promoter and the transcriptional factors determine which RNA polymerase binds and which genes can be transcribed. A complex consisting of TFIIA, IID, IIB, IIE, and RNA polymerase II is capable of giving only a **basal level** of transcription (Fig. 4-8b). The complex of transcriptional factors, specificity factors, and RNA polymerase will be referred to in this book as the **preinitiation complex.**

4.40. How do enhancer sites affect transcription?

Generally, a preinitiation complex consisting only of transcriptional factors, tissue specificity factors, and the RNA polymerase yields only basal levels of RNA. Induction of the operon to higher levels requires the presence of other activators binding to distinct initiator sites many hundreds or thousands of base pairs from the promoter site. These distant initiator sites are known as enhancer sites (Fig. 4-8c). Activator proteins like **Ctf** (NF-1) binding to CAAT boxes (GGC*CAAT*CT), **Sp1** binding to G boxes (*GGGCGG*), **Spi-1** attaching to PU boxes (GAG*GAA*), and **Oct-1** interacting with AT-CAT boxes (*ATTTCAT*) cause the DNA to loop back upon itself, so that the activator proteins either directly or indirectly interact with the preinitiation complex. When they do, this signals the RNA polymerase to begin synthesizing high levels of RNA. Some of these enhancer sites are fairly close to the promoter region. The CAAT box is usually about -75 base pairs before the base pair where transcription initiates.

Transcription generally requires **coregulators (coactivators)** that bind the transcriptional activators to the transcriptional factors (Fig. 4-8d). The interaction of TAs with TFs causes the DNA to enter into a conformation that can be unwound and translated.

> **Example 4.5.** One coactivator has been discovered that binds Sp1 and TFIID together. Some coactivators that bind to RNA polymerase are believed to have helicase (DNA unwinding) activity.

4.41. What is the nature of operons in eukaryotic cells?

One or more structural genes and the controlling sites that regulate the rate of transcription of these genes are called operons. Most operons in eukaryotic cells consist of a large number of controlling sites and a single structural gene. The average structural gene in eukaryotes is about 10,000 base pairs long and specifies a pre-mRNA. Generally, large stretches of the pre-mRNA are excised while various small regions of the pre-mRNA are spliced together, yielding one or more different messenger RNAs (mRNAs). These different mRNAs may specify similar proteins or very different proteins with dissimilar enzyme activities. Some eukaryotic viruses produce mRNA molecules that specify polyproteins. **Polyproteins** are cleaved into separate proteins with various functions.

The large number of controlling sites in animals and plant cells reflects the fact that differentiated cells working together have to respond to more signals and be shielded from more influences than do bacterial cells.   In bacteria, one promoter site (for binding of RNA polymerase), one or two operator sites (for binding of repressor), one or two initiator sites (for binding of activator), and an attenuator site (for destabilizing RNA polymerase attachment to DNA) are usually sufficient to control the most complicated bacterial operons.

4.42.   What role does the structure of the chromosome play in controlling DNA replication and transcription?

DNA replication and transcription in eukaryotes are affected by the degree to which the chromosomes are condensed.   Replication and transcription are inhibited by highly condensed chromosomes, apparently because DNA and RNA polymerases as well as accessory proteins essential for replication and transcription are unable to bind to the DNA.   On the other hand, decondensed DNA can be transcribed.

4.43.   What chemical modification of DNA is known to be important in eukaryotic gene regulation?

The degree of **DNA methylation** has been shown to control transcription.   In general, highly methylated DNA inhibits transcription, whereas **demethylated DNA** allows transcription.   It is thought that cellular methylases add methyl groups to guanines and adenines located in activator binding sites.   The methyl groups sterically hinder activator binding at controlling sites.   Since many different activator binding sites usually control gene expression, methylation of the controlling sites may reduce transcription to basal levels.

4.44.   What is gene imprinting?

When a gene's activity is affected by its history in a parent, the gene is said to demonstrate **gene imprinting.**   Typically, imprinted genes are altered by methylation while in the gonads of one sex but not in the other.   Gene methylation generally blocks the expression of the altered gene even when the gene is transferred to another host.

> **Example 4.6.**   If a gene is imprinted (methylated) in the female's gonads but not in the male's, the gene inherited from the female is not expressed, but the one from the male is expressed in both female and male progeny.

Gene imprinting often produces maternal and paternal genomes that complement each other rather than provide duplicate products.   For example, a chromosome from a female might express genes A and B but not C, D, or E, whereas the homologous chromosome from a male would express C, D, and E but not A or B.   Thus, diploid animals that derive both sets of chromosomes from one parent are more likely to develop abnormally in some manner.   Even imprinted genes on just one chromosome can have a major effect.   Generally, if both copies of certain chromosomes come from the same parent, the fetus develops abnormally or dies.

> **Example 4.7.**   Mice embryos containing two complete sets of chromosomes from their mother (but no genes from a father) initially develop almost normally, but because the placental tissue is barely present, the embryo dies.   In contrast, mice embryos with two complete sets of chromosomes from their father (but no genes from a mother) generally do not develop, and there is an abnormally large development of placental tissue.

4.45.   How is gene imprinting implemented?

DNA methylation is believed to be the most important mechanism in gene imprinting.   Scientists have followed the methylation patterns of a foreign gene in mice for a number of generations and found that the methylation pattern depends upon the sex of the parent.

> **Example 4.8.**   When the mouse gene was inherited from the female it was highly methylated, but much less so when inherited from the male.   The highly methylated gene inherited from the female was inactive in the sons, but would become demethylated in the son's gametes.   Thus, the son passed on a demethylated, active gene to both grandsons and granddaughters.   The gene was active in both the grandsons and the granddaughters.   The gene was found to be highly methylated, however, in the granddaughter's eggs.   Methylation of transcriptional factor binding sites, promoter sites, or sites

within the structural genes that lead to premature termination of transcription could be responsible for gene imprinting.

It has been found that there is more methylation of DNA when it is condensed than when it is decondensed. Thus, the structure of DNA may regulate its methylation, or methylation may affect DNA condensation.

RNA Processing

4.46. After transcription, RNA in eukaryotes undergoes significant processing. How might this affect the expression of rRNA genes?

Much of the ribosomal RNA (rRNA) is encoded by very large genes that in humans are as much as 13 kilo-base pairs long with a Svedberg value of 45S (Fig. 4-9a). The 13-kilobase-pair transcript is processed by endonu-cleases in the nucleolus. An endonuclease, possibly consisting of a small nuclear ribonucleoprotein (U3 snRNP), cleaves the **spacer RNA** (a noncoding RNA) from the rRNAs. The 45S pre-rRNA yields an 18S rRNA, a 5.8S rRNA, and a 28S rRNA. These rRNAs are structural components of the 40S ribosomal subunit (18S rRNA) and the 60S ribosomal subunit (5.8S rRNA and 28S rRNA). The 60S ribosomal subunit also contains a 5S rRNA that is a transcript from another gene.

4.47. How does RNA processing affect the expression of genes coding for proteins?

Transcripts that specify proteins are modified in the nucleus by the addition of **caps** at their 5′ ends and poly-A tails at their 3′ termini (Fig. 4-9b). The RNA caps consist of a modified base called **7-methylguanine** (Fig. 4-10). The poly-A tail, formed by adding one adenine nucleotide at a time, is usually between 100 and 250 adenines long (Fig. 4-9c). The resulting RNA molecule is referred to as pre-mRNA. The pre-mRNA is con-verted into mRNA by the excision of introns and the splicing of exons (Fig. 4-9d); the mRNA is subsequently transported to the cytoplasm, where it is translated into protein.

The 7-methylguanosine cap that is attached to the phosphate at the 5′ end of pre-mRNA through a 5′ → 5′ linkage is necessary for the completion of a nascent pre-mRNA (Fig. 4-10c). The cap is also apparently neces-sary for the translation of most eukaryotic mRNAs.

All eukaryotic mRNAs, except for those that code for the histone proteins, must have a poly-A tail to enter the cytoplasm. It is believed that mRNAs lacking poly-A tails are digested in the nucleus. Thus, poly-A tails are thought to protect most mRNAs from digestion. The mechanism that adds poly-A tails (or not) effectively regu-lates whether a mRNA will survive to be translated. Since poly-A tails may be added to a number of exons, the mechanism that adds poly-A tails also determines which exons will end a particular protein.

> **Example 4.9.** The pre-mRNA of adenovirus late gene expression contains five potential poly-A ad-dition sites. The exon that has a poly-A tail added to it is the last one to be spliced to the processed RNA. As it turns out, there are several splicing sites for each of the exons, so that as many as 15 dif-ferent proteins can be synthesized. Thus, the mechanism that determines where the poly-A tail will be attached determines what the gene product will be like.

The largest RNA molecules isolated from a cell's nucleus includes primary transcripts and partly processed transcripts. These RNA molecules are known as **heterogeneous nuclear RNA (hnRNA).** Primary transcripts are referred to as **pre-mRNA** and **pre-rRNA.**

4.48. What are the separate functions of spacer DNA, introns, and exons?

In more complex eukaryotes such as humans, it is estimated that over 95% of the genome consists of noncod-ing DNA. Much of the noncoding DNA in such "higher" eukaryotes is found between genes. This DNA is re-ferred to as **spacer DNA.** Spacer DNA is found *between* the "genes" for 18S rRNA, 5.8S rRNA, and 28S rRNA (Fig. 4-9a). Transcription of these genes results in a polycistronic 45S rRNA. Endonucleases cut away the spacer RNA, releasing the three ribosomal RNAs.

Within genes specifying proteins, 90 to 95% of DNA may be noncoding also. This DNA separates genes into noncoding regions called introns and coding regions called exons. RNA polymerases normally transcribe an en-tire gene, producing RNA transcripts that contain introns and exons. The introns of various primary transcripts specifying cellular proteins are removed and various exons are spliced together to produce the final mRNA. In-tron excision and exon splicing require a special complex of enzymes called the **spliceosome.** Spliceosomes are

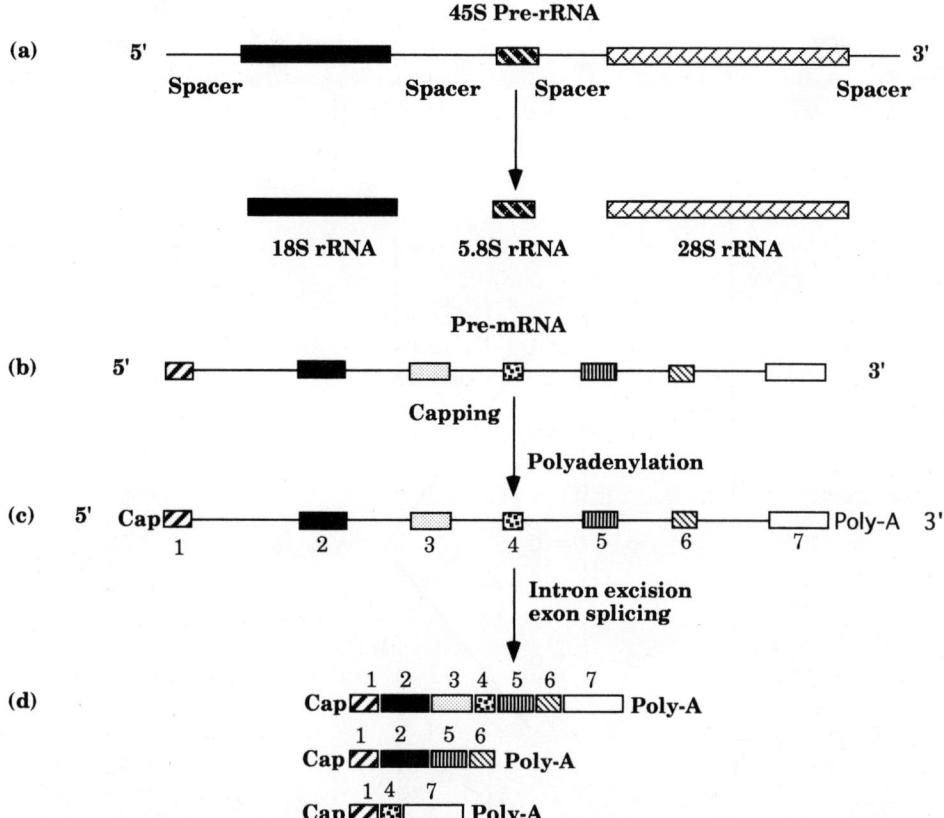

Fig. 4-9. Posttranscriptional processing of eukaryotic RNA. (*a*) The 45S rRNA transcript is processed to ribosomal RNA by an endonuclease that cuts out the spacer material between the ribosomal RNAs. Usually, there are no introns to be removed or exons to be spliced together in rRNA; see Question 14.8 for an exception. (*b*) to (*c*) Pre-mRNA is capped at the 5′ end and has a poly-A tail synthesized at its 3′ end. (*c*) to (*d*) Spliceosomes cut out the introns and splice together the exons (numbered rectangles). DNA sequences between exons are introns. (*d*) Many different mRNAs can be produced by different exon splicing and different poly-A additions (3 are shown). These different mRNAs may specify proteins with totally different activities.

discussed in Question 4.50. In summary, almost all mature, cytoplasmic mRNA consists of a 5′ cap, spliced exons, and a poly-A tail.

4.49. How can splicing result in the production of more than one gene product from a single gene?

Sometimes not all the exons in a transcript of a gene are used to make the mRNA. For example, suppose that the initial transcript of a gene consists of exons 1, 2, 3, 4, 5, 6, and 7 (Fig. 4-9*c*). Splicing together exons 1, 2, 3, 4, 5, 6, and 7 would produce mRNA that would code for a slightly different protein than mRNA with exons 1, 2, 5, and 6 (Fig. 4-9*d*). Proteins with similar functions but different amino acid sequences are called **isozymes** and are frequently found in different tissues. The consequence of splicing different exons together is that some eukaryotic genes may each code for more than one protein.

4.50. How does splicing occur and how might it be regulated?

Most splicing is carried out by enzyme complexes, called spliceosomes, in the nucleus. Spliceosomes consist of 4 different small nuclear ribonucleoprotein particles (snRNPs) that work together to bring the ends of exons

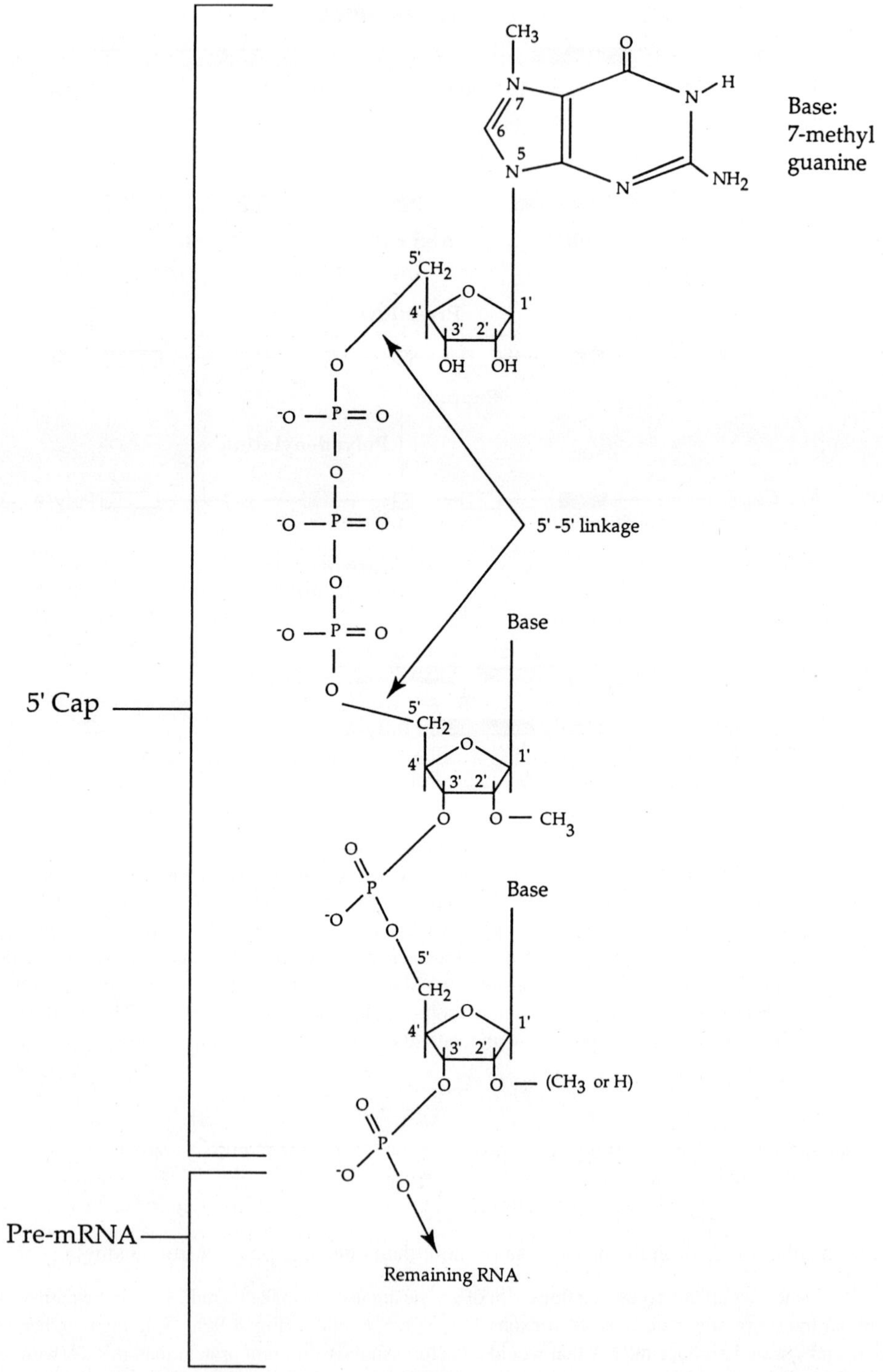

Fig. 4-10. Messenger RNA 5′ cap. The mRNA 5′ cap contains the unusual ribonucleotide 7-methylguanosine (m7G) triphosphate attached at the 5′ end through an unusual 5′ → 5′ linkage. Most nucleotides are attached by bonds that connect the 3′ → 5′ carbons. The two nucleotides following the m7G may also be methylated on the 2′ hydroxyl groups. These modified nucleotides are considered part of the 5′ cap.

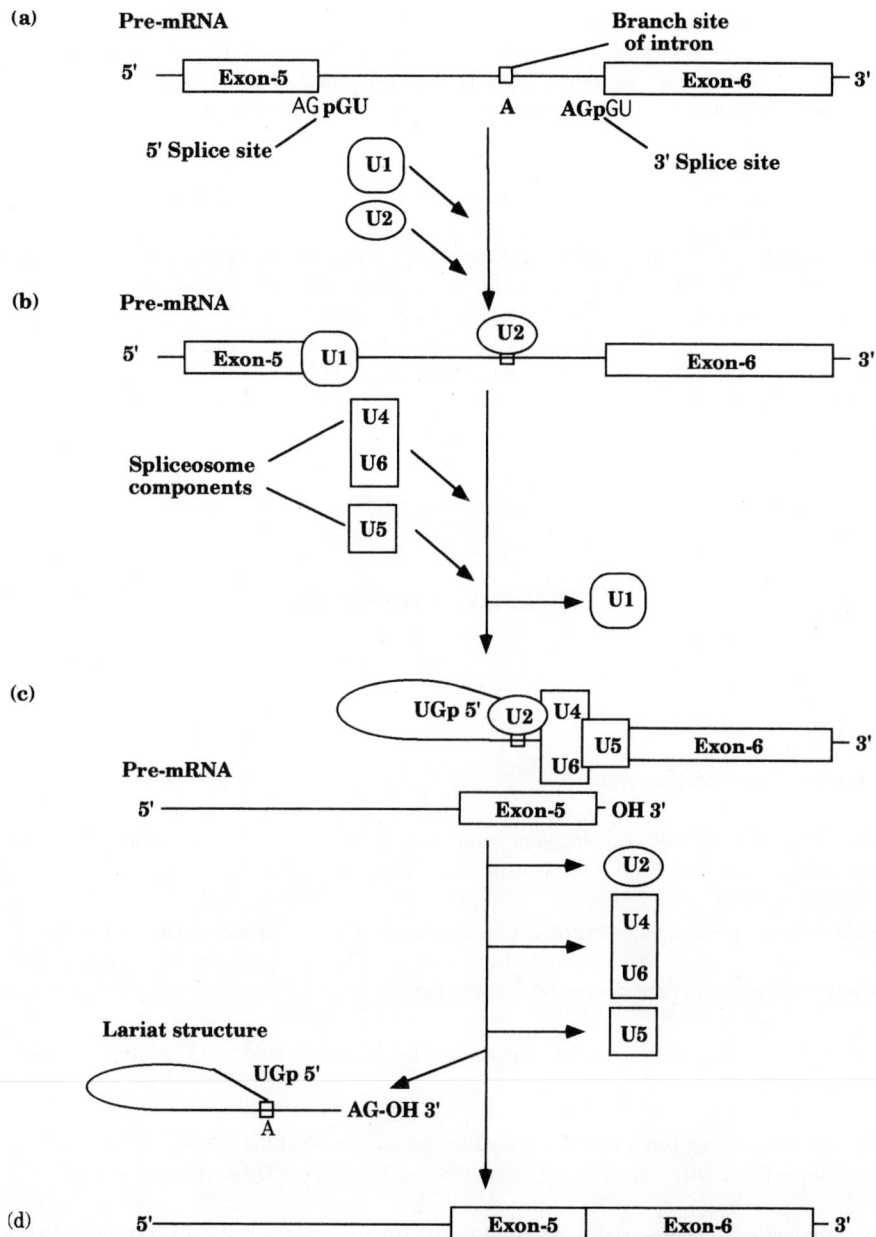

Fig. 4-11. RNA splicing of pre-mRNA. (*a*) to (*b*) Small nuclear ribonucleoprotein particles U1 and U2 attach to the 5′ splice site and branch site, respectively. The 5′
splice site is to the left of the phosphate in the sequence 5′AGpGU3′, whereas
the 3′ splice site is to the right. (*b*) to (*c*) Other small nuclear ribonucleoprotein
particles U4-U6 and U5 form a spliceosome with U1 and U2. The spliceosome
cleaves the 5′ exon from the intron and attaches the intron's 5′ end to the branch
site, creating a lariat structure. The 3′ end of the "free" exon (Exon 5, in this example) is in position to react with the 5′ end of the next exon (Exon 6, is this example). (*c*) to (*d*) The spliceosome catalyzes the reaction between the −OH at
the 3′ end of the "free" exon (Exon 5) and the 5′ end of the second exon (Exon
6). The spliceosome depolymerizes into individual ribonucleoprotein particles.
The lariat, now consisting only of intron material, is released.

in a primary transcript near each other (Fig. 4-11). The snRNPs are constructed from six to ten proteins and one or two of the five **small nuclear RNAs** (snRNAs) designated U1, U2, U4, U5, and U6. U4 and U6 are found in the same snRNP. The snRNPs are generally designated by the snRNAs they contain.

U1 snRNP binds to the 5′ exon-intron junction, U5 snRNP attaches near the 3′ intron-exon junction, whereas U4-U6 snRNP binds near U5, and U2 associates where a **lariat branch point** will form (Fig. 4-11*b,c*). The spliceosome, in particular U1 snRNP, cuts at the 3′ end of an exon (#5 as an example in Fig. 4-11*c*). U2 snRNA catalyzes the formation of the lariat, whereas U5 catalyzes the cut at the 5′ end of exon 6 and the splicing of exon 5 to exon 6 (Fig. 4-11*d*).

In the simplest case, a spliceosome promotes excision of an intron between two exons and the splicing together of the two exons. In more complicated cases, a spliceosome may promote the excision of introns and exons between two exons and then the splicing together of the targeted exons.

In some cases, splicing may not require snRNPs or spliceosomes. The pre-mRNA acts as an **autocatalyst,** splicing itself. All splicing discovered to date is negatively regulated (inhibited) by proteins that have not been characterized. These proteins are believed to bind to the RNA and inhibit the splicing capability of the pre-mRNA.

Objective Questions

Multiple Choice

Directions: Choose the one best answer.

1. The proteins that bind to the TATA box in the promoter region are known as (*a*) coregulators. (*b*) coactivators. (*c*) enhancers. (*d*) transcriptional activators. (*e*) transcriptional factors.

2. A gene inherited from a female is not expressed in either male or female offspring, but the same gene is expressed in both male or female offspring when inherited from a male. This can be an example of gene (*a*) recombination. (*b*) repression. (*c*) deletion. (*d*) imprinting. (*e*) transcription.

3. Steroid hormone receptors, when bound by an appropriate hormone, bind to (*a*) protein. (*b*) rRNA. (*c*) mRNA. (*d*) snRNA. (*e*) DNA.

4. DNA that is located *between* genes and is not used to specify a product but is discarded during RNA processing is known as (*a*) intron DNA. (*b*) exon DNA. (*c*) spacer DNA. (*d*) regulatory DNA. (*e*) either *a* or *b*.

5. DNA that is located *within* genes and is not used to specify a product but is discarded during RNA processing is known as (*a*) intron DNA. (*b*) exon DNA. (*c*) spacer DNA. (*d*) regulatory DNA. (*e*) either *a* or *b*.

6. Which of the following sites may be thousands of base pairs from a promoter, yet regulate transcription? (*a*) Operators (*b*) initiators (*c*) enhancers (*d*) attenuators (*e*) all of the above

7. What is the chemical basis of gene imprinting? (*a*) Phosphorylation of DNA (*b*) methylation of DNA (*c*) oxidation of DNA (*d*) glycosylation of DNA (*e*) none of the above

8. Eukaryotic RNA that specifies proteins is modified before translation by (*a*) addition of a 5′ cap. (*b*) splicing of exons. (*c*) addition of a 3′ poly-A tail. (*d*) all of the above. (*e*) none of the above.

9. Spliceosomes consist of 4 different (*a*) signal recognition particles (SRPs). (*b*) signal recognition particle receptors (SRPRs). (*c*) signal sequence receptors (SSRs). (*d*) small nuclear ribonucleoprotein particles (snRNPs). (*e*) small nuclear RNAs (snRNAs) devoid of proteins.

10. What is the characteristic form introns have after being cut from a pre-mRNA? (*a*) Linear structure (*b*) circular form (*c*) lariat shaped (*d*) theta structure (*e*) chi or χ structure

11. Initiation of protein synthesis from most eukaryotic mRNA requires (*a*) a 3′ poly-A tail. (*b*) a 5′ poly-A tail. (*c*) a 3′ cap. (*d*) a 5′ cap. (*e*) both a poly-A tail and a cap.

12. On the average, structural genes in bacteria are how many base pairs long? (*a*) 15 (*b*) 50 (*c*) 100 (*d*) 500 (*e*) 1,000

13. On the average, controlling sites in bacteria are how many base pairs long? (*a*) 15–30 (*b*) 50–100 (*c*) 300–500 (*d*) 750–1,000 (*e*) more than 2 kbp

14. When a number of genes are transcribed as one mRNA, the mRNA is said to be (*a*) multimeric. (*b*) polymeric. (*c*) polycistronic. (*d*) polyclonal. (*e*) polysomal.

15. Enzyme-III in its phosphorylated form (E-IIIP) (*a*) stimulates adenylcyclase. (*b*) opens cAMP pores. (*c*) blocks the lactose permease. (*d*) stimulates phosphorylase. (*e*) all of the above.

16. Which of the following characterize cAMP? (*a*) It is a nucleotide. (*b*) It is synthesized from ATP. (*c*) It is an effector molecule. (*d*) It binds to CRP. (*e*) All of the above.

17. In bacteria, which of the following is *not* an effector? (*a*) cAMP (*b*) L-arabinose (*c*) allolactose (*d*) lactose (*e*) IPTG

18. In bacteria, which of the following is *not* an inducer of an operon? (*a*) Tryptophan (*b*) allolactose (*c*) IPTG (*d*) L-arabinose (*e*) more than one of the above

19. The tryptophan biosynthetic operon is an example of a (*a*) positively controlled repressible operon. (*b*) negatively controlled inducible operon. (*c*) positively controlled inducible operon. (*d*) negatively controlled repressible operon. (*e*) none of the above.

20. The tryptophan regulatory operon consists of a single structural gene *trpR* that codes for an inactive regulatory protein called the (*a*) corepressor. (*b*) holorepressor. (*c*) aporepressor. (*d*) coactivator. (*e*) attenuator.

21. The corepressor of the tryptophan biosynthetic operon is a (*a*) protein. (*b*) polysaccharide. (*c*) tRNA. (*d*) tryptophan. (*e*) cAMP.

22. What is the effector for the tryptophan regulatory protein? (*a*) protein (*b*) polysaccharide (*c*) tRNA (*d*) tryptophan (*e*) cAMP

23. The leader region *(trpL)* of the tryptophan biosynthetic operon codes for RNA that may function as the operon's (*a*) corepressor. (*b*) holorepressor. (*c*) aporepressor. (*d*) coactivator. (*e*) attenuator.

24. Attenuation of the tryptophan operon can occur (*a*) if tryptophan and all other amino acids are present in excess. (*b*) if synthesis of the leader peptide is not initiated. (*c*) if the leader peptide is completed. (*d*) if methionine or *N*-formylmethionine is not present. (*e*) with all of the above.

25. If both the leader region *(trpL)* and the regulatory protein are deleted *(trpR)*, the tryptophan biosynthetic operon will (*a*) be expressed constitutively. (*b*) be attenuated maximally. (*c*) not be transcribed. (*d*) be inhibited by the corepressor. (*e*) none of the above.

26. The growth kinetics that result from metabolizing one sugar before another is referred to as (*a*) exponential growth. (*b*) diphasic growth. (*c*) diauxic growth. (*d*) chemotaxis. (*e*) autotrophy.

True–False

1. The initiation of RNA replication in eukaryotes may require transcriptional activators such as Fos and Jun.

2. Signal receptor proteins consist of an RNA (srpRNA) and a number of proteins.

3. Phosphorylation is an important mechanism for gene imprinting.

4. If a gene is "imprinted" in females but not in males, this imprinted gene can be expressed in males.

5. Steroid hormone receptors, when bound by an appropriate hormone, are known to function as transcriptional activators.

6. Most pre-mRNAs are modified by the addition of a $5'$ poly-A tail and a $3'$ cap.

7. Exons in pre-mRNAs can be spliced together in a number of different ways, resulting in a number of different mRNAs from a single gene.

8. Another term for a structural gene is attenuator.

9. When a number of genes are transcribed as one mRNA, the mRNA is said to be polymeric.

10. Catabolite repression decreases cAMP levels within a cell and inhibits the lactose permease.

11. E-III (enzyme-III in its unphosphorylated state) does not stimulate adenylcyclase, the enzyme that catalyzes the synthesis of cAMP from ATP.

12. Starvation for any one of at least 4 amino acids and found at the beginning of the leader peptide causes attenuation of the tryptophan operon, even when tryptophan is limiting.

13. Transcriptional activators (TAs) bind to enhancer sites many hundreds of base pairs from the promoter region.

14. Coactivators bind to RNA polymerases and help them attach to the correct promoter region.

15–46. The lactose regulatory operon and the lactose catabolic operon of wild-type cells are symbolized as $I^+P^+O^+Z^+Y^+A^+$. A plus superscript indicates that the controlling site or structural gene functions normally. A negative superscript indicates that the controlling site or structural gene is defective or missing. Other symbols such as I^{-d}, I^s, Z^{-ns}, and O^c are explained in the text. A genotype with a slash line such as $I^+P^+O^+Z^+Y^+A^+/I^+P^+O^+Z^+Y^+A^+$ represents a partially diploid cell. For each question number in the table below, indicate whether β-galactosidase (β-gal) and permease are produced for the genotypes listed. Notice that inducer is absent in some cases, but present in others. Mark T if the protein is produced; mark F if it is not produced.

Genotype	Inducer Absent		Lactose Present	
	β-Gal	Permease	β-Gal	Permease
$I^+P^+O^+Z^+Y^+A^+$	15	16	17	18
$I^sP^+O^+Z^+Y^+A^+$	19	20	21	22
$I^-P^+O^+Z^+Y^+A^+$	23	24	25	26
$I^+P^-O^+Z^+Y^+A^+$	27	28	29	30
$I^+P^+O^cZ^+Y^+A^+$	31	32	33	34
$I^sP^+O^+Z^+Y^+A^+/I^+P^+O^cZ^+Y^+A^+$	35	36	37	38
$I^-P^-O^+Z^+Y^+A^+/I^-P^-O^+Z^+Y^+A^+$	39	40	41	42
$I^-P^+O^cZ^{-ns}Y^+A^+/I^sP^+O^+Z^+Y^+A^+$	43	44	45	46

47–66. The L-arabinose regulatory operon and the L-arabinose catabolic operons are symbolized as $O^+C^+P^+O^+CRP^+I^+P^+B^+A^+D^+$. A plus superscript indicates that the controlling site or structural gene functions normally. A negative superscript indicates that the controlling site or structural gene is defective or missing. Other symbols such as P^c and C^c are explained in the text. The genotype $O^+C^+P^+O^+CRP^+I^+P^+B^+A^+D^+/O^+C^+P^+O^+CRP^+I^+P^+B^+A^+D^+$ represents a partially diploid cell. Indicate whether isomerase and kinase are produced for the genotypes listed below. Notice that inducer is absent in some cases, but present in others. Mark T if the enzyme is produced; mark F if it is not.

Genotype	Inducer Absent		Inducer Present	
	Isomerase	Kinase	Isomerase	Kinase
$O^+C^+P^+O^+CRP^+I^+P^+B^+A^+D^+$	47	48	49	50
$O^+C^-P^+O^+CRP^+I^+P^+B^+A^+D^+$	51	52	53	54
$O^+C^cP^+O^+CRP^+I^+P^+B^+A^+D^+$	55	56	57	58
$O^+C^-P^+O^+CRP^+I^+P^+B^+A^+D^+/$ $O^+C^+P^+O^+CRP^+I^+P^+B^-A^-D^-$	59	60	61	62
$O^+C^-P^+O^+CRP^+I^+P^+B^-A^+D^+/$ $O^+C^-P^+O^cCRP^+I^+P^+B^+A^-D^+$	63	64	65	66

Matching

Directions: In each of the following matched sets, not every term in **Column B** may be matchable to the numbered descriptions in **Column A.** Also, a term may apply to more than one description.

Column A	Column B
1. The enzyme that synthesizes pre-mRNA	A. RNA polymerase I
2. The enzyme that synthesizes snRNA	B. RNA polymerase II
3. The enzyme that synthesizes 45S rRNA	C. RNA polymerase III
4. The enzyme that synthesizes srpRNA	D. RNA polymerase IV
5. The enzyme that synthesizes 5S rRNA	
6. The enzyme that synthesizes tRNA	

Column A	Column B
7. The DNA sequence to which RNA polymerase binds	A. operator site
8. The DNA sequence to which sigma factors bind	B. promoter site
9. The DNA sequence that codes for a protein	C. leader region
10. The DNA sequence to which repressors bind	D. structural gene
11. The DNA sequence to which activators bind	E. none of the above

Terms

Directions: Unless otherwise specified, each answer is a single word.

1. The protein that binds to TATA boxes. (3 words)

2. The proteins that are required for the initiation of transcription by associating with TBP and RNA polymerase. (2 words)

3. Controlling sites that are hundreds to thousands of base pairs from promoters. (1 or 2 words)

4. A region of a chromosome that contains the genes for 45S rRNA. (3 words)

5. A body in the nucleus (consisting of rDNA, rRNA, and protein) where ribosomal subunits (40S and 60S) begin to be assembled.

6. The term used to describe the proteins that connect transcriptional activators to transcriptional factors.

7. RNA molecules, including primary transcripts, partially processed RNA, discarded introns, and small nuclear RNAs. (3 words)

8. The DNA located between genes. (2 words)

9. The DNA located within genes that is not used to specify a product. (2 words)

10. The enzyme complex that splices exons together.

11. Protein complexes containing small nuclear RNAs (snRNAs) U1, U2, U4, U5, and U6 as structural components. (2 words)

12. A compound that, when hydrolyzed, phosphorylates glucose and allows it to be transported by the phosphotransferase system.

13. The state of phosphorylation that enzyme-III (E-III) is in when glucose is transported by the phosphotransferase system and the lactose permease is inhibited.

14. The terminology that describes operators, promoters, attenuators, and terminators. (2 words)

15. A unit consisting of one or more structural genes that are under the control of one or more controlling sites, one of which must be a promoter.

16. The enzyme of the glucose phosphotransferase system that stimulates a phosphatase and directly phosphorylates glucose.

17. The growth kinetics that result from completely metabolizing first one sugar and then another. (2 words)

Answers to the Objective Questions

Multiple Choice

1. *e*   2. *d*   3. *e*   4. *c*   5. *a*   6. *c*   7. *b*   8. *d*   9. *d*   10. *c*   11. *d*
12. *e*   13. *a*   14. *c*   15. *a*   16. *e*   17. *d*   18. *a*   19. *d*   20. *c*   21. *d*
22. *d*   23. *e*   24. *e*   25. *a*   26. *c*

True–False

1. T   2. T   3. F (methylation)   4. F (imprinted gene cannot be expressed in either sex)   5. T
6. F (5′ cap and 3′ poly-A tail)   7. T   8. F (cistron)   9. F (polycistronic)   10. T
11. T   12. T   13. F (TAs bind to sites in the promoter region)   14. F (coactivators do not help RNA polymerases to bind to promoter regions. This is the job of transcriptional factors. Coactivators connect TFs and TAs, stimulating transcription above basal levels)   15. F   16. F   17. T   18. T   19. F   20. F
21. F   22. F   23. T   24. T   25. T   26. T   27. F   28. F   29. F   30. F
31. T   32. T   33. T   34. T   35. T   36. T   37. T   38. T   39. F   40. F
41. F   42. F   43. F   44. F   45. F   46. F   47. F   48. F   49. T   50. T
51. F   52. F   53. F   54. F   55. T   56. T   57. T   58. T   59. F   60. F
61. T   62. T   63. F   64. F   65. F   66. F

Matching

1. *B* **2.** *C* **3.** *A* **4.** *C* **5.** *B* **6.** *C* **7.** *B* **8.** *B* **9.** *D* **10.** *A* **11.** *E* (an activator may bind to initiator site, transcriptional factor binding site, or enhancer site)

Terms

1. TATA binding protein **2.** transcriptional factors **3.** enhancers of enhancer sites **4.** nucleolar organizing region **5.** nucleolus **6.** coregulator or coactivator **7.** heterogeneous nuclear RNA (hnRNA) **8.** spacer DNA **9.** intron DNA **10.** spliceosome **11.** ribonucleoprotein particles (RNPs) **12.** PEP (phosphoenolpyruvate) **13.** unphosphorylated **14.** controlling sites **15.** operon **16.** enzyme-III **17.** diauxic growth

Chapter 5

Translation

INTRODUCTION

In the previous chapter, transcription (the process by which all RNA molecules are produced) was outlined. The present chapter considers translation, the process by which various types of RNAs and other molecules interact in the synthesis of proteins. Following an explanation of the genetic code, translation will be discussed, first in prokaryotes and then in eukaryotes. It will be assumed that the reader has knowledge of the structures of proteins and nucleic acids equivalent to that presented in Chapter 2.

THE GENETIC CODE

5.1. Even before the nature of the genetic code was elucidated, it was hypothesized that the simplest genetic code would be a triplet code. What was the rationale for this hypothesis?

Given that each ribonucleotide has one of the four kinds of bases (A, U, G, C) and that there can be as many as 20 kinds of amino acids in biological proteins synthesized on ribosomes, the simplest genetic code would be a triplet of nucleotides. A group of three adjacent nucleotides in DNA would be transcribed into three complementary RNA nucleotides, which in turn would be translated into a single amino acid within a polypeptide chain. A singlet code (one nucleotide per amino acid) could only code for four amino acids. A doublet code (two nucleotides per amino acid) could only form $4 \times 4 = 16$ combinations and thus code for 16 amino acids. A triplet code, however, could form $4^3 = 64$ combinations (many more than needed to code for 20 amino acids). Experimental evidence later proved that the genetic code is indeed a triplet code, and that 61 of the 64 possible triplets specify various amino acids. Each coding triplet is referred to as a **codon**. However, because proteins are synthesized from mRNA templates rather than from DNA templates, codon tables are always displayed as mRNA triplets and their corresponding amino acid specifications (Table 5.1). Each mRNA codon in Table 5.1 is conventionally written with the 5′ nucleotide at the left and the 3′ nucleotide at the right, because protein synthesis begins at the 5′ ends of mRNA molecules and proceeds toward their 3′ ends. The 5′ end of the mRNA specifies the free amino end of the polypeptide chain and the 3′ end of the mRNA specifies the free carboxyl end of the polypeptide chain (see Question 2.29). Under optimal conditions in bacterial cells, translation may occur at a rate of 20 to 40 amino acids per second.

5.2. What is meant by code degeneracy? What implications does code degeneracy have for the expression of mutations?

The code is highly **degenerate** in that more than one codon can specify the same amino acid. Methionine and tryptophan are each specified by a single codon. The codon 5′AUG3′ near the 5′ end of a mRNA molecule is the usual **start codon (initiation codon)** that places methionine at the beginning (amino end) of all nascent eukaryotic polypeptide chains. Some amino acids, such as phenylalanine and histidine, are each specified by two codons. Isoleucine has three codons, proline has four codons, and serine has six codons. Different codons that specify the same amino acid are said to be **synonymous**. Because of code degeneracy, many changes (mutations) can occur in a gene that will have no effect on the amino acid composition of the gene product. Such changes are referred to as **samesense mutations** or **silent mutations**. Most of the degeneracy in the genetic code exists in the third position (3′ end) of the codon. The complementary base pairing between an mRNA codon and its anticodon in a transfer RNA (tRNA) molecule is usually much less restrained at the third position than in the other two positions of the triplet. For example, in the case of alanine, the first two bases in its mRNA codon (GC) completely specify that amino acid; any of the four nucleotides can be in the third position. If the first two nucleotides of a codon are UU, then either pyrimidine U or C can be in the third position to specify phenylalanine, whereas either purine A or G in that same third position specifies leucine. This phenomenon, called **wobble,** allows the same tRNA to recognize more than one mRNA codon in many cases.

5.3. Do the noncoding triplets perform any functions?

There are three codons that are not recognized by any tRNA: UAA, UAG, and UGA. These are termed **nonsense codons, termination codons,** or **stop codons,** because they act as periods at the end of a sentence; i.e., they provide part of the signal that protein synthesis should stop at that point. The completed polypeptide can then be released from its cognate tRNA and from the ribosome to do whatever job its role may be. The other 61 mRNA codons are **sense codons** that specify some amino acid.

5.4. Is the genetic code universal?

The genetic code, corresponding to the mRNA codons shown in Table 5.1, at first seemed to apply to all genes in both prokaryotes and eukaryotes. It was thus initially thought to be a **universal code.** However, some exceptions were later found in mitochondrial DNA of some organisms. For example, in yeast mitochondria, CUA codes for threonine instead of leucine, and in mammalian mitochondria, AUA codes for methionine instead of isoleucine.

Table 5.1. Codons (displayed as mRNA triplets)

First Letter		Second Letter				Third Letter
		U	C	A	G	
	U	UUU ⎫ Phe UUC ⎬ UUA ⎫ Leu UUG ⎭	UCU ⎫ UCC ⎬ Ser UCA UCG ⎭	UAU ⎫ Tyr UAC ⎬ UAA ⎫ Nonsense UAG ⎭	UGU ⎫ Cys UGC ⎬ UGA Nonsense UGG Trp	U C A G
	C	CUU ⎫ CUC ⎬ Leu CUA CUG ⎭	CCU ⎫ CCC ⎬ Pro CCA CCG ⎭	CAU ⎫ His CAC ⎬ CAA ⎫ Gln CAG ⎭	CGU ⎫ CGC ⎬ Arg CGA CGG ⎭	U C A G
	A	AUU ⎫ Ile AUC ⎬ AUA ⎭ AUG Met	ACU ⎫ ACC ⎬ Thr ACA ACG ⎭	AAU ⎫ Asn AAC ⎬ AAA ⎫ Lys AAG ⎭	AGU ⎫ Ser AGC ⎬ AGA ⎫ Arg AGG ⎭	U C A G
	G	GUU ⎫ GUC ⎬ Val GUA GUG ⎭	GCU ⎫ GCC ⎬ Ala GCA GCG ⎭	GAU ⎫ Asp GAC ⎬ GAA ⎫ Glu GAG ⎭	GGU ⎫ GGC ⎬ Gly GGA GGG ⎭	U C A G

The three-letter symbols, names, and one-letter symbols (in parentheses) of the amino acids are as follows: ala = alanine (A), arg = arginine (R), asn = asparagine (N), asp = aspartic acid (D), cys = cysteine (C), glu = glutamic acid (E), gln = glutamine (Q), gly = glycine (G), his = histidine (H), ile = isoleucine (I), leu = leucine (L), lys = lysine (K), met = methionine (M), phe = phenylalanine (F), pro = proline (P), ser = serine (S), thr = threonine (T), trp = tryptophan (W), tyr = tyrosine (Y), val = valine (V).
From *Schaum's Outline of Theory and Problems of Genetics, Third Edition,* 1991, by W. D. Stansfield, McGraw-Hill, Inc.

5.5. How was the genetic code cracked?

The first break in deciphering the genetic code was made in the early 1960s by M. W. Nirenberg and J. H. Matthei. They used a bacterial **cell-free system** containing all the necessary machinery for protein synthesis (tRNAs, ribosomes, amino acids, ATP, GTP, etc.) but devoid of intact cells and mRNA molecules. Using the enzyme **polynucleotide phosphorylase,** which randomly links ribonucleotide monomers together without a template, they were able to make synthetic mRNA molecules. At first they synthesized **homopolymers** consisting of a single kind of ribonucleotide. For example, addition of a synthetic mRNA consisting of only uracil monomers (poly-U) to such a cell-free system resulted in the production of polypeptide chains of variable lengths made entirely of phenylalanine. This suggested that some sequence of U's codes for that amino acid; later it was proven to be the triplet UUU. Similarly, the synthetic homopolymer poly-C codes for proline, poly-A codes for lysine, and poly-G codes for glycine.

5.6. How was most of the genetic code deciphered?

In another laboratory, H. G. Khorana began synthesizing mRNA **heteropolymers** consisting of alternating ribonucleotides such as . . . ACACACAC . . . and found that they specified polypeptide chains consisting of alternating threonine and histidine residues. Because protein synthesis on these artificial mRNAs in a cell-free system is indiscriminate as to where it begins to translate, it was not known which amino acid was being specified by codons ACA or CAC; translation is also much slower on synthetic mRNAs than on natural bacterial mRNAs (possessing specific ribosome binding sites). To resolve this problem, mRNAs consisting of repeating sequences of AAC (. . . AACAACAACAAC . . .) were synthesized and introduced into a cell-free system. This resulted in three kinds of homopolymeric polypeptide chains: all asparagine, all threonine, and all glutamine. Thus the mRNAs were being read in any of the three possible **reading frames** AAC, ACA, and CAA. Since the only codon in common between mRNAs consisting of alternating ribonucleotides A and C (. . . ACACAC . . .) and mRNAs containing repeating AAC triplets (. . . AACAACAAC . . .) is ACA, and since the only amino acid their polypeptide products have in common is threonine, ACA is logically the codon for threonine. This type of analysis allowed most of the codons to be deciphered.

5.7. How was the remainder of the genetic code resolved?

The remainder of unassigned codons was resolved by ribosome-binding tests. It was discovered that all synthetic RNA triplets (except those of stop codon sequences) would pair with the anticodon region of their respective tRNA–amino acid complexes; such complexes would then bind to ribosomes. Twenty different systems were examined. Each system contained a different radioactive amino acid; its other 19 amino acids were unlabeled. When these complexes were poured onto nitrocellulose filters, only ribosomes (with or without bound tRNAs) stuck to the filter; all free tRNAs (whether attached to their cognate amino acids or not), free synthetic mRNA trinucleotides, and free amino acids passed through the filter. The system that produced a radioactive filter thus revealed the amino acid that was specified by the mRNA triplet under test.

TRANSLATION IN PROKARYOTES

Before discussing the translation process, we must be aware of several facts. Ribosomes contain two sites for binding tRNA molecules: the **aminoacyl site (A site),** where each tRNA molecule first attaches, and the **peptidyl site (P site),** where a tRNA holds the growing polypeptide chain. The bacterial ribosome has two major subunits: a 30S subunit to which the mRNA and tRNAs become bound, and a 50S subunit to which tRNAs also bind. Each tRNA molecule becomes **charged (loaded)** by an enzyme that attaches a specific amino acid to each species of tRNA. Each tRNA molecule has a loop containing a triplet of ribonucleotides, called the anticodon (Question 2.60), that can base pair with a comparable triplet codon in mRNA.

5.8. There are three major stages in the translation process of both prokaryotes and eukaryotes: (1) initiation, (2) elongation, and (3) termination. What happens during the initiation stage of translation in bacteria?

Translation in prokaryotes begins at the methionine mRNA codon 5'AUG3' with the help of some protein **initiation factors** (Fig. 5-1). A tRNA loaded with the amino acid *N*-formylmethionine (tRNA^{fMet}) binds to a 30S-mRNA complex. The 3'UAC5' anticodon of the tRNA pairs (in antiparallel fashion; Question 2.55) with the complementary 5'AUG3' codon in the mRNA. Next, a 50S ribosomal subunit joins the 30S-mRNA-tRNA^{fMet} complex to form a complete 70S **initiation complex.** The tRNA^{fMet} ends up in the P site of the ribosome. This completes the initiation phase of translation. The ribosome acts like a jig to hold all the reactants in proper alignment during the translation process.

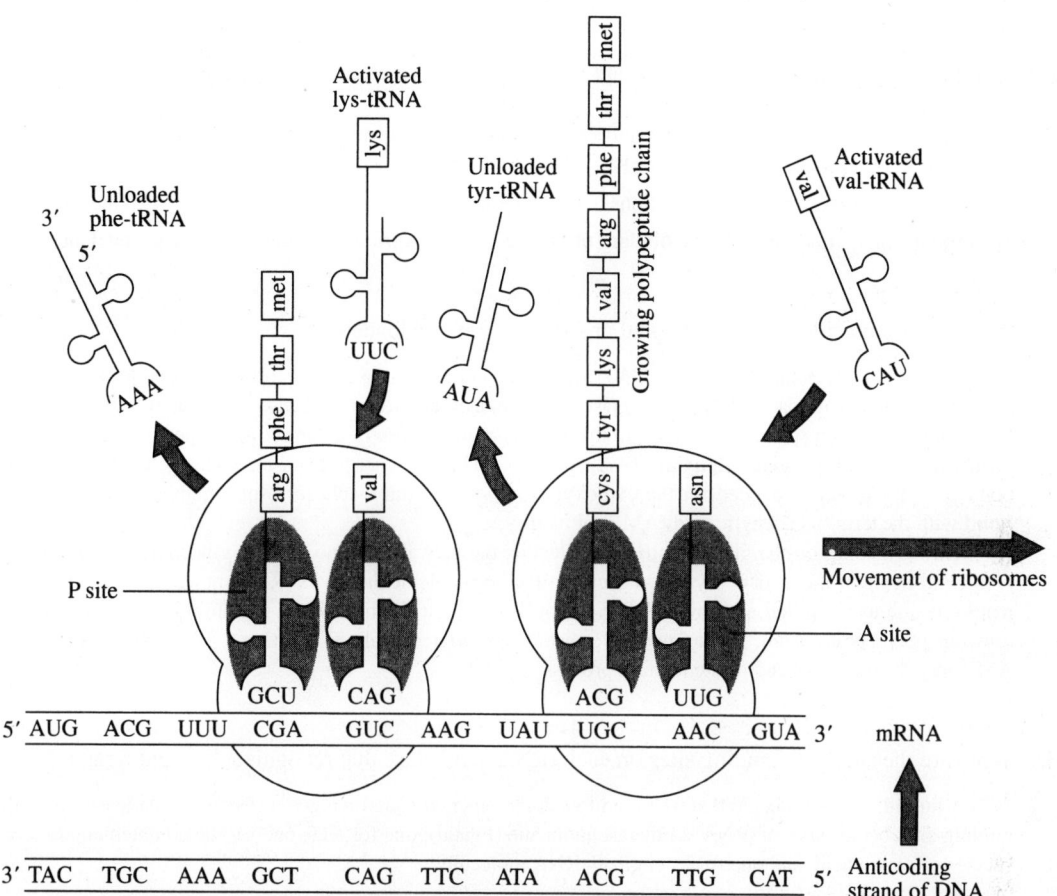

Fig 5-1. Diagram of protein synthesis. In bacteria, *N*-formylmethionine (fMet) initiates all polypeptide chains instead of methionine (Met) as in eukaryotes. (*From Schaum's Outline of Theory and Problems of Genetics, Third Edition, W. D. Stansfield, McGraw-Hill, 1991.*)

5.9. What happens during the elongation phase of translation in bacteria?

The elongation phase of translation proceeds with the help of a group of protein **elongation factors** when a second activated tRNA (e.g., one loaded with threonine) enters the A site (again by specific codon-anticodon base pairing). This places *N*-formylmethionine and threonine next to one another so that a peptide bond can be formed between them by action of an enzymatic, ribosomal component called **peptidyl transferase.** The amino-acyl bond that held the *N*-formylmethionine to the 3′ end of its tRNA is broken simultaneously with formation of the peptide bond. The now "unloaded" tRNA^{fMet} leaves the P site (usually to become reactivated with another *N*-formylmethionine). The mRNA shifts **(translocates)** three nucleotides through the ribosome. A new open codon is now positioned in the A site, and the fMet-Thr-tRNA^{Thr} (a dipeptide complex) is in the P site. The third tRNA (e.g., one loaded with phenylalanine) enters the A site; a peptide bond forms between the second amino acid (threonine) and the third one (phenylalanine); the second uncharged tRNA^{Thr} exits the P site; translocation of the mRNA through the ribosome positions the next codon (e.g., for arginine) in the A site, while simultaneously shifting the fMet-Thr-Phe-loaded tRNA^{Phe} (a tripeptide complex) from the A to the P site. This process is repeated until a termination codon is encountered in the A site.

5.10. How is translation terminated in bacteria?

The termination phase of translation (Question 5.46) occurs when the system reaches one or more mRNA stop codons (UAA, UAG, or UGA). At this point, with the help of proteins called **release factors** (see Question 5.46), the polypeptide is released from the tRNA; the mRNA and the last tRNA are released from the ribosome as the ribosome dissociates into its 30S and 50S subunits.

Transfer RNA (tRNA)

5.11. List three classes of proteins that interact with specific tRNAs and specify their functions.

1. Enzymes that activate amino acids and couple them with their respective tRNAs, referred to as **aminoacyl-tRNA synthetases.** See Question 5.12.
2. Enzymes that trim or add modifications to tRNA molecules during their synthesis (e.g., see Question 5.33).
3. Proteins that are required in the binding of charged tRNAs to ribosomal complexes (e.g., see Question 5.38).

5.12. How do amino acids become activated and tRNAs become loaded?

Each tRNA molecule ends with CCA3′ where its cognate amino acid becomes attached to the terminal adenine. The same aminoacyl-tRNA synthetase that activates an amino acid also loads or transfers the activated amino acid to its own species of tRNA. In the process of **activation** of an amino acid, these enzymes catalyze the formation of an **amino acid adenylate** (AA $\sim$ AMP). The free energy from the cleavage of the pyrophosphate bond (P $\sim$ P) in ATP is needed for the carboxyl group of the amino acid to form a high-energy ($\sim$) **aminoacyl bond** with the terminal adenylic acid (AMP) of its tRNA.

During loading or charging of a tRNA molecule, the activated amino acid may become attached by its carboxyl end to either the 2′ or 3′ hydroxyl group of the terminal adenine in its tRNA molecule, and it readily shifts from one site to the other, as shown in Fig. 5-2. However, at the time when the amino acid is transferred to a growing polypeptide chain, the amino acid must be adenylated at its 3′ position. The synthetase enzyme and AMP are released at the end of the loading process.

5.13. What are the substrate binding sites on each tRNA synthetase that recognize different ligands?

A **ligand** is any molecule that will bind to a complementary site on a given structure. At least three different binding sites occur on each tRNA synthetase molecule (ligand): one for ATP, one for the complementary (cognate) species of amino acid, and one for its cognate tRNA molecule.

5.14. What is the source of the energy used in formation of a peptide bond?

The energy in the amino acid $\sim$ tRNA bond (an aminoacyl bond) is used in the formation of a lower-energy peptide bond.

5.15. How many different kinds of tRNA molecules are required for translation?

Since there are 20 kinds of amino acids in proteins synthesized on ribosomes, there must be at least 20 different species of tRNA molecules. Of the 64 possible mRNA codons, 61 make sense (i.e., specify an amino acid). Because of wobble in the third position of a codon, however, a maximum of 61 kinds of tRNA molecules is not needed. The numbers actually used vary from one species to another. There may be as few as 30 or 40 different tRNA molecules used in bacteria and about 50 in animal and plant cells.

5.16. How many aminoacyl-tRNA synthetase enzymes are required for translation?

All of the tRNAs that become loaded with the same amino acid are called **isoacceptors.** All of the isoacceptors that bind to synonymous codons are activated by the same aminoacyl-tRNA synthetase. Thus, only 20 synthetases are required.

5.17. In what respects are aminoacyl-tRNA synthetases similar in structure? In function?

Although all these synthetases perform the same functions (they activate amino acids and charge tRNAs), structurally they are an extremely diverse group of proteins with respect to both the size and number of their polypeptide subunits.

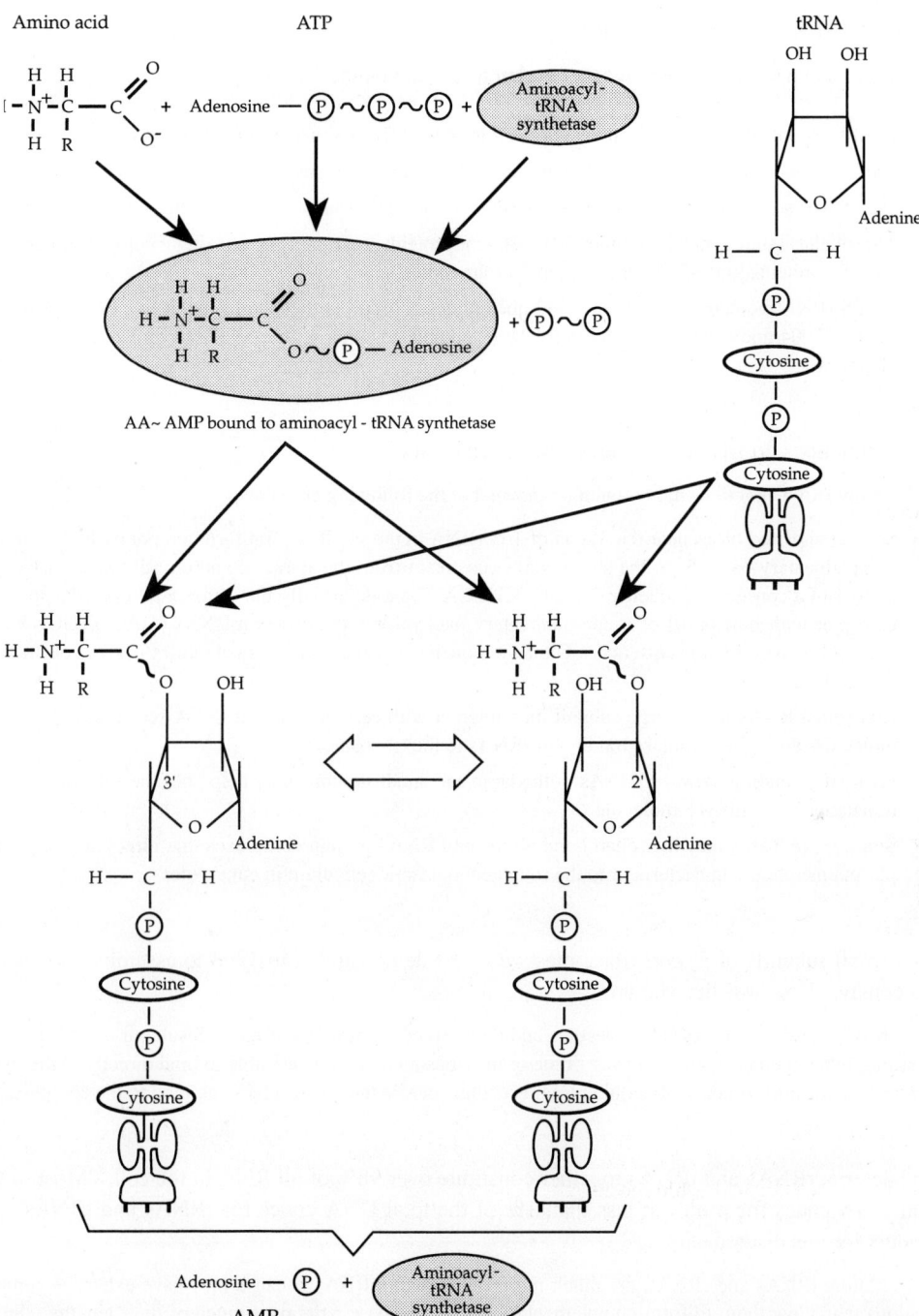

Fig 5-2. Activation of an amino acid and loading of a tRNA molecule are carried out by the same aminoacyl-tRNA synthetase. Attachment of the amino acid can be at either the 2′—OH or 3′—OH positions of the terminal adenine and can alternate rapidly between these two sites.

Ribosomal RNA (rRNA) and Ribosomes

5.18. What are some of the characteristics of a bacterial ribosome?

1. Each bacterial ribosome has a molecular weight of slightly less than 3 million daltons.

2. There are two major subunits: the larger one is 50S; the smaller one is 30S; together they are 70S.

3. The rRNA: protein molecular weight ratio is about 2:1 in bacteria; in most other organisms, the ratio is about 1:1.

4. The small subunit generally contains 21 different proteins, each present in a single copy; the large subunit contains 34 proteins, one of which is present in four copies.

5. A 16S rRNA molecule containing 1,542 nucleotides is found in the small subunit; a 23S rRNA molecule containing 2,904 nucleotides is found in the large subunit, together with a 5S rRNA molecule containing 120 nucleotides.

5.19. What functions do rRNA molecules perform in bacteria?

Many rRNA functions may remain unknown, but the following are known:

1. Several unpaired bases near the 3′ end of 16S rRNA in the small subunit form temporary hydrogen bonds with complementary bases 5′ to the start codons (upstream from the start codons) on mRNAs. These rRNA sites consist of a consensus sequence 3′AUUCCUCCA5′ that is virtually the same in all bacterial species and can base-pair with part or all of a complementary nucleotide sequence in mRNAs (5′AGGAGGU3′) called the **Shine-Dalgarno** (S-D) **sequence.** The S-D sequence is usually 8–13 nucleotides in front of the AUG initiation codon.

2. Ribosomal RNAs in the large subunit may interact with certain invariant tRNA sequences of different tRNA molecules during preliminary binding of tRNAs to the ribosome.

3. Hydrogen bonding between rRNAs in the large and small subunits may help hold the subunits together during translation of an mRNA molecule.

4. Some rRNA bases may hydrogen bond with small RNA–protein complexes that direct nacent proteins to the plasma membrane in bacteria or to the rough endoplasmic reticulum in eukaryotes.

5.20. The small subunits of *E. coli* ribosomes were first demonstrated in 1968 to assemble themselves spontaneously. How was this shown?

Mixing purified 16S rRNA molecules and the purified 21 proteins of the 30S subunit *in vitro* produced particles that functioned normally *in vivo*. Some of the ribosomal proteins are able to bind directly to the rRNA. Others require the prior binding of certain protein subunits before they themselves can bind to the complex.

5.21. In bacteria, rRNAs and tRNAs together constitute over 98% of all RNA in the cell. Most of the bacterial DNA codes for proteins; less than 1% of the total DNA codes for rRNAs and tRNAs. What accounts for this discrepancy?

Mature tRNAs and rRNAS are much more stable than mRNAs because they have extensive complementary regions that allow them to form complementary intrastrand base pairs over much of their lengths. Such double-stranded regions are less subject to cleavage by ribonuclease enzymes; the latter tend to digest mRNAs, which are single-stranded over most of their lengths. Also, only a small number of genes encoding proteins are expressed at any given time in any one cell. In rapidly growing cells, rRNA genes are fully expressed. Amplification of rDNA is explained in Question 3.10.

5.22. Are the transcripts of bacterial genes all made in the same way?

All RNAs are synthesized by enzymes called RNA polymerases. Although mRNA transcripts can function without modification in bacteria, transcripts for rRNAs and tRNAs must be enzymatically cleaved from larger precursor molecules (pre-rRNA, pre-tRNA).

5.23. What names are given to the RNA regions in rRNA and tRNA transcripts that do not function as rRNAs or tRNAs in bacteria?

The expendable segment at the 5′ end of an RNA precursor molecule is called a **leader;** the discarded region at the 3′ end is called a **trailer;** the nonfunctional RNA segments between these end regions are called **spacers.**

5.24. What is the advantage of all rRNAs being produced from the same precursor molecule?

Since each bacterial ribosome contains one each of 23S, 16S, and 5S rRNA molecules, it is advantageous to ensure that these three molecules are produced in equal amounts.

5.25. Is there any advantage for tRNAs to be produced from the same precursor molecule?

As many as seven different tRNAs may be contained in a common pre-tRNA transcript. It is not obvious why some tRNAs should be made from the same transcript, because tRNAs are rarely needed in equal amounts.

5.26. How are tRNAs produced from their precursor molecules?

An endonuclease called **ribonuclease P (RNase P)** removes the 5′ flanking nucleotides by a single cut next to each tRNA in the pre-tRNA molecule. An exonuclease called **RNase D** trims away the extra 3′ nucleotides. What signals the exonuclease to stop when it reaches the CCA end of each tRNA is not known.

5.27. What is unusual about RNase P?

This enzyme is unusual in that it consists of both protein and RNA covalently connected. The RNA molecule contains 377 nucleotides; the small protein has a molecular weight of approximately 20,000. Under physiological conditions, both components are required for the enzyme's activity. However, in nonphysiological buffers containing high concentrations of magnesium ions, the RNA component alone can carry out cleavage as accurately, at nearly the same rate, and without being degraded as the intact enzyme **(holoenzyme).** Thus, the RNA component of RNase P represents a new class of catalytic molecules known as **ribozymes.** The significance of ribozymes in theories about the origin of life are discussed in Chapter 14.

5.28. How many genetic loci code for the tRNAs and rRNAs in *E. coli?*

Usually only one or two specific regions of the bacterial chromosome code for each of the 30 to 40 different pre-tRNA molecules. In *E. coli,* seven complex loci (each containing coordinately controlled sets of genes called operons) code for pre-rRNA molecules. These seven loci differ in only about 1% of their nucleotide sequences. The redundancy of these loci is needed to produce the large numbers of ribosomal RNA molecules required for rapid cell growth. Genetic redundancy can be extensive for various DNA sequences in eukaryotes (especially in the regions that do not code for proteins), but it is uncommon in prokaryotes, this case of rRNAs being an extreme example. Some pre-rRNA molecules also contain tRNAs. Either one or two tRNA genes are located in the spacer between the 16S and 23S sequences; in the trailer region beyond the 5S rRNA, there may be zero, one, or two tRNA molecules (Fig. 5-3).

5.29. How are rRNAs derived from the primary transcripts in *E. coli?*

Complementary nucleotide sequences flanking the 16S, 23S, and 5S rRNAs allow rRNA precursors to form double-stranded regions that are cut by an endonuclease called **ribonuclease III (RNase III).** The double-stranded regions form the stems of "stem and loop" structures. Each loop contains one of the three molecular weight classes of rRNA. Some ribosomal proteins attach to the pre-rRNA before it is completely transcribed. Further 5′ and 3′ trimming of the RNase III cleavage products by other enzymes requires the presence of ribosomal proteins. Thus, the processing of pre-rRNA and the assembly of ribosomal proteins occurs simultaneously.

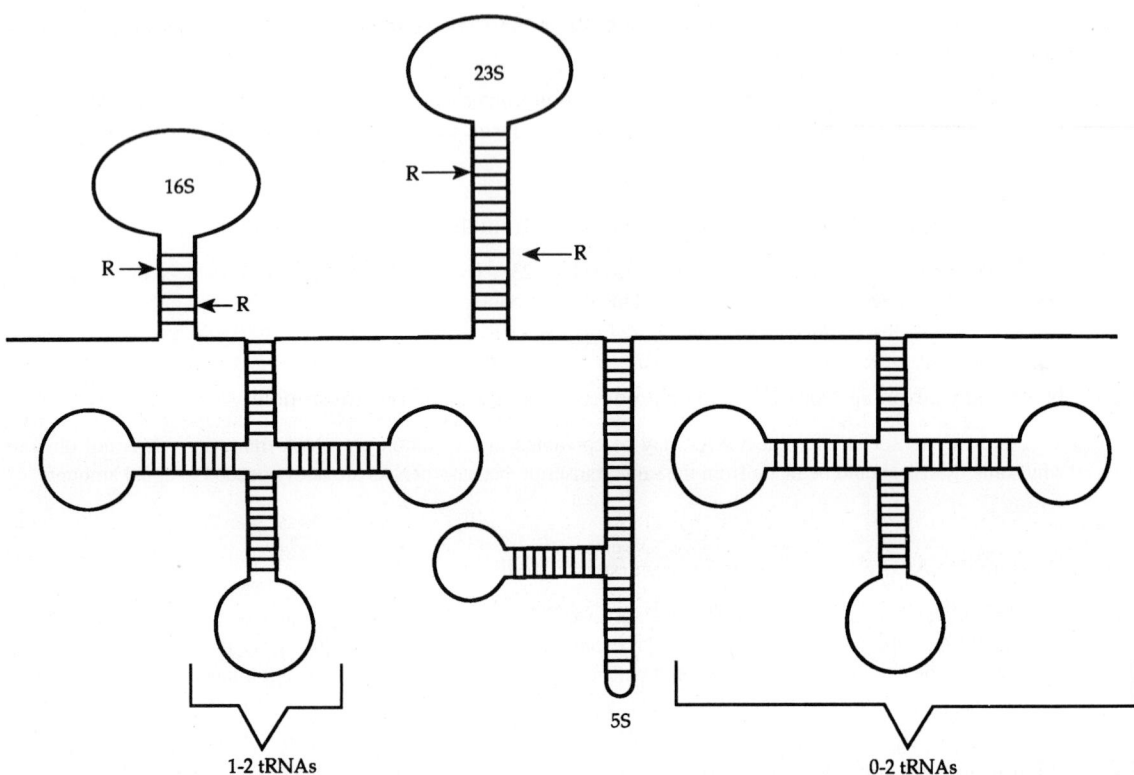

Fig 5-3. An example of a pre-rRNA molecule from *E. coli* that may also contain one or more pre-tRNA molecules. R = sites of primary processing by enzyme ribonuclease III (RIII). *(After Watson, J. D., et al., Molecular Biology of the Gene, Fourth Edition, The Benjamin/Cummings Publishing Co., Inc., 1987.)*

Messenger RNA (mRNA)

5.30. Bacterial mRNA molecules differ considerably in size. Is this also true of tRNAs and rRNAS?

Bacterial tRNA molecules have molecular weights of about 2.5×10^4; rRNA molecules have three defined sizes (4×10^4, 5×10^5, and 10^6 for 5S, 16S, and 23S molecules, respectively). In contrast to tRNAs and rRNAs, mRNAs vary greatly in their molecular weights, depending on two major factors: one is the length of each structural gene, and the other is the number of structural genes encoded by the mRNA molecule. Some bacterial mRNAs are **monocistronic,** i.e., the mRNA molecule contains a single structural gene that codes for a single polypeptide chain. Many bacterial mRNAs, however, are **polycistronic** (containing two or more structural genes coding for an equivalent number of polypeptide chains).

5.31. Unlike rRNA and tRNA molecules, bacterial mRNA molecules are not cleaved from larger precursor molecules and so do not leave leader, spacer, and trailer fragments behind. Do mRNA molecules contain leader, spacer, and trailer sequences that are not translated into proteins?

The 5′ end of each bacterial mRNA has a nontranslated leader region containing a Shine-Dalgarno nucleotide sequence that serves as a recognition site for binding to the 3′ end of the 16S rRNA in the small 30S ribosomal subunit. The 3′ end of each mRNA contains a nontranslated trailer region, the function of which is unknown in bacteria; in eukaryotes the trailer region contains the binding site for the enzyme that adds a poly-A tail to the mRNA before it leaves the nucleus. Most (if not all) of the polypeptides encoded by the same bacterial mRNA molecule are usually involved in the same metabolic pathway. For example, the five enzymes required for synthesis of the amino acid tryptophan from chorismic acid are encoded by nucleotide sequences in the same polycistronic mRNA; each of these sequences is separated from the others in the same mRNA by noncoding segments called spacers.

5.32. Suppose that hemoglobin-synthesizing reticulocytes (immature red blood cells) are briefly exposed to radioactively labeled amino acids, and this is followed by immediate isolation of newly completed (free, unattached to ribosomes) hemoglobin chains. The time of labeling is short, relative to the time required to synthesize a complete chain. When fragments of these chains are assayed, a gradient of increasing radioactivity is observed from their amino ends to their carboxyl ends. What do these results indicate?

　　Most hemoglobin chains were already at least partly synthesized before the brief exposure to radioactive amino acids. Relatively few chains would have been initiated during the brief interval from radiation exposure to the isolation and analysis phases. The fact that more radioactivity is found at the carboxyl ends of completed chains indicates that translation proceeds from the amino terminus toward the carboxyl end of these polypeptide chains.

5.33. How can the bacterial mRNA translation initiation codon (AUG) encode *N*-formylmethionine, whereas internal AUG codons specify methionine?

　　There are two types of bacterial methionine adaptors (tRNA molecules). A formyl group (COH) is added by a **transformylase** enzyme after the methionine has become attached to the adaptor molecule, symbolized tRNA^{fMet}. The name of the aminoacyl tRNA loaded with its cognate amino acid is usually given as fMet-tRNA^{fMet}. The other charged adaptor, symbolized Met-tRNA^{Met}, is not recognized by transformylase. The symbol Met-tRNA^{Met} is sometimes used to indicate that the tRNA has been correctly loaded with its cognate amino acid (methionine) rather than incorrectly loaded with some other amino acid such as aspartic acid (Asp-tRNA^{Met}). During the initiation phase of translation, a protein translation initiation factor recognizes only the loaded fMet-tRNA^{fMet}, thus accounting for the specificity of formylated methionine at the 5′ ends of all bacterial primary translation products. Translation elongation factors promote the use of tRNA^{Met} (but not tRNA^{fMet}) at internal AUG codons.

5.34. How can mRNA 5′GUG3′ occasionally function as a translation initiation signal?

　　The 5′AUG3′ mRNA codon is normally used to initiate translation by complementary base pairing with the 3′UAC5′ anticodon of tRNA^{fMet}. Wobble is normally restricted to the 3′ end of mRNA codons and the 5′ end of anticodons. An unusual wobble exists at the 5′ end of the initiation codon that necessarily involves the 3′ base of the anticodon. The valine codon 5′GUG3′ may therefore be able to serve as the initiation (start) codon in bacteria, but it is used about tenfold less than 5′AUG3′. Even the 5′UUG3′ and 5′CUG3′ codons may initiate translation in bacteria, but they are used even more rarely than 5′GUG3′. The ability of tRNA^{fMet} to participate in 5′mRNA wobble may be due to the fact that an unmodified adenine lies adjacent to the 3′ end of its anticodon. All other tRNAs have an alkylated derivative at that site (e.g., 1-methylinosine in alanine tRNA; see Figs. 2-16 and 2-17).

5.35. Despite the fact that all bacterial proteins initiate synthesis with fMet at their N termini, many functional (fully formed) proteins start with either a nonformylated methionine or some other amino acid. What is the explanation for these facts?

　　In some cases, a **deformylase** enzyme removes the formyl group from the growing polypeptide chain soon after chain synthesis commences. In other cases, a Met-specific aminopeptidase enzyme removes the N-terminal methionine, and the processed polypeptide begins with the "second" amino acid introduced into the growing polypeptide chain.

5.36. List all of the components of the 70S translation initiation complex.

　　(1) 30S ribosomal subunit, (2) fMet-tRNA^{fMet}, (3) mRNA, (4) 50S ribosomal subunit, (5) three protein initiation factors (IF1, IF2, IF3), (6) guanosine triphosphate (GTP; Fig. 11-2*b*).

5.37. How does a bacterial ribosome get started at the 5' end of a mRNA molecule and position the initiator codon 5'AUG3' in the correct orientation for complementary base pairing with the anticodon of the initiator tRNA?

Each bacterial mRNA molecule has one ribosome binding site for each of its independently synthesized polypeptide products. Ribosomes binding to mRNA molecules protect a terminal segment around the AUG start codon from ribonuclease digestion. The protected segment is about 30 nucleotides long. The AUG start (initiator) codon is approximately in the middle of this protected segment. Centered approximately 8 to 13 nucleotides upstream from the start codon is the Shine-Dalgarno (S-D) sequence—rich in purines and usually containing AGGA or GAGG—that constitutes the ribosome binding site. The S-D sequence of mRNA pairs with a complementary pyrimidine-rich sequence at the 3' end of a 16S rRNA chain in the 30S ribosomal subunit. This interaction places the start codon of the mRNA in the proper register for hydrogen bonding with the anticodon of the initiator tRNA. These two sets of base pairings distinguish a true initiator region from other sites containing internal AUG codons.

5.38. There are four major steps in the formation of the 70S initiation complex. What events happen in each step?

In the first step (Fig. 5-4a,b), three protein initiation factors (IF1, IF2, IF3) and GTP bind to the 30S ribosomal subunit. IF3 assists the 30S ribosomal subunit in binding to the Shine-Dalgarno (S-D) sequence of the mRNA. Binding of the 50S subunit to the mRNA is inhibited by IF3. Then (Fig. 5-4b) GTP binds to F2, and is stabilized by EF1. The N-(fMet-tRNA)-F2-GTP complex binds to the 30S-F3-mRNA complex. IF3 is released (Fig. 5–4b,c). This completes the **30S initiation complex.** In the final step (Fig. 5-4c,d), hydrolysis of GTP releases IF1, IF2, GDP and phosphate, allowing the binding of the 50S subunit. The final aggregation is a **70S initiation complex** (Fig. 5-4d).

5.39. Which subunit of a ribosome forms the A and P sites? Does the initiation tRNA first enter the A site or the P site?

Each ribosome has two tRNA-binding sites (Fig. 5-5a): one is the peptidyl (P) site; the other is the aminoacyl (A) site. Each of these sites is formed partially by the 30S and 50S subunits. The mRNA codon in the A site makes the tRNA binding site specific for a unique tRNA molecule.

It is not known if the initiating tRNA^{fMet} enters the A site or the P site. However, all AA $\sim$ tRNAs must be in the P site before the second AA $\sim$ tRNA can bind to the ribosome (Fig. 5-5b).

5.40. When a peptidyl-tRNA translocates from the A to the P site on a ribosome, it leaves the A site empty. What events must occur before an AA $\sim$ tRNA enters the vacant A site?

Elongation factor Tu (EF-Tu) reacts with GTP and the AA $\sim$ tRNA (loaded with glycine as an example) to form a complex (Fig. 5-5b). EF-Tu can bind to every AA $\sim$ tRNA except tRNA^{fMet}. The AA $\sim$ tRNA fraction enters the A site with release of a free EF-Tu-GDP complex and phosphate. **Elongation factor Ts (EF-Ts)** displaces GDP from EF-Tu by itself combining with EF-Tu. When the (EF-Tu)–(EF-Ts) complex combines with GTP, and (EF-Tu)–GTP complex is formed, releasing EF-Ts. Thus (EF-Tu)–GTP is regenerated to bind to another AA $\sim$ tRNA, and the cycle repeats.

5.41. How does a peptidyl-tRNA (AA $\sim$ tRNA) move from the A site to the P site during the elongation phase of translation?

The process whereby a peptidyl-tRNA (and the mRNA codon to which it is bound) move from the A site to the P site is termed translocation. It occurs subsequent to the formation of the peptide bond (Fig. 5-5c,d). The amino acid attached to the tRNA in the A site acts as the receptor for the peptide held by the tRNA in the P site. Then the peptidyl-tRNA in the A site moves to the P site with the help of the enzyme peptidyl transferase (an integral component of the 50S subunit), **elongation factor G (EF-G; also known as translocase),** and the hydrolysis

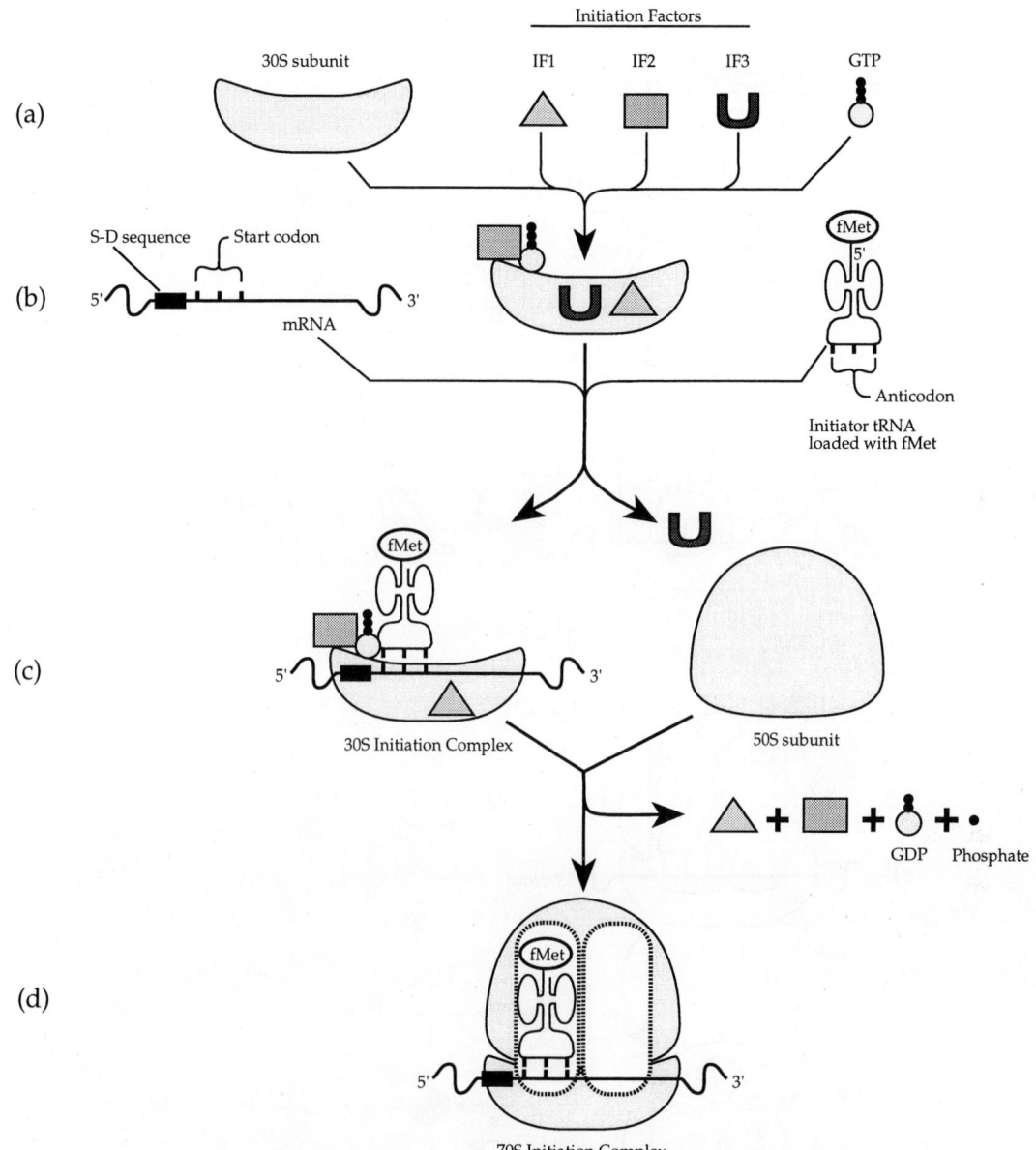

Fig 5-4. Diagram of the initiation phase of translation in bacteria.

of GTP. Coinciding with translocation, the mRNA also moves one codon through the 30S ribosomal subunit and the unloaded tRNA exits the P site.

5.42. How is EF-G (translocase) recycled?

After the P site becomes vacant (Fig. 5-5c), an (EF-G)−GTP complex binds to the 70S ribosome. This stimulates the hydrolysis of GTP to GDP and phosphate (Fig. 5-5d). Upon translocation of the peptidyl-tRNA to the P site, EF-G and GDP are released. EF-G becomes reactivated by union with GTP to form a new (EF-G)−GTP complex.

5.43. Is there a way by which polypeptide chain extension can be easily disrupted at random to produce incomplete polypeptide chains of varying lengths?

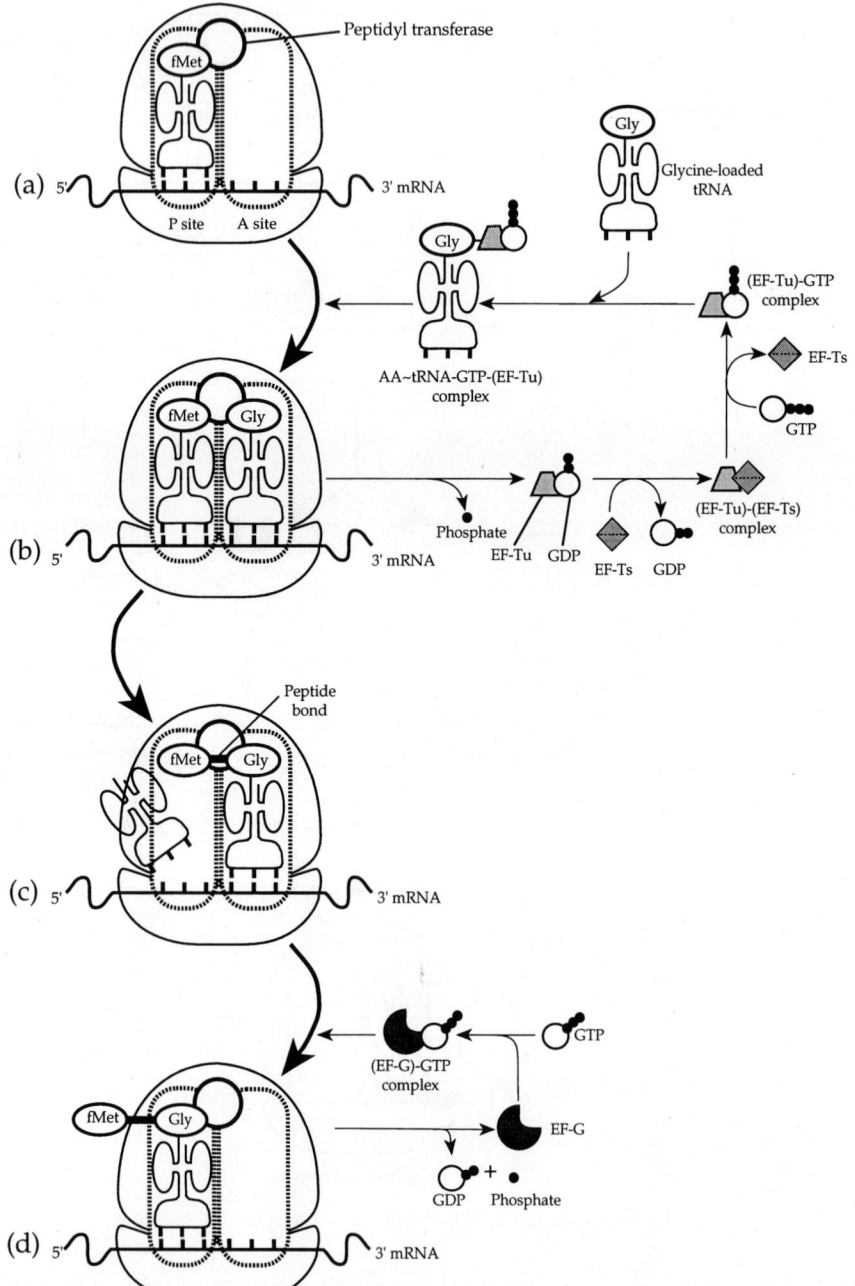

Fig 5-5. Diagram of the elongation phase of translation in bacteria. (*a*) fMet-tRNA in the P site; A site vacant. (*b*) Entry of gly-tRNA (as an example) into the A site and regeneration of the (EF-Tu)–GTP complex by EF-Ts. (*c*) Peptide bond formation and departure of unloaded fMet-tRNA from the P site. (*d*) Peptidyl-tRNA translocation from the A site to the P site, aided by EF-G and GTP; ready for the third AA~tRNA complex to enter the A site for another round of elongation.

The antibiotic **puromycin** is a powerful inhibitor of growth of all cells. Since it is not specific against microbes, it cannot be used as a therapeutic antibiotic. But it can be used to study translation. Part of its structure resembles the amino end of an amino acid. It also binds to the A site of a ribosome and prevents the entry of a normal AA $\sim$ tRNA by **competitive inhibition.** Peptidyl transferase transfers nascent polypeptide chains in the P site to puromycin in the A site. Puromycin is smaller than an AA $\sim$ tRNA molecule and binds only weakly to the A site. Thus the puromycin-terminated chains dissociate from the ribosome, creating incomplete polypeptide chains of varying lengths. When these chains are analyzed, puromycin is found at their carboxyl ends, providing additional proof that translation proceeds from the amino end toward the carboxyl end.

5.44. What are some other examples of how antibiotics that block protein synthesis affect translation in bacteria? How is it possible for certain antibiotics to inhibit or kill bacteria without harming eukaryotic cells?

    Fusidic acid prevents the release of the (EF-G)–GTP complex after it has functioned in translocation, thereby blocking subsequent translocations. **Chloramphenicol** inhibits peptidyl transferase from forming the peptide bond. **Streptomycin** binds to the 30S subunit and interferes with chain initiation.

    The complexity of the translation process offers many points at which various drugs and antibiotics can interfere to abort or mutate the process. Furthermore, there are several differences between prokaryotic and eukaryotic translation components and processes, so that some of these antibiotics can be directed specifically against bacteria while leaving eukaryotic cells relatively unaffected.

5.45. When do polypeptide chains fold up into their tertiary shapes?

    Nascent polypeptide chains begin folding during the elongation phase of their synthesis. The final shape of a polypeptide chain may be almost completed before the chain has terminated its synthesis. If the completed polypeptide will function as an enzyme, some nearly completed chains may exhibit catalytic activity even before they are released from their ribosomes.

5.46. Describe the process of polypeptide chain termination in bacteria.

    At least one of three mRNA stop codons must be present to terminate translation in bacteria. A stop codon immediately follows the codon specifying the last amino acid in the polypeptide chain. No tRNA recognizes any of these stop codons. Instead, they are recognized by protein release factors (RFs). The stop codons UAG and UAA are recognized by RF1; stop codons UGA and UAA are recognized by RF2. These two RFs induce peptidyl transferase to transfer the growing polypeptide chain to water, thereby releasing the polypeptide chain from the tRNA in the P site of the ribosome. A third release factor (RF3), believed to be identical to IF3, appears to stimulate the 30S and 50S subunits to dissociate from the mRNA and from one another. The uncharged tRNA is released as the P site disappears due to the dissociation of ribosomal subunits.

5.47. How do the roles of ATP and GTP differ in translation?

    ATP is used to generate the high-energy aminoacyl bond that connects an activated amino acid to its tRNA adaptor molecule. By contrast, GTP is not used in the formation of any covalent bonds. Instead, it plays a role in the formation of noncovalent binding of various translation factors (IF2, EF-G, EF-Tu). It is thought that the presence of GTP induces the translation factors to undergo a change of shape that is necessary for their binding to ribosomes. Cleavage of GTP to GDP allows these translation factors to return to their original shape and to dissociate from ribosomes.

5.48. What order of magnitude is the error rate in translation?

    About one amino acid is misincorporated into polypeptide chains for every 10^4 amino acids polymerized. Thus, only about one in every 25 protein molecules of average size (400 amino acids) is likely to contain an error (which may or may not affect the protein's function).

5.49. What mechanisms contribute to the fidelity of translation?

One recognition site of an aminoacyl tRNA synthetase has a high (but not absolute) degree of specificity for the side chain of only one kind of amino acid. Another site on that same synthetase molecule has specificity for all of the isoadaptors that carry the same amino acid. If an incorrect activated amino acid becomes attached to a tRNA molecule, the synthetase will normally remove it by hydrolysis before it has a chance to combine with an elongation factor. If the wrong AA ~ tRNA–(EF-Tu)–GTP complex manages to get into the A site of a ribosome, abnormal codon recognition causes the GTP component to be hydrolyzed to GDP and phosphate, thus allowing EF-Tu to dissociate from the complex. The weak hydrogen bonding of an incorrect tRNA to a mRNA codon would usually cause that tRNA to leave the A site rather quickly. The short delay between codon-anticodon base pairing and formation of a peptide bond provides an opportunity for an incorrect AA ~ tRNA to exit without having extended the polypeptide chain. This process is referred to as "proofreading." **Proofreading** is defined as any mechanism for correcting an error in protein or nucleic acid synthesis during polymerization.

5.50. How many high-energy bonds must be broken for each amino acid added to a polypeptide chain?

At least three high-energy bonds must be broken if no editing is required. Activation of an amino acid must occur before it can become attached to its tRNA adaptor. The energy for this step is produced by cleavage of ATP to ADP. One molecule of GTP is hydrolyzed in the release of the (EF-Tu)–GDP complex. Finally, one GTP molecule is hydrolyzed from the EF-G complex following translocation of the peptidyl-tRNA from the A site to the P sites of the ribosome.

5.51. Is there a mechanism to prevent cells from wastefully producing more ribosomes than are needed?

When bacterial cells are starved for amino acids, large amounts of two unusual guanine nucleotides (guanosine tetraphosphate, pp5'-G-3'pp, also called "magic spot," and guanosine pentaphosphate, ppp5'-G-3'pp) accumulate. An enzyme called **stringent factor** uses ATP as a pyrophosphate (P ~ P) donor to convert GDP (pp5'-G-3'OH) into magic spot when an uncharged tRNA molecule occupies an A site. Normally, only aminoacyl-tRNAs are placed in the A site by EF-Tu. But when aminoacyl-tRNAs are not available, uncharged tRNAs can enter the A site. Magic spot specifically causes a 10- to 20-fold reduction in the synthesis of tRNA and rRNA molecules and about a three-fold reduction in mRNA synthesis in what is called the **stringent response.** The rate of protein degradation is increased, and reduced synthesis of nucleotides, carbohydrates, and lipids occurs. These unusual guanine nucleotides specifically inhibit the initiation of transcription at the promoters of bacterial operons coding for rRNAs. They also inhibit the elongation phase of transcription for many templates, but the specificity of such inhibitions remains elusive.

5.52. How many ribosomes can simultaneously be translated on a mRNA molecule?

The collection of ribosomes bound to a given mRNA molecule is called a **polysome** or **polyribosome.** The size of a polysome depends on the length of the mRNA and the frequency with which ribosomes can attach to the Shine-Dalgarno (S-D) ribosome binding site at the 5' end of the bacterial mRNA. Variations in nucleotide sequences within the S-D region make some mRNAs very attractive to ribosomes; other sequences are only weakly attracted to ribosomes. Even though a single ribosome makes contact with only about 30 mRNA nucleotides, the maximum utilization of a mRNA chain is only about one ribosome for every 80 mRNA nucleotides.

5.53. Messenger RNA constitutes only about 1 to 2% of the total cellular RNA. Yet some proteins need to be produced in large amounts. How can this be accomplished?

Since each mRNA can be simultaneously translated by multiple ribosomes, many polypeptides can be synthesized from one mRNA molecule. Also, the more ribosomes that are bound to a mRNA molecule, the less likely it is to be degraded by cellular RNases. The average half-life of a mRNA molecule is only about 1–2 minutes. If an average bacterial protein contains about 300 amino acids, it could be synthesized in about 20 seconds (15 amino acids per second). The rate of bacterial transcription under ideal culture conditions (14 codons per second) is about the same as the translation rate. If ribosomes can follow one another on the same mRNA molecule as closely as 80 nucleotides, then a mRNA consisting of 900 nucleotides could be simultaneously translated by about a dozen ribosomes.

5.54. How do proteins reach their correct destinations in bacterial membranes and in the periplasmic space?

Most proteins destined for the plasma membrane or beyond are synthesized with 15 to 30 additional amino acids at their amino termini that are not present in mature or properly positioned proteins. This end sequence is rich in uncharged (usually hydrophobic) amino acids and is referred to as the **signal sequence.** It is hypothesized that the signal sequence and an adjacent region each fold into short helical segments that bend into an antiparallel helical pattern. This structure becomes stuck in the hydrophobic lipid bilayer of the inner membrane.

The ribosome that translates a protein bearing a signal sequence (peptide) would be bound to the membrane because hydrophobic sequences tend to avoid water by burying themselves in lipid membranes. Once the N-terminus has anchored the polypeptide in the membrane, the remainder of the chain is extruded through the membrane to the other side. If a second hydrophobic region is translated, it too may become stuck in the membrane. The signal sequence is usually trimmed off by a specific **signal peptidase** enzyme. If no hydrophobic region is synthesized other than that at the N terminus, the trimmed polypeptide will be released free into the periplasm. It is not known how proteins destined for the outer membrane are specifically targeted.

TRANSLATION IN EUKARYOTES

The process of translation in eukaryotes is essentially the same as that in bacteria, but differs in several important ways.

5.55. How do the ribosomal subunits of eukaryotes differ in molecular weight from those of prokaryotes?

Structurally, each eukaryotic ribosome consists of a 40S and a 60S subunit that together form an 80S complex. Thus, eukaryotic ribosomes are larger than their bacterial counterparts (30S + 50S = 70S).

5.56. Are transcription and translation in eukaryotes coupled processes, as they are in bacteria?

Transcription and translation are coupled processes in prokaryotes, mitochondria, and chloroplasts. That is to say, as soon as the 5′ end of a bacterial mRNA is transcribed, ribosomes can attach and begin translation. However in eukaryotes, the primary transcript must be processed into a functional mRNA molecule and then must be transported from the nucleus to the cytoplasm before translation can begin (Chapter 4). Thus, transcription and translation are not coupled processes for eukaryotic nuclear (chromosomal) genes.

5.57. Can a eukaryotic mRNA molecule code for more than one protein?

Although most bacterial mRNAs specify multiple proteins before protein processing occurs, all eukaryotic mRNAs code for a single nascent polypeptide chain. However, some newly synthesized polypeptide chains may subsequently be enzymatically cleaved into two or more functional protein components. A multicomponent protein such as this is termed a **polyprotein.**

> **Example 5.1.** A polyprotein called pro-opiomelanocortin is synthesized by the anterior lobe of the pituitary gland. A cut near the carboxyl (C) terminus first produces β-lipotropin. Then a cut near the amino terminus produces adrenocorticotropic hormone (ACTH). In the intermediate lobe of the pituitary, β-lipotropin is further digested, releasing the C-terminal peptide β-endorphin; the ACTH is cleaved to release α-melanotropin.

5.58. How does the initiation of translation in eukaryotes differ from that in prokaryotes?

Whereas only three well-defined initiation factors are required for translation of *E. coli* mRNAs, many more are needed in eukaryotes. Eukaryotic initiation factors are designated **eIFs** to distinguish them from their bacterial counterparts (IFs). During the initiation phase, a special initiator tRNA (tRNA^{iMet}) brings an unformylated methionine into the first position on the ribosome. In eukaryotes there is no specific ribosomal binding site comparable to the Shine-Dalgarno sequence of prokaryotes. Instead, the 40S ribosomal subunit is thought to attach to the mRNA at its capped 5′ terminus, and then it slides along (consuming ATP) until it reaches the first AUG start codon and begins translation. Only AUG is an efficient initiator in eukaryotes, whereas GUG, UUG, and CUG may sometimes be used in *E. coli,* albeit at much lower frequencies. In prokaryotes, IF1, IF2, IF3, and mRNA all bind to the 30S ribosomal subunit before the initiator aminoacyl-tRNA is bound. In eukaryotes, however, the initiator aminoacyl-tRNA, GTP, and eIF2 form a complex before being bound to the 40S ribosomal subunit.

5.59. How do the elongation and termination phases of eukaryotic translation differ from those of prokaryotes?

Three different elongation factors in eukaryotes replace those found in bacteria. However, a single termination factor or release factor (RF) in eukaryotes replaces RF1 and RF2 of bacteria. RF recognizes all three stop codons (UAG, UAA, and UGA).

5.60. What are molecular chaperones? What are some of the functions they may perform in eukaryotes?

Many proteins spontaneously fold into their proper tertiary shapes by themselves. Some proteins, however, need to interact with various polypeptides to attain their correct folded shapes. These helper peptides are called **molecular chaperones.** Some of these chaperones are **heat-shock proteins** (Hsps) that are synthesized by eukaryotic cells in response to stresses such as an increase in temperature. However, some Hsps are present in cells under normal, nonstressful conditions, and they may be essential for life. Chaperones may also prevent polypeptide chains from making premature or nonproductive associations in multisubunit proteins. Some chaperones bind to nascent polypeptide chains while they are being synthesized on ribosomes, and they may help the chains move out of the tunnel of the eukaryotic 60S ribosomal subunit. Other chaperones may keep the translating polypeptide in an unfolded conformation, thus aiding its passage across membranes, as into the endoplasmic reticulum or into mitochondria. Once across the appropriate membrane, the polypeptide is passed off to a different chaperone that will help it fold correctly into a functional protein.

Objective Questions

Multiple Choice

Directions: Choose the one best answer.

1. During the initiation phase of translation in bacteria, which of the following is first to dissociate from the 30S ribosomal subunit? (*a*) IF1 (*b*) IF2 (*c*) IF3 (*d*) GTP (*e*) tRNA

2. The maximal rate of protein synthesis in bacterial cells (amino acids per second) is (*a*) approximately 0.5. (*b*) 3–4. (*c*) 12–15. (*d*) 30–40. (*e*) 75–100.

3. How many polypeptide chains can be formed simultaneously by a given ribosome? (*a*) One (*b*) about a dozen (*c*) up to 30 (*d*) variable, depending on the length of the mRNA (*e*) variable, depending both on length of mRNA and temperature

4. Which of the following amino acids is specified by only a single codon? (*a*) Glutamine (*b*) tryptophan (*c*) asparagine (*d*) isoleucine (*e*) none of the above

5. Without referring to Table 5.1, answer the following question. Which of the following single nucleotide mutations in a mRNA codon is most likely to be a "silent" mutation? (*a*) 5′CAU3′ → CUU (*b*) 5′UUU → UUC (*c*) 5′AGA → CGA (*d*) 5′GUG → GGG (*e*) 5′UAU → UAG

6. Using Table 5.1, answer the following question. What amino acids would be coded for by the repeating synthetic mRNA heteropolymer 5′ . . . CGACGACGA . . . 3′ in a bacterial cell-free system? (*a*) Only poly-arginine (*b*) only poly-asparagine (*c*) only poly-threonine (*d*) three homopolymers of arginine, asparagine, and threonine (*e*) none of the above

7. The drug chloramphenicol blocks (*a*) cell-wall formation. (*b*) transcription. (*c*) translation termination release factors. (*d*) polypeptide chain elongation. (*e*) polypeptide chain initiation.

8. What percentage of RNA in most bacterial cells is mRNA? (*a*) 75–85% (*b*) 50–60% (*c*) 35–40% (*d*) 10–15% (*e*) less than 5%

9. An antibiotic that resembles the 3′ end of a charged tRNA molecule is (*a*) streptomycin. (*b*) penicillin. (*c*) sparsomycin. (*d*) puromycin. (*e*) tetracycline.

10. Which of the following is *not* characteristic of eukaryotes? (*a*) Ribosomes are larger than those of prokaryotes. (*b*) All mRNAs specify a single protein or a polyprotein. (*c*) No specific ribosome binding sites exist comparable to those of the S-D sequences in prokaryotes. (*d*) Methionine rather than *N*-formylmethionine initiates all nascent polypeptide chains. (*e*) More release factors are required than those in prokaryotes.

True–False

1. Each amino acid must become activated before it can be loaded onto its cognate tRNA. Both the activation and loading steps are catalyzed by the same enzyme.

2. At no time during translation does an amino acid side chain group make contact with the mRNA template.

3. Every tRNA ends with CCA5′.

4. The bond between a tRNA and its specific amino acid is a covalent, high-energy bond.

5. Peptide bonds are formed by an enzyme that is part of the 30S ribosomal subunit.

6. Isoacceptor tRNA molecules that bind to different codons are activated by the same aminoacyl-tRNA synthetase.

7. When transferred to the growing polypeptide chain, each amino acid must leave its tRNA from the 2′ position of ribose of the terminal A.

8. Three different rRNA molecules are present in every bacterial ribosome.

9. The various rRNAs and protein components of a ribosome are assembled with the aid of several different enzymes.

10. In bacterial cells, transcription and translation are coupled processes; i.e., a mRNA can be translated at the same time it is being transcribed.

11. The 3′ end of a mRNA molecule specifies to the N-terminus of its polypeptide product.

12. The 30S and 50S ribosomal subunits must be assembled before mRNA can be bound.

13. Messenger RNAs are enzymatically degraded more rapidly than rRNAs or tRNAs.

14. The 16S, 23S, and 5S rRNA molecules are each synthesized as individual larger transcripts that are later trimmed to their functional size.

15. Some pre-rRNA molecules also contain tRNA molecules.

16. Proteins are the only macromolecules that can bind a substrate and catalyze a biochemical reaction.

17. Genetic redundancy is rare within a prokaryotic genome.

18. A regulatory system is required to ensure that 23S, 16S, and 5S rRNA molecules are each made in the precise ratios needed for ribosome synthesis.

Matching

Directions: Match each item in **Column A** with the one in **Column B** to which it is most closely associated. Each item in **Column B** can be used only once.

Column A	Column B
1. homopolymer of phenylalanine	*A.* third position of codons
2. peptidyl transferase	*B.* initiation methionine
3. aminoacyl-tRNA synthetase	*C.* binding to 30S ribosomal subunit; interfering with chain initiation
4. ribonuclease P	
5. puromycin	*D.* amino acid activation
6. streptomycin	*E.* 3′ end of tRNA molecules
7. wobble	*F.* prevents entry of AA ~ tRNA into A site of ribosome
8. transformylase	
9. ribonuclease III	*G.* peptide bond formation
10. CCA	*H.* removal of tRNAs from their precursor molecules
	I. removal of rRNAs from primary transcripts
	J. poly-U

Terms

Directions: Unless otherwise specified, each answer is a single word.

1. The process of attaching an amino acid to its tRNA.

2. The process of converting a mRNA nucleotide sequence into an amino acid sequence of a polypeptide chain.

3. Structurally different transfer RNA molecules that recognize the same amino acid.

4. The nucleotide sequence of a pre-rRNA preceding the 16S rRNA.

5. The nucleotide sequences between the 16S and 23S rRNAs or between the 23S and the 5S rRNAs in pre-rRNA transcripts.

6. The nucleotide sequence following the 5S rRNA in pre-rRNA transcripts.

7. An adjective describing any mRNA molecule that codes for more than a single polypeptide chain.

8. The nucleotide sequence on a bacterial mRNA molecule that is recognized by ribosomes. (2 names, hyphenated)

9. The translational steps that reduce the error rate of incorporation of incorrect amino acids into a polypeptide chain.

10. An adjective describing the response that occurs when cells are starved for amino acids and when large amounts of the unusual nucleotide pp5′-G-3′pp or "magic spot" accumulate.

11. RNA molecules with catalytic ability.

12. A series of hydrophobic amino acids at the N terminus of a protein that targets that protein for delivery to the cell membrane or beyond. (2 words)

13. The high-energy compound required during the initiation and elongation phases of translation. (2 words or 3-letter acronym)

14. A synonym for elongation factor G (EF-G).

The Translation Story

Directions: Specify the appropriate term, word, number, or letter at each parenthetical number that best completes the story.

The monomer used in the formation of proteins is called an **(1)**. The template on which proteins are made is **(2)**. Each **(1)** must first become **(3)** by cleavage of a molecule of **(4)**. Each kind of **(1)** becomes specifically attached to its own adaptor molecule, called a **(5)**. The attachment is catalyzed by an enzyme called an **(6)**. All **(5)**s can be represented by a model in the form of a **(7)**, and each contains **(8**; how many?**)** major loops. One of these loops contains a **(9**; how many?**)**-nucleotide sequence called the **(10)**, which base-pairs with a complementary sequence in the **(2)** called a **(11)**. Each **(3)** **(1)** is attached by its **(12)** group to the **(13)** nucleotide at the **(14)** end of its **(5)** molecule by a process called **(15)**. Structures called **(16)** act as jigs to hold the **(2)** and the **(5)** together during protein synthesis. Bacterial **(16)** consist of about two-thirds **(17)** and one-third protein. The size of the larger subunit of bacterial **(16)** is **(18**; a number**)**S and that of the smaller subunit is **(19**; a number**)**S. The symbol S is in honor of **(20)**, the inventor of the **(21)**. Each polypeptide chain grows from its **(22)**-terminal end toward its **(12)** end. Individual **(16)** have two **(5)** binding sites. Each **(1)** ~ **(5)** complex first enters the site designated by the letter **(23)**, allowing the subsequent formation of a covalent **(24)** bond between adjacent **(1)**s, with the growing chain held in the site designated by the letter **(25)**. Formation of the **(24)** bond transfers the nascent chain to the **(5)** in the **(23)** site. A process called **(26)** moves the **(5)** carrying the nascent chain from the **(23)** site to the **(25)** site, advances the **(2)** to its next **(11)**, and ejects the unloaded **(5)** from the **(25)** site. Another cycle can then commence. The translational process consists of three major stages. The first stage is called **(27)**, the second stage is called **(28)**, and the third stage is called **(29)**. During the first two stages, the high-energy compound **(30)** is converted to **(31)**. All three stages require the help of more than one protein translation factor. All bacterial polypeptide chains begin with the **(1)** **(32)** that has been enzymatically modified by the addition of a **(33)** group. The **(29)** stage requires the help of at least two proteins called **(34)** factors, each of which recognizes **(35**; how many?**)** of the **(9)** possible **(36)** **(11)**s. The term **(37)** refers to the complex of a **(2)** molecule being simultaneously translated by multiple **(16)**.

Answers to the Objective Questions

Multiple Choice

1. *c*   **2.** *c*   **3.** *a*   **4.** *b*   **5.** *b*   **6.** *d*   **7.** *d*   **8.** *e*   **9.** *d*   **10.** *e*

True-False

1. T   **2.** T   **3.** F (CCA3′)   **4.** T   **5.** F (50S subunit)   **6.** T   **7.** F (3′ position)   **8.** T **9.** F (self-assembly)   **10.** T   **11.** F (5′ end)   **12.** F (mRNA binds to 30S subunit before the addition of the 50S subunit)   **13.** T   **14.** F (they are all synthesized as parts of a single transcript)   **15.** T   **16.** F (e.g., the RNA component of RNase P can do this)   **17.** T   **18.** F (one molecule of each class is cleaved from a common pre-rRNA molecule)

Matching

1. *J*   **2.** *G*   **3.** *D*   **4.** *H*   **5.** *F*   **6.** *C*   **7.** *A*   **8.** *B*   **9.** *I*   **10.** *E*

Terms

1. loading or charging   **2.** translation   **3.** isoacceptors   **4.** leader   **5.** spacers   **6.** trailer
7. polycistronic   **8.** Shine-Dalgarno   **9.** proofreading   **10.** stringent   **11.** ribozymes
12. signal sequence   **13.** GTP, or guanosine triphosphate   **14.** translocase

The Translation Story

1. amino acid   **2.** messenger RNA (mRNA)   **3.** activated   **4.** adenosine triphosphate (ATP)
5. transfer RNA (tRNA)   **6.** aminoacyl-tRNA synthetase   **7.** cloverleaf   **8.** three (3)   **9.** three (3)
10. anticodon   **11.** codon   **12.** carboxyl (COOH)   **13.** adenine   **14.** 3′   **15.** loading or charging
16. ribosomes   **17.** rRNA   **18.** 50   **19.** 30   **20.** Svedberg   **21.** ultracentifuge
22. amino or N   **23.** A   **24.** peptide   **25.** P   **26.** translocation   **27.** initiation   **28.** elongation
29. termination   **30.** guanosine triphosphate (GTP)   **31.** guanosine diphosphate (GDP)   **32.** methionine
33. formyl (CHO)   **34.** release   **35.** two (2)   **36.** stop (termination; nonsense)
37. polysome or polyribosome

Chapter 6

Mutations

INTRODUCTION

Lewis Thomas, writing about DNA in his book *The Medusa and the Snail,* penned the following passage: "The capacity to blunder slightly is the real marvel of DNA. Without this attribute we still would be anaerobic bacteria and there would be no music." Mutations are the *sine qua non* of evolution. The ability of organisms to adapt to environmental changes is ultimately due to the accumulations of mutations in the gene pool of a population. Genetic diversity in the population enhances the chance for survival of the species. Many species of organisms, possibly even the dinosaurs, have become extinct because the gene pool was not sufficiently heterogeneous to allow at least a few members of the population to adapt to a changing environment. In this chapter we will discuss the most important types of mutations, how mutations occur, and how they are detected.

TYPES OF MUTATIONS

6.1. What are mutations?

Mutations are heritable changes in the genetic material that give rise to alternate forms of any gene. These alternate forms of the gene are called **alleles.** Generally, scientists recognize two broad types of mutations, those that affect the gene, also known as **gene mutations,** and those that affect whole chromosomes or segments of chromosomes. These latter type of mutations are called **chromosomal aberrations.** Gene mutations at the nucleotide level are generally described as **point mutations.**

6.2. What is the molecular basis of gene mutations?

The sequence of nucleotides determines the structure of rRNA, mRNA, and tRNA, as well as the myriad of proteins that are required in a living cell. The genetic information must be accurately replicated and passed on to future generations for the survival of the species. Any errors in replication of a gene within the DNA molecule resulting in the insertion, deletion, or substitution of one or more bases will give rise to a mutation. Even though the cell has mechanisms to improve the fidelity of a DNA replication process, every once in a while (1 in 10^4 to 10^7 replications) a spontaneous mistake is made leading to a heritable change in the sequence of a DNA molecule. In the laboratory, mutation rates can be greatly increased by exposure to chemicals or physical agents (e.g., UV and ionizing radiation) called **mutagens.**

Many mutations are due to the instability of the nucleotide bases in the DNA. When the nucleotide bases absorb sufficient energy from bombarding water molecules, they undergo structural changes called **tautomeric shifts** (Fig. 6-1). A tautomeric shift causes the redistribution of electrons and protons in the bases so that they no longer pair normally (i.e., A≡T and G≡C). In the unstable, high-energy state, G pairs with T and A with C. Spontaneous tautomeric shifts can result in heritable changes in the nucleotide sequence of the affected region of the DNA molecule, if such a molecule is replicated before the base can shift back to its normal state (Fig. 6-2).

> **Example 6.1.** Suppose that as the DNA molecule is replicated, a base (e.g., A) undergoes a tautomeric shift just as the DNA polymerase reaches it. If this happens, a C will be incorporated into the new copy strand of DNA rather than a T. The other template strand, when replicated, will exhibit normal base pairing. When the DNA molecule with the mispaired region is replicated, a perfectly matched mutant DNA molecule will result and a G-C pairing will appear where an A-T was before.

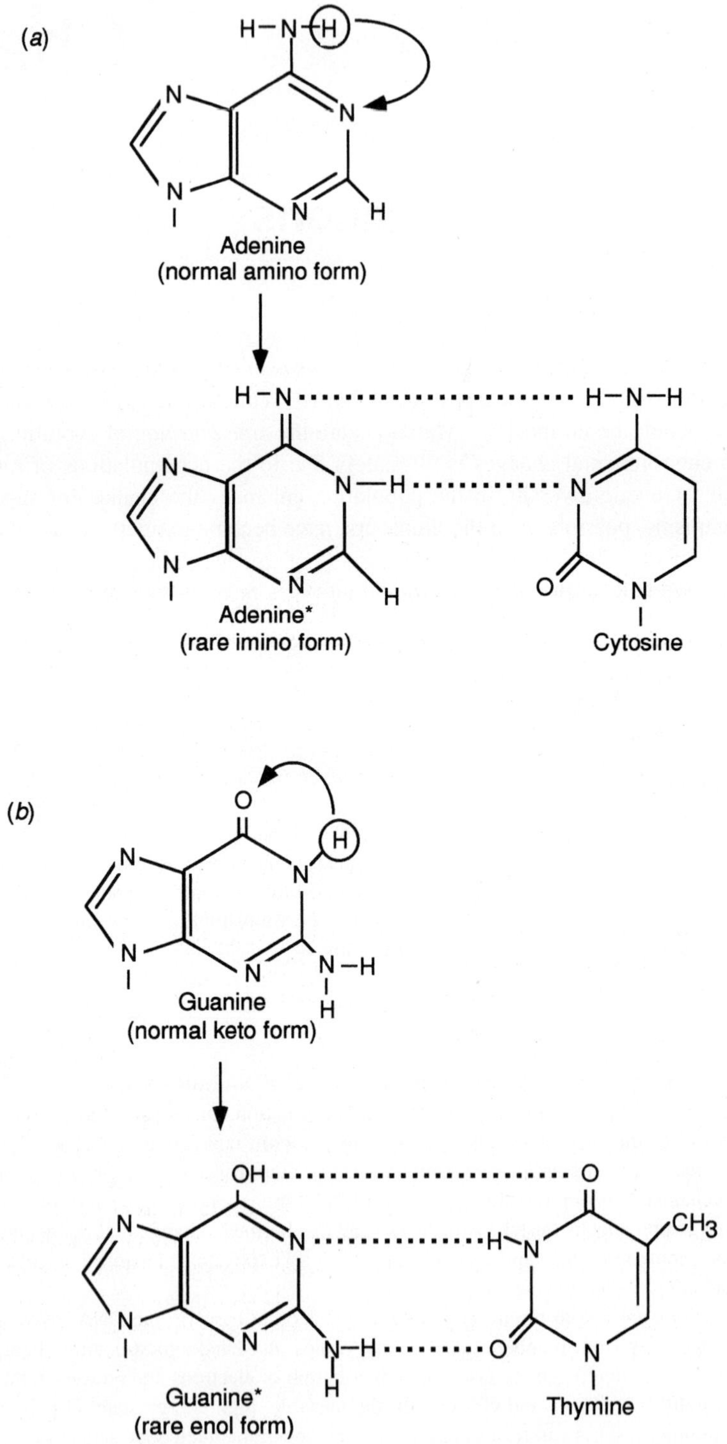

Fig 6-1. Tautomeric shifts and abnormal base pairing. A tautomeric shift occurs when molecules redistribute their electrons and protons. Tautomeric shifts can cause the bases in nucleotides to become unstable and pair abnormally with each other. (*a*) The unstable (*) adenine forms two hydrogen bonds (dotted lines) with cytosine (A*═C). (*b*) The unstable guanine pairs by three hydrogen bonds to thymine (G*≡T). Conversely, unstable thymine bonds with guanine, and the unstable cytosine pairs with adenine.

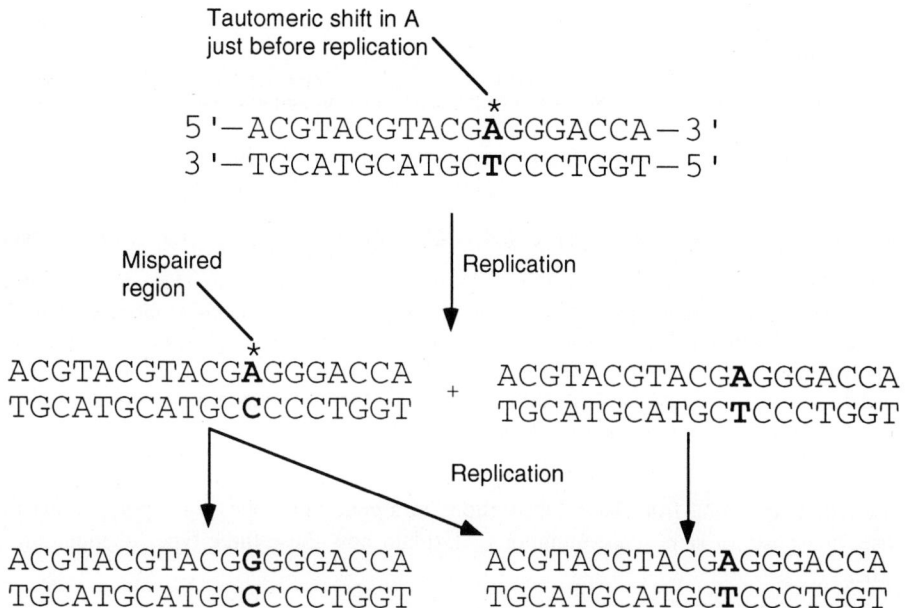

Fig 6-2. Consequence of abnormal base pairing as a result of a tautomeric shift. A tautomeric shift in the template strand during DNA synthesis causes the wrong nucleotide to be incorporated into the strand of DNA being synthesized. When the DNA is replicated again, one of the progeny DNA molecules contains a mutated base pair.

Other types of genetic changes can occur when chemical mutagens or high-energy radiation cause the removal (deletion) or insertion (addition) of one or more nucleotides into the DNA molecule. **Insertions** and **deletions** in protein-coding genes can cause **frameshift** mutations. By inserting or removing a base from a coding region, it is possible to shift the reading frame in a mRNA molecule.

6.3. What are the tautomeric forms of purines and pyrimidines?

The spontaneous movement of a hydrogen atom from one position to another within a base converts it to an alternative isomeric form called a **tautomer** (Fig. 6-1). The process itself involves a tautomeric shift. All purines and pyrimidines can exist as tautomers. Thymine and guanine are normally in **keto** forms, but when in the rare **enol** forms, they can join by three hydrogen bonds with keto forms of guanine or thymine, respectively. Likewise, adenine and cytosine are normally in the **amino** forms, but when in the rare **imino** forms, they can join by two hydrogen bonds with amino forms of cytosine and adenine, respectively. When a tautomeric shift occurs, the altered base tends to pair with the wrong complementary base, thus creating a mutation that may give rise to a different phenotype. One of the major impediments encountered by Watson and Crick in their attempts to solve the structure of DNA was that the biochemistry textbook they used displayed DNA bases in their rare tautomeric forms.

6.4. What type of base substitutions take place as a result of tautomeric shifts?

Base substitutions resulting from tautomeric shifts can be either transitions or transversions. **Transitions** occur when the mispairing caused by the tautomeric shift results in the replacement of one purine for another (e.g., A ⟷ G) or one pyrimidine for another (T ⟷ C). **Transversions** result when a purine is replaced by a pyrimidine or vice versa (A ⟷ C or T ⟷ G). Because the structural changes leading to transitions are relatively small (since only minor atomic rearrangements are required for the change), they occur more frequently than transversions (which require more substantial modifications of the molecule).

6.5. Can the number of transitions and transversions give an indication about the evolutionary relationship between organisms?

   The number of transitions and transversions in a given gene or genome can be used to compare the degree of evolutionary divergence of one organism from another. The number of base changes along the length of a given gene can give an indication of genetic divergence from the common ancestor.

6.6. Can DNA polymerases, during the process of DNA replication, correct some of the mismatched bases?

   Some DNA polymerases possess 3′ to 5′ exonuclease activity. This characteristic endows these polymerases with "proofreading" properties. When the DNA polymerase adds a mismatched nucleotide to the growing chain, structural changes occur in the DNA molecule that are recognized by one of the subunits of the DNA polymerase (e.g., the ε subunit of bacterial DNA polymerase III). This leads to the excision of the mismatched base at the 5′ end of that base, thus leaving a free 3′-OH group at the end of the growing chain with which the incoming base can form a phosphodiester bond.

6.7. Point mutations resulting from base substitutions in a gene that codes for a polypeptide may result in missense, nonsense, or samesense mutations. Explain how these three types of mutations differ from each other.

   Missense mutations result in the replacement of one sense codon for another.

   Example 6.2. The disease known as sickle cell anemia is a good example of the consequences of missense mutations. The normal hemoglobin A molecule, which is responsible for the transport of oxygen from the lungs to all parts of the body, is composed of two alpha (α) and two beta (β) chains. The first seven amino acids of the normal β chain are val-his-leu-thr-pro-glu-glu. Individuals afflicted by sickle cell anemia have a mutant β chain in which a valine (val) has replaced the glutamine (glu) as the sixth amino acid of the polypeptide. This mutation is due to a transversion in the second base of the sixth codon coding for glutamine (GAG) from an A to a U (Fig. 6-3). The sixth codon now has the sequence GUG that codes for valine (val). This missense mutation results in a hemoglobin molecule called HbS. Cells having only HbS tend to become crescent (sickle) shaped under low oxygen tension. These abnormal cells rupture easily and do not move through capillaries as easily as normal cells, leading to anemia, and possibly death.

Hb A	1	2	3	4	5	6	7
DNA	CAC	GTG	GAC	TGA	GGA	CTC	CTC
mRNA	GUG	CAC	CUG	ACU	CCU	GAG	GAG
Amino acids	val	his	leu	thr	pro	glu	glu

Hb S	1	2	3	4	5	6	7
DNA	CAC	GTG	GAC	TGA	GGA	CAC	CTC
mRNA	GUG	CAC	CUG	ACU	CCU	GUG	GAG
Amino acids	val	his	leu	thr	pro	val	glu

Fig 6-3. Single-base substitution leading to a missense mutation in the β-globin gene.

   A **nonsense mutation** creates one of the three terminating codons (UGA, UAA, UAG). Nonsense mutations result in polypeptides that are shorter than the normal ones. These shortened polypeptides generally are unable to function properly and therefore represent a loss of a biological characteristic.

   Example 6.3. The β-globin variant known as McKees Rocks is missing the last two amino acids in the chain (at amino acid positions 145 and 146 from the amino terminus). This occurs as a result of substitution of a U at the third position of the tyrosine-inserting codon UAU for an A. This base substitution creates the terminating codon UAA, causing the premature stoppage of polypeptide elongation.

Samesense mutations alter the third nucleotide of those synonymous codons that can be read by the same wobbling tRNA.

> **Example 6.4** If a mutation changes a glycine codon from GGA to GGC, there would be no noticeable change in the polypeptide's amino acid composition since both of the codons call for the insertion of glycine.

6.8. Why is the observed mutation rate of a gene likely to be an underestimate of the "true" mutation rate?

Since mutations arise spontaneously, any given base (say an A at position 125) within the coding region of the gene can change into any other base (say a C). When this happens, a mutation occurs. This mutation may be detected as a phenotypic change in some aspect of the cell. If another mutational event at that same location changes the C into an A, the phenotype of the cell will revert back to that of the wild type (a **back mutation**). Even though two mutational events have taken place within the gene, none will be detected if the phenotype of the cell was assessed only after the second mutational event took place. Samesense mutations and mutations in a different codon that compensate for another mutation, also known as mutation suppression (Question 6.11), can result in an underestimation of actual mutation rates.

6.9. What are hot spots?

Even though mutations arise spontaneously, some genes contain regions that are more likely to mutate than others. These sites are called **hot spots.** Hot spots were first demonstrated by Seymour Benzer during his landmark study of the topology of the *rII* locus of bacteriophage T4. Benzer mapped a total of 1,612 spontaneous mutations in the *rII* locus and found 60 sites within the locus exhibited a larger-than-expected number of spontaneous mutations. In particular, he located two sites within which more than 100 mutations were mapped. He also found 129 locations within the *rII* locus in which no mutations were found at all. The causes of these variations in mutation rates are largely unknown.

6.10. What are frameshift mutations?

Frameshift mutations are the result of nucleotide or base insertions or deletions within the coding region of a gene. The genetic code is translated by the protein synthesis apparatus by reading sequential groups of three bases that make up a codon, beginning from the start codon (AUG). If a single base is added (inserted) or removed (deleted) from the coding region, all of the codons from that point on will likely be changed.

> **Example 6.5.** Take the following sentence as an illustration of what happens when an insertion (or deletion occurs): THE FAT CAT ATE HIS HAT. If an I is inserted at the second position of the third word, the sentence would then read THE FAT CIA TAT EHI SHA T. The new sentence does not make sense because the insertion has altered the reading frame. In a genetic context, the new sets of codons generated and the resulting polypeptides would not make sense to the cell at all. Similarly, if the letter F is deleted from the correct sentence it would read THE ATC ATA TEH ISH AT. It, too, does not make sense.

> **Example 6.6.** The normal α-globin mRNA has 141 amino acids, with the last five codons in the mRNA being as follows:

Codon number:	138	139	140	141	142
Codon sequence:	UCC	AAA	UAC	CGU	UAA
Amino acid coded for:	ser	lys	tyr	arg	stop

The abnormal α-globin variant known as Hb Wayne 1 has a total of 146 amino acids, 138 of which are identical to those in the normal α-globin molecule and the remaining eight amino acids are different. This mutant α globin results from the deletion of an adenine (A) in position 3 of codon 139.

This event is illustrated below:

Codon number:	138	139	140	141	142	. . . 147
Normal codon sequence:	UCC	AAA	UAC	CGU	UAA	
Mutant codon sequence:	UCC	AAU	ACC	GUU	AAG	. . . UAA
Amino acid coded for:	ser	asn	thr	val	lys	. . . stop

The single base deletion caused a shift in the reading frame so that the normal terminating codon UAA at 142 was split up into two different codons, causing eight new amino acids to be added to the polypeptide chain until a new terminating codon was reached (position 147).

6.11. What are suppressor mutations?

Some suppressor mutations are able to compensate for a mutation in a different gene. Another type of suppressor mutation may be located at a different site from that mutation which it nullifies, yet still be present within the same gene as the mutation it suppresses. Even though suppressor mutations cause an apparent "reversal" of a mutant phenotype to wild type, this phenotypic reversal does not represent a true back mutation, since crosses between the wild type and the revertant can yield progeny carrying the original (unsuppressed) mutation. For example, a base deletion causes a frameshift mutation in a cell that leads to the premature termination of a polypeptide involved in amino acid metabolism, converting the cell into an auxotrophic mutant (see Question 6.22). If a subsequent base addition occurs (or a deletion removes a short sequence including the stop codon), it may place the gene sequence back in the proper reading frame, allowing the synthesis of the polypeptide. The resulting polypeptide may possess one or more amino acids that are different from the native polypeptide, but if these do not affect the active site of the protein, a wild type phenotype may develop. This second mutation would compensate for the first mutation, making the cell prototrophic (see Question 6.22) once again.

6.12. What are mutator genes?

Cells have an array of different repair systems aimed at reducing the number of potentially lethal mutations in the genetic material. As is the case for all cell processes, repair mechanisms are expressed by proteins, all of which are encoded in the genome of the cell and therefore subject to mutations. When a mutation occurs in any of the genes coding for a repair protein, thus rendering the gene product nonfunctional, a higher-than-normal mutation rate is noted in the cell. These genes are commonly known as **mutator genes.**

> **Example 6.7.** Strains of the bacterium *Escherichia coli* that are *ung⁻* lack the enzyme uracyl-*N*-glycosylase. These mutants are unable to repair cytosine deaminations and therefore have high levels of C → T transitions.

CAUSES OF MUTATIONS

6.13. Can mutations be induced with chemicals or physical agents?

Yes. Chemical and physical agents that cause mutations are called **mutagens.** Mutagens can cause mutations by replacing one base with another in the DNA molecule (e.g., base analogs), causing structural changes in a base so that it causes it to mispair (e.g., alkylating agents); causing insertions or deletions (e.g., intercalating agents such as acridine dyes); or damaging a base so much that it is unable to pair with any other normal base (e.g., ultraviolet light).

6.14. How do base analogs cause mutations?

Base analogs are sufficiently similar to the normal nitrogen bases in DNA that they can be incorporated into a replicating DNA molecule by DNA polymerases. Once incorporated, however, base analogs have abnormal base-pairing properties so that they produce mutations during subsequent DNA replication cycles. Two common base analogs are 5-bromouracil (5-BU) and 2-amino-purine (2-AP).

> **Example 6.8.** The analog 5-BU is an analog of thymine and can become incorporated into a DNA molecule instead of thymine by DNA polymerases. The 5-BU molecule has a bromine group at the C-5 position of the base instead of the normal CH_3 found in thymine. The enol form of this analog, which occurs at a high rate in DNA molecules with the incorporated analog, pairs with guanine instead of cytosine, giving rise to transitions.

6.15. How do alkylating agents cause mutations?

Alkylating agents cause mutations by chemically altering normal nitrogen bases so that they pair up with a specific base other than the normally complementary base.

> **Example 6.9.** Hydroxylamine (HA) causes GC → AT transitions. This molecule adds OH groups to the NH$_2$ nitrogen at position C-4 of the cytosine molecule. The resulting molecule, known as *N*-4-hydroxycytosine, pairs with adenine, causing the transitional mutation of GC → AT.

6.16. How do intercalating agents cause mutations?

Intercalating agents like **proflavin, acridine orange,** and **ethidium bromide** are planar molecules that can insert themselves between the stacked nitrogen bases within the DNA double helix. These agents alter the molecule of DNA in such a way that DNA polymerases may insert or skip one or more bases during replication. These types of mutagens often cause frameshift mutations.

6.17. How does ultraviolet light cause mutations?

When **ultraviolet light** is absorbed by adjacent pyrimidines in one strand of a dsDNA molecule they form a cyclobutane-pyrimidine photodimer. These dimers on one strand of the DNA molecule interfere with proper base pairing during DNA replication. The impact on DNA replication is so extensive that the normal replicative process is stopped until these dimers are removed or bypassed. Under these circumstances, a repair mechanism known as the **SOS repair system** is induced. The SOS system is an emergency repair process aimed at bypassing "replication blocks" such as those caused by photodimers. The mechanism of repair is not fully known, but it is thought that SOS proteins interact with DNA polymerase III and lower the strict specificity of base pairing required by the DNA polymerase. Under SOS mediation, stalled DNA polymerases bypass the stoppage site by incorporating one or more bases at random until the polymerase passes through the damaged site. When this occurs, it is possible that a base other than those normally complementary to the dimerized pyrimidines are inserted, and a mutation results. For this reason, ultraviolet irradiation normally causes transitions and transversions, although frameshift mutations are possible but unlikely.

6.18. Can UV-damaged DNA be repaired before the SOS system is activated?

Most bacterial cells have DNA repair mechanisms aimed at excising the UV-damaged DNA and replacing the excised segment with an undamaged one. Fig. 6-4 illustrates this process. The absorption of UV light by adjacent pyrimidines in the same strand will cause them to react with each other and form a dimer. The most common lesions produced are thymidine dimers. The thymidine dimers distort the DNA, causing it to bulge and inhibit polymerase activity. The repair system utilizes the *uvrABC* endonuclease, which recognizes DNA distortion. This endonuclease binds on either side of the distortion, nicking the damaged strand of DNA. DNA polymerase I associates with the nick at the 5′ side of the damaged DNA, excising the damaged region and replacing it with new DNA. Once the damaged DNA has been replaced with new DNA by the polymerase, DNA ligase repairs the remaining nick.

DETECTION OF MUTATIONS

6.19. How are bacterial mutants detected?

Bacterial mutants are detected by comparing the phenotype of the bacterial strain (or isolate) under observation with that of the wild type. The wild-type strain is that strain which is commonly isolated from nature. The characteristics of the wild type reflect the usual or native form of a gene. For example, wild-type strains of *E. coli* are capable of growing in a medium containing only inorganic salts and glucose as a source of carbon and energy. They do so because they have the biochemical pathways to synthesize all the necessary amino acids, vitamins, and nucleotides. Mutant strains require that one or more amino acids be provided in the medium in order for them to grow. This change in phenotype reflects a mutation in that amino acid's synthetic pathway.

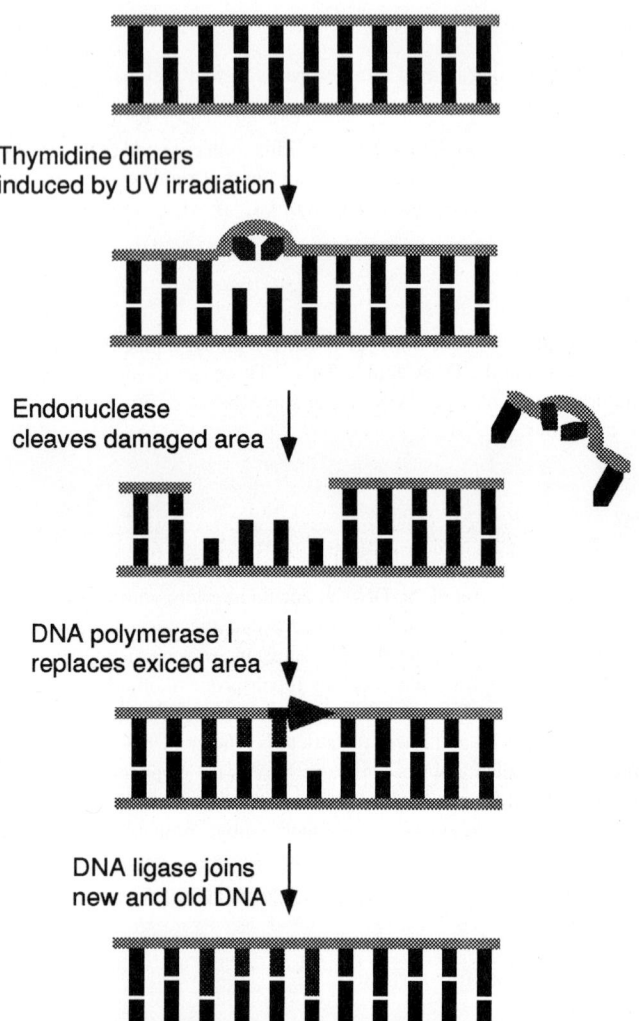

Thymidine dimers
induced by UV irradiation

Endonuclease
cleaves damaged area

DNA polymerase I
replaces exiced area

DNA ligase joins
new and old DNA

Fig 6-4. The repair of pyrimidine dimers by the cell. When two pyrimidines lie next to each other in the same DNA strand, they can react with each other to form a dimer if they are destabilized by the absorption of ultraviolet light. Pyrimidine dimers distort the DNA molecule and prevent DNA replication and transcription. Most bacteria have a repair system that removes these dimers and replaces them with the appropriate nucleotides. The repair process involves the excision of the damaged DNA by the gene product of the gene *uvr*ABC and the replacement of the damaged DNA by the enzyme DNA polymerase I. The repair process is completed when a DNA ligase joins the newly synthesized strand with the old one.

6.20. Bacteria undergo spontaneous or induced mutations. These mutations are passed on from one generation to another as a result of cell division. Much of what we know about mutations comes from the study of spontaneous and laboratory-induced mutations, using bacteria as experimental models. Do bacterial mutations arise spontaneously, as a result of exposure to specific environments, or both?

   This question, unanswered until 1943, is of fundamental importance to the understanding of genetics. Salvador Luria and Max Delbrück, working with bacteriophages of the bacterium *E. coli,* designed an experiment—later known as the **fluctuation test**—providing evidence that bacterial mutations arise spontaneously rather than adaptively in response to specific environmental agents (such as phage).

6.21. What is the Luria-Delbrück fluctuation test?

   The Luria-Delbrück fluctuation test was the first experimental demonstration that bacterial mutations arise spontaneously. To appreciate the value of this test, let us examine the phenotypic change from susceptibility (sensitivity) to bacteriophage T1 (normally found in nature) to a resistant state as seen in *E. coli.* The wild-type strain of *E. coli* is sensitive to the bacteriophage T1; that is, upon infection by the bacteriophage, the bacterial cells lyse. Some cells in the population, however, are resistant to infection and are said to be bacteriophage resistant. Did these mutants arise *as a result* of the bacteriophage's presence in the bacterial environment (acquired resistance) or were they *already present* (spontaneous) and proliferated preferentially because the phage acted as a selective agent in the environment? The fluctuation test, as described below, provides an answer to that question:

1. Seed 10 plates of nutrient agar with a culture of bacteriophage T1, then spread approximately 10^9 *E. coli* cells from an overnight nutrient broth **stock culture** over the surface of the bacteriophage-seeded plates. Incubate the plates at 35°C overnight. Count and record the number of colonies appearing on each of the plates.

2. Prepare 10 subcultures of *E. coli* in nutrient broth from the same stock culture used to inoculate the plates above. Incubate overnight under the same culture conditions. Repeat step 1 above for each of the 10 subcultures.

    The difference between the control experiment (step 1) and the test experiment (step 2) is simply that the latter were allowed about 12 hours of growth in liquid medium before being plated.

3. Analyze the results statistically, comparing the number of mutants present in the stock culture with those obtained from each of the 10 subcultures. Table 6.1 summarizes the results obtained when a fluctuation test is performed.

Table 6.1.   Results Obtained from a Representative Fluctuation Test

Stock Culture		Subcultures	
Plate Number	Number of T1-Resistant Colonies	Plate Number	Number of T1-Resistant Colonies
1	23	1	2
2	18	2	7
3	19	3	24
4	25	4	0
5	20	5	43
6	19	6	6
7	23	7	20
8	29	8	0
9	17	9	1
10	16	10	3

The results shown in Table 6.1 support the hypothesis that mutations arise spontaneously. If phage-resistant cells arise as a result of exposure to the same concentration of selective agent (phage T1), all of the plates, both seeded from stock or from subcultures, would be expected to exhibit the same approximate number of colonies due only to sampling error. On the other hand, if random mutations to T1 resistance arise spontaneously, variation in the number of phage-resistant clones would be expected to fluctuate widely from one subculture to another. Those mutations that happened to occur early after subculturing would give rise to many resistant colonies, whereas mutants that happened to occur late in the incubation phase preceding plating would produce relatively few resistant colonies.

6.22.   What are auxotrophic mutants?

Auxotrophic mutants are those mutants that exhibit one or more *nutritional* requirements not exhibited in the wild type; wild-type strains in this regard are designated as **prototrophic** strains. For example, prototrophic strains of *E. coli* do not require tryptophan in their medium in order for them to grow, because they have complete, functional biosynthetic pathways for tryptophan synthesis. Mutations in any of the tryptophan genes that lead to the synthesis of inactive tryptophan-pathway enzymes will give rise to auxotrophic mutant bacteria requiring tryptophan.

6.23.   What is the replica-plating method?

The replica-plating method (Fig. 6-5) is a common technique employed for the detection of auxotrophic mutants. This procedure can be used to select almost any type of nutritional mutant. Let us say that we want to detect tryptophan-requiring mutants in a population of *E. coli* cells exposed to ultraviolet irradiation (a mutagenic agent that should increase the mutation rate). To do this by replica plating, a large sample (e.g., 1×10^7 cells) of the *E. coli* population is spread over the surface of a glucose-salts medium supplemented with tryptophan, so that individual colonies appear after a suitable incubation period. This plate, called the **master plate,** will usually contain both prototrophs and tryptophan auxotrophs. From this master plate, exact replicas of the colonies are made onto two agar plates, one with glucose salts alone (minimal medium) and the other with glucose-salts supplemented with tryptophan (complete medium).

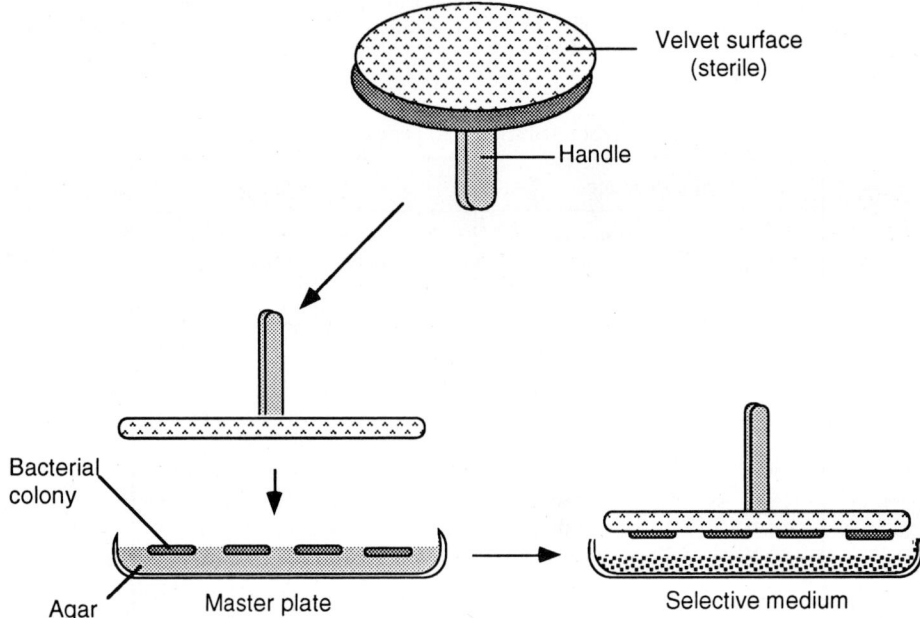

Fig 6-5. Replica-plating. The replica-plating technique permits the detection and selection of specific mutants in a mixed microbial population. An applicator disk of a diameter slightly smaller than the interior of a Petri dish is coated with sterile velvet cloth. The disk is pressed lightly over the surface of a plate containing the desired bacterial colonies, so that the velveteen fibers pick up representative cells from each colony. The applicator is then pressed over the surface of a selective culture medium that will allow the growth of those bacterial cells with the desired phenotype.

The replicas are made by pressing a cylinder covered with a sterile velveteen cloth onto the surface of the master plate so that some of the cells making up the colonies are transferred to the velveteen fibers. The velveteen cloth, bearing an imprint of the master plate, is then pressed onto the surface of two agar plates, one containing a minimal medium and the other a complete medium. During this process, exact replicas of the master plate are transferred onto the two agar plates. Auxotrophic mutants in the master plate will grow in complete medium but not in minimal medium. Upon examination of both plates, colonies appearing in the supplemented medium but not present in the minimal medium are likely to be tryptophan auxotrophs.

6.24. The results of the fluctuation test gave only statistical evidence to support the conclusion that at least one type of bacterial mutation arises spontaneously. How can replica plating provide more direct evidence for the same conclusion?

Using the replica plating technique they invented, Joshua and Esther Lederberg, in 1952, provided more direct supporting evidence to the conclusion (based on statistics from fluctuation experiments) that mutations arise spontaneously. Colonies of *E. coli* grown on an agar plate were transferred by replica plating onto two or more plates containing a culture medium and a high concentration of the bacteriophage T1. After overnight incubation, only those colonies made up of T1-resistant cells appeared on the plates. Each of the replica plates showed exactly the same number of colonies in identical geometric positions. These observations make sense if resistant mutants existed on the master plate before any exposure to phage.

6.25. Mutations in bacteria can be classified into two categories: those that are selectable mutations and those that are not selectable. What are selectable mutations?

Selectable mutations are those which afford the microorganism a growth advantage over the wild-type strain. For example, the mutation that endows a bacterium with resistance to the antibiotic streptomycin is a selectable mutation. Why? Because when a mixture of wild-type and mutant cells are grown in a culture medium containing streptomycin, only those cells that are resistant to the antibiotic will grow. Hence, the presence of streptomycin in the culture environment *selects* for those mutants which are resistant to streptomycin. Can you think of additional selectable mutations?

6.26. What are nonselectable mutations?

Nonselectable mutations are those which do not afford the microorganism a growth advantage. These mutations can only be recognized by visual or biochemical examination of many cells. Mutations in colony pigmentation are common types of nonselectable mutations. Wild-type *Serratia marscescens,* for example, forms red-colored colonies when grown at room temperature. When the gene responsible for the pigment production mutates, the colony becomes cream-colored. Because they are rare, mutants of *S. marscescens* can only be detected by examining a large number of plates, each with numerous colonies.

6.27. What is genetic complementation?

Genetic complementation is a phenomenon that may occur in cells that are diploid or partially diploid (having only some homologous chromosome segments in duplicate rather than entire genomes) and carry two mutations. The two mutations may reside on the same DNA molecule or on different DNA molecules. If one chromosome has two mutant sites and the comparable sites on the homologous chromosome are completely normal (unmutated), the mutations are said to be in *cis* position (Fig. 6-6a). If two homologous DNA molecules each carry a mutation at different sites (either in the same gene or in different genes), the mutations are said to be in *trans* position (Fig. 6-6b,c). If diploid cells bearing two mutations in *trans* position display a normal or nearly normal (wild) phenotype, genetic complementation is said to have occurred.

6.28. What is the genetic basis of complementation?

When a diploid or partially diploid cell carries two mutations in *cis* position, the normal, undamaged chromosome makes one or more functional proteins that allows the cell to display a normal or near normal (wild) phenotype, despite the fact that the other chromosome carries two mutations. It does not matter whether the two mutations lie within the same DNA segment coding for a single polypeptide chain (the segment is called a gene or cistron) or whether they lie in different cistrons; complementation should occur under either condition. Methodologically, therefore, the *cis* condition just described establishes a control for a *trans* experiment. The complete experiment is called a ***cis-trans* test.**

If a diploid or partially diploid cell, having two mutations in *trans* position, displays a wild or nearly normal phenotype, again, complementation will have occurred. But now we can infer from the *cis-trans* test that the two mutations probably lie in different cistrons, because the defective protein made by one mutant gene has been compensated for or nullified by a functional protein made by a normal gene in the homologous DNA molecule (chromosome). It is as if you had two pairs of pants with a hole (defect) in each pair on a different leg. If you put on both pairs of pants, the hole in one pants is covered by the undamaged leg in the other pants. If, on the other hand, two mutations in *trans* position fail to complement one another, a mutant (abnormal) phenotype will appear. From this we may infer that the two mutations were most likely in the same cistron. Neither DNA molecule has a normal gene by which to complement the defect in a comparable segment on the homologous DNA molecule.

If complementation occurs in both *cis* and *trans* positions of a *cis-trans* test, we would not know whether the polypeptides produced by two different cistrons function as independent proteins (e.g., different enzymes) or whether they are part of the same heteromeric protein (e.g., containing two different chains as heterodimers or heterotetramers). For example, in *E. coli* the enzyme tryptophan synthetase is a tetrameric aggregate of two α and two β chains. This enzyme catalyzes an essential step in the biosynthetic pathway for the production of the

(a) *Cis* configuration = Complementation occurs

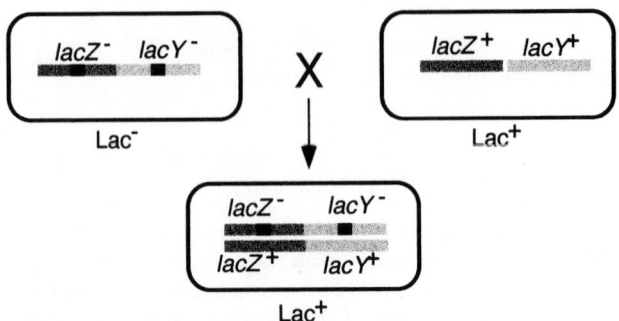

(b) *Trans* configuration = Complementation occurs

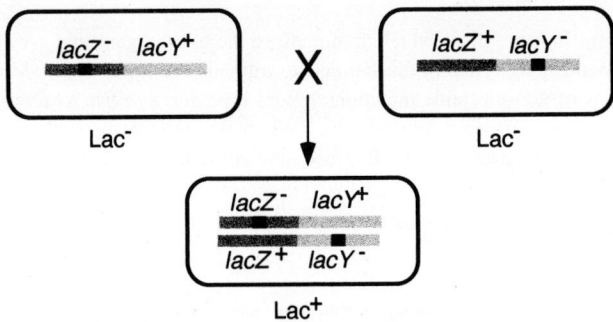

(c) *Trans* configuration = No complementation occurs

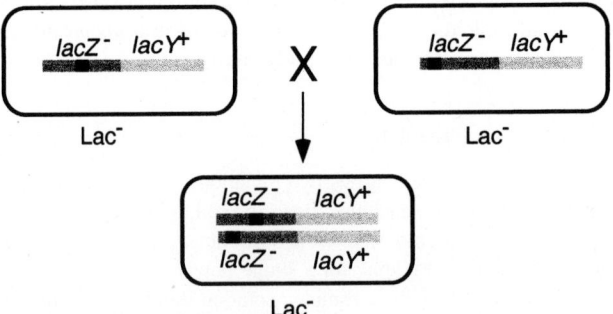

Fig 6-6. Genetic complementation. (*a*) *Cis* configuration. The *E. coli* cell at left contains a mutant *lacZ⁻* gene (dark stripe) that cannot produce the enzyme β-galactosidase. It also contains a mutant *lacY⁻* gene (dark stripe) that cannot produce the transport protein β-galactoside permease. Either mutation would prevent the cell from metabolizing lactose (Lac⁻ phenotype). In the cell at right, both loci contain normal genes (*lacZ⁺* and *lacY⁺*), and the cell is able to metabolize lactose (Lac⁺ phenotype). When crossed, the resulting partial diploid has one DNA molecule containing both functional *lac* genes and is able to utilize lactose (Lac⁺). Complementation has occurred. (*b*) *Trans* configuration. The cell at left is defective only in its ability to make β-galactosidase. The cell at right is defective only in its ability to make β-galactoside permease. Neither cell can utilize lactose, but for different reasons. When crossed, the resulting hybrid cell contains one unmutated copy of each gene and therefore can metabolize lactose. Complementation has occurred. (*c*) If complementation had not occurred in the hybrid, we would have had to conclude that the two mutations were in the same gene (cistron). But we would not have known whether that cistron was *lacZ* or *lacY*.

amino acid tryptophan. A diploid cell having two mutations in *trans* position in the gene for the α chain cannot produce the functional enzyme. The same is true if the two mutations are in the gene for the β chain. Either way, the cell cannot synthesize tryptophan (Trp$^-$ phenotype). But if one mutation is in the gene for the α chain and the other mutation is in the gene for the β chain, the functional genes in the homologues should allow the production of some functional enzymes containing four normal chains, and thus at least partial complementation should occur (development of near normal phenotype). Both normal and abnormal α and β chains may be produced, and they may spontaneously aggregate in various combinations into heterotetramers, but tryptophan is unlikely to be synthesized (Trp$^+$ phenotype) by the cell unless some tetramers consist of two normal α and two normal β chains.

6.29. Is complementation the same as recombination?

No. Recombination represents the formation of new gene combinations resulting from the breakage and reunion of chromosomes. As a result, the progeny represent new genotypes not found in the parental strains. Complementation, on the other hand, does not involve the formation of new genotypes, but rather a mixture of gene products. Complementation can be distinguished from recombination experimentally by the massive effect of the former. All hybrid cells are expected to show wild-type phenotype. Recombination occurs far less frequently, especially if the two mutants are tightly linked.

6.30. What is a cistron?

A **cistron** is a gene defined on the basis of a *cis-trans* complementation test. Two mutations in the same cistron (*cis* configuration) cannot complement each other, whereas two mutations in different cistrons may complement each other. A normal cistron is commonly thought of as a segment of DNA that codes for a complete polypeptide chain.

CHROMOSOMAL ABERRATIONS

6.31. Two types of chromosomal aberrations can occur in cells: changes in chromosome structure and changes in chromosome number. What are the most common types of changes in chromosome structure and how do they occur?

Common chromosomal structural changes are deletions, duplications, inversions, and translocations. **Deletions** are chromosomal changes in which one or more genes or chromosomal segments are lost. **Duplications** occur when one or more copies of a given gene are present on the same or different chromosomes. Deletions and duplications can occur in the same mutational event when two homologous DNA strands overlap, break at the same time at two different (nonhomologous) points, and then rejoin with the wrong strand. One of the strands will be missing one or more genes, and the reciprocal strand will have an extra copy of one or more genes (Fig. 6-7). **Inversions** occur when a breakage in one of the chromosomes occurs and the segment rotates 180° before it rejoins. **Translocations** take place when nonhomologous chromosomes break and exchange segments.

6.32. How can mutants resulting from chromosomal deletions be distinguished from nucleotide substitutions?

Mutants that result from deletion of a segment of chromosome cannot back mutate, since an entire segment of chromosome would have to be acquired for the cell to revert to the wild type. Base substitutions, however, can back mutate to the wild type at a rate that is characteristic for the gene and the locus.

6.33. In diploid organisms, there are two major types of chromosomal aberrations that are the result of changes in chromosome number. These are polyploidy and aneuploidy. What are polyploid mutants?

The gametes of diploid (2n) organisms normally contain a haploid (n) set of chromosomes. Some organisms, like male bees and fungi, normally have a single set of chromosomes in their body cells (somatic cells) and are commonly known as haploid organisms. However, since the haploid number refers specifically to gametes,

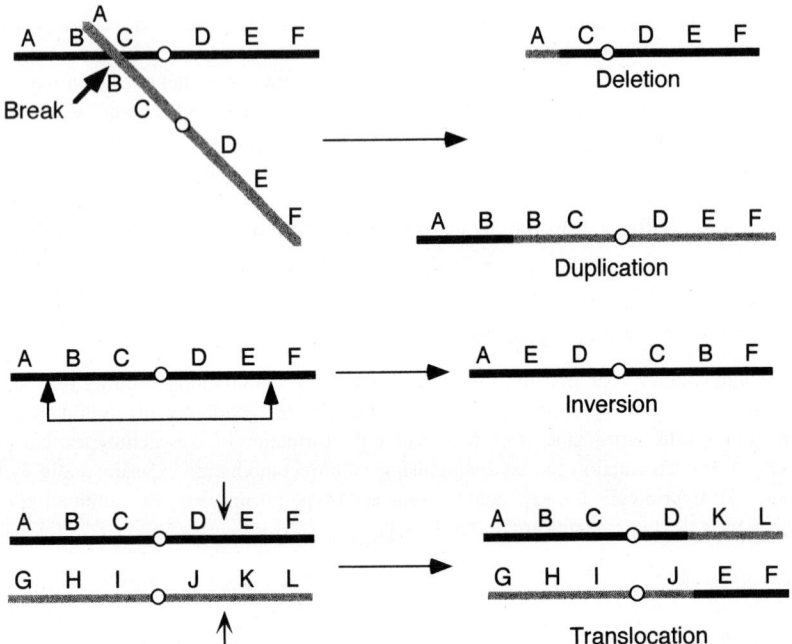

Fig 6-7. Types of chromosomal aberrations leading to changes in chromosome structure.

the term **monoploid** (one set of chromosomes) should be applied to those organisms with one set in their somatic cells.   Humans and most other eukaryotes, on the other hand, have two sets of chromosomes in their somatic cells.   These organisms are said to be diploid.   For example, the haploid number of chromosomes in human egg or sperm cells is 23, but the number of chromosomes in somatic cells (e.g., skin cells) is 46.   When cells acquire one or more sets of chromosomes beyond the "normal" number of sets, they are said to be **polyploids.**   Common polyploids include **triploids** (3n) and **tetraploids** (4n).   Triploids, like all other odd-number ploids, are characteristically sterile because they can rarely produce genetically balanced gametes by meiosis.   Tetraploids, on the other hand, may be fertile if meiosis regularly produces genetically balanced 2n gametes.   In certain types of crop plants, tetraploids have been found to have the added advantage of being generally larger and more productive plants.

6.34.   What are aneuploid mutants?

Aneuploid mutants are the result of changes in the individual number of homologous chromosomes in a set. During meiosis, homologous chromosomes pair along the metaphase plane and segregate so that four haploid cells are produced.   If, during one of the divisions of meiosis, one of the homologous chromosomes in meiosis I (or chromatids in meiosis II) does not migrate to the opposite pole, an uneven distribution of that particular chromosome will take place.   Depending upon where in meiosis this **nondisjunction** occurs, some gametes will end up with one or more extra chromosomes and others will end up with fewer chromosomes than a complete set. This phenomenon is illustrated in Fig. 6-8.

The aneuploid condition that results in three copies of a given chromosome is known as **trisomy** (2n + 1). This condition is a chromosomal aberration that can lead to abnormality or even death, depending on which chromosome is present in triplicate.

> **Example 6.10.**   A condition known as Kleinfelter syndrome is a trisomic condition in which a human male possesses two X chromosomes in addition to the normal Y chromosome.   These individuals are usually sterile, mentally retarded, and have a lanky build.

> **Example 6.11.**   Possibly the most common type of trisomic condition in humans is **Down syndrome.**   It occurs in three out of 2,000 births.   Down syndrome is the result of a nondisjunction of chromosome 21.   When an abnormal gamete possessing two copies of chromosome 21 is fertilized by

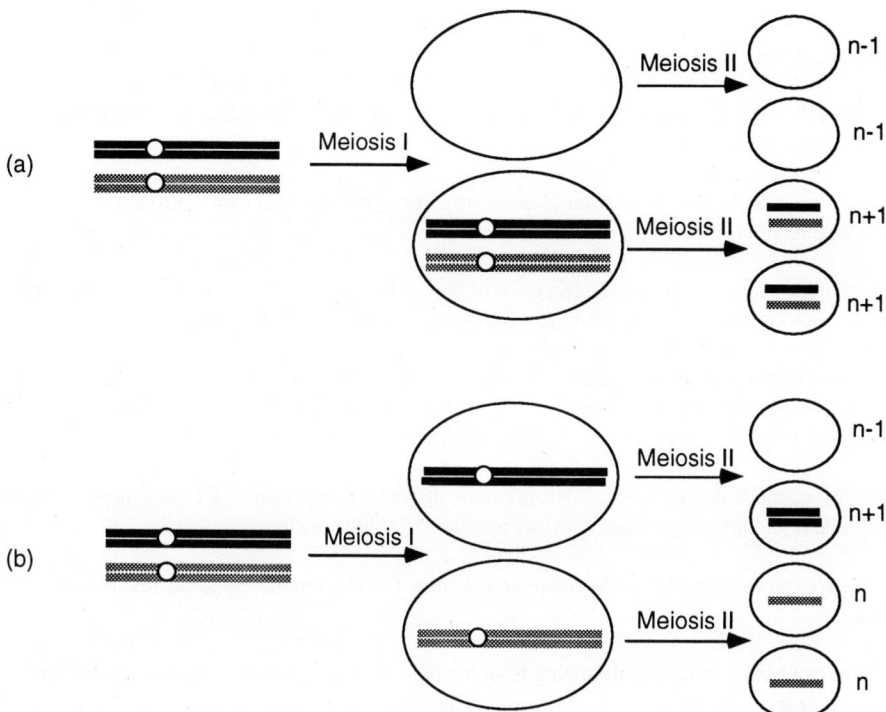

Fig 6-8. Mechanisms of formation of aneuploid gametes during meiosis. (*a*) Chromosomal nondisjunction. (*b*) Chromatidal nondisjunction.

a normal haploid gamete, the resulting zygote is trisomic for chromosome 21. Although a proportion of trisomic 21 fetuses die before birth, many grow up to be adults, but few live past 35–40 years of age. Abnormalities in Down syndrome include mental retardation (IQ 20–60), short stature, stubby hands with a crease across the middle, a large, wrinkled tongue, and a flat face. Female Down syndrome patients may be fertile and have normal (or trisomic) offspring, but males are infertile.

Objective Questions

Multiple Choice

Directions: Choose the one best answer.

1. Alternate forms of a gene are known as (*a*) tautomers. (*b*) isomers. (*c*) alleles. (*d*) aneuploids. (*e*) polyploids.

2. Mutations often occur as a result of base substitutions. The most common cause of base substitutions is (*a*) tautomeric shifts. (*b*) mitotic errors. (*c*) meiotic errors. (*d*) base insertions. (*e*) base deletions.

3. In the imino form, adenine pairs with (*a*) C. (*b*) G. (*c*) U. (*d*) T. (*e*) A.

4. In the keto form, thymine pairs with (*a*) C. (*b*) G. (*c*) U. (*d*) T. (*e*) A.

5. In the enol form, guanine pairs with (*a*) C. (*b*) G. (*c*) U. (*d*) T. (*e*) A.

6. In the enol form, thymine pairs with (*a*) C. (*b*) G. (*c*) U. (*d*) T. (*e*) A.

7. The McKees Rocks β-globin variant results from *(a)* a base substitution. *(b)* a translocation. *(c)* an inversion. *(d)* a nondisjunction. *(e)* a frameshift.

8. A tautomeric shift causing the substitution of one purine for another is called a(n) *(a)* transversion. *(b)* translocation. *(c)* transition. *(d)* deletion. *(e)* inversion.

9. A tautomeric shift causing the substitution of one purine for a pyrimidine is called a(n) *(a)* transversion. *(b)* translocation. *(c)* transition. *(d)* deletion. *(e)* inversion.

10. When a mutational event leads to the replacement of one codon for another calling for the same amino acid, the resulting mutation is known as *(a)* frameshift. *(b)* transition. *(c)* transversion. *(d)* deletion. *(e)* silent.

11. Strains of *Neisseria gonorrhoeae,* as usually isolated from the genito-urinary tract of humans suffering from gonorrhea, oxidize glucose, but not lactose, and are catalase and oxidase positive. Strains with such characteristics are considered to be *(a)* auxotrophic. *(b)* prototrophic. *(c)* wild type. *(d)* mutants. *(e)* avirulent.

12. Replica-plating methods can be used to detect strains of bacteria that exhibit all the following phenotypes *except* *(a)* auxotrophic. *(b)* antibiotic-resistant. *(c)* virus-resistant. *(d)* pigmentation. *(e)* pilated.

13. The term cistron means essentially the same as the term *(a)* auxotroph. *(b)* gene. *(c)* mutant. *(d)* transformant. *(e)* conjugant.

14. A large number of hybrid diploid cells arising from the mating of two alanine-requiring mutants of *Escherichia coli* can synthesize their own alanine. This observation tells us that the new phenotype most likely arose as a result of *(a)* back mutations. *(b)* cis complementation. *(c)* *trans* complementation. *(d)* transduction. *(e)* transposable elements.

15. Which of the following chemical mutagens are likely to cause frameshift mutations? *(a)* 5-Bromouracil *(b)* 2-amino-purine *(c)* acridine orange *(d)* hydroxylamine *(e)* catalase

16. Which of the following chemical mutagens is likely to cause GC → AT transitions? *(a)* 5-Bromouracil *(b)* 2-amino-purine *(c)* acrydine orange *(d)* hydroxylamine *(e)* nitrous acid

17. Which of the following chemical mutagens is likely to cause base substitutions after incorporation into a replicating DNA molecule? *(a)* 5-Bromouracil *(b)* 2-amino-purine *(c)* acrydine orange *(d)* hydroxylamine *(e)* more than one of the above

18. The SOS repair mechanism is activated by which of the following? *(a)* 5-Bromouracil *(b)* 2-amino-purine *(c)* acrydine orange *(d)* hydroxylamine *(e)* thymidine dimers

19. Sites where mutations occur at rates higher than normal are known as *(a)* suppressor sites. *(b)* hot spots. *(c)* mutator sites. *(d)* cistrons. *(e)* promoters.

20. When chromosome breakage occurs and the broken segment rotates by 180° before reunion, the resulting chromosomal aberration is known as a(n) *(a)* insertion. *(b)* deletion. *(c)* inversion. *(d)* translocation. *(e)* frameshift.

True–False

1. Mutations are inheritable changes in the genetic material and take place exclusively as a result of chromosomal aberrations.

2. Tautomeric shifts within the bases that make up the nucleic acids are the most common cause of point mutations.

3. Insertions and deletions most often result in missense mutations.

4. Bacterial mutants can be detected by comparing their phenotypes to that of the wild-type strain.

5. Auxotrophic mutants exhibit resistance to antimicrobial agents.

6. Mutations in bacteria arising spontaneously are generally not selectable.

7. Resistance to antibiotics is a selectable mutation.

8. Two mutants usually complement each other only when their corresponding mutations are in different cistrons.

9. The most common tautomeric form of adenine in cells is the enol form.

10. Another name for a gene is cistron.

11. Point mutations can lead to nonsense mutations.

12. Hot spots are sites where mutations occur at very high rates.

13. Frameshift mutations are caused by base substitutions.

14. Aneuploidy and polyploidy are common chromosomal aberrations.

15. Trisomy is a term that describes a cell with three sets of chromosomes.

16. SOS repair mechanisms involve cutting and replacing damaged DNA segments.

17. Mutator genes are found in mutants with nonfunctional DNA repair mechanisms.

18. Intercalating agents cause primarily transitions and transversions.

19. Observed mutation rates are normally higher than true mutation rates.

20. Proofreading properties of DNA polymerases rely on their $5' \rightarrow 3'$ exonuclease activity.

Matching

Directions: Match each item in **Column A** with the one in **Column B** to which it is most closely associated. Any item chosen from **Column B** can be used only once.

Column A	Column B
1. transversion	A. replica plate technique
2. keto form	B. nutritional mutant
3. prototroph	C. fluctuation test
4. transition	D. chromosomal structural change
5. nonsense	E. trisomy
6. inversion	F. incorrect amino acid substitution
7. cistron	G. high mutation rate
8. missense	H. substitution of one purine for another
9. tautomer	I. Hb S
10. sickle cell anemia	J. alleles
11. velveteen	K. translation termination
12. aneuploidy	L. enol form
13. alternative forms of a gene	M. alkylating agent
14. proof of spontaneous mutations	N. pyrimidine to purine
15. auxotroph	O. isomer
16. 5-bromouracil	P. wild type with regard to nutritional requirements
17. hydroxylamine	Q. imino form
18. pyrimidine dimers	R. one complete polypeptide
19. ethidium bromide	S. polyploidy
20. hot spot	T. SOS repair
	U. base analog
	V. intercalating agent

Terms

Directions: Unless otherwise specified, each answer is a single word.

1. Inheritable changes in the genetic material

2. The procedure used to provide the first experimental evidence supporting the hypothesis that mutations occur spontaneously. (2–4 words)

3. An adjective descriptive of bacterial mutants with nutritional requirements that are not present in the wild type.

4. Variant forms of the same gene

5. Many mutations are due to the instability of nucleotide bases. The conformational changes that these bases undergo when hydrogen atoms are moved to new locations in the base are called. (2 words)

6. This mutational event results from addition of a base to the coding region of a gene.

7. This mutational event results from the removal of a base.

8. This type of mutation results from insertions and deletions.

9. An alternative isomeric form of a base due to the movement of one hydrogen atom from one position to another.

10. Thymine and guanine are normally in this tautomeric form.

11. This form of cytosine is the most common isomer in nature.

12. This type of mutation results when a base substitution in a codon leads to the formation of another codon calling for the insertion of a different amino acid.

13. Mutations that lead to premature termination of the polypeptide chain.

14. The test used by Joshua Lederberg and Esther Lederberg to demonstrate that mutations arise spontaneously. (2 words)

15. Name given to those strains (phenotypes) most common in nature. (2 words)

16. The rate of mutation at these sites is much greater than at other locations in the same gene. (2 words)

17. These types of mutations cause the reversion of a mutant phenotype to the wild type even though they are not back mutations.

18. This term defines a gene and was originally coined from the application of the complementation test.

19. This term describes chromosomal deletions, duplications, inversions, and translocations.

20. Two major types of chromosomal abnormalities that result in changes in chromosome number. (2 words)

21. Mutagens that cause base mispairing to occur by chemically modifying normal bases. (1 or 2 words)

22. Mutagens that are incorporated by DNA polymerases during synthesis but which cause mispairing of the bases. (2 words)

23. Acridine orange is an example of this type of mutagen. (2 words)

24. These kinds of genes are the result of inactive DNA repair systems.

25. This repair system is induced by DNA damage caused by UV irradiation.

Answers to the Objective Questions

Multiple Choice

1. *c*   **2.** *a*   **3.** *a*   **4.** *e*   **5.** *d*   **6.** *b*   **7.** *a*   **8.** *c*   **9.** *a*   **10.** *e*   **11.** *c*
12. *e*   **13.** *b*   **14.** *c*   **15.** *c*   **16.** *d*   **17.** *a*   **18.** *e*   **19.** *b*   **20.** *c*

True–False

1. F (also result from base substitutions and other "point" alterations)   **2.** T   **3.** F (result in frameshift mutations)   **4.** T   **5.** F (mutants that exhibit nutritional requirements)   **6.** F (whether they are selectable depends on the phenotype of the mutated gene and the function it performs)   **7.** T   **8.** T   **9.** F (keto form)
10. T   **11.** T   **12.** T   **13.** F (are caused by insertions or deletions)   **14.** T   **15.** F (triploid is the term, 2n + 1 represents trisomy)   **16.** T   **17.** T   **18.** F (cause frameshift mutations)   **19.** F (lower)
20. F $(3' \rightarrow 5')$

Matching

1. *N*   **2.** *L*   **3.** *P*   **4.** *H*   **5.** *K*   **6.** *D*   **7.** *R*   **8.** *F*   **9.** *O*   **10.** *I*   **11.** *A*
12. *E*   **13.** *J*   **14.** *C*   **15.** *B*   **16.** *U*   **17.** *M*   **18.** *T*   **19.** *V*   **20.** *G*

Terms

1. mutations   **2.** Luria-Delbrück fluctuation test   **3.** auxotrophic   **4.** alleles   **5.** tautomeric shifts
6. insertion   **7.** deletion   **8.** frameshift   **9.** tautomer   **10.** keto   **11.** amino   **12.** missense
13. nonsense   **14.** replica plating   **15.** wild type   **16.** hot spots   **17.** suppressor   **18.** cistron
19. aberrations   **20.** aneuploidy, polyploidy   **21.** alkylating agents   **22.** base analogs   **23.** intercalating agent (frameshifter)   **24.** mutator   **25.** SOS

Chapter 7

The Genetics of Bacteria and Their Viruses

INTRODUCTION

The survival of any species depends on its continued ability to remain adapted (live and reproduce) in its environment. This is largely determined by the size and diversity of the genetic pool of the population. The spread of a spontaneous mutation throughout a population by natural selection would require the replacement of most of the population and the extensive proliferation of the mutant cell carrying the beneficial gene. Many bacteria have evolved parasexual mechanisms to enhance their genetic variability beyond that possible through spontaneous mutations. These mechanisms, known as transformation, transduction, and conjugation, involve the transfer of genetic information from one bacterium to another. The transferred DNA may then be incorporated into the recipient's genome by recombination. These processes allow the propagation of many different genes and alleles among the members of a population and thus enhance the gene pool for that population.

Bacterial viruses, or bacteriophages, are found in many bacteria and in some instances they can impact the phenotype and genotype of their bacterial hosts. To understand bacterial genetics it is important that we familiarize ourselves with bacterial viruses and their characteristics. We will look at these first.

BACTERIOPHAGES

7.1. What is a bacteriophage?

A **bacteriophage,** or **phage,** is a virus that infects bacteria. These viruses, like all other viruses, are obligate intracellular parasites, devoid of protein-synthesizing machinery and energy-conversion systems; they contain only one kind of nucleic acid enclosed in a protein coat, or **capsid.** Bacteriophages require a living bacterial host in which to carry out their reproductive cycle. When a suspension of these viruses is spread over the surface of a **bacterial lawn** (confluent layer of bacteria on the surface of a nutrient agar plate) the viruses appear to "eat" (Greek: *phag, -e, -o*; eat) part of the lawn, leaving "holes" (cleared areas) on the surface. These clear areas, known as **plaques,** represent a "viral colony" of thousands of identical phage particles. When a bacteriophage reproduces inside a bacterium, it destroys the cell and releases a few hundred viral progeny, which, in turn, infect adjacent bacteria. When these viral infection–cell lysis cycles are repeated over and over again, eventually the amount of cell lysis is visible to the unaided eye, giving rise to a plaque. In liquid media, a turbid bacterial suspension will become clear as lysis of all the cells becomes complete.

7.2. Not all bacteriophages reproduce in the same way. The reproductive cycle of bacteriophages can either be lytic or lysogenic. Describe the lytic life cycle.

The **lytic** or **vegetative life cycle** culminates in the lysis (rupture) of the host cell and the release of numerous viral progeny. Bacterial viruses exhibiting a lytic life cycle only are also known as **virulent bacteriophages** (or lytic phages) because they inevitably cause the death and destruction of the host bacterium. Examples of virulent phages are the T-even phages (e.g., T2, T4, T6). The life cycle of T4 is illustrated in Fig. 7-1.

The lytic life cycle consists of five steps. These are attachment, penetration, synthesis, assembly, and release. The cycle begins by the **attachment** of the bacteriophage to specific host cell molecules (receptors) on the bacterial cell wall. After the virus attaches to its host, it introduces its genetic material into the cell **(penetration).** In the case of T4, the phage physically "drills" a hole through the bacterial cell wall and membrane and then pushes the core of its tail into the cytoplasm through contraction of a sheath around the core. The viral DNA is then injected into the cell by conformational changes in the phage head. Once the bacteriophage genome en-

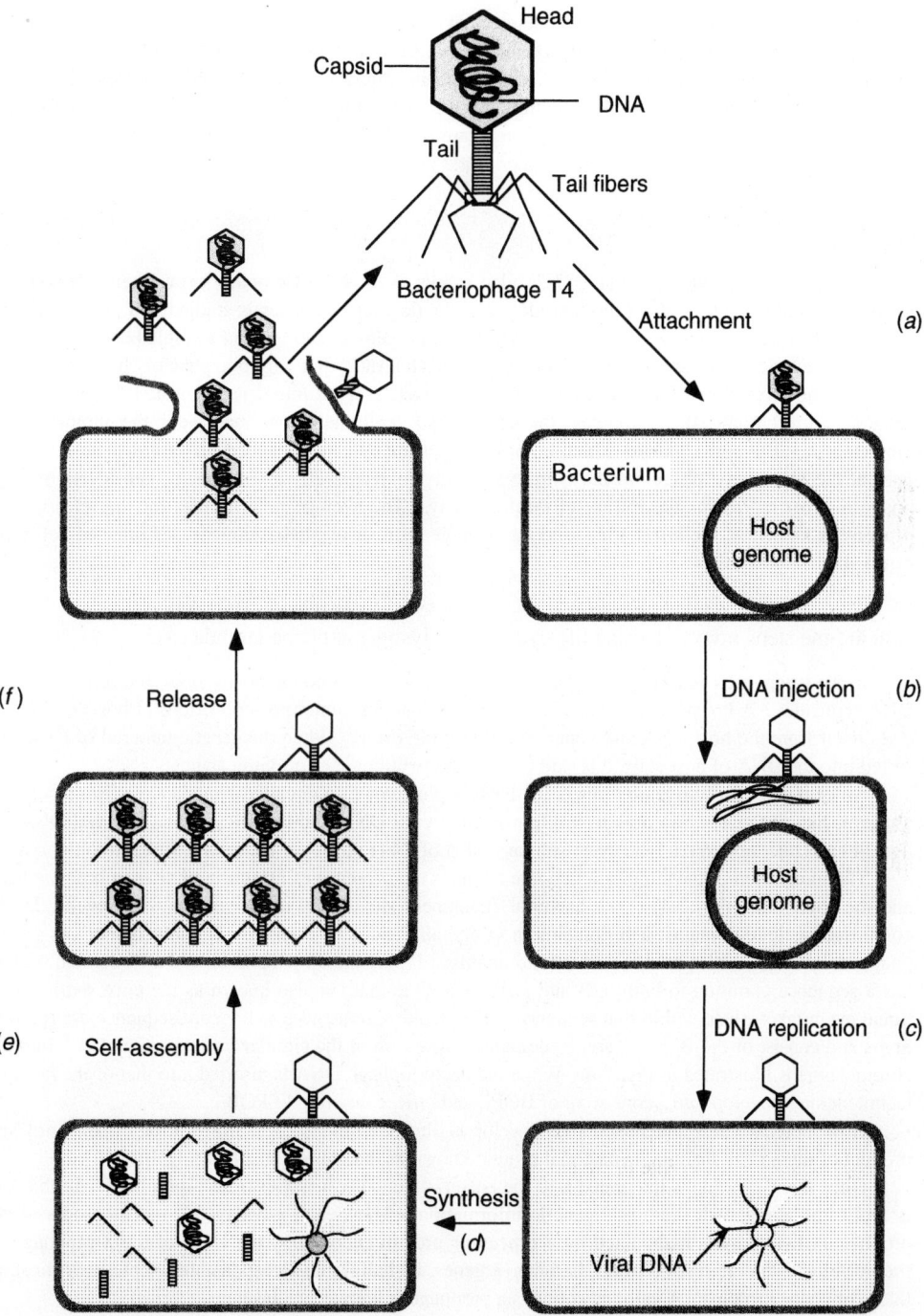

Fig 7-1. The lytic cycle of bacteriophage T4. The lytic cycle of bacteriophage T4 begins (*a*) with the attachment of a viral particle to a susceptible host cell (*E. coli*). (*b*) The virus injects its DNA (located within the head) into the host cell's cytoplasm. Early T4 proteins appear 2–3 minutes after infection. (*c*) After 5 minutes, the host DNA is completely degraded by viral enzymes. This is immediately followed by viral DNA replication via a rolling circle mechanism. (*d*) Other viral components begin to be synthesized. (*e*) Self-assembly begins after about 10 minutes. Complete bacteriophages are assembled within the cell 15–20 minutes after infection. (*f*) Late viral proteins cause the lysis of the cell, and viral progeny are released from the cell after 25 minutes at 37°C. Each infectious particle released can infect another cell and begin another cycle of replication.

ters the cytoplasm, it subverts the host's nucleic acid and protein synthesis apparatus and initiates the **synthesis** of viral proteins and DNA. As the virus proteins are synthesized, they self-assemble into viral components such as the head (containing the phage DNA), tail, and tail fibers (Fig. 7-1*e*). The **assembly** process results in the formation of numerous intact phage particles within the cell. After the assembly step is completed, viral proteins cause the lysis of the host cell, and all the viral progeny are **released** into the environment. Many of the progeny infect other adjacent bacterial cells, initiating another round of replication.

7.3.   What is the one-step growth curve?

The **one step growth curve** (Fig. 7-2), also known as the **Ellis-Delbrück experiment,** demonstrates the life cycle of a viral particle. When the reproduction of lytic bacteriophages is studied and plotted (Fig. 7-2), it appears to take place in steps, each step representing the progeny resulting from a single bacteriophage. In this experiment bacteriophages are mixed with bacteria in such a ratio that only one phage is likely to adsorb to a bacterium. The mixture is then incubated at 37°C, sampled every 5 minutes, and tested for the presence of viral progeny. For the first 20–25 minutes after T4 infects *E. coli,* there is no increase in the number of progeny that are detected. The period during which no new viruses are detected is called the **latent period.** After the latent period, the number of phages in the culture suddenly increases 70–250-fold and then levels off. The rise in phage number represents the release of viral progeny. The average time it takes for the release of new phage progeny is known as the **burst time,** and the average number of phage progeny per infected cell represents the **burst size.**

7.4.   What are the steps involved in the life cycle of the lysogenic phage lambda (λ)?

In the **lysogenic cycle** (Fig. 7-3), the viral DNA is inserted into the host's DNA and reproduces as the host's DNA replicates. A lysogenic virus can remain in this state for numerous replications of host cell DNA until it excises itself from the host DNA and undergoes a lytic life cycle. When the genetic material of these phages is inserted into the DNA of host cells, it is said to be in the **prophage,** or proviral, state.

An example of a lysogenic virus is the bacteriophage of *E. coli* called lambda (λ). The lysogenic bacteriophage λ, like phage T4, also uses its tail to inject the viral DNA into the host. Once the viral DNA is inside the host, it can either undergo a lytic cycle similar to that of T4 (Fig.7-1) or incorporate itself into the host DNA.

The integration of the circular λ genome involves the presence of **attachment sites** in both the bacteriophage and the bacterium (Fig. 7-4). The bacterial attachment site, called *att*B, consists of a specific DNA sequence composed of three domains: a B domain, an O domain, and a B$'$ domain. The attachment site of the bacteriophage, called *att*P, is also made up of three domains: a P domain, an O domain, and a P$'$ domain. The O domain has a sequence common to both *att*B and *att*P. The O domain is also known as the **core sequence.** An insertional event takes place within that sequence. The flanking sequences to the core sequence are referred to as the **arms** and consist of the B, B$'$, P, and P$'$ domains. Insertion of the circular bacteriophage DNA into the bacterial chromosome is illustrated in Fig. 7-4. When the bacteriophage DNA is inserted into that of the host, two new attachment sites develop: *att*L, consisting of BOP$'$, and *att*R, consisting of POB$'$.

How these two new attachment sites develop is illustrated in Fig. 7-5. Integration (*att*B $\times$ *att*P) requires the product of the λ gene *int* and the bacterial protein known as integration host factor (IHF).

Once the λ genome is integrated into the bacterial genome, it is known as a prophage. In this latent state it is replicated along with the host's genetic information. Thus, each progeny cell that is produced contains the prophage. Lysogeny is maintained by a **repressor** protein encoded by the DNA of the λ prophage. The λ repressor blocks the expression of all the other λ genes, including those responsible for phage replication and synthesis of phage proteins. A cell that contains a prophage is known as a **lysogen.**

Under certain conditions (DNA damage to a lysogenic cell, nutrient starvation, or UV irradiation), the λ repressor is inactivated and the prophage state is no longer maintained, and the excision process begins. **Excision,** the removal of the phage DNA from the host DNA, involves recognition of sites *att*L and *att*R by phage enzymes and requires the λ gene *xis,* in addition to *int* and IHF. Excision culminates in a lytic life cycle (Fig. 7-1).

7.5.   Can virulent phages become prophages?

No. By definition a virulent bacteriophage can only undergo the lytic cycle and therefore cannot incorporate its DNA into the host's genome and become a prophage. A **temperate phage,** on the other hand, can exist within the host bacterium as a prophage, allowing the lysogenized bacterium to carry out a seemingly normal life cycle.

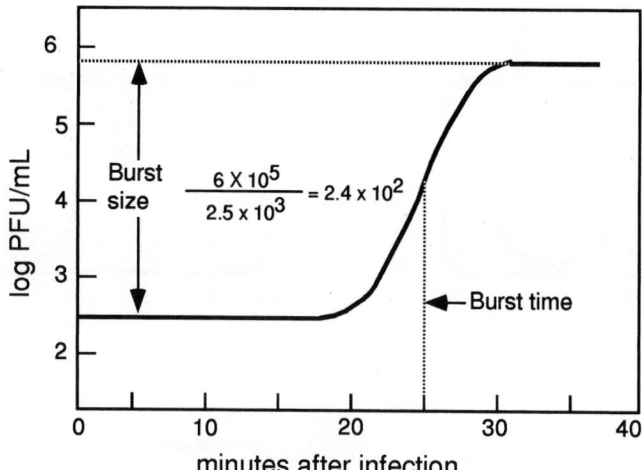

Fig 7-2. One-step growth curve of bacteriophage T4. The one-step growth curve reflects the reproductive process of the lytic bacteriophage (see text for explanation of the various steps). In a typical experiment, a bacterial broth culture is infected with a suspension of bacteriophages at a concentration that will result in infection of a bacterial cell by a single bacteriophage. The infected culture is incubated at 37°C and sampled every few minutes in order to enumerate the viral progeny resulting from that sampling period. The sampled culture is plated on an agar-containing medium containing a lawn of host bacteria. Viral progeny are measured as the number of plaque-forming units (PFU), which represents the number of viral colonies on the plate. The burst size indicates the mean number of viral progeny released per infected bacterial cell and is calculated by dividing the number of PFUs at the plateau of the sigmoidal (S-shaped) curve (e.g., 30 minutes after infection) by the number of PFUs before exponential growth begins (e.g., 10 minutes after infection). The burst time, by convention, is measured as the half-way point of the sigmoidal curve.

GENETIC RECOMBINATION AND GENETIC TRANSFER

7.6. What is genetic recombination?

Genetic recombination in bacteria is a nonreciprocal process whereby segments of genetic material from two different sources are brought together into a single DNA molecule.

7.7. Is there more than one type of genetic recombination?

Yes, there are two basic mechanisms of genetic recombination. These are **homologous** or **general recombination** and **site-specific recombination.** Homologous recombination was discussed in Chapter 3.

7.8. What is site-specific recombination?

Site-specific recombination involves the recombination of two DNA molecules at specific locations variously called insertion sequences (ISs), long terminal repeats (LTRs), and **attachment (*att*) sites.** The integration of the bacteriophage λ into the *E. coli* chromosome is a common example of site-specific (specialized) recombination involving *att*.

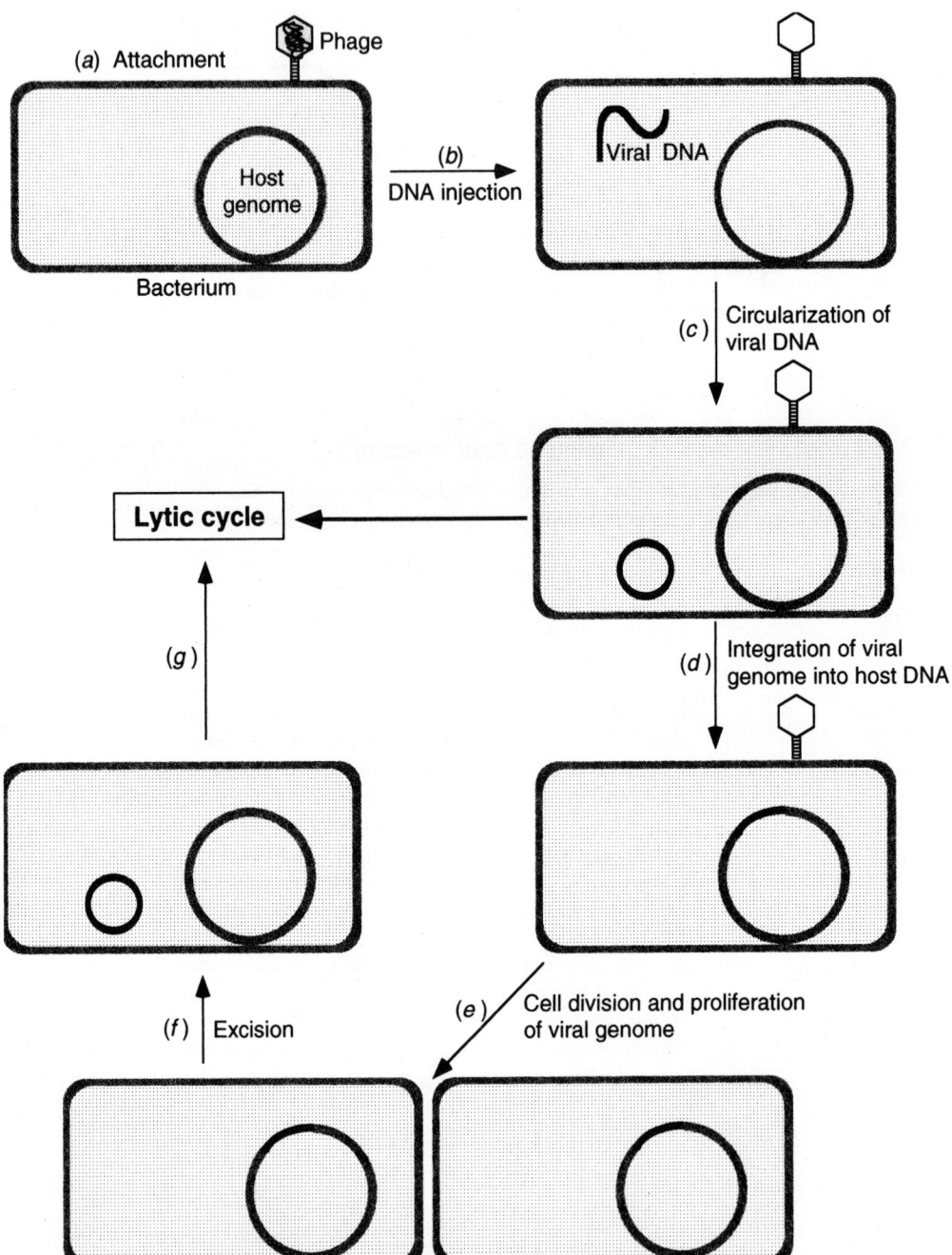

Fig 7-3. The lysogenic life cycle of λ bacteriophage. Lambda (λ) bacteriophage can undergo a lysogenic life cycle that begins with (a) the attachment (adsorption) of the viral particle to the susceptible host cell. After adsorption, the linear viral DNA is injected (b) into the cell and becomes circularized (c). The circularized form of the λ bacteriophage becomes incorporated (d)
into the host genome by a process called site-specific recombination (see Figs. 7-4 and 7-5), becoming a prophage. The bacterium possessing an inserted copy of the λ genome is called a
lysogen. In the prophage state, the virus is replicated during bacterial cell division (e). (f)
When exposed to UV light or nutrient deprivation, the λ prophage is excised from the bacterial
genome, leading to a lytic cycle (g) similar to that illustrated in Fig 7-1.

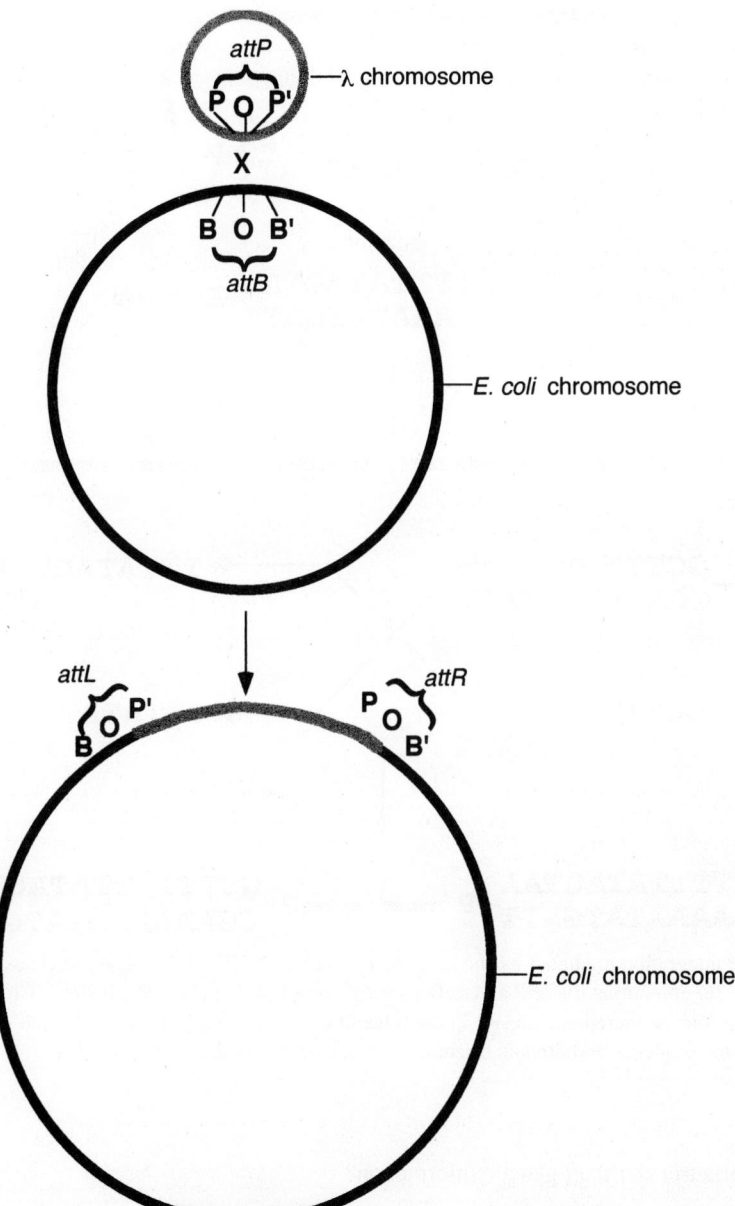

Fig 7-4. Integration of bacteriophage λ into the chromosome of *E. coli*. Bacteriophage λ and *E. coli* both possess attachment sites in their chromosomes. These attachment sites are recognized by the λ integration and excision enzymes. Both chromosomes share a short region of homology indicated by "O." This region of homology is flanked by short DNA sequences that are unique to the organism. The flanking *E. coli* regions are indicated by B and B′, while those of the bacteriophage λ are indicated by P and P′. Thus, the bacterial attachment site is represented by BOB′ and the bacteriophage attachment site is represented by POP′. After integration of the phage into that of *E. coli* by site-specific recombination, the λ chromosome is flanked by the sequences BOP′ and POB′.

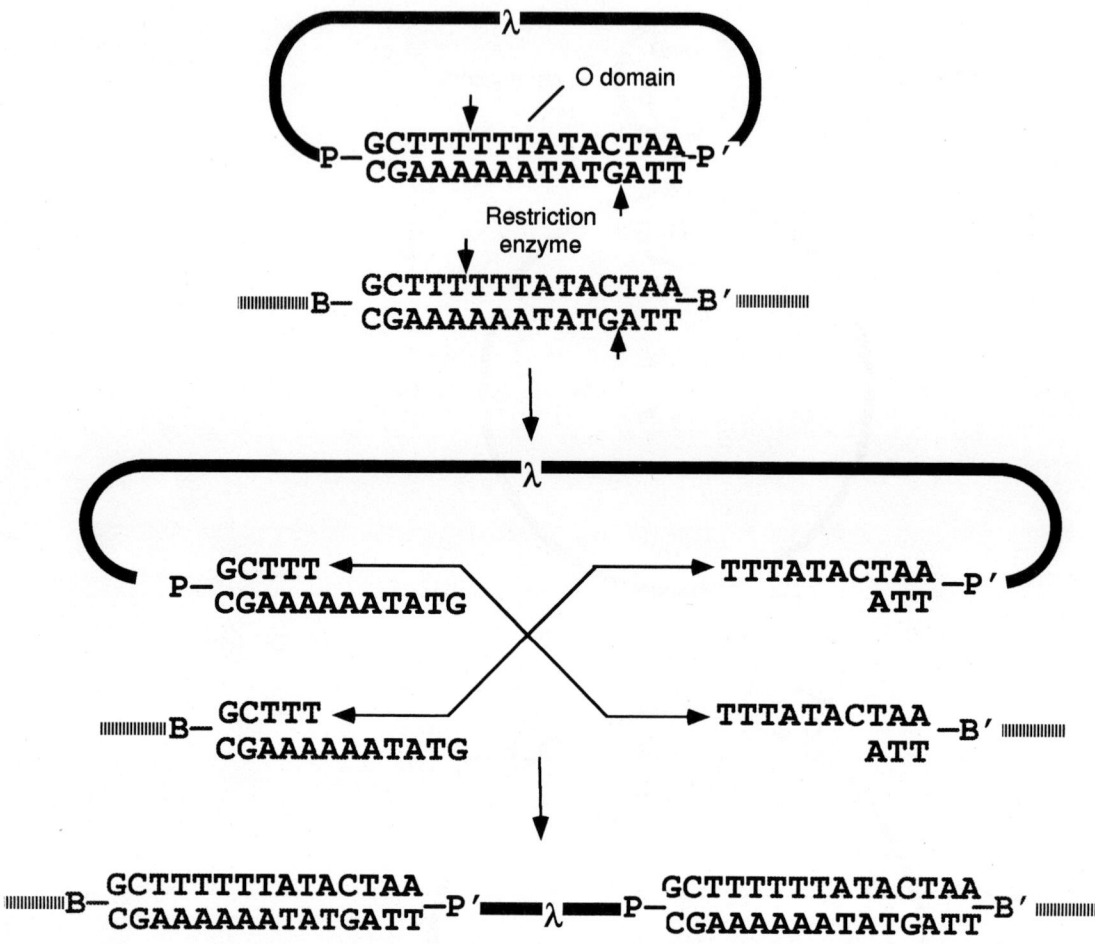

Fig 7-5. Details of integration mediated by the *int* gene product and IHF. The staggered cleavages within the core sequences (O domain) at the *att*P and *att*B sites are caused by a restriction enzyme. Crosswise reunions between opposite staggered ends result in the integration of the phage genome into that of the bacterial host. Compare the sequence and domain orientation shown here with the one seen in Fig. 7-4.

7.9. How do bacteria transfer their genetic information?

There are three recognized mechanisms of genetic transfer between bacteria. These are transformation, transduction, and conjugation. They differ from each other in the source of the donor DNA and the vehicle or mechanism of transfer.

Transformation

7.10. What is transformation?

Transformation (Question 2.17) is a mechanism of genetic transfer between bacteria in which the donor DNA is cell-free DNA in the recipient bacterium's immediate environment. DNA can be naturally released into the environment when cells die and subsequently lyse. Experimentally, known amounts of DNA containing genes of interest can be introduced into the environment in order to transform bacterial cells. Once the free DNA is brought into the cell, it may become integrated into the recipient's genome by homologous recombination. Recipient cells that undergo this process and acquire a new phenotype as a result are said to be **transformed.** This phenomenon allows only the uptake of DNA from the same or related species. The process of transformation in *E. coli* is illustrated in Fig. 7-6.

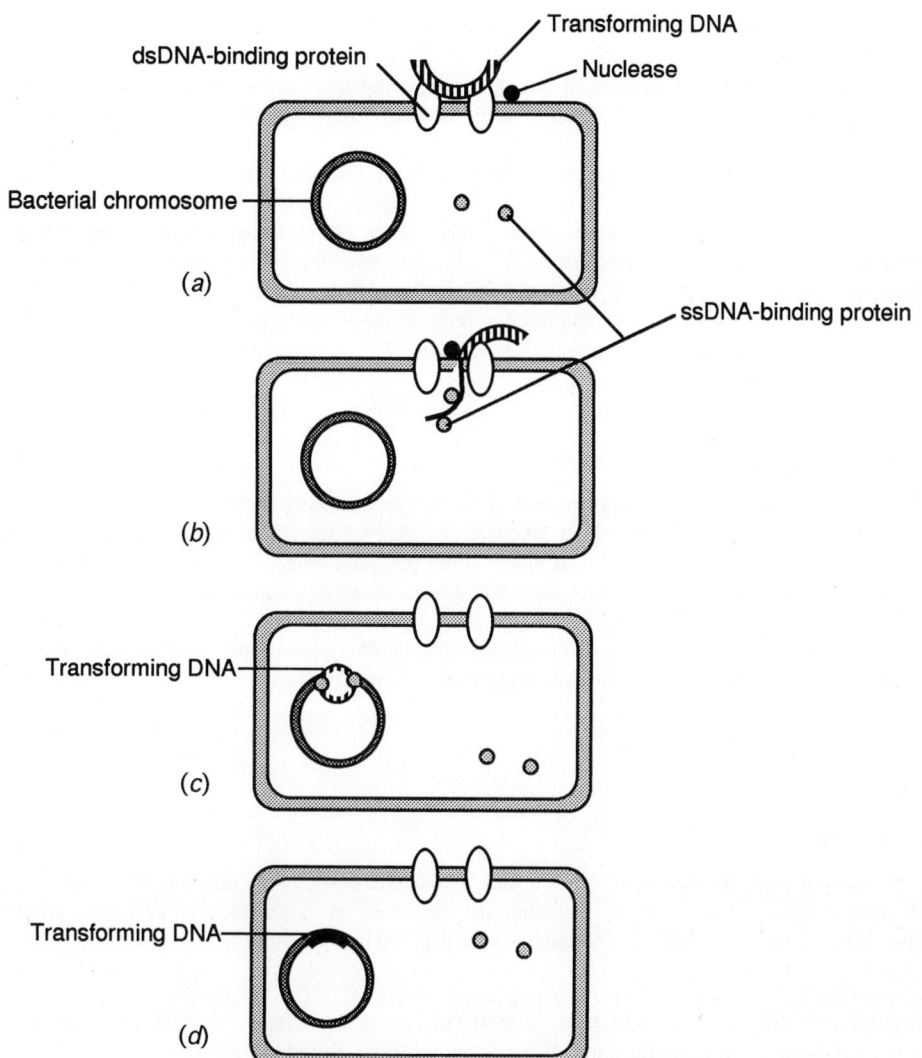

Fig 7-6. Mechanism of DNA transfer in bacteria by transformation. (*a*) Double-stranded DNA (dsDNA) attaches to membrane-bound dsDNA-binding protein. (*b*) One of the two strands of the transforming DNA passes into the cell while the other strand is degraded by a nuclease. (*c*) The single-stranded DNA (ssDNA) within the cell is bound by ssDNA-binding proteins. During this process, homologous recombination takes place and the transforming DNA is incorporated into the bacterial chromosome. The integrated ssDNA is converted into dsDNA by DNA polymerase I. (*d*) The transformed cell contains the segment of transforming DNA incorporated into its genome.

7.11.   What is competence?

The ability of a recipient bacterium to take up free DNA and become transformed is known as **competence.** Competence is an inheritable characteristic of certain strains of bacteria. Some species of bacteria are "naturally" competent; transformation can occur at a high frequency since most cells in such a population can take up environmental DNA at any time. In these bacteria, which include *Streptococcus pneumoniae, Bacillus subtilis,* and *Haemophilus influenzae,* linear, double-stranded DNA (dsDNA) enters the cell, but one strand is degraded while the other is integrated into the chromosome by homologous recombination mediated by RecA protein (see Fig. 3-10).

In many other species of bacteria, including *E. coli,* competence is a brief physiological state during the exponential growth phase, and transformation occurs at a low frequency. In these bacteria, Ca^{2+} ions enhance somewhat the level of competence.

In certain bacteria, competence is associated with the production of double-stranded DNA-binding proteins on the plasma membrane of competent cells and single-stranded DNA-binding proteins in the cytoplasm of the cell. These proteins are present only at specific stages of the life cycle for certain species of bacteria (e.g., *E. coli*) or constitutively for other species (e.g., *Neisseria gonorrhoeae*). Competence-associated proteins function in binding environmental DNA and protecting it from digestion by the cell's endonucleases. In the case of *N. gonorrhoeae* or *H. influenzae,* some competency proteins associated with the plasma membrane bind specific short segments of DNA sequences (e.g., 5'-AAGTGGGTCA-3') that are highly repetitive in these bacterial genomes.

7.12. How many genes can be cotransformed?

Only a few genes, usually less than 15, can be incorporated in a single transformational event. There are several reasons for this observation. First, when donor cells lyse, the DNA that is released usually breaks up into fragments of about 10 million daltons. If one assumes that an average gene consists of 1,000 nucleotide pairs, and estimating that each nucleotide pair weighs 650 daltons, it can be calculated that there are 15 average genes in each 1×10^7 D fragment. Second, it is doubtful that such a large piece of DNA can be taken up intact into the cell because the cell wall and plasma membrane of bacteria offer a permeability barrier to large molecules. It is more reasonable to assume that cells usually take up and incorporate 3–5 average-sized genes in a single transformational event.

Transduction

7.13. What is transduction?

Transduction is a mechanism of DNA uptake by bacteria in which the donor DNA, consisting of fragments of the bacterial chromosome, is introduced into the recipient's cell via a bacterial virus (bacteriophage) vector. During this process, new genetic information may be acquired by the host cell.

7.14. There are two different types of transduction mechanisms. These are generalized transduction and specialized (restricted) transduction. What is generalized transduction?

During **generalized transduction** virtually any bacterial gene can be transferred (Fig. 7-7). The genetic transfer is mediated by a virulent or lytic bacteriophage. Lytic bacteriophages are bacterial viruses that upon infection of a host bacterial cell destroy the DNA of the host and ultimately lyse the cell, releasing numerous viral progeny. Upon infection of a host cell by a lytic bacteriophage, viral enzymes degrade the host's DNA into fragments. Viral DNA is not degraded because some of the bases in the bacteriophage genome are modified so that they are not recognized by viral enzymes (see the section on restriction enzymes in the next chapter). The viral DNA is then replicated and viral proteins are synthesized. The newly replicated DNA is packaged into the coat proteins, and then infectious viral particles are assembled (Fig. 7-1). When the viruses are fully assembled, viral enzymes degrade the cell's envelopes, lysing the cell and releasing the viral progeny.

Infrequently, however, some of the host's DNA is packaged into the virus along with an incomplete viral genome. When this happens, a **generalized transducing phage** is formed, which, although capable of initiating an infection, is unable to replicate itself or lyse the host cell. Some phage genes must be given up to accommodate the bacterial genes within the confines of the virus head. These defective transducing phages serve as vehicles for the transfer of the host DNA (incorporated during viral assembly) from one cell to another. Because the packaging of host DNA into the viral particle is a random event, any given bacterial gene has an equal chance of being packaged and transferred to a recipient cell. Upon infection of a bacterial host cell by the transducing phage, the transducing DNA is introduced into the host's cytoplasm and becomes incorporated into the bacterial genome by homologous recombination. The infected cell is not destroyed because the transducing phage is defective in that it does not possess a full complement of genes.

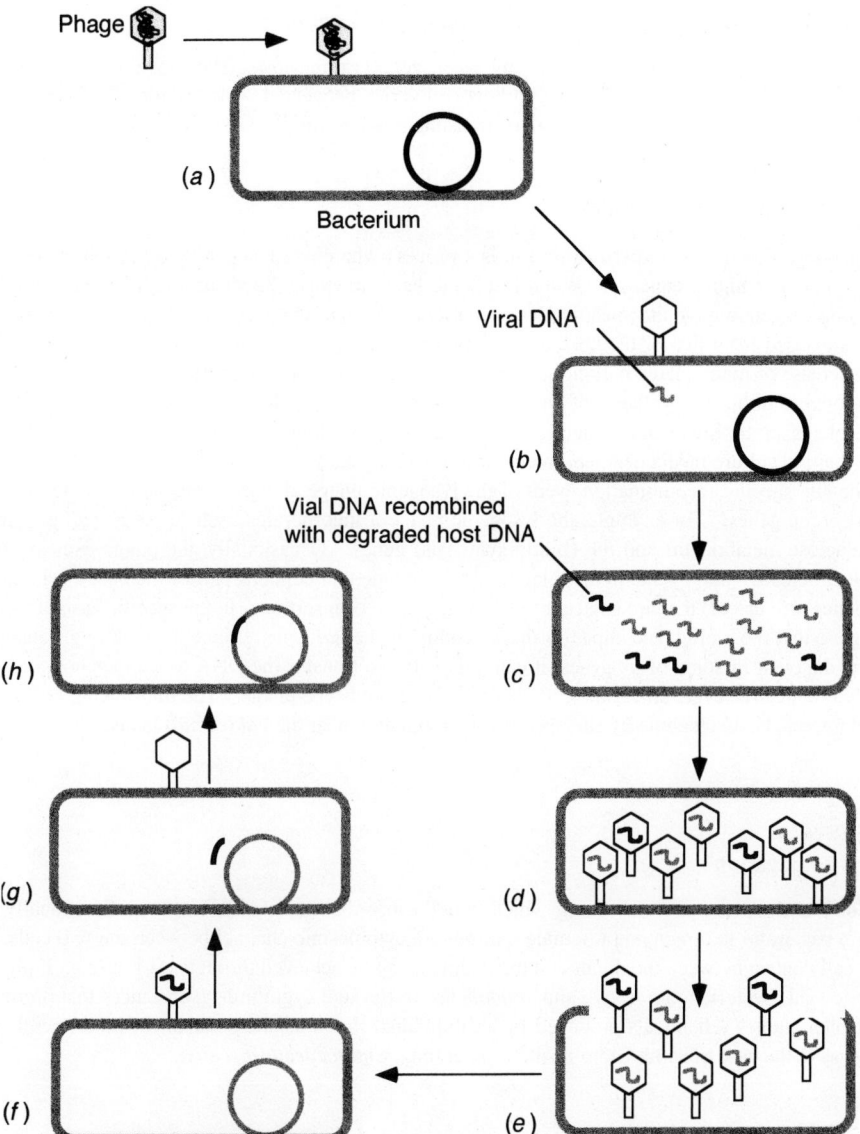

Fig 7-7. Generalized transduction. Generalized transducing phages like T4 invade bacteria by adsorbing to the cell wall (*a*) and injecting their genetic material (*b*).
Shortly after infection, the host's DNA is degraded into small fragments (*c*).
During viral assembly (*d*), some of the viral particles take up some of the small
fragments of host DNA (transducing DNA). After cell lysis (*e*), these viral particles can infect other bacterial cells (*f*) and inject their genome with transducing
DNA (*g*). Homologous segments of transducing DNA can then become incorporated into the host's DNA (*h*) by homologous recombination and impart one or a
few new genes to the host's genome.

7.15. What is the frequency of recombination via generalized transduction?

The frequency of recombination for any given gene is quite low. The probability that any given gene is transferred to and incorporated into a given host cell varies between 1×10^{-6} and 1×10^{-8}. Generally only a few genes (<5) are transferred during a given transductional event.

7.16. What is specialized transduction?

Specialized or **restricted transduction** is a process whereby a lysogenic bacteriophage serves to transfer a specific gene at a high frequency. When lysogenic bacteriophages infect host cells, their DNA is incorporated into the host's genome by site-specific recombination. Through the process of **induction,** the prophage genome becomes excised from that of the host and undergoes a lytic life cycle, which results in the lysis of the host cell and the release of many phage particles. Induction occurs when the λ repressor protein becomes inactivated and can no longer inhibit the synthesis of enzymes involved in the excision of the prophage from the host's genome and the phage's lytic life cycle. Induction can be initiated by exposing lysogenized cells to UV light or by transferring them to culture media deficient in one or more amino acids.

The site-specific recombination event of the lysogenic phage always occurs at a specified location and adjacent to certain genes. For example, the lysogenic phage λ incorporates itself in the *E. coli* genome between the *gal* (galactose metabolism) and *bio* (biotin synthesis) genes. Occasionally, the phage excision from the host's genome is defective and results in the release of a λ genome that contains part of the host's genome; in particular, those genes (*bio* or *gal*) that are adjacent to the phage's site of insertion. In the specific case of the bacteriophage λ, the gene that most often accompanies the λ genome is the *gal* gene. Since the resulting transducing bacteriophage is deficient in some of its genes, it can only serve to transfer the DNA to another host cell, but cannot go through a replicative cycle unless the cell also contains a wild-type phage. The transducing DNA becomes incorporated into the host's genome by site-specific recombination near the *gal* (or *bio*) locus.

Conjugation

7.17. What is conjugation?

Conjugation is a process during which genetic information is transferred unidirectionally from a donor ("male") bacterium to a recipient ("female") through a cytoplasmic channel between the two cells. The required cell-to-cell contact between the donor and the recipient can be achieved through sex pili (e.g., F pili) as in the case of enteric (gut) bacteria (e.g., *E. coli* and *Salmonella*) or through agglutinins (substances that promote cell clumping) or pheromones (chemicals produced by an individual that alter the behavior of other members of the same species) as in the case of some Gram-positive cocci (e.g., *Enterococcus faecalis*).

7.18. How many genes can be transferred during conjugation?

During conjugation, whole plasmids, some containing 50 or more genes, and large portions of the bacterial chromosome can be transferred. In theory, the entire bacterial chromosome can be transferred during conjugation, but this rarely happens because the cytoplasmic channel usually breaks before transfer is complete.

7.19. Are there different types of donor bacteria?

Yes, there are at least three different types of donor bacteria. These are $\mathbf{F^+}$, $\mathbf{F'}$, and **Hfr** bacteria. All these three types of bacteria possess sex pili, which form a cytoplasmic bridge with recipient bacteria during conjugation. Recipient bacteria, which lack the sex pilus, are symbolized $\mathbf{F^-}$.

7.20. What are plasmids?

Plasmids (Question 1.6) are relatively small, autonomous pieces of hereditary material distinct from the main genome. In bacteria, these are circular DNA molecules containing 10 to 100 genes; they vary tremendously in size from about 2,500 base pairs to more than 80,000 base pairs in length. Plasmid replication is often independent of the chromosome's replication.   Depending upon the type of plasmid, there may be 2 to 250 copies of a plasmid in each bacterial cell.   In general, most of a plasmid's genes are involved in the control of its own replication, but some plasmids provide the cell with one or more useful functions (Table 7.1).   Frequently, the function of a plasmid is unknown and the plasmid is said to be **cryptic.**   The genetic information contained in plasmids generally is not essential for cell function.   Plasmids, however, contain genes that afford the cell a survival advantage in certain environments (such as in the presence of antibiotics) over cells lacking the plasmid.

Table 7.1.   Some Functions Attributed to Bacterial Plasmid Genes

Promote the transfer of genetic material from one cell to another
Provide resistance to antimicrobial agents (e.g., penicillin and streptomycin)
Provide resistance to heavy metals (e.g., mercury, lead, arsenic)
Production of certain toxins (e.g., coagulase, hemolysin, enterotoxins)
Nutrient assimilation and metabolism (e.g., lactose, nitrogen fixation, urea)
Induce tumors in plants (e.g., nodules and crown galls)
Degradation of toxic organic materials (e.g., camphor, octane, naphthalene)
Production of pigments
Production of antimicrobial substances called colicins (B, I, V, E1, E2, E3, D, K)

7.21.   What are F$^+$ bacteria?

These are bacterial strains that donate only plasmid genes during conjugation.   The symbol F$^+$ comes from the fact that these bacteria contain a "Fertility factor" that allows them to conjugate.   The fertility factor, we now know, is a type of plasmid known as a **conjugative plasmid.**   It carries genetic information that codes for a protein called pilin, used to make a sex pilus (Question 1.10) known as the F pilus, as well as for proteins that promote the transfer of plasmid DNA from the donor to a recipient.   F$^-$ cells lack the F factor and thus are not pilated.   These F$^-$ bacteria, however, become F$^+$ after conjugation with F$^+$ bacteria for the reason discussed in the next question.

The F factor (plasmid) of the bacterium *E. coli* is a circular double-stranded DNA molecule.   It measures 94.5 kilobase pairs (kbp) and is capable of directing its own (autonomous) replication in the cytoplasm of the cell.

7.22.   How are plasmids transferred during conjugation?

Conjugation between an F$^+$ and an F$^-$ bacterium begins with the attachment of the two cells via a sex pilus (Fig. 7-8) made by the F$^+$ cell.   The sex pilus then retracts, pulling the two cells into intimate contact with each other.   Two plasmid-encoded proteins (TraY and TraZ) cut the plasmid at an origin of replication (*ori*T) site. The plasmid DNA then replicates by the rolling circle mechanism explained in Chapter 3.   Only one strand of plasmid DNA is transferred from the donor to the recipient cell through the cytoplasmic bridge with the help of a **pilot protein.**   Both the donor and recipient cells synthesize a DNA strand complementary to that of the single-stranded F plasmid, forming double-stranded F plasmids.   When conjugation is complete, the exconjugant cells separate.   Since each has at least one copy of the plasmid, they both are F$^+$ cells.

7.23.   What are Hfr strains?

Hfr strains were so named because they exhibited a *H*igh *f*requency of *r*ecombination for certain chromosomal genes.   We now know that this high frequency of transfer by conjugation is due to the fact that the conjugative (F) plasmid has become incorporated into the bacterial chromosome.   Genetic elements (like plasmid F and λ phage) that can exist either as autonomously replicating entities or as chromosomal entities are called **episomes.**

7.24.   How is the F plasmid inserted into the bacterial chromosome?

The insertion of the F plasmid into the bacterial chromosome takes place by site-specific recombination (Fig. 7-9).   The regions or sites at which recombination occur are known as **insertion sequences** (IS) and consist of

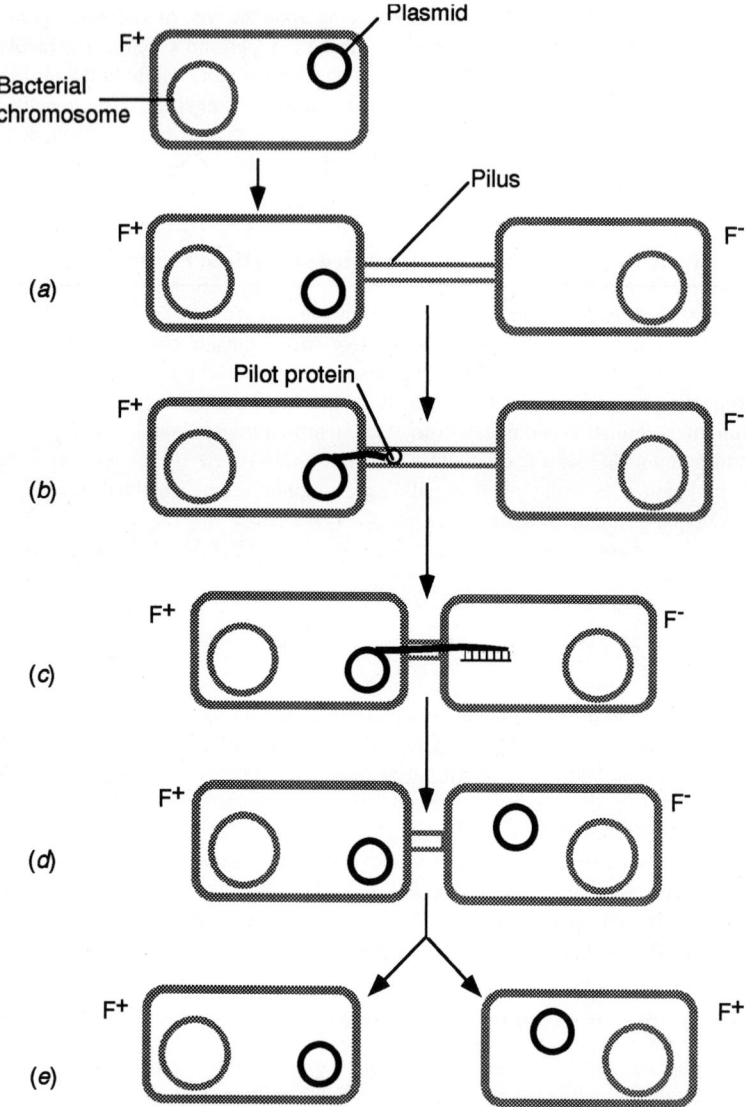

Fig 7-8. Bacterial conjugation. In bacterial conjugation, the F⁺ bacterium
(containing the F plasmid) is pilated and attaches to an F⁻ bacterium
through its pilus (*a*). The pilus provides a cytoplasmic channel
through which plasmid DNA can be transferred (*b*) with the aid of a
pilot protein (tiny sphere). The plasmid DNA replicates by a process
known as rolling circle replication. During this process, a single-
stranded DNA molecule is transferred through the pilus to the F⁻
bacterium. As DNA transfer takes place, the two cells come closer
together as the pilus retracts (*c*). Eventually, the host DNA uses the
single-stranded DNA transferred from the F⁺ bacterium as a template
to synthesize a complementary strand. At the end of this replicating
cycle (*d*) both cells are F⁺, since each has a copy of the F plasmid.
The process of conjugation ends when the two conjugants come
apart (*e*).

homologous DNA sequences between the conjugative plasmid and the chromosome. These IS sites are distributed at eight different locations throughout the *E. coli* chromosome, and thus the F plasmid can be integrated into the chromosome at any one of these sites in a given cell. The frequency of recombination and the order in which the genes are transferred depend upon the orientation and site of insertion of the F plasmid in the chromosome.

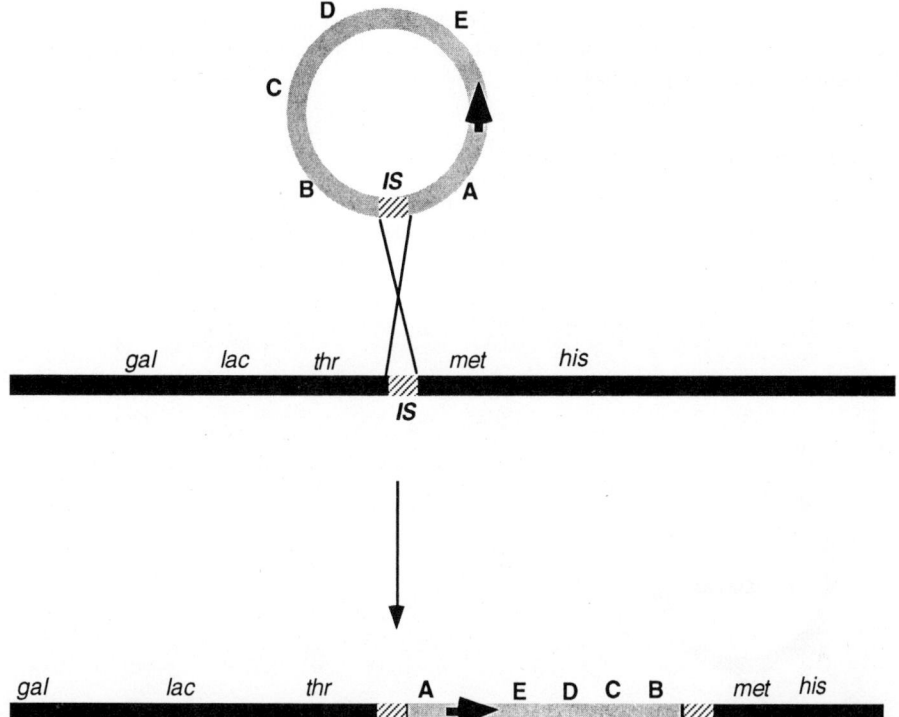

Fig 7-9. Insertion of the F plasmid into the bacterial chromosome. The capital letters indicate arbitrary plasmid genes. The *italic* abbreviations indicate chromosomal genes. The **bold** arrow shows the orientation or direction of transfer. IS indicates homologous insertion sequences. During conjugation between Hfr and F⁻ strains, chromosomal genes are transferred before most of the plasmid genes.

7.25. How are chromosomal genes transferred by Hfr strains?

The process of DNA transfer between Hfr strains and F⁻ strains is essentially the same as that between F⁺ and F⁻ strains. There is, however, one significant difference. Whereas the recipient F⁻ bacterium always becomes F⁺ after conjugation with an F⁺ strain, it rarely becomes F⁺ during conjugation with an Hfr strain. This is because during a conjugation between an Hfr and an F⁻ strain, all the essential plasmid genes (some of which are at the end of the transferred chromosome) are not transferred until all the chromosomal genes are transferred. It is a rare event indeed when the entire chromosome is transferred without disruption of the conjugation bridge between the conjugants. Since the entire chromosome is rarely transferred, it follows that some F-plasmid genes (such as the sex pilus gene) are likewise rarely transferred.

7.26. Can we map the location of genes in the chromosome with Hfr strains?

Conjugation between different Hfr and F⁻ strains can be used to map the relative positions of genes in the bacterial chromosome. The mapping procedure is based on the fact that genes closer to the origin of transfer will be transferred at a higher frequency than genes farther away from the origin. In a typical mapping experiment, 1×10^7 cells/ml of a prototrophic wild-type Hfr strain (sensitive to the antibiotic streptomycin, Str^s) are mixed with 1×10^8 cells/ml of a multiple auxotrophic F⁻ mutant (F⁻, Thr⁻, Lac⁻, Gal⁻, His⁻, Str^r) of *E. coli*. At specified time intervals after mixing, samples are taken and violently agitated in a kitchen blender to break apart mat-

ing pairs. The samples are then spread onto plates containing various chemically defined media (including streptomycin) and either lacking one of the required amino acids (threonine, histidine) or containing galactose or lactose as the only sugars. After an incubation period the number of colonies on the media are counted. Colonies
appearing on these plates are recombinants because the donor bacteria can transfer their genes but are sensitive to
streptomycin and cannot grow (wild-type *E. coli* are streptomycin sensitive). The recipient strain requires two
amino acids and cannot catabolize galactose or lactose. The frequency of different recombinants after various
time intervals indicates the order in which the genes are located on the chromosome. Since the rate of DNA
transfer from donor to recipient is fairly constant, the time of appearance of prototrophic markers (genes) is in-
dicative of distance between those markers.

7.27. What are F′ (pronounced F prime) strains?

   F′ strains are those bacteria that possess an F plasmid containing one or more chromosomal genes. These
strains of bacteria are easily identified because the only chromosomal genes they transfer during conjugation are
those that have been incorporated in the F plasmid plus the F plasmid gene(s) that confers the ability to form F
pili on the recipient cell.

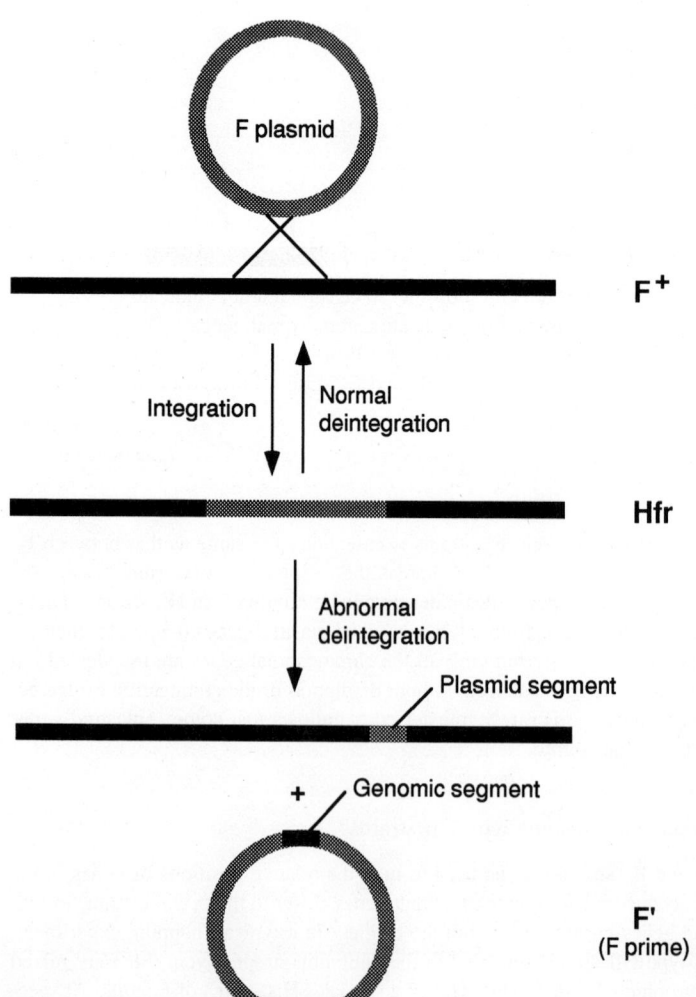

Fig 7-10. Formation of an F′ strain. An F′ strain forms when there is an abnormal deintegration or excision of the F plasmid from an Hfr. During the excision process, a chromosomal gene remains attached to the F plasmid, becoming part of the F plasmid. The end result is a plasmid with an additional bit of genetic information that can be expressed.

7.28. How are F′ strains formed?

The F plasmid can exist in a population of cells either in the integrated or free state. As was discussed previously, the process of integration is by site-specific recombination. The integrated F plasmid, on occasion, becomes abnormally excised (deintegrated) from the chromosome (Figure 7-10). When the excision is imperfect and the excised plasmid takes along some chromosomal genes, the resulting cell is said to be an F′ strain. These cells transfer the excised chromosomal gene at a very high frequency. This process of genetic transfer is called **F-duction** or **sexduction.** The transferred genes need not be integrated into the recipient's chromosome to be expressed because the F′ plasmid retains its autonomy, including independent reproduction, and therefore can express all of its genes. An F′ cell is thus a partial diploid for these genomic genes that have alleles in the plasmid.

Transposable Elements

7.29. What are transposable elements?

There are pieces of DNA measuring 700 to 20,000 base pairs in length that "jump" from one region of the genome to another. In both prokaryotes and eukaryotes, the jumping can occur from one location on the chromosome to another or, in the case of bacteria, from the chromosome to a plasmid, or vice versa. These "jumping genes" are known as **transposable elements** or **transposons.** Transposable elements are important because they may be involved in turning genes on or off when they move from one location to the control region of another gene. Additionally, transposable elements, because they insert themselves at various locations in the genome, can cause mutations if they are integrated into a gene.

Transposable elements have been found in many different organisms, including bacteria, fungi, protozoa, insects, and plants, and perhaps in all types of cells. There are at least three types of transposable elements: insertion sequences, complex transposons, and retrotransposons.

7.30. What are insertion sequences?

An **insertion sequence** is a transposable element possessing only the genes required for its own transposition (Fig. 7-11). A **transposase** enzyme recognizes its own inverted repeats, cuts, and ligates the DNA during the transposition process. A **resolvase** enzyme is also required for the excision of an IS from its chromosomal or plasmid site. Insertion sequences range in size from 700 to 5,000 base pairs. They are the simplest transposition modules and are normal components of bacterial chromosomes and plasmids. Insertion sequences are given the prefix **IS** followed by a number that identifies the type of insertion sequence. Hfr strains of *E. coli* are formed when site-specific recombination occurs at insertion sequences in the bacterial chromosome and the F plasmid. Short regions of DNA, called **inverted terminal repeats (IR),** are at the ends of each IS. Next to each IR there is a **direct** (target) **repeat** of identical nucleotide sequences that serves as an insertion site for transposable elements. Direct repeats are so named because the two copies of the sequence are repeated in the same orientation.

7.31. What are complex transposons?

Complex (composite) **transposons** carry one or more genes other than those required for their own transposition. Complex transposons, ranging in size from 2,000 to 20,000 base pairs, are bordered by insertion sequences that may be identical or closely related and may be oriented in the same or inverted direction (Fig. 7-12). Complex transposons synthesize replicas of themselves by DNA replication. When a replica transposes itself from one site to another, it leaves the original transposon behind. Complex transposons frequently carry genes that make the cell resistant to antibiotics (Table 7.2), confer resistance to heavy metals, or genes for enterotoxins, but any cellular gene may be found between the insertion sequences.

7.32. What are retrotransposons?

Retrotransposons (retroposons) are many thousands of base pairs long and are also bordered by insertion sequences. Retrotransposons characteristically carry genes for the enzymes **reverse transcriptase** and an **integrase.** Retrotransposons make copies of themselves from RNA transcripts of the DNA from which these transposons originated. Thus, the original retrotransposon is conserved whereas all the copies of the DNA are totally new. Because recombination places the DNA copies at new sites, the retrotransposons appear to jump from site to

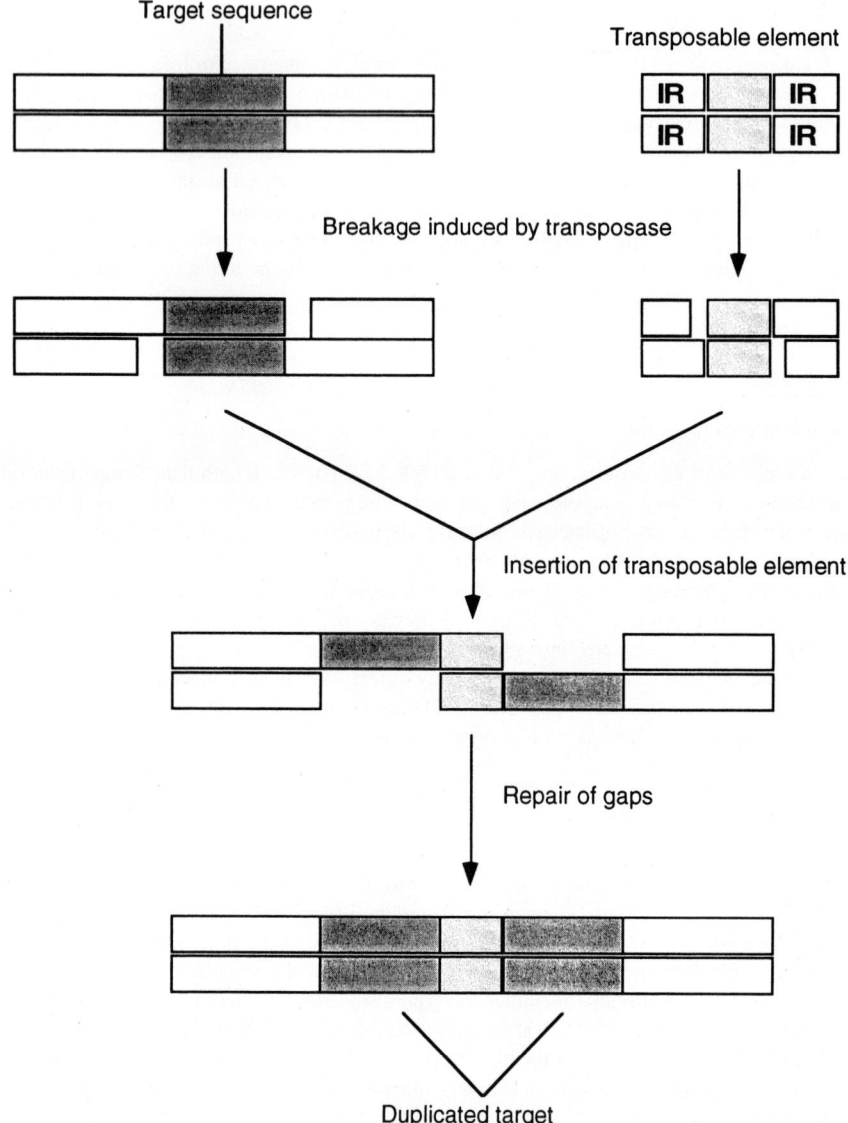

Fig 7-11. The transposition process. A transposase, coded for by a transposable element, causes breaks in both the element itself and at sites adjacent to the target sequence in the bacterial genome. The transposable element inserts itself by site-specific recombination into the bacterial genome. Enzymes repair the gaps created by the insertion of the transposable element in the bacterial genome. This repair process duplicates the target sequence.

Table 7.2.   Description of Representative Complex Transposons

Transposon	Organism	Length (bp)	IS Number	Resistance
Tn5	*Staphylococcus aureus*	5,700	IS50	kanamycin
Tn9	*Escherichia coli*	2,500	IS1	chloramphenicol
Tn10	*Escherichia coli*	9,300	IS10	tetracycline
Tn903	*Escherichia coli*	3,100	IS903	kanamycin

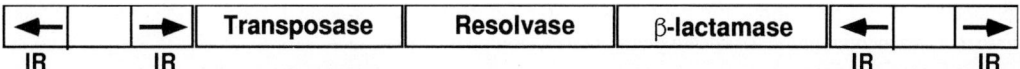

Fig 7-12. A complex transposon. Complex transposons are large transposable elements measuring 2,000 to 20,000 base pairs in length. They are flanked by insertion sequences that have retained their corresponding inverted repeats (arrows). Additionally, they carry a transposase and a resolvase, which allow this transposable element to duplicate itself and insert the duplicate into another region of the chromosome or into a plasmid. Some complex transposons have additional genes that impart resistance to one or more antibiotics. The gene coding for the enzyme β-lactamase is indicated in the figure.

site within and among chromosomes or plasmids. The best understood retroposons are the retroviruses (e.g., human immunodeficiency virus associated with AIDS).

Objective Questions

Multiple Choice

Directions: Choose the one best answer.

1. Small circular DNA molecules that are capable of self replication are called (*a*) introns. (*b*) exons. (*c*) plasmids. (*d*) transposable elements. (*e*) none of the above.

2. When a population of bacteria capable of conjugation transfers a specific chromosomal gene (say a gene coding for galactose metabolism) at a very high frequency, but no other genes, regardless of how long the bacteria are allowed to mate, these bacteria are said to be (*a*) F$^+$. (*b*) F$^-$. (*c*) Hfr. (*d*) F'. (*e*) none of the above.

3. A mechanism of genetic exchange that does not occur in a culture medium containing the enzyme DNase is (*a*) specialized transduction. (*b*) conjugation. (*c*) transformation. (*d*) generalized transduction. (*e*) homologous recombination.

4. During conjugation between an Hfr strain and an F$^-$ strain, the time it takes for a given gene to pass from the donor to the recipient (*a*) is independent of the distance of the gene from the origin of the plasmid. (*b*) depends on the competence of the cell. (*c*) differs from time to time. (*d*) is inversely proportional to the distance of the gene from the origin of transfer site. (*e*) is directly proportional to the distance of the gene from the origin of transfer site.

5. If two populations of bacteria cultured in a U-shaped tube are separated by a membrane filter (which does not allow phage particles to pass), but recombination takes place anyway, the mechanism of genetic exchange is (*a*) specialized transduction. (*b*) site-specific recombination. (*c*) conjugation. (*d*) transformation. (*e*) generalized transduction.

6. The Ellis-Delbrück experiment demonstrates (*a*) the presence of a latent period in bacteriophage reproduction. (*b*) the presence of an eclipse period in bacteriophage reproduction. (*c*) that most bacterial mutants are auxotrophic mutants. (*d*) that wild-type bacteria rarely mutate to auxotrophy. (*e*) that bacteriophages can lysogenize bacteria.

7. Plasmids generally code for genetic traits that are (*a*) not essential for the survival of the species. (*b*) essential for the survival of the species. (*c*) also present in the chromosome. (*d*) mostly involved in imparting resistance to heavy metals. (*e*) involved in conjugation.

8. Bacterial cells that can be lysogenized by λ possess in their genome (*a*) 80S ribosome genes. (*b*) *att*B sites. (*c*) *att*P sites. (*d*) *att*L sites. (*e*) *att*S sites.

9. Competence in *E. coli* is limited to the following stage in the population growth curve: (*a*) lag phase. (*b*) stationary phase. (*c*) exponential phase. (*d*) death phase. (*e*) division phase.

10. Plaques, which can be seen on the surface of bacterial lawns infected by lytic viruses, indicate the presence of (*a*) bacterial colonies. (*b*) viral colonies. (*c*) auxotrophic mutants. (*d*) F'. (*e*) antibiotic-producing bacteria.

11. A prophage is (*a*) an auxotrophic mutant. (*b*) a gene. (*c*) a phage DNA incorporated into the host genome. (*d*) host DNA packed into viral heads. (*e*) DNA of lytic phages.

12. The protein that guides the transfer of an F factor through the pilus is called a (*a*) chaperone. (*b*) pilot. (*c*) leader. (*d*) signal. (*e*) slave.

True–False

1. Bacterial cells generally lack plasmids.

2. Bacteria reproduce by a process known as parthenogenesis.

3. Lysogens are cells that contain prophages.

4. Bacterial lawns infected by lytic viruses can be detected by the presence of plaques.

5. Virulent phages can assume a prophage state.

6. The bacteriophage T4 infects bacterial cells by injecting its DNA through a pilus.

7. All plasmids insert themselves into bacterial chromosomes by site-specific recombination.

8. The integration of the phage λ genome is dependent upon the presence of attachment sites in both the host DNA and the virus DNA.

9. Site-specific recombination is the mechanism by which the bacteriophage λ incorporates itself into the host chromosome.

10. Lysogens can acquire viral genes.

11. The insertion of the F factor into the chromosome to form Hfr strains is by homologous recombination.

12. Calcium ions enhance competence in certain bacteria.

13. Competence is an inheritable characteristic.

14. Lysogenic bacteriophages can mediate specialized or restricted transduction.

15. Pili function as cytoplasmic bridges during transformation.

16. Cryptic plasmids have no known function.

17. UV light can cause a lysogenic bacterium to lyse.

18. During conjugation, a molecule of ssDNA resulting from a rolling circle replication is transferred from the donor to the recipient, which then synthesizes the complementary strand.

19. Insertion of the F plasmid into the bacterial chromosome takes place at an IS.

20. Transposable elements carry terminal inverted repeats.

For questions 21–25 answer true or false by using the following experimental results: At time zero, Hfr strain D1 (*ala*+ *bio*+ *cis*+ *dra*+ *eco*+ *his*+ *pro*+ *try*+ *kan*ˢ) was mixed with the F⁻ strain R1 (*ala*⁻ *bio*⁻ *cis*⁻ *dra*⁻ *eco*⁻ *his*⁻ *pro*⁻ *try*⁻ *kan*ʳ), and at various time intervals, small aliquots were removed and mixed violently to separate mating pairs (interrupted mating). The samples were then plated onto selective media to measure the frequency of *try*+ *kan*ʳ recombinants that had received certain genes from the Hfr strain. The results obtained are illustrated in the figure below.

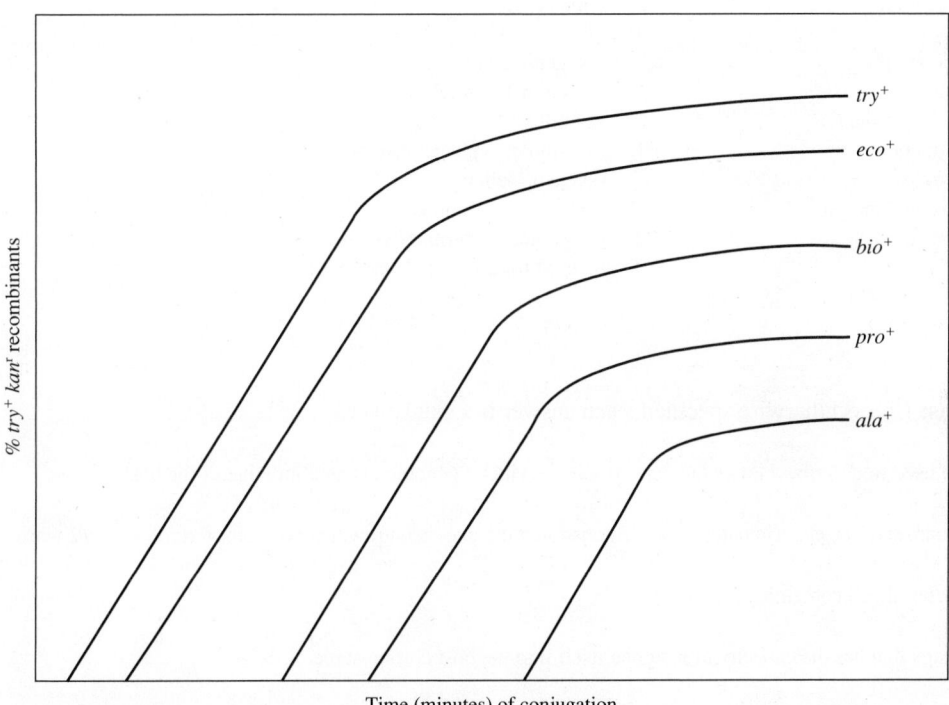

Time (minutes) of conjugation

21. All F⁻ cells that received the *try*+ allele from D1 also received the *bio*+ gene.

22. The order of the genes in the Hfr chromosome is *ala* → *pro* → *bio* → *eco* → *try*.

23. Most *try*+ *kan*ʳ recombinant cells are likely to be Hfr cells.

24. We can tell the order of genes on the *E. coli* chromosome by the time interval between mixing and disruption of the conjugants and the phenotype of the exconjugant cells.

25. If the earliest transfers of the *try*+ and *eco*+ genes are 15 and 20 minutes, respectively, it is possible to detect *try*+*eco*+ recombinants after 21 minutes.

Matching

Directions: Match each item in **Column A** with the one in **Column B** to which it is most closely associated. Each item in **Column B** can be used only once.

	Column A		Column B
1.	direct repeat	*A.*	specialized transduction
2.	insertion sequence	*B.*	reverse transcriptase
3.	transformation	*C.*	bacterial attachment site for λ phage
4.	burst size	*D.*	F factor + chromosomal gene
5.	capsid	*E.*	one-step growth curve
6.	Ellis-Delbrück experiment	*F.*	simplest type of transposable element
7.	lysogen	*G.*	average number of phage progeny per infected cell
8.	plaque	*H.*	F factor or λ phage
9.	Hfr	*I.*	insertion site for IS
10.	F′	*J.*	bacterium with prophage
11.	conjugation	*K.*	competence
12.	episome	*L.*	viral colony
13.	temperate phage	*M.*	lysogenic phage
14.	*att*B	*N.*	viral transfer of DNA
15.	*xis* gene product	*O.*	sex pilus
16.	transduction	*P.*	excision of λ phage from host DNA
17.	retrotransposon	*Q.*	phage protein coat
18.	interrupted mating	*R.*	F factor in chromosome
19.	lambda phage	*S.*	nonpilated bacterium
20.	F⁻	*T.*	method for mapping bacterial genes

Terms

Directions: Unless otherwise specified, each answer is a single word.

1. Term used to describe a temperate bacteriophage with its genome inserted into that of the host.

2. The names of the experimenters who demonstrated the one-step growth curve of bacteriophages. (2 words)

3. A bacterial cell containing a prophage.

4. A phage that has the capacity to integrate itself into the host chromosome.

5. The clear zones that appear on the surface of bacterial lawns infected by T4.

6. The recombination process involved in the integration of F factors into the bacterial chromosome. (2 words)

7. The stage of the lytic life cycle of an individual phage particle from the time it enters a host cell until the cell releases phage progeny. (2 words)

8. During site-specific recombination, this λ gene product and this bacterial protein participate in the integration of the λ genome into that of the bacterial host. (2 words)

9. The mechanism of genetic transfer during which DNA is transferred from a donor to a recipient via a virus.

10. The mechanism of genetic transfer in which a host cell takes up naked DNA.

11. The mechanism of genetic transfer during which DNA is transferred from a donor to a recipient via a cytoplasmic bridge.

12. This condition imparts to bacteria the ability to take up environmental DNA.

13. During this type of transduction, a large number of genes at low frequency are normally transferred.

14. During this type of transduction, one or a few genes are normally transferred at a high frequency.

15. Competence in many bacteria appears during this stage of their growth cycle.

16. Extrachromosomal, autonomous DNA molecules found in the cytoplasm of many bacteria.

17. The following three *E. coli* strains are donor strains during conjugation.   (3 designators)

18. An adjective applied to any plasmid (such as the fertility factor of *E. coli*), that can promote the transfer of genetic information by conjugation.

19. *E. coli* donor strains possess protein structures that allow them to carry out conjugation while the recipients do not. What is the name of these structures?

20. The mechanism of DNA replication during conjugation. (2 words)

21. The name given to the simplest transposable element. (2 words)

22. Another name given to transposable elements.

23. Transposable elements insert themselves into host DNA at these specific sites.   (3 words)

24. These transposable elements have a gene that codes for a reverse transcriptase.

Answers to the Objective Questions

Multiple Choice

1. *c*   **2.** *d*   **3.** *c*   **4.** *e*   **5.** *d*   **6.** *a*   **7.** *a*   **8.** *b*   **9.** *c*   **10.** *b*   **11.** *c*
12. *b*

True–False

1. F (plasmids are commonly found in bacterial cells)   **2.** F (most bacteria reproduce by binary fission)   **3.** T
4. T   **5.** F (only lysogenic viruses can assume a prophage state)   **6.** F (T4 injects its DNA through the cell wall)   **7.** T   **8.** T   **9.** T   **10.** T   **11.** F (site-specific recombination)   **12.** T   **13.** T
14. T   **15.** F (pili are only required for conjugation)   **16.** T   **17.** T   **18.** T   **19.** T   **20.** T
21. F (only those that have been conjugating for a long time would have received the *bio*[+] allele *AND ONLY AFTER* they have received the *eco*[+] allele)   **22.** T   **23.** F (the entire donor chromosome is rarely transferred)   **24.** T
25. T

Matching

1. *I*   **2.** *F*   **3.** *K*   **4.** *G*   **5.** *Q*   **6.** *E*   **7.** *J*   **8.** *L*   **9.** *R*   **10.** *D*   **11.** *O*
12. *H*   **13.** *M*   **14.** *C*   **15.** *P*   **16.** *N*   **17.** *B*   **18.** *T*   **19.** *A*   **20.** *S*

Terms

1. prophage   **2.** Ellis-Delbrück   **3.** lysogen   **4.** temperate or lysogenic   **5.** plaques   **6.** site-specific   **7.** latent period   **8.** int, IHF   **9.** transduction   **10.** transformation   **11.** conjugation   **12.** competence   **13.** generalized   **14.** specialized   **15.** exponential   **16.** plasmids   **17.** F$^+$, F$'$, Hfr   **18.** conjugative   **19.** pili   **20.** rolling circle   **21.** insertion sequence   **22.** transposons   **23.** direct terminal repeats   **24.** retrotransposons (retroposons)

Chapter 8

Genetic Engineering/Recombinant DNA Technology

INTRODUCTION

Recombination between DNA molecules from different organisms is a common phenomenon in nature. Viruses such as λ are able to insert their genomes into the *E. coli* chromosome. In the process, they may change the genetic makeup of the cell. Sometimes a bacterium can become pathogenic (disease-causing) when receiving new genetic information from a virus.

> **Example 8.1.** When the bacterium *Corynebacterium diphtheriae* is infected by the β virus, a toxin is produced that is responsible for the signs and symptoms of diphtheria. The viral gene that codes for the toxin is inserted in the bacterial chromosome as a provirus. This genetic change in the bacterium brought about by the virus, known as **lysogenic conversion,** is an example of genetic engineering in nature.

Discoveries in molecular biology have allowed scientists to duplicate these natural phenomena in the laboratory and develop methods to introduce almost any type of genetic information into an organism. Most of the developments involve the genetic manipulation of bacteria such as *E. coli* and *B. subtilis,* and the yeast *S. cerevisiae* for the production of consumer goods, especially those that are very expensive or virtually impossible to make by traditional manufacturing methods (Table 8.1).

Considerable progress is being made toward the introduction of genes into animals and plants to cure genetic disorders, to enhance productivity of plants and livestock, and to make crops more resistant to diseases. For example, numerous genes have been introduced into plants to make them resistant to insects and to weed killers. It is hoped that the genes coding for the β-globin subunit of hemoglobin can be introduced into people suffering from the degenerative and deadly diseases sickle-cell anemia and β-thalassemia, which are due to defective β-globin genes. Many scientists are attempting to introduce into cereal crops such as wheat, rice, barley, and corn the bacterial genes for converting atmospheric nitrogen into ammonia. This could allow the cereal crops to grow without the expensive nitrogen fertilizers they now need. Additionally, investigators at the National Institutes of Health are conducting pioneering studies in gene therapy of an immunodeficiency disease known as **adenosine deaminase deficiency (ADA).** This disease is caused by a defective gene coding for adenosine deaminase. Absence of this enzyme, which removes amine groups from adenine, results in the accumulation of metabolites that are toxic for B and T cells (lymphocytes involved in the expression of immunity in mammals). The marked reduction in immune cells as a result of toxic death leads to an immune deficiency.

GENETIC ENGINEERING

8.1. What is genetic engineering?

> **Genetic engineering** is the creation of new DNA, usually by the recombination of DNA from different organisms by artificial means using enzymes known as **restriction enzymes,** and the production of many copies of the recombined DNA by a process known as **cloning.** A typical cloning experiment is illustrated in Fig. 8-1. The amplification of a specific cloned gene or genes, coupled with a marked increase in production of their protein products, makes it relatively easy to extract and purify these proteins in the laboratory.

Table 8.1.   Some Goods and Services Produced Through Biotechnology

Hybridoma	Antibodies
Proteins	Insulin
	Interferon
	Human serum albumin
	Human growth hormone
	Tissue plasminogen activator
	Antithrombin
	Blood coagulation factors
	Luciferase (firefly)
	Lymphokines
	Tumor necrosis factor
	Superoxide dismutase
	Human gonadotropin
Agricultural	Disease-resistant crops
	Hydroponic tomatoes
	Pesticide resistance
	Bioinsecticides
	Nitrogen fixation
Vaccines	Hepatitis B
	Herpes
	Influenza
	Malaria
Bioremediation	Grease/oil digesters
	Herbicide removers
	Pesticide degraders

8.2.   What is cloning?

   Cloning is a process whereby a section of DNA is inserted into a plasmid or a phage chromosome and then allowed to replicate to produce numerous copies of the DNA. The replication normally takes place when the plasmid or phage chromosome is inserted into a suitable host (e.g., a bacterium or a yeast cell) and the host DNA synthesis apparatus replicates the inserted DNA in the host cell.

8.3.   What components are involved in cloning?

   A cloning experiment involves five major components. These are listed in Table 8.2, along with their functions.

Donor DNA

8.4.   Why is donor DNA necessary?

   The goal of any cloning experiment is to isolate a desired gene or segment of DNA from an organism and introduce it into a suitable host cell to obtain large quantities of the DNA. The DNA that is to be cloned is referred to as **donor DNA.**   Often, the donor DNA is used for the large-scale production of important proteins (Table 8.1). But, the DNA may also be used in the detection of infectious agents or abnormal cells.

8.5.   How is donor DNA obtained for cloning experiments?

   Normally, the donor DNA is a small portion of the genome of a cell, and it is present as one or two copies in each cell. Therefore, before donor DNA can be extracted, a sufficient number of cells containing the desired DNA must be obtained, either from a small segment of tissue or by culturing the cells. After a sufficient number

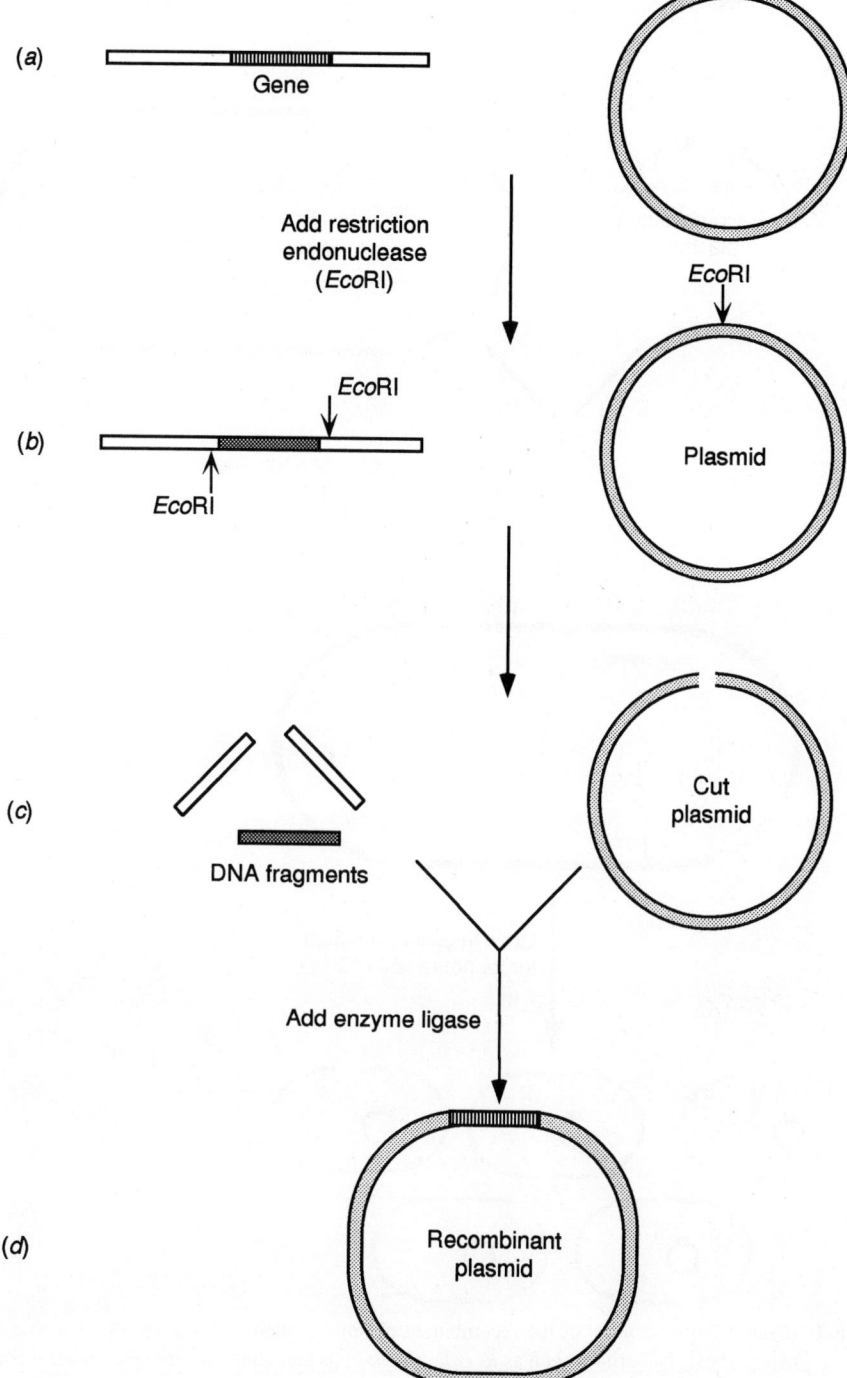

Fig 8-1. A typical cloning experiment. (*a*) A suitable plasmid (vector) is selected in which to insert a desired gene (donor DNA). Both donor DNA and vector have specific DNA sequences (4−6 base pairs) where restriction endonucleases nick the DNA strands. (*b*) In separate test tubes, both the DNA fragment containing the gene to be cloned and the plasmid in which the gene is to be inserted are mixed with the chosen restriction endonuclease (in this case, *Eco*RI) and the appropriate buffer. The enzyme chosen must cleave the plasmid at only one position. The DNA-enzyme mixtures are incubated, resulting in the cleavage and release of the desired donor DNA fragment. The cut plasmid can then receive an additional piece of DNA. (*c*) The DNA fragments and the cut-open plasmid are incubated with the enzyme ligase, which joins some of the DNA fragments and plasmids. (*d*) The result of this ligation is the formation of a recombinant plasmid, which contains the desired DNA fragment.

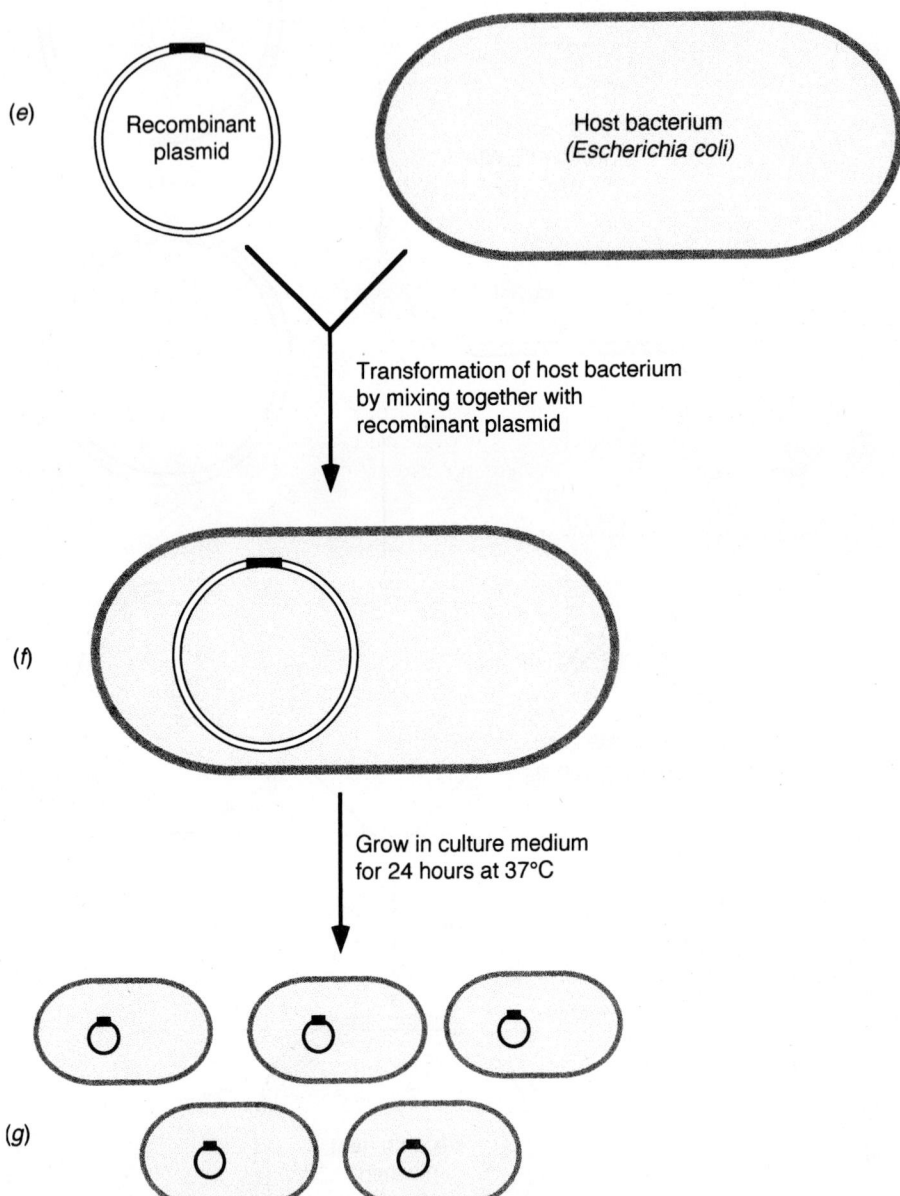

Fig 8-1. (*Cont.*) (*e*) A solution of the recombinant plasmid is then incubated with a suitable receptor (host) bacterium, such as *E. coli,* whose cells can attach to and take up the recombinant plasmid. Host DNA is not shown. (*f*) Cells that take up plasmids are said to be
transformed. (*g*) The transformed cells are incubated overnight in a suitable culture
medium, so that they can multiply. Each cell that is transformed may contain as many
as 200 copies of the recombinant plasmid.

of cells containing the desired genetic material has been obtained, each cell must be disrupted and the genetic material must be extracted. This genetic material might be present in chromosomes or in plasmids.

8.6. How is chromosomal DNA extracted from cells?

The chromosomes of eukaryotic cells are located within the nucleus of the cell (see Chapter 3). Each eukaryotic chromosome consists of a single DNA molecule wrapped around proteins called histones. All the DNA

Table 8.2. Components of a Cloning Experiment

Cloning Component	Function
Donor DNA (insert)	Source of the DNA or gene to be cloned.
Restriction endonuclease	Enzyme used to cut both donor and vector DNA at specified locations, so that the donor DNA can be spliced into the vector.
Vector	Plasmid or bacteriophage used to introduce the gene to be cloned into a suitable host cell.
DNA ligase	Enzyme used to join the spliced ends of vector and donor DNA and thus form a recombinant vector.
Host cell	Usually a bacterium or a yeast cell. Recombinant vectors are introduced into host cells to obtain larger quantities of the recombinant DNA molecule.

molecules that make up the chromosomes of a cell are referred to as the **genomic DNA** of the cell. In order to release the genomic DNA from a cell, the membranous envelopes (plasma membrane and nuclear membrane) that enclose and protect the genetic material must be dissolved and the histones and other chromosomal proteins digested. The membranes are dissolved using detergents that solubilize the phospholipids that make up these membranes. Very common detergents used for the extraction of genomic DNA are sodium dodecyl sulfate (SDS), Triton X-100®, and Tween®. Once the membranes are dissolved, the resulting cell **lysate** (consisting of cytoplasmic and nuclear material, organelles, and chromosomes) is treated with **proteases** (e.g., protease K) to digest the proteins away from the chromosomes. After the protease treatment, any DNA will be in aqueous solution. The cell lysate is treated with organic solvents such as a mixture of phenol and chloroform to remove proteins and other cell debris, leaving the genomic DNA in the aqueous interphase between the phenol and chloroform.

Collection of this partially purified DNA is the next and final step. When absolute (100%) ethanol is layered on top of the aqueous DNA solution, an interphase develops between the two liquids due to their difference in densities. Genomic DNA in the aqueous phase will precipitate in this interphase as long, thin, white threads gathered into an opaque-looking mass. Genomic DNA can be **spooled** from the interphase with a glass rod. To do this, a glass rod is inserted into the interphase and twisted to wrap around it the strands of DNA. This spooled DNA can be dried, redissolved in an aqueous solution, and used in a variety of procedures.

8.7. How is plasmid DNA extracted from bacteria?

The genome of a bacterial cell consists of a single, circular DNA molecule ("chromosome"). Plasmids are small, self-replicating DNA molecules that may exist independently of the bacterial chromosome. Plasmids carry the genetic information to direct their own replication and a few genes, many of which impart some selective advantage to the cell. Examples of these genes include resistance to antimicrobial agents, resistance to heavy metals, and the production of antibiotics. Since many bacterial cells possess both the bacterial chromosome and several plasmids, the isolation of plasmid DNA involves a step in which the desired plasmid DNA is separated from the bacterial chromosome (genomic) DNA.

As with eukaryotic cells, bacterial cells are treated with detergents in order to disrupt the plasma membrane. Since bacteria possess a cell wall, treatment with the enzyme **lysozyme** is also required to remove this outer envelope. This treatment, along with that of detergents, releases both plasmid and chromosomal DNA from the cell. For the selective isolation of plasmid DNA from cells, sodium hydroxide (NaOH) is also added to the detergent solution to **denature** the DNA (i.e., cause the two strands to separate). The denaturation of bacterial DNA with NaOH and detergents is called **alkaline lysis**. Chromosomal DNA and many of the proteins and membrane components are removed by adding potassium (usually in the form of potassium acetate) to the mixture, followed by centrifugation. The selective precipitation of chromosomal DNA is based on the much larger size of the genomic DNA; the larger size makes it more vulnerable to precipitation by potassium acetate than the small plasmid DNA molecule. After this step, plasmid DNA remains in solution along with cellular RNA. This solution is also known as a **cleared lysate.** The RNA can be eliminated by treatment with the enzyme **RNase,** which can then be removed with organic solvents (phenol-chloroform). Plasmid DNA is then precipitated with ethanol, dried, redissolved in an aqueous solution, and used in a variety of applications.

Restriction Endonucleases

8.8.   How were the first restriction enzymes discovered and used to create recombinant DNA molecules?

In 1962 Werner Arber at the University of Geneva in Switzerland reported that the bacterial virus λ grown on one strain of *E. coli* did not infect a second strain. He found that 100% of the virus grown on an *E. coli* strain known as K could reproduce well within that strain, but was unable to reproduce in a similar strain known as B. This evidence led Arber and his colleagues to believe that certain bacteria possess enzyme systems that digest and destroy foreign DNA, in this case, viral DNA. They called these enzymes **restriction endonucleases.** Since the isolation of the first restriction endonuclease in 1970 by Hamilton Smith and his coworkers, scientists have discovered and characterized more than 200 different restriction endonucleases (Table 8.3). In 1971, Paul Berg, Herbert Boyer, and their colleagues employed an endonuclease to create the first recombinant DNA molecule by using DNA from simian virus 40 (SV40) and the bacterial virus λ.

8.9.   What are restriction endonucleases?

Restriction endonucleases are bacterial enzymes that recognize specific nucleotide sequences within a double-stranded DNA molecule and cleave the DNA at those locations. These enzymes cut DNA into fragments of various lengths, depending upon the number of times the enzyme's recognition site is repeated within the molecule. Restriction endonucleases type II, which are used in cloning experiments, recognize base pair sequences 4 to 8 nucleotides long and cut within these sequences. For example, the restriction endonuclease known as *Sma*I recognizes the sequence 5′-CCCGGG-3′ and cuts between the adjacent C and G, whereas the enzyme *Eco*RI recognizes the sequence 5′-GAATTC-3′ and cuts between the G and the A. These sites are called **recognition sequences.** Hence, any DNA molecule that is cut (or restricted) by a restriction endonuclease type II must have the recognition sequence at every site where the DNA is cut. Many restriction enzymes have recognition sequences known as **palindromes.** Palindromes are 4 to 6 base (nucleotide) sequences that are the same on both strands of the DNA molecule when read in the same direction (5′ → 3′). For example, the recognition sequence for the enzyme *Bam*HI is

5′-GGATCC-3′

3′-CCTAGG-5′

Notice that the sequence reads GGATCC in the 5′ → 3′ direction on both strands. That is an example of palindrome. Restriction endonucleases may cut the DNA to produce fragments with **cohesive ("sticky") ends** (Fig. 8-2*a*) or **blunt** ends (Fig. 8-2*b*).

Restriction endonucleases are named after the bacterial species from which they are isolated. For example, the restriction endonuclease *Eco*RI is obtained from the bacterium *Escherichia coli* strain R, and the enzyme *Hind*III, from the bacterium *Haemophilus influenzae* strain **D.** The Roman numeral indicates the order in which the particular restriction enzyme was isolated from the microorganism. For example, *Eco*RI was the first restriction enzyme isolated from *E. coli* strain R while *Hind* III was the third restriction enzyme isolated from *H. influenzae* strain D.

8.10.   Can two different restriction enzymes cut within the same recognition sequence?

Yes, different microorganisms produce restriction endonucleases that have the same recognition sequence. These are know as **isoschizomers.** Table 8.3 cites some of these enzymes.

8.11.   How can the products of a restriction enzyme digestion of a DNA molecule be detected?

The resulting fragments of a restriction enzyme digestion can be visualized by a procedure known as electrophoresis. Electrophoresis involves the movement of charged molecules or ions in a stabilized matrix support medium under the influence of an electrical field. Agarose gels are common media for the electrophoresis of

(a) Cohesive ends are formed when *Bam*HI cleaves the DNA

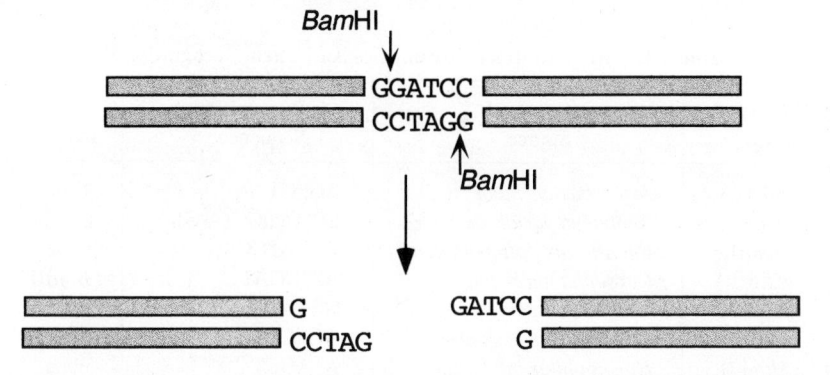

(b) Blunt ends are formed when *Hae*III cleaves the DNA

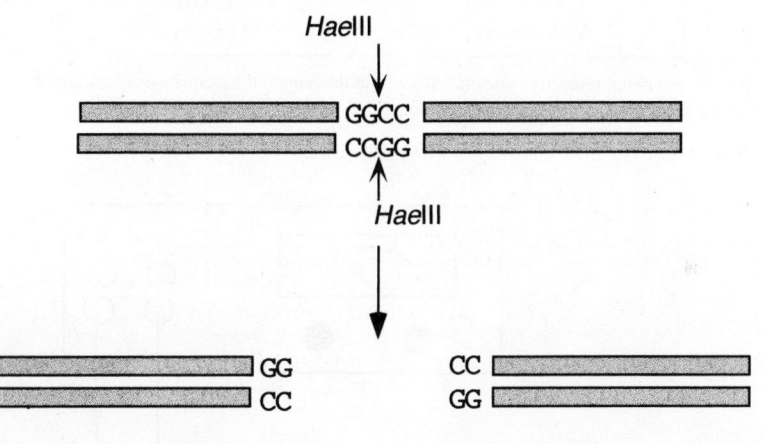

Fig 8-2.   Restriction enzymes can cut DNA to form (*a*) cohesive ends or (*b*) blunt ends, depending on where they cut within the recognition sequence.

DNA. In a typical apparatus, an electrophoretic chamber holds the gel and the buffer. The direct current power supply must be capable of reaching a potential difference of 50 to 300 volts (Fig. 8-3).

The agarose gel, cast as a thin slab in a mold with sample wells at one end, is submerged in the buffer solution with the sample well side toward the negative pole (cathode). The samples are dispensed into the wells with a micropipet and a current from the power supply is applied to the system. Since nucleic acids have a negative charge at pH of about 8, they will migrate within the agarose gel matrix from the negative pole to the positive pole (anode). The agarose gel serves as a molecular sieve to separate molecules of DNA based on their size and the total charge of the molecule. DNA solutions are normally loaded along with a colored dye (e.g., **bromophenol blue**) so that the electrophoretic migration of the DNA through the gel can be visualized and stopped before the DNA migrates out of the gel and into the buffer solution.

The migration rate of charged macromolecules (such as DNA) in an applied electric field is proportional to the total negative charge of the macromolecule and inversely proportional to its mass. The electrophoretic mobility of DNA is mainly affected by the size and shape of the fragment. Small fragments of DNA migrate faster than larger ones. So, through electrophoresis, it is possible to sort DNA fragments based on size. Figure 8-4 illustrates how the genome of the bacteriophage λ, when cut with the enzyme *Hin*dIII, appears on a gel after electrophoresis. The bands represent the sorted DNA fragments. Larger fragments are seen as bands near the origin of the gel, and small fragments are seen as bands lower in the gel.

Table 8.3.   Some Restriction Enzymes And Their Recognition Sequence

Name of Enzyme	Bacterial Source	Recognition Sequence*	Isoschizomer†
*Alu*I	*Arthrobacter luteus*	AG®CT	none
*Ava*I	*Anabaena variabilis*	C®(T/C)CG(A/G)G	*Eco*47 I, *Sin*I
*Bam*HI	*Bacillus amyloliquefaciens*	G®GATCC	none
*Cla*I	*Caryophanon latum*	AT®CGAT	*Ban*III
*Eco*RI	*Escherichia coli*	G®AATTC	none
*Hae*III	*Haemophilus aegyptius*	GG®CC	*Pal*I
*Hind*III	*Haemophilus influenzae*	A®AGCTT	none
*Kpn*I	*Klebsiella pneumoniae*	GGTAC®C	*Dpn*I
*Not*I	*Nocardia otidis-caviarum*	GC®GGCCGC	none
*Pst*I	*Providencia stuartii*	CTGCA®G	none
*Xba*I	*Xanthomonas badrii*	T®CTAGA	none
*Xho*I	*Xanthomonas holcicola*	C®TCGAG	*Pae*R7 I

*The site where the restriction enzyme cuts within the recognition sequence is indicated by "®".
†See Question 8.10.

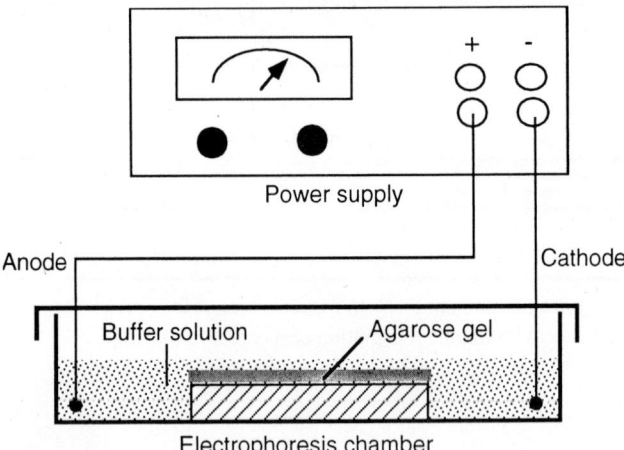

Fig 8-3.   Horizontal electrophoresis apparatus.

8.12.   How can we visualize the DNA within the gel?

DNA molecules are invisible to the naked eye, but can be seen in gels by staining them with a solution of a dye called **ethidium bromide.** Ethidium bromide is a molecule that becomes intercalated between the stacked bases of the DNA molecule and fluoresces. Consequently, a DNA molecule treated with ethidium bromide fluoresces orange when excited with ultraviolet light. The photograph in Fig. 8-3 was taken while the gel was illuminated with ultraviolet light.

8.13.   What is a restriction map and how is it constructed?

Since restriction endonucleases cut DNA at specified locations, a given DNA molecule from an organism can be differentiated from a different DNA molecule by the number of times a restriction endonuclease cuts the DNA.

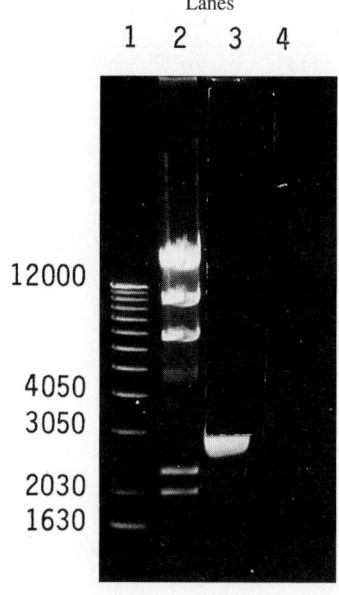

Fig 8-4. Restriction digest of bacteriophage λ DNA and the plasmid pBR322. This photograph illustrates the electrophoretogram of the restriction endonuclease digestion of two different DNA molecules with the enzyme *Hin*dIII. Even though *Hin*dIII produces seven fragments from the digestion of bacteriophage λ DNA (lane 2), only five are seen here because two of the smallest fragments have migrated off the gel during electrophoresis. The DNA samples were loaded at the top (cathode side) of the gel and migrated downward (toward the anode). Plasmid pBR322 digestion is in lane 3. Lane 1 contains the DNA of a reference molecular-weight marker, so that the exact molecular weights of the restriction digests can be measured. Lane 4 is empty. The numbers on the left represent the number of base pairs of the corresponding bands.

For example, the restriction endonuclease *Hin*dIII cuts the DNA of the bacteriophage λ at six different locations along the viral genome, yielding seven fragments (Fig. 8-5). On the other hand, the plasmid pBR322 has only one recognition site for *Hin*dIII. When the enzyme cuts this plasmid, it converts the circular DNA molecule into a linear one, with the remains of a *Hin*dIII recognition site at each end of the molecule. The graphical representation of recognition sites for two or more restriction endonucleases is known as a **restriction (endonuclease) map** for that molecule. Figure 8-5 illustrates a restriction map of the bacteriophage λ.

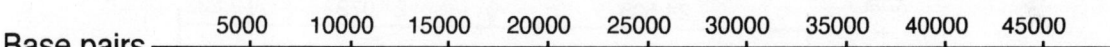

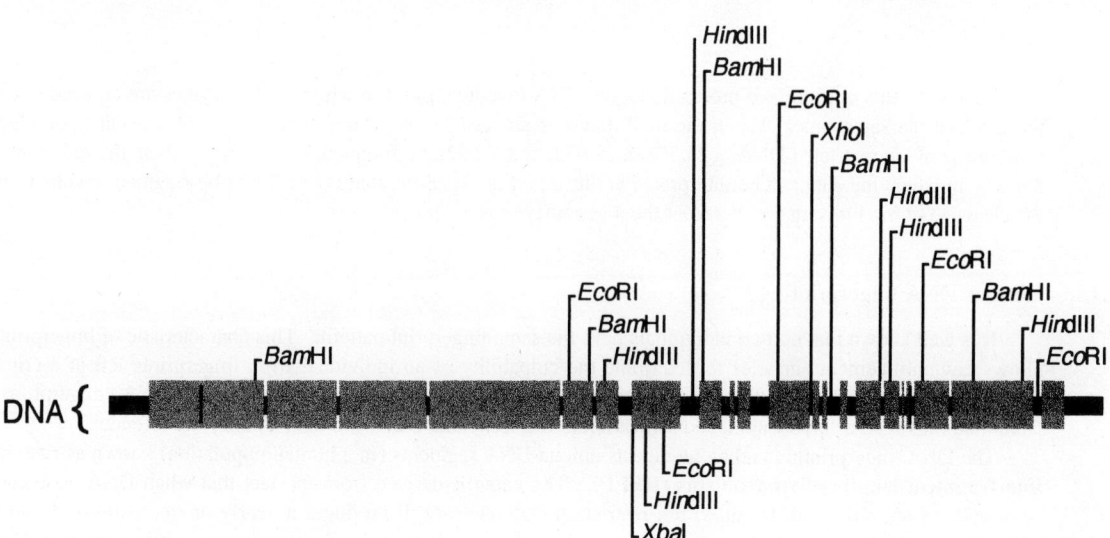

Fig 8-5. Restriction endonuclease map of bacteriophage λ. The restriction endonuclease map was constructed using the enzymes *Bam*HI, *Eco*RI, *Hin*dIII, *Xba*I, and *Xho*I. The number of base pairs from a standard reference point on the λ genome are displayed at the top of the figure. Thus, the first cut by *Bam*HI, at the left, occurs about 5,000 base pairs from the reference point.

8.14. What is the strategy of restriction mapping?

The strategy of restriction mapping can be illustrated with the following example. Restriction endonuclease *Xba*I cuts λ DNA into two fragments; a 5′-24,508 base pair (bp) fragment and a 23,994-3′ bp fragment (Fig. 8-6*a*). Restriction endonuclease *Nar*I cuts DNA also into two fragments; fragments 45,680 and 2,822 bp long (Fig. 8-6*b*). However, relative to enzyme *Xba*I, there are two possible ways (left to right) in which *Nar*I could have cut the λ DNA; a 5′-45,680 and a 2,822-3′ bp fragment or a 5′-2,822 and a 45,680-3′ bp fragment. Which of the two possibilities is the correct cleavage site for *Nar*I with respect to *Xba*I?

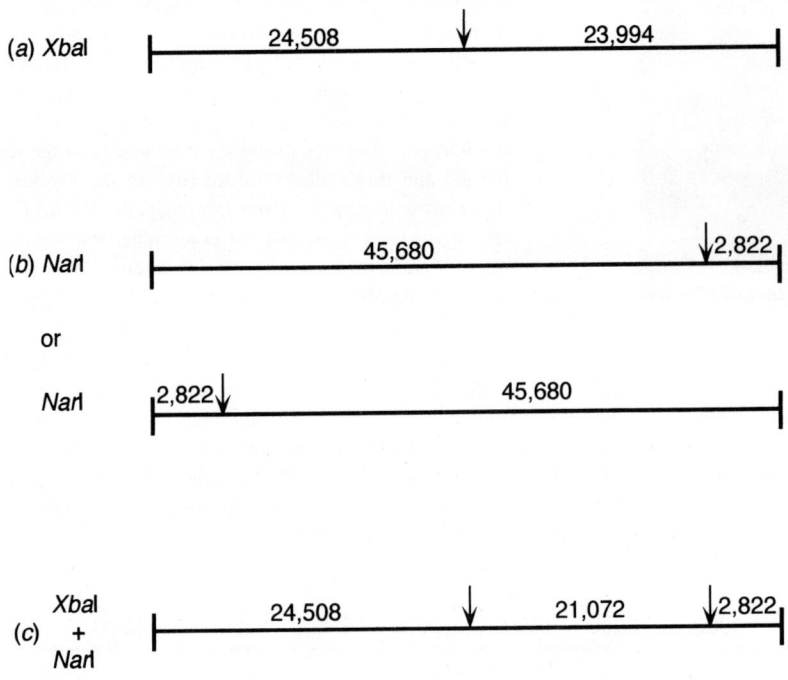

Fig 8-6. Strategy for restriction mapping. See text for an explanation.

To answer this question we must subject the DNA to a dual digest in which both enzymes are allowed to cut the DNA at the same time. When the dual digest is allowed to take place, three fragments appear upon electrophoresis of the restricted DNA, a 24,508, a 21,072, and a 2,822 bp fragment (Fig. 8-6*c*). Then, the only possible way in which the data can be interpreted is illustrated in Fig. 8-6*c*; that is, the 2,822 bp fragment results from the cleavage of *Nar*I toward the 3′ end of the λ genome.

8.15. What is DNA fingerprinting?

It is well known that no two individuals have the same fingerprint pattern. This characteristic of fingerprints allows law enforcement agencies to determine the culpability of an individual from fingerprints left at a crime scene. Molecular biology techniques are now available that allow forensic scientists to make DNA "fingerprints" from materials such as blood, skin, and semen left at the scene of a crime and thus identify the offender.

The DNA fingerprinting technique detects unique DNA segments (in a human population) known as **restriction fragment length polymorphisms (RFLP).** The name is derived from the fact that when DNA, extracted from cells of an individual, is cut with a restriction endonuclease, it produces a nearly unique pattern of bands when the cut DNA is subjected to electrophoresis. Different alleles arise from mutations (insertions and deletions) in the DNA that give rise to alterations in the sizes of the fragments cut by the restriction endonuclease. The pattern of restricted DNA fragments (different lengths of DNA) that appears after electrophoresis resembles a bar code. The pattern of bands from cells of one individual can be very different from that of another. This is due to the presence of hypervariable regions in human chromosomes.

As pointed out in Chapter 3, much of the DNA in human chromosomes does not code for RNA or proteins. Among the noncoding regions of a chromosome there are sequences many thousands of base pairs long consisting of a short segment of DNA that repeats itself over and over (Fig. 8-7). These segments are known as **variable number of tandem repeats (VNTR).** The number of repeating segments varies greatly from individual to individual. In human chromosomes, many VNTR regions are bordered by restriction endonuclease sites on either side (Fig. 8-7). Alleles of a given VNTR differ from each other in the size of the segment cut by the restriction endonuclease, giving rise to nearly unique RFLP patterns.

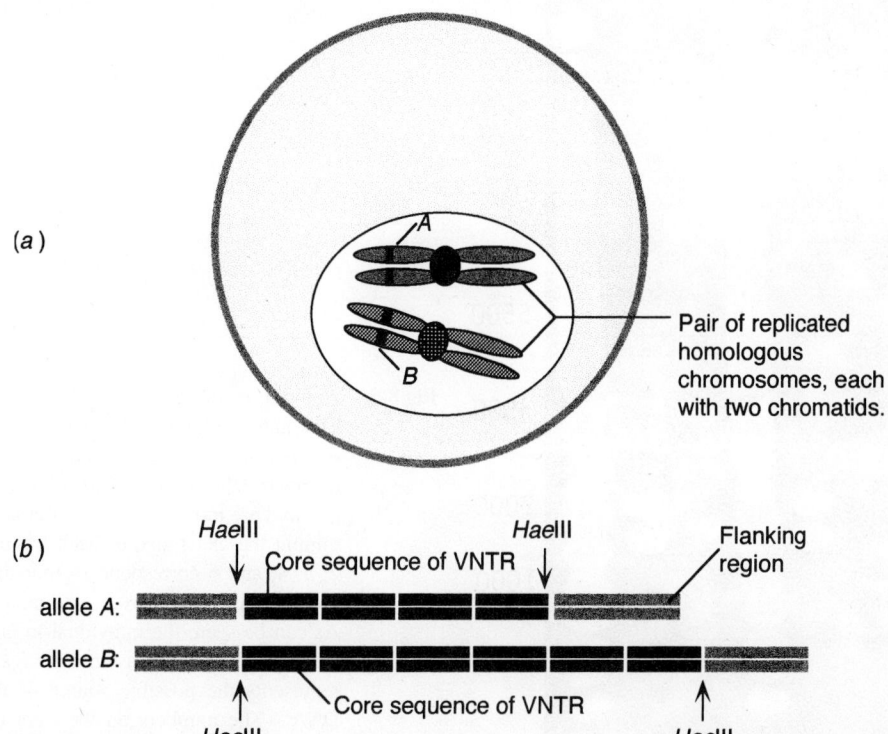

Fig 8-7. Structure of a human variable number of tandem repeats (VNTR). Each diploid human cell contains 23 pairs of homologous chromosomes. Each chromosme contains an allele (alternative form) for every VNTR segment in the chromosome. Figure (*a*) illustrates a cell with a pair of chromosomes bearing two alleles (*A* and *B*) at the same locus where a VNTR is found. (*b*) Each allele differs from the other by the number of end-to-end (tandem) repeats of the core sequence. The VNTR is defined by the recognition site of the restriction enzyme *Hae*III. Since each allele differs from the other by the number of repeats of the core sequence, they can be seen as two different bands in an electrophoresis or a Southern blot (Fig. 9-3). There are numerous VNTR alleles in a population. Therefore, they can be used to distinguish between cells of two different individuals, because chances are the two individuals will not share the same pattern of repeats in their pairs of alleles.

The procedure used to obtain a DNA fingerprint involves first the isolation of the cell's DNA. This DNA is cut with just one restriction endonuclease such as *Pst*I or *Hae*III and then electrophoresed to sort the different fragments based on their lengths (molecular weights). Since humans have more than 3 billion base pairs (6 billion nucleotides) in their genome, after electrophoresis all that can be seen is a smear because all the resulting bands overlap each other.

To visualize a fingerprint pattern for a specific VNTR site, the DNA fragments corresponding the VNTR must be detected by *DNA hybridization,* a technique discussed in detail in Chapter 9. After electrophoresis, the DNA is denatured by alkali and transferred to a piece of nitrocellulose (or nylon) membrane. The DNA on the membrane is then made to hybridize with a radioactive DNA molecule (called a probe) made in the laboratory that has a sequence complementary to that of the core sequence. After hybridization, autoradiography is performed and the x-ray film developed. The pattern of bands appearing on the resulting autoradiogram is the DNA fingerprint for that individual (Fig. 8-8).

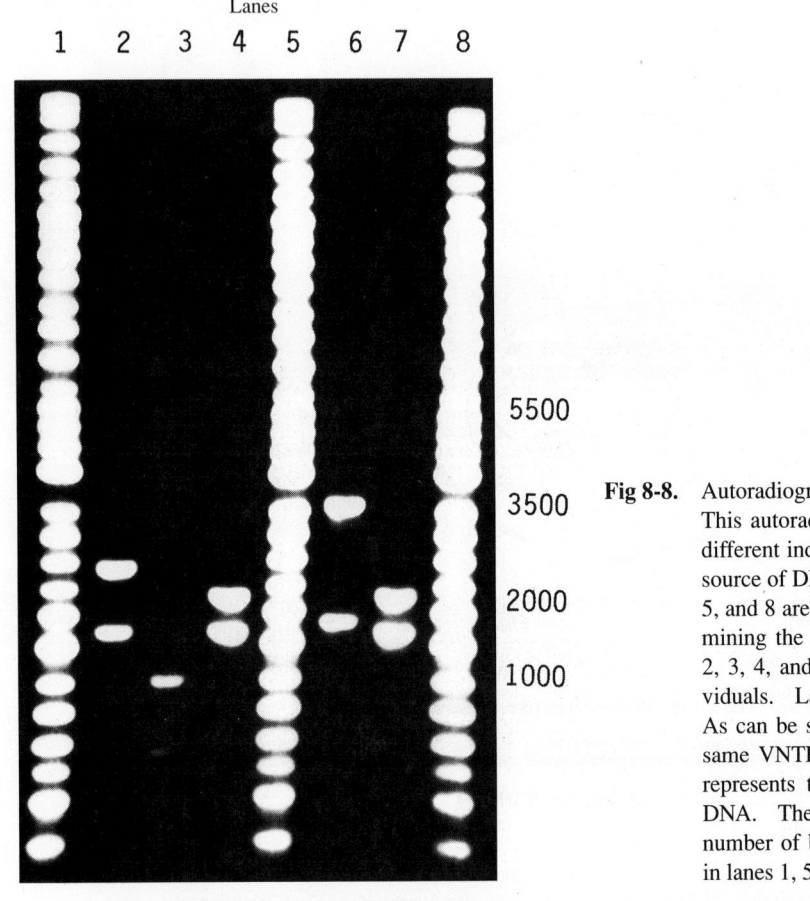

Lanes

1   2   3   4   5   6   7   8

5500

3500

2000

1000

Fig 8-8. Autoradiograph of a human VNTR site (D2S44). This autoradiograph illustrates the study of four different individuals in an attempt to identify the source of DNA of an unknown sample. Lanes 1, 5, and 8 are molecular-weight markers, for determining the exact size of each fragment. Lanes 2, 3, 4, and 6 correspond to four different individuals. Lane 7 is from the unknown source. As can be seen, the individual in lane 4 has the same VNTR pattern as in lane 7 and therefore represents the possible source of the unknown DNA. The numbers on the right represent the number of base pairs of the corresponding band in lanes 1, 5, and 8.

With this technique, forensic scientists have been able to find missing persons, determine whether or not an individual is the parent of a child, and establish beyond reasonable doubt that a person is guilty of a crime for which he or she has been accused.

8.16. How is the fragment containing a gene of interest identified and isolated for cloning?

One common way of identifying genes of interest within the DNA molecule of an organism is by a **Southern blot hybridization.** This technique, discussed in detail in Chapter 9, can be used to identify specific DNA sequences after restriction enzyme digestion and electrophoresis of genomic DNA. To do this, a small portion of the restricted DNA is electrophoreses on an agarose gel and a Southern blot performed. Once the precise location of the restriction fragment with the desired gene has been determined by the Southern blot, electrophoresis is performed; the portion of the gel containing the desired gene is excised from the gel, and the DNA is eluted. This DNA can be used in subsequent cloning experiments.

8.17. What happens in a cloning experiment after the desired donor DNA is cleaved from the genome with restriction endonucleases?

After the desired segment of donor DNA is cut away from the donor's genome with restriction endonucleases it is spliced, or **ligated,** into a **vector** DNA molecule.

Vectors

8.18. What is a vector?

A vector is a DNA molecule into which foreign DNA molecules can be spliced and inserted into cells so that the "recombinant" DNA can be replicated. Vectors can be introduced into host cells by transformation or transduction (see Chapter 7).

8.19. What characteristics must a cloning vector have?

There are several different types of vectors used in cloning genes. Plasmids and bacterial viruses (bacteriophages), however, are the most commonly employed because they are easily introduced into cells and can be easily manipulated in the laboratory. Regardless of the type, vectors must have certain characteristics to be suitable for cloning. Table 8.4 summarizes the most important characteristics that a cloning vector must have.

Table 8.4.   Characteristics of Cloning Vectors

Characteristic	Functions Served
Are stable in the host cell	Permits replication
Control their own replication	Enables them to multiply to high copy numbers within the cell
Are small in size	Permits ready introduction into the cell by transformation, electroporation, or transduction
Are cut at a single site by a restriction endonuclease	Permits the donor DNA to be inserted and the plasmid recircularized
Are not transferred by conjugation	Prevents recombinant DNA escape into natural populations of bacteria
Have easily detected traits	Makes it possible to distinguish between transformed and untransformed cells
Are readily isolated from cells	Enhances yields of recombinant plasmids

Most cloning vectors are relatively small (2,000 to 6,000 base pairs), and their replication is not strictly controlled by the host cell. The most frequently used vectors (e.g, plasmids and phages) direct their own replication and form many copies (50 to 200) within each cell. On the other hand, plasmids whose replication is regulated by the host cell normally produce only 2 to 3 copies per cell, a low copy number. The advantage of a high copy number vector is that it is possible to get many copies of the desired DNA per cell.

Cloning vectors must possess restriction endonuclease recognition sites where genetic information can be inserted. As shown in Fig. 8-9, the cloning vector pGEM® has single sites for each of the restriction endonucleases *Eco*RI, *Hin*dIII, *Bam*HI, and several others. It is at these sites that DNA to be cloned (foreign, insert, or donor DNA) can be inserted. Several cloning vectors have been engineered to contain a **multiple cloning site (MCS),** each containing several different restriction sites.

Most modern cloning vectors have the MCS within the *lac*Z gene. This gene, in the presence of an inducer such as IPTG (isopropylthiogalactoside), produces the enzyme β-galactosidase, which converts lactose into galactose and glucose. If a segment of donor DNA is inserted at the cloning site, it alters the sequence of the *lac*Z gene so that it no longer produces a functional enzyme. This phenomenon is called **insertional inactivation.** It is routinely used to detect recombinants by a screening method known as **blue-white colony screening.** Cells with intact *lac*Z genes that produce β-galactosidase will catabolize **X-gal** (5-bromo-4-chloro-3-indol-β-D-galactoside; or more simply β-D-galactoside), producing a blue pigment that colors the colonies. Blue colonies contain cells that have an intact *lac*Z gene. Recombinant cells will have inactive *lac*Z genes and will not metabolize the X-gal. These white colonies will contain the desired cloned gene.

A cloning vector must also be easily selected and maintained in cells. Vectors that carry antibiotic resistance genes can be easily selected and maintained by including the antibiotic in the growth medium. When these vectors are introduced into suitable host cells, the cells become resistant to the antibiotic and can grow in culture me-

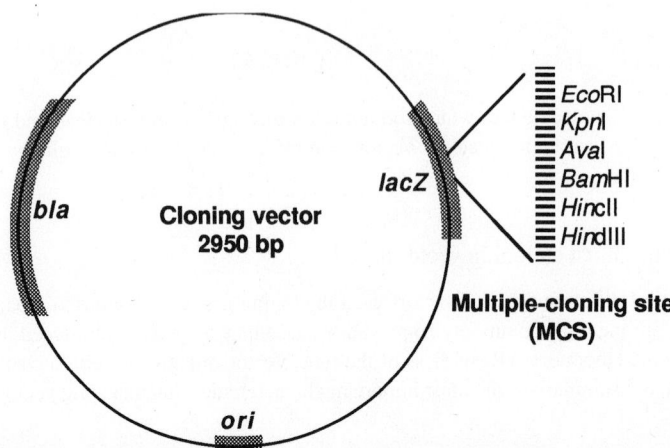

Fig 8-9. Salient features of the cloning vector pGEM®. Cloning vectors such as pGEM® are small (2,500–3,500 bp), with an origin of replication (*ori*) site that can be recognized by the host bacterium, thus permitting the replication of the cloning vector within the host. In addition, cloning vectors possess a multiple-cloning site (MCS) consisting of a short segment of DNA with single recognition sites for two or more restriction endonucleases. The MCS is found within an expressible gene such as the *lac*Z so that recombinant molecules can be detected by insertional inactivation. Cloning vectors also possess selectable genes such as the *bla* gene, which codes for a β-lactamase and confers resistance to ampicillin (Amp^r) to the bacteria transformed by the cloning vector.

dia containing the antibiotic. Cells that have not been transformed cannot multiply in the antibiotic-containing medium.

8.20. What are expression vectors?

One of the principal goals of cloning is to produce large quantities of a desired gene or its product. The production of a desired protein can only be achieved if the cloning vectors have regulatory sites for the initiation of transcription of its gene and translation of its mRNA by the host cell (Fig. 8-10). **Expression vectors** are vectors that carry a gene that can be efficiently transcribed and translated by the host cell.

For transcription, a promoter site (P) and a terminator site (TER) are necessary. Transcription of the desired gene begins at the promoter site and ends at the terminator site. A ribosome binding site (RBS) upstream from the start codon is also present in many of the expression vectors. This site is required for the efficient initiation of translation in bacteria (see Chapter 5).

DNA Ligase

8.21. How are insert (donor) DNA and vector (recipient) DNA molecules spliced together?

Donor and vector DNA molecules are joined or spliced together by enzymes known as **DNA ligases** or **polynucleotide ligases**. DNA ligases are enzymes that catalyze the formation of a phosphodiester bond between the 3'-hydroxyl group of a segment of donor DNA and the 5'-phosphate group of the vector DNA. In a typical cloning experiment in which a cut (restricted) vector and a donor segment of DNA cut by the same enzyme are brought together in a solution, DNA ligase carries out the actual splicing of the DNA into the vector. DNA ligases commonly used in cloning experiments are those obtained from *E. coli* or from the bacteriophage T4.

Normally, vectors are prevented from self-ligating by removing their 5'-phosphates with **alkaline phos-**

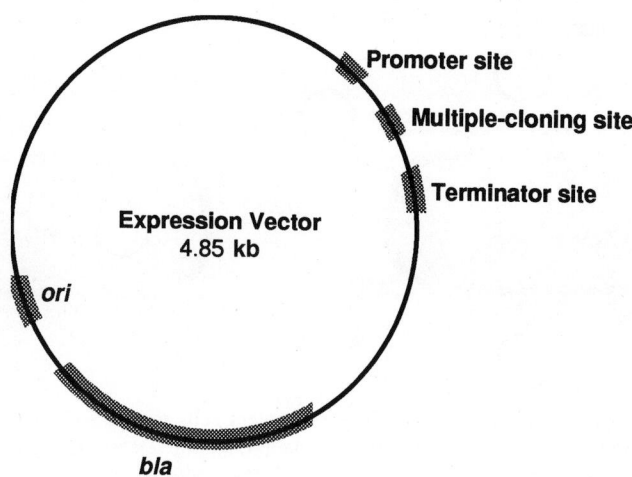

Fig 8-10. An expression vector. This diagram illustrates the
principal components of an expression vector.

phatase. Without 5′-phosphates, vectors cannot form the required covalent bond. Donor DNA can ligate to the vector because it still retains its 5′-phosphate group intact.

Host Cells

8.22. What attributes should host cells possess for genetic engineering purposes?

A number of bacterial and yeast strains have been developed especially for recombinant DNA experiments. Many of these host cells have a corresponding group of vectors that must be used in order to clone a gene successfully. This is because not all cloning vectors have the same origin of replication site *(ori)*. In order for a given plasmid to be replicated by the cell, the cell must recognize its own *ori* site and begin DNA replication there. For safety reasons, the host cells must not be able to reproduce in nature. This decreases the chances of accidental infection of laboratory workers and the general population. Additionally, transfer of DNA from one host cell to another must be avoided to prevent the dissemination of recombinant DNA in natural populations of organisms.

8.23. How is a recombinant DNA molecule introduced into host cells?

Recombinant plasmid vectors are normally introduced into cells by a process similar to transformation (Fig. 8-11). Cells are made competent by treatment with a 0.1 M (molar) solution of $CaCl_2$, then incubated with the vector. During the incubation period a few cells become transformed by the recombinant vector and are subsequently selected using appropriate culture media.

A method now used by many laboratories for introducing recombinant DNA molecules into host cells is **electroporation.** In this method, a suspension of exponentially growing host cells is mixed with a solution of recombinant DNA molecules and exposed to a high electric field (up to 2,500 volts) for a new milliseconds. The high voltage alters the structure of the membrane so that pores are temporarily formed, allowing plasmid DNA to enter the cell. This method is gaining popularity because it is fast and very efficient, especially when DNA is to be introduced into cells that do not transform readily.

> **Example 8.2.** On warm summer nights in the South, one can see the flashing of the lanterns on the abdomen of the firefly *(Photinus pyralis)*. These light flashes are produced by the chemical reaction catalyzed by the enzyme **luciferase** in the presence of oxygen, ATP, and a substance called luciferin. The luciferase gene *(luc)* has been used as an experimental tool to determine when a gene has been inserted into plants along with other genes. The luciferase gene product acts as a marker (reporter) to indicate when genes are active in the plant, because when they are, the plant lights up. Developmental biologists believe that they can use the luciferase gene to study when genes are turned on and off during the development of an organism. In addition, luciferase is used extensively in clinical chemistry and microbiology laboratories in biochemical assays.

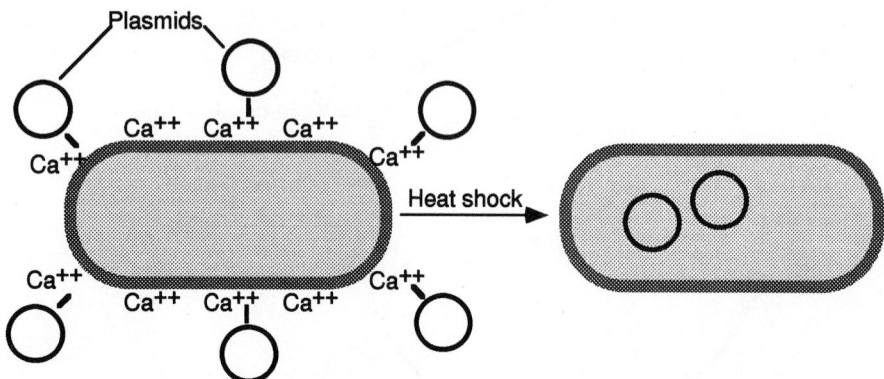

Fig 8-11. Introduction of plasmid DNA into competent cells treated with calcium chloride. The CaCl$_2$ around the cells promotes the binding of DNA to the cell's surface and makes the cell more permeable to DNA. Upon heat shock (40°C for 60–90 sec), some of the bound DNA enters the cell.

Until recently, the only source of firefly luciferase was extracts of firefly abdomens. The extraction process is tedious and the supply of fireflies is sometimes unpredictable. A team of scientists at the University of California at San Diego isolated the mRNA coding for the firefly luciferase from fireflies, converted it into DNA, and cloned it into an *E. coli* plasmid. The steps involved in the isolation and cloning of the firefly luciferase gene are outlined in Fig. 8-12. A functional firefly luciferase gene has also been introduced into plant cells. The plant cells express the luciferase gene, and the plant glows in the dark when luciferin, ATP, and oxygen are provided. This type of experiment is being used to evaluate procedures for introducing into crop plants useful genes such as those that make plants fix nitrogen or become resistant to herbicides.

8.24. Bacteria do not possess mechanisms for processing eukaryotic pre-mRNA molecules. How is it possible then to clone eukaryotic genes in bacteria so that a sensible mRNA can be made and functional proteins produced?

Bacteria are commonly used to express eukaryotic genes, but the genetic information must be devoid of introns and contain only coding sequences. To do this, it is necessary to isolate the mRNA (which already has been processed and all the introns removed) from donor eukaryotic cells. The conversion of ssRNA to dsDNA is achieved with the enzyme **reverse transcriptase**. This enzyme uses an RNA template for the synthesis of DNA. The resulting DNA molecules, known as cDNA (complementary DNA), can then be used for cloning in bacteria since they possess only intron-free protein-coding genetic information.

8.25. Can cloning experiments be carried out with microorganisms other than bacteria?

Yes. Although bacteria, particularly *E. coli,* are very desirable cloning hosts because they multiply rapidly in common laboratory media and have well-understood genomes, they are not always suitable for all types of cloning experiments. Frequently, the cloning and expression of eukaryotic genes are best done in an eukaryotic host. The yeast *S. cerevisiae* is the organism of choice for such cloning experiments.

8.26. Why is *S. cerevisiae* such a well-suited host for cloning of eukaryotic genes?

This yeast is probably the best characterized of all eukaryotic organisms. It has been used for centuries in the making of bread, wine, and beer, and it grows very well under laboratory conditions in liquid cultures as individual cells. *S. cerevisiae* has a genome of approximately 2×10^7 base pairs contained in 17 linear chromosomes, and some strains possess a type of plasmid known as the **2-micron circle.** This plasmid has 6,318 base pairs and is present in a copy number of about 50 copies per cell.

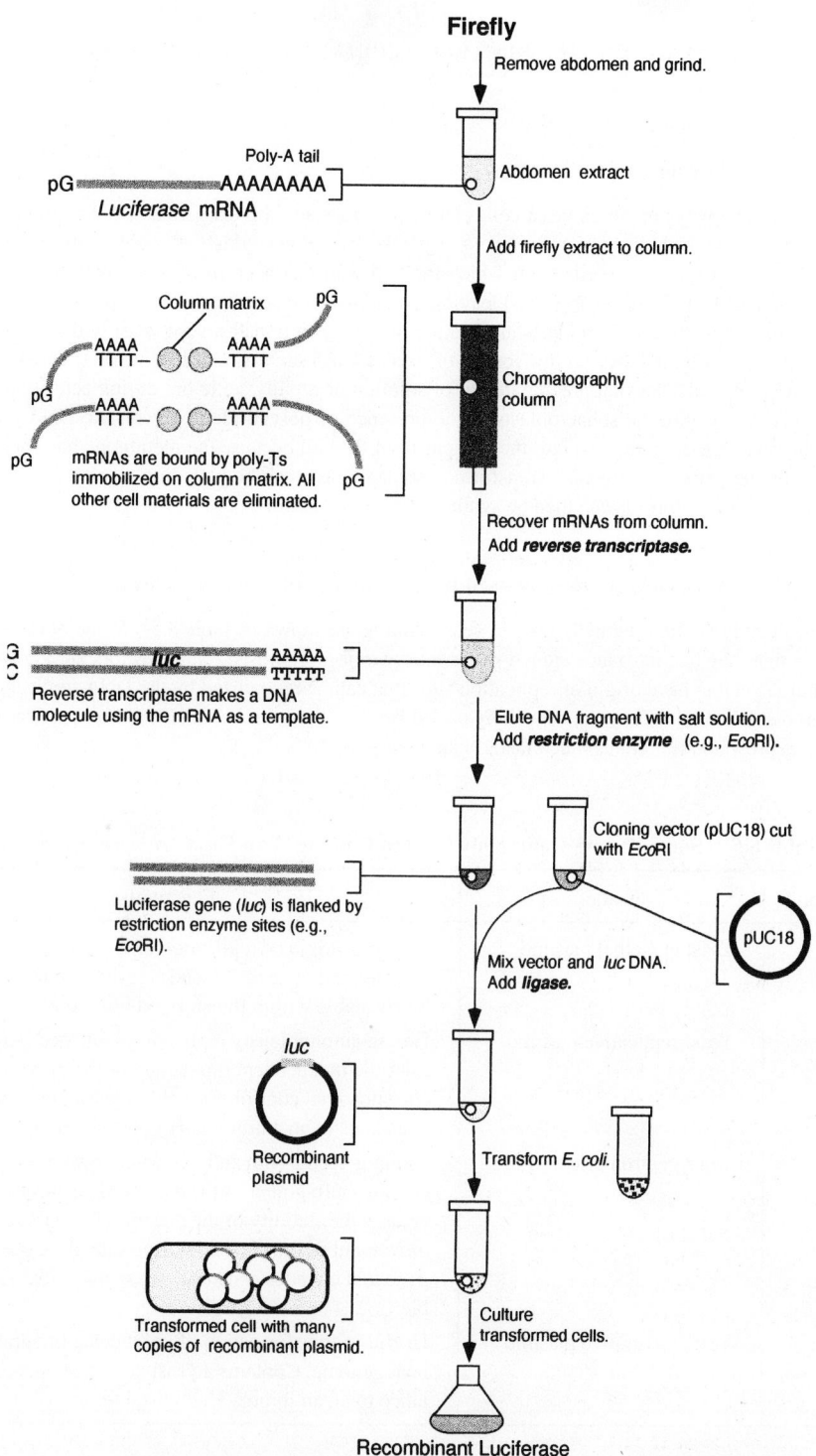

Fig 8-12. Cloning of the luciferase (*luc*) gene in *Escherichia coli*. The cloning of the *luc* gene in *E. coli* involves several steps. Luciferase mRNA, which like all other eukaryotic mRNA molecules contains a poly-A tail, is captured in a chromatography column. Immobilized on the column matrix is a poly-T sequence. The mRNA becomes bound to the poly-T and is subsequently eluted from the column. The mRNA molecules are then converted into ds-DNA using the enzyme reverse transcriptase, and the DNA molecules are subjected to digestion by the enzyme *Eco*RI. The restricted DNA is then inserted into plasmid vectors. Cells transformed by the vectors are assayed for luciferase production. Luciferase is extracted from the cells by treating the cells with detergents and the enzyme purified using special chromatography columns.

199

8.27. How can yeasts be transformed?

There are two ways in which yeast cells can be transformed with exogenous DNA: uptake by intact cells (at low efficiency through the porous cell wall) and uptake by **sphaeroplasts,** the latter being the most commonly employed. Sphaeroplasts are cells from which the cell wall has been mostly removed by enzymatic treatment (e.g., with **chitinases** in the case of yeasts) leaving a spherical cell surrounded by its plasma membrane and bits of cell wall remnants. Once sphaeroplasts are formed, it is easy to transform the yeast with exogenous donor DNA by first mixing the donor DNA with a source of excess Ca^{2+} such as $CaCl_2$ or $CaPO_4$. A Ca–DNA complex forms microscopic particles that precipitate out of solution or rapidly settle out during centrifugation. When this precipitate is mixed with the sphaeroplasts in the presence of **polyethylene glycol,** the DNA moves quite easily through the plasma membrane and into the cytoplasm of the cell because the membrane has been made more permeable by the polyethylene glycol. Transformed sphaeroplasts can be plated on a suitable culture medium with 3% agar so that new cell walls can then be synthesized around them.

8.28. What kinds of vectors are commonly used for the cloning of genes in yeasts?

Some cloning vectors routinely used in yeast cloning are shown in Table 8.5. Some of these vectors are constructed so that they can introduce cloned genes into two or more hosts. Such vectors are known as **shuttle vectors.** Shuttle vectors have origin of replication sites that can be recognized by the DNA replication mechanism of more than one host cell. They are especially useful because they can be transferred into prokaryotic cells, which are easier to grow and reproduce much faster than eukaryotic cells.

Table 8.5.   Some Vectors Commonly Used in Cloning With *Saccharomyces Cerevisiae*

Designation	Name	Function(s)
YEp	Yeast episomal plasmid	Contains origin of replication of 2-micron circle. Contains a yeast gene that encodes resistance to an antibiotic. Fairly stable within transformed host cells.
YRp	Yeast replicative plasmid	Has an autonomously replicating sequence (ARS). Allows the independent replication of the plasmid without chromosomal control. Contains a yeast gene that encodes resistance to an antibiotic. Highly stable.
YCp	Yeast centromeric plasmid	Cloning vector with an inclusion known as **cen**, which contains a fragment of the eukaryotic centromere that increases the stability of the plasmid. Appears to direct the movement of the plasmid during mitosis. Contains a yeast gene that encodes resistance to an antibiotic. Highly stable.
YIp	Yeast integrative plasmid	Useful in transformation of yeast cells. Integrates into the host genome. Contains a yeast gene that encodes resistance to an antibiotic. Highly stable.

Objective Questions

Multiple Choice

Directions: Unless otherwise specified, choose the one best answer.

1. Genetic engineering involves which of the following? (*a*) Conjugation (*b*) deamination (*c*) cloning (*d*) nucleotide synthesis (*e*) mutations

2. Cloning requires all of the following *except* (*a*) donor DNA. (*b*) DNA ligase. (*c*) a vector. (*d*) methylases. (*e*) restriction endonucleases.

3. DNA containing a genetic sequence that is to be cloned is referred to as (*a*) vector DNA. (*b*) donor DNA. (*c*) single-stranded DNA. (*d*) RFLP. (*e*) host DNA.

4. For the extraction of genomic DNA from cells, all of the following are required *except* (*a*) detergent. (*b*) proteases. (*c*) potassium ions. (*d*) organic solvents. (*e*) ethanol.

5. The selective isolation of plasmid DNA from total genomic DNA involves (*a*) detergent. (*b*) proteases. (*c*) potassium ions. (*d*) organic solvents. (*e*) ethanol.

6. Bacterial enzymes that cut DNA at specific sites *within* the DNA molecule are called (*a*) exonucleases. (*b*) methylases (*c*) transferases. (*d*) kinases. (*e*) restriction endonucleases.

7. The sites at which a restriction endonuclease cuts the DNA molecule is known as (*a*) palindromic DNA. (*b*) recognition sequences. (*c*) sticky sites. (*d*) blunt ends. (*e*) restriction maps.

8. Sequences on the DNA molecule that are the same on both strands when read in the same direction (e.g., $3' \rightarrow 5'$) are known as (*a*) palindromes. (*b*) recognition sequences. (*c*) sticky sites. (*d*) blunt-end sites. (*e*) restriction maps.

9. Restricted DNA fragments can be isolated using the following technique: (*a*) gas chromatography. (*b*) electrophoresis. (*c*) liquid chromatography. (*d*) restriction mapping. (*e*) centrifugation.

10. The movement of DNA at pH 8 in an electrical field during electrophoresis in agarose gels is (*a*) from the positive to the negative pole. (*b*) from the negative pole to the positive pole. (*c*) from an area of high concentration to one of lower concentration. (*d*) from an area of low concentration to one of high concentration. (*e*) dependent on the ionic strength of the buffer.

11. Analyses of restriction fragment length polymorphisms involve which of the following two techniques: (*a*) electrophoresis, (*b*) chromatography, (*c*) restriction enzyme digestion, (*d*) methylation, (*e*) electroporation.

12. All of the following are descriptive of cloning vectors *except* (*a*) plasmids. (*b*) bacteriophages. (*c*) genomic DNA. (*d*) shuttle. (*e*) expression.

13. Which of the following characteristics is *undesirable* in cloning vectors? (*a*) control their own replication (*b*) readily isolated from cells (*c*) high copy number (*d*) vulnerable at several sites to a restriction enzyme (*e*) small in size

14. Which of the following enzymes "splice" together two pieces of DNA that have been cut by the same restriction endonuclease? (*a*) Ligase (*b*) kinase (*c*) methylase (*d*) polymerase (*e*) lyase

15. Polynucleotide ligase joins two DNA molecules together by forming a covalent bond between (*a*) two OH groups. (*b*) two phosphate groups. (*c*) complementary nucleotides on opposite strands. (*d*) two carbon atoms of adjacent nucleotides on the same strand. (*e*) a $3'$-OH group and a $5'$-PO_4 group.

16. Expression vectors contain a sequence, *not normally found* in other vectors that is known as (*a*) a ribosome-binding site. (*b*) an *ori* site. (*c*) a multiple-cloning site. (*d*) an antibody-resistant marker. (*e*) a recognition sequence.

17. All of the following are *common* methods of experimentally introducing DNA into host cells *except* (*a*) transformation. (*b*) transduction. (*c*) conjugation. (*d*) electroporation. (*e*) more than one of the above.

18. One important characteristic of host cells for cloning is (*a*) thick cell walls. (*b*) ability to grow without growth factors. (*c*) exacting growing requirements. (*d*) ability to conjugate with wild-type strains. (*e*) inability to multiply in natural environments.

19. Which of the following eukaryotic genera contain common cloning host cells? (*a*) *Paramecium* (*b*) *Saccharomyces* (*c*) *Penicillium* (*d*) *Spirogyra* (*e*) *Euglena*

20. Which of the following ions is essential for the introduction of DNA into yeast cells? (*a*) Sodium (*b*) potassium (*c*) magnesium (*d*) calcium (*e*) zinc

21. Sphaeroplasts lack the following cellular component: (*a*) endoplasmic reticulum. (*b*) ribosomes. (*c*) intact cell wall. (*d*) cytoskeleton. (*e*) nucleus.

22. Cloning vectors that can be used to transfer DNA from eukaryotic to prokaryotic cells are called (*a*) expression vectors. (*b*) mobile elements. (*c*) shuttle vectors. (*d*) bacteriophages. (*e*) transposons.

23. Which of the following is a yeast plasmid that has an autonomously replicating sequence? (*a*) YEp (*b*) YRp (*c*) YCp (*d*) YIp (*e*) pUC

24. Which of the following is a yeast plasmid that has a section of the eukaryotic centromere? (*a*) YEp (*b*) YRp (*c*) YCp (*d*) YIp (*e*) pUC

25. Which of the following is a yeast plasmid that can integrate into the host's chromosome? (*a*) YEp (*b*) YRp (*c*) YCp (*d*) YIp (*e*) pUC

True–False

1. Genetic engineering usually involves the introduction of native DNA into the same species of host from which the DNA originated.

2. Cloning is involved mainly in introducing a segment of recombinant DNA into a host bacterium.

3. To clone donor DNA, the DNA must be double-stranded.

4. The donor DNA to be inserted into a cloning vector usually is excised from the donor's genome by restriction endonucleases.

5. Detergents are used in the extraction of genomic DNA in order to dissolve membranes.

6. Organic solvents (e.g., phenol or chloroform) are used to precipitate plasmid DNA for cloning.

7. The selective precipitation of genomic DNA to obtain a pure preparation of plasmid DNA involves the use of potassium ions.

8. Ethanol is used to precipitate DNA from aqueous solutions.

9. Any enzyme that cuts DNA anywhere except at its ends is a called a restriction endonuclease.

10. Restriction endonucleases of type II often cut at specific recognition sites that are palindromes.

11. Palindromes are segments of DNA molecules that read the same in the $5' \rightarrow 3'$ direction as in the $3' \rightarrow 5'$ direction of the same strand.

12. Restriction mapping can be accomplished by using a single restriction endonuclease.

13. Products of restriction enzyme digestion can be isolated by electrophoresis.

14. Identification of VNTR patterns involves the use of restriction enzymes and agarose gel electrophoresis.

15. DNA molecules, because they have an overall negative charge along their backbone, move toward the anode in electrophoresis.

16. Under the same electrophoretic conditions, larger DNA molecules move faster than smaller ones.

17. Restriction mapping involves the use of single and double digestions in order to determine the relative sites at which each restriction endonuclease cuts.

18. VNTRs are polymorphic sites in noncoding regions of the chromosomal DNA.

19. Vector DNA is used as a vehicle to introduce the foreign DNA into a host cell, which can then replicate the DNA.

20. Cloning vectors include bacteriophages, shuttle vectors, expression vectors, and plasmids.

21. Yeasts are frequently used to clone eukaryotic genes.

22. Plasmids are good cloning vectors if they regulate their own replication to produce high copy numbers.

23. A good cloning vector is readily transferred by conjugation to wild-type bacteria.

24. Expression vectors lack both a ribosome binding site and a promoter site.

25. The ability of YIp yeast plasmids to integrate themselves into a host chromosome is an undesirable trait.

Matching

Directions: Match each item in **Column A** with the one in **Column B** to which it is most closely associated. Each item in **Column B** can be used only once.

Column A		Column B
1. plasmid	A.	recognition sequence
2. vector	B.	dissolves plasma membranes
3. restriction enzyme (RE) site	C.	RE cuts middle of recognition sequence
4. shuttle vector	D.	selective isolation of plasmid DNA
5. cleared lysate	E.	self-replicating DNA molecule
6. Triton X-100®	F.	noncoding region polymorphism
7. K+	G.	restriction mapping
8. palindromes	H.	remains after removal of K+ and cell debris
9. blunt ends	I.	possesses ribosome binding site
10. electrophoresis	J.	covalently unites two DNA molecules
11. double digests	K.	vehicle to introduce DNA into host cell
12. VNTRs	L.	same sequence ($5' \rightarrow 3'$) on both strands
13. expression vector	M.	moves DNA between two hosts
14. high copy number	N.	sorts DNA molecules based on size
15. DNA ligase	O.	desirable quality of cloning vector

Terms

Directions: Unless otherwise specified, each answer is a single word.

1. The process whereby a section of DNA is inserted into a plasmid or bacteriophage and then replicated to make numerous copies of the insert.

2. In a cloning experiment, the DNA to be cloned.

3. Substances used to dissolve membranes during the extraction of genomic DNA from cells.

4. The name of the disrupted cell mixture, once the membranes are dissolved.

5. The process of removing genomic DNA from a water-alcohol interphase by twirling a glass rod.

6. The resulting solution after the addition of potassium acetate and centrifugation to remove genomic DNA and cell debris. (2 words)

7. Enzymes that cut DNA at specified sites. (2 words)

8. Sites within which restriction enzymes cut double-stranded DNA. (2 words)

9. Sequences of double-stranded DNA that read the same (e.g., $5' \rightarrow 3'$) on both strands of the DNA molecule.

10. The graphical representation of recognition sites for two or more restriction enzymes. (2 words)

11. Restriction fragments can be sorted using this technique.

12. In an electrical field, DNA molecules move toward this pole.

13. Name given to DNA molecules made from mRNA templates using reverse transcriptase.

14. DNA fingerprinting is a technique that detects unique DNA chromosomal segments known as: (4 words or 4-letter acronym).

15. Restriction enzymes isolated from different bacterial species, but with the same recognition sequence.

16. DNA molecules employed to introduce foreign genes into host cells.

17. An adjective given to vectors having ribosome binding sites and promoter sites.

18. An adjective given to vectors that transfer spliced DNA into eukaryotic or prokaryotic host cells.

19. The process whereby a readily detected phenotype (such as the digestion of lactose) is eliminated when a recombinant cell is formed. (2 words)

20. This type of vector (referring to the number of vector molecules per cell) is commonly used in cloning. (3 words)

21. Plasmid cloning vectors can be transferred from one cell to another by this process, either naturally or experimentally.

22. This enzyme splices two DNA molecules by forming covalent bonds between the $3'$-hydroxyl group of one molecule and the $5'$-phosphate group of the other.

23. High voltage can be used to introduce recombinant DNA molecules into cells.

24. A type of plasmid found in yeast cells. (2–3 words)

25. The segment of a YRp plasmid that allows the independent replication of the plasmid without chromosomal control. (3 words)

Answers to the Objective Questions

Multiple Choice

1. *c*   **2.** *d*   **3.** *b*   **4.** *c*   **5.** *c*   **6.** *e*   **7.** *b*   **8.** *a*   **9.** *b*   **10.** *b*   **11.** *a & c*
12. *c*   **13.** *d*   **14.** *a*   **15.** *e*   **16.** *a*   **17.** *c*   **18.** *e*   **19.** *b*   **20.** *d*   **21.** *c*
22. *c*   **23.** *b*   **24.** *c*   **25.** *d*

True–False

1. F (It could, but normally the DNA originates from a foreign cell.)   **2.** F (Cloning is done to replicate an introduced segment of DNA.)   **3.** T   **4.** T   **5.** T   **6.** F (used to remove proteins)   **7.** T   **8.** T
9. F (Not all endonucleases are restriction endonucleases.)   **10.** T   **11.** F (that read in the same direction on opposite strands)   **12.** F (Two or more enzymes are needed for restriction mapping.)   **13.** T   **14.** T
15. T   **16.** F (slower)   **17.** T   **18.** T   **19.** T   **20.** T   **21.** T   **22.** T   **23.** F (Conjugative transfer is an undesirable property of cloning vectors.)   **24.** F (They possess both promoters and ribosome binding sites.)   **25.** F (It is indeed a desirable trait.)

Matching

1. *E*   **2.** *K*   **3.** *A*   **4.** *M*   **5.** *H*   **6.** *B*   **7.** *D*   **8.** *L*   **9.** *C*   **10.** *N*   **11.** *G*
12. *F*   **13.** *I*   **14.** *O*   **15.** *J*

Terms

1. cloning   **2.** donor   **3.** detergents   **4.** lysate   **5.** spooling   **6.** cleared lysate   **7.** restriction endonucleases   **8.** recognition sequences   **9.** palindromes   **10.** restriction map   **11.** electrophoresis
12. positive (anode)   **13.** cDNA   **14.** restriction fragment length polymorphism (RFLP)
15. isoschizomers   **16.** vectors   **17.** expression   **18.** shuttle   **19.** insertional inactivation
20. high copy number   **21.** transformation   **22.** ligase   **23.** electroporation   **24.** two (2) micron circle   **25.** autonomously replicating sequence

Chapter 9

Nucleic Acid Manipulations

INTRODUCTION

Over the past two decades, molecular biology applications have spread into almost every field of endeavor in the sciences. Techniques such as DNA hybridization, nucleic acid sequencing, and the polymerase chain reaction have revolutionized the basic and applied sciences, from evolution to forensics, to biotechnology and medical diagnostics. This chapter will discuss the theory and applications of some of the most important manipulation techniques of nucleic acids.

NUCLEIC ACIDS HYBRIDIZATION

9.1. What is DNA hybridization?

From developments in the area of genetic engineering and molecular biology, a powerful tool known as **DNA hybridization** has emerged. In the laboratory, this technique is used to detect the presence of DNA from pathogens in clinical specimens and to locate specific genes in cells. DNA hybridization takes advantage of the ability of nucleic acids to form stable, double-stranded molecules when two single strands with complementary bases are brought together under favorable conditions of temperature, pH, and ionic concentration.

In DNA hybridization assays, DNA from a virus or a cell is denatured with alkali to separate the strands. The single strands of DNA are then attached to a solid support such as a nitrocellulose or nylon membrane so that the strands do not reanneal (Fig. 9-1). The denatured DNA is attached to the membrane by its sugar-phosphate backbone with the nitrogenous bases projecting outward. To characterize or identify the target DNA, a single-stranded DNA or RNA molecule of known origin, called a **probe,** is added to the membrane in a buffered solution. This allows the formation of hydrogen bonds between complementary bases. The probe, so called because it is used to seek or probe for DNA sequences, is labeled with a **reporter** group. A reporter group can be a radioactive atom or an enzyme whose presence can be easily detected.

The probe is allowed to react with the target DNA; then any unreacted probe is removed by washing in buffered solutions. After the washes, all that remains on the nitrocellulose is the target DNA and any probe molecules that have attached to complementary sequences in the target DNA, forming stable hybrids.

Hybridization between target and probe DNAs is detected by assaying for the probe's reporter group. If the reporter group is detected, hybridization has taken place. If no reporter group is detected, it can be assumed that the target molecule does not have sequences that are complementary to those of the probe, and hence, the gene or DNA segment sought is not present in the sample.

9.2. What are the components of a DNA hybridization assay?

There are four major components in a DNA hybridization assay. These are the target DNA, the probe, the detection system, and the hybridization format.

9.3. What characteristics are important for a target DNA molecule in a hybridization assay?

DNA hybridization assays are aimed at detecting the presence of unique nucleic acid sequences in the genome of organisms or in DNA molecules of interest. These nucleic acid sequences represent the target DNA molecule. Target molecules can be segments of DNA or the entire molecule in a chromosome, a plasmid, or even a mRNA or rRNA. Target DNA molecules must have unique, contiguous sequences along the length of the molecule that can be discriminated from other, similar sequences.

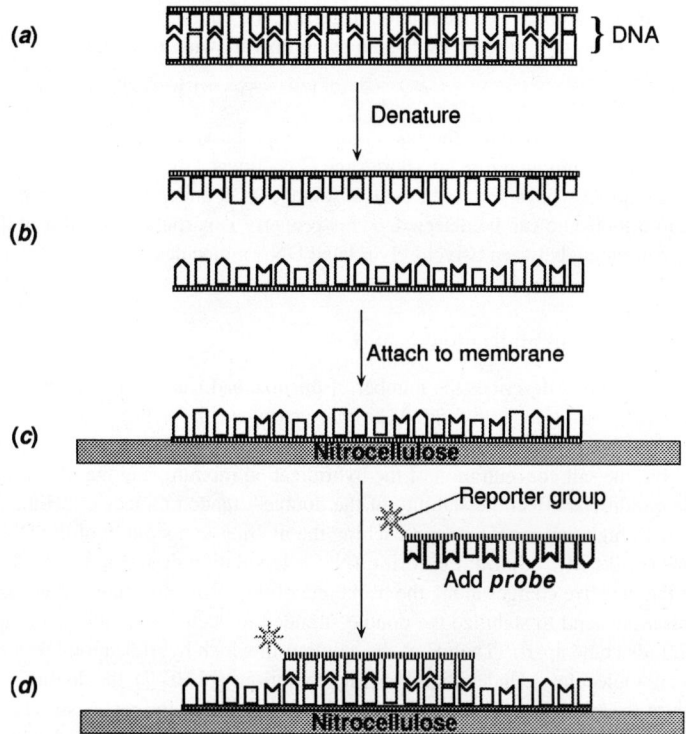

Fig. 9-1. DNA hybridization. When a DNA molecule (*a*) is heated above its melting temperature or is placed in an alkaline environment, it becomes denatured and the strands separate (*b*). The single strands of the DNA are then attached to a solid support such as a nitrocellulose or nylon membrane (*c*) by baking or by using UV light. A probe bearing a reporter molecule is added and allowed to react with the target DNA on the membrane for a few hours (*d*). After unreacted probe is washed away, the reporter molecule, which is attached to the hybridized probe, is detected enzymatically, fluorometrically, or by autoradiography.

A minimum number of target DNA molecules must be present for the hybridization to be detected. For this reason, DNA targets that appear in more than one copy in each cell, such as those coding for ribosomal RNA, insertion sequences, or those found in plasmids, are prime candidates for targets in hybridization assays. If the desired target DNA exists as a single copy per haploid cell, then alternative methods of hybrid detection have to be used to optimize the level of detection.

9.4.   What are nucleic acid probes?

Nucleic acid probes are segments of DNA or RNA that have a reporter group attached (Fig. 9-1) and can bind with high specificity to complementary sequences of target DNA. Probes can be complementary to either DNA or RNA and can be from as few as 20 bases to many thousands of bases long. **Oligonucleotide probes** (20 to 50 bases in length) can be chemically synthesized in any nucleotide sequence desired.

9.5.   What advantages do oligonucleotide probes have?

Because oligonucleotide probes are relatively short, they hybridize rapidly (sometimes within a few minutes) with target DNA molecules. The time it takes for a probe molecule to hybridize with target DNA (known as

$C_ot_{1/2}$) is dependent on the probe (x) concentration in μg/ml, the complexity of the probe (y), and the reaction volume (z) in milliliters. $C_ot_{1/2}$ times can be calculated using the following formula:

$$C_ot_{1/2} = [2(1/x)(y/5)(z/10)]$$

As can be seen from the formula, the less complex a probe is (that is, the fewer the number of bases) the smaller the number will be and therefore the shorter the $C_ot_{1/2}$ times.

Oligonucleotide probes have the additional advantage that under high **stringency** conditions, single nucleotide changes in a molecule can be detected. This property is particularly useful in detecting point mutations in genes or in discriminating between two closely related DNA molecules.

9.6. What is the stringency of a hybridization reaction?

Stringency is a term that describes the number of mismatched bases that are allowed in a hybridization reaction and still have a double-stranded hybrid. The more stringent the hybridization conditions are, the fewer the mismatched bases that are allowed before the two strands come apart. Stringency is influenced by the temperature of hybridization, the salt concentration of the hybridization mixture, and the pH of the hybridization reaction. All three of these conditions affect the stability of the double-stranded molecule. High stringency conditions are usually achieved with high temperatures (approaching the melting temperature of the DNA-DNA hybrid) and low salt concentrations (< 0.1 M). Sodium chloride (NaCl) in solution dissociates to form Na^+ and Cl^- ions. The Na^+ ions bind to the negative charges along the backbone of the DNA molecule and weaken their repulsive forces. High salt concentrations tend to stabilize the double-stranded molecule and allow more mismatches to take place before the two strands come apart. The pH of the solution in which hybridization takes place also affects the stability of hybridizing molecules. Under slightly acidic conditions (pH 6–7), the double-stranded DNA molecule is more stable because the increased H^+ concentration also helps reduce the repulsive forces between the two backbones of the DNA molecule. Much higher H^+ concentrations destroy the DNA molecule by removing the purines (depurination).

9.7. What are reporter molecules and what is their function?

Reporter molecules are chemical entities covalently attached to the nucleic acid probe that serve to indicate the presence of the hybridized probe with the target DNA. Reporter molecules play a very important role in the assay since it is their detection that heralds the presence or absence of hybridization.

Early investigators used radioactive phosphorus (^{32}P) as the reporter molecules and detected their presence by either autoradiography or liquid scintillation. Radioactive reporter molecules have fallen out of favor because of the hazard of handling radioactive isotopes and the waste disposal problem that they present. Nonradioactive reporter molecules include enzymes such as alkaline phosphatase, **fluorescent** molecules such as fluorescein isothiocyanate, chemiluminescent moeities, **antigenic** groups, and **affinity labels** such as biotin. They have the advantage of being nonhazardous and can be disposed of safely, but the level of sensitivity of hybridization assays using these molecules is generally lower than that obtained when using radioactive reporter molecules.

Reporter molecules that are part of a nucleotide triphosphate can be attached to nucleic acid probes using a variety of DNA-modifying enzymes—for instance, **terminal transferase** (adds nucleotides to the 3′ end of the probe molecule), **polynucleotide kinase** (adds nucleotides to the 5′ end of the probe molecule); DNA polymerase I (removes a few bases from nicked DNA and replaces them with a labeled nucleotide)—and photochemical reactions. Enzymes are covalently bound to the 3′ and/or 5′ end of the probe.

9.8. How are nonradioactive reporter molecules detected?

Affinity labels such as biotin by themselves are not detectable. To detect a biotinylated probe, the target-probe hybrid is first mixed with a solution of **streptavidin** (a molecule with high binding affinity for biotin) that has been linked to an enzyme molecule, usually alkaline phosphatase. A colorigenic or fluorogenic substrate for the enzyme is then added to detect the presence of hybrids. When the enzyme bound to the target DNA-probe hybrid reacts with the substrate, it produces a colored substrate that can be readily detected by visual observation or with a colorimeter (Fig. 9-2).

Chemiluminescent moeities such as acridinium esters are chemical groups that emit light when exposed to certain excitor substances. In a typical assay, after the hybridization is completed, the excitor is added and an x-

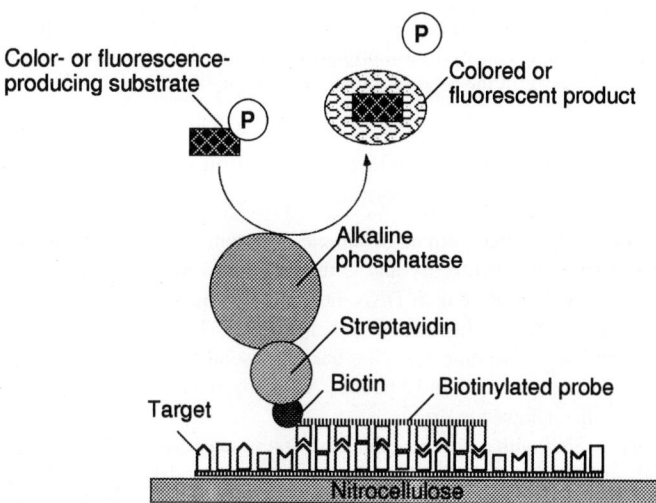

Fig. 9-2. Components of a detection system using biotin as a probe label and alkaline phosphatase as a reporter molecule. Single-stranded target DNA molecules are bound to a nitrocellulose membrane. Biotin-labeled probes are added so that they hybridize with complementary sequences in the target DNA molecules. After a suitable period of hybridization (e.g., 12 hours), a solution containing alkaline phosphatase complexed to streptavidin (SA-AP) is added and incubated for 30–60 minutes. During this time the streptavidin binds to the biotin on the probes, forming a biotin-SA-AP complex. After several washes aimed at removing unbound SA-AP, a color- or fluorescence-producing substrate for alkaline phosphate is added. When alkaline phosphate reacts with this substrate, it produces a colored or fluorescent product, providing visual evidence of hybridization.

ray film is layered over the membrane where the hybridization took place. Any light given off by the chemiluminescent reporter molecule exposes the film and a photograph is formed. Alternatively, luminometers can be used to quantify the amount of light given off by the reporter molecule.

Fluorescent reporter molecules are usually detected using fluorometers that can measure the amount of fluorescence after excitation of the reporter molecules by UV light. These reporters can be used in quantification assays. Since the amount of fluorescence emitted is directly proportional to the amount of target DNA-bound probe, it is possible to make crude estimations of the amount of DNA present or the degree of hybridization that has taken place.

9.9. The format in which DNA hybridizations are performed affects the time of the overall assay and how the hybrids are detected. What types of formats are commonly used in nucleic acids hybridization assays?

Nucleic acids hybridization assays are done either in solutions (liquid phase) or on solid supports. In **liquid-phase hybridizations,** both the target nucleic acid and the probe are in solution and are free to react with each other. The probe must be single stranded and not self-complementary. The rate of hybridization under these conditions is very rapid. The detection of hybridized probes requires the selective capture of double-stranded DNA (probe hybrids) using columns of **hydroxyapatite** (which specifically binds double-stranded DNA). The entire solution is subjected to treatment by **S1 nuclease,** which digests single-stranded DNA. All that remains after S1 digestion is the double-stranded probe-target hybrid. These double-stranded molecules are captured by the hydroxyapatite and eluted from the column, and the reporter molecule on the probe is detected.

In **solid-phase hybridization** assays, either the target nucleic acid or the probe is bound to a solid support such as a nitrocellulose or nylon membrane or polystyrene surfaces of microwell plates or beads. One of the most common assays involves the binding of the target DNA to the surface of a nitrocellulose membrane followed by alkali denaturation (dsDNA → ssDNA) of the target molecules. The membrane-bound DNA is made to react with a probe. Probe hybrids are detected by first thoroughly washing the membrane to remove any unbound probe, then by assaying for the presence of the reporter molecules on the membrane.

9.10. Three common formats are used in solid-phase hybridization assays: dot blot, Southern blotting, and *in situ* hybridization. Describe the dot blot format.

In the **dot blot** assay, a specified volume of sample or specimen is spotted onto a small area of a nitrocellulose membrane. The spot is allowed to dry and is then treated with an alkaline solution (e.g., 0.2N NaOH) to break open microbial cells and denature their DNA molecules. After this treatment, the sample is neutralized and then baked for approximately 20 minutes in an oven set at 80°C. The baking is to ensure that the single-stranded DNA is firmly attached to the nitrocellulose. After baking, a solution of a radioactively labeled probe is added to the membrane and allowed to react for 2 to 12 hours. After this, the membrane is washed several times to remove any unreacted probe and then allowed to dry.

The membrane is covered with a photographic film that is sensitive to radioactivity (e.g., x-ray film) and then placed in a darkroom for 12 to 24 hours. After this time, the film is developed and examined for the presence of dark spots. A dark spot indicates that the specimen contained DNA sequences that hybridized with the radioactive probe and, therefore, that the sought target DNA is present in the specimen. If no dark spot appears on the film, it can be inferred that no DNA hybridization has occurred.

This type of assay is becoming increasingly popular in the clinical laboratory because in theory (and frequently in practice) hybridization assays are specific for the DNA sequence and very sensitive. **Sensitivity,** a measure of the lowest level of target that can be detected by the assay, is very important. Small numbers of cells or viruses are extremely difficult to detect by biochemical, staining, or immunological assays. In these situations, it is necessary to employ sensitive assays, such as DNA hybridization, for the detection of the target DNA.

9.11. Describe the Southern blot hybridization assay.

Southern hybridization (blot) assays (Fig. 9-3) involve the restriction enzyme digestion and agarose gel electrophoresis of the target DNA prior to the hybridization assay. The different bands on the agarose gel are transferred by capillary action onto a nitrocellulose or nylon membrane in a blotting apparatus. During the transfer, each of the DNA bands is transferred onto the membrane in the same relative position that it had in the gel. After the transfer, the target DNA is probed and detected, as in the dot blot assay.

9.12. Describe the in situ hybridization assay.

In situ hybridization assays involve the probing of intact cells or tissue sections affixed to a microscope slide. This type of solid-phase assay has the advantage that one cannot only detect the presence of target DNA in intact cells, but also determine the location of such target DNA within a tissue. An important application of in situ hybridization is for the detection of viruses and certain types of bacteria within infected cells.

THE POLYMERASE CHAIN REACTION

9.13. What is the polymerase chain reaction?

The replication of the genetic material is carried out by enzymes called DNA polymerases. These enzymes initiate the synthesis of DNA starting from a primer bound to a template (Chapter 3). The primers are generally 9 to 20 bases in length and establish the site where DNA replication begins. With the **polymerase chain reaction (PCR),** any particular stretch of genetic material can be pinpointed and replicated numerous times simply by selecting a pair of primers that flank the desired stretch of DNA. The PCR is predicated on the annealing of two oligonucleotides (primers) of known composition to a target sequence of interest and the extension of the oligonucleotides with a DNA polymerase. Each reaction is repeated subsequent to a denaturation step, thus allowing for exponential amplification.

The polymerase chain reaction (Fig. 9-4) involves three temperature incubations or steps that are repeated 20 to 50 times. One repetition of the three steps is called a **cycle.** In the first step of the chain reaction, called **denaturation,** the two strands of the target DNA molecule are separated (denatured) by heating the DNA to 94°C. This breaks the hydrogen bonds between the bases, yielding two separate strands. In the second step, called **an-**

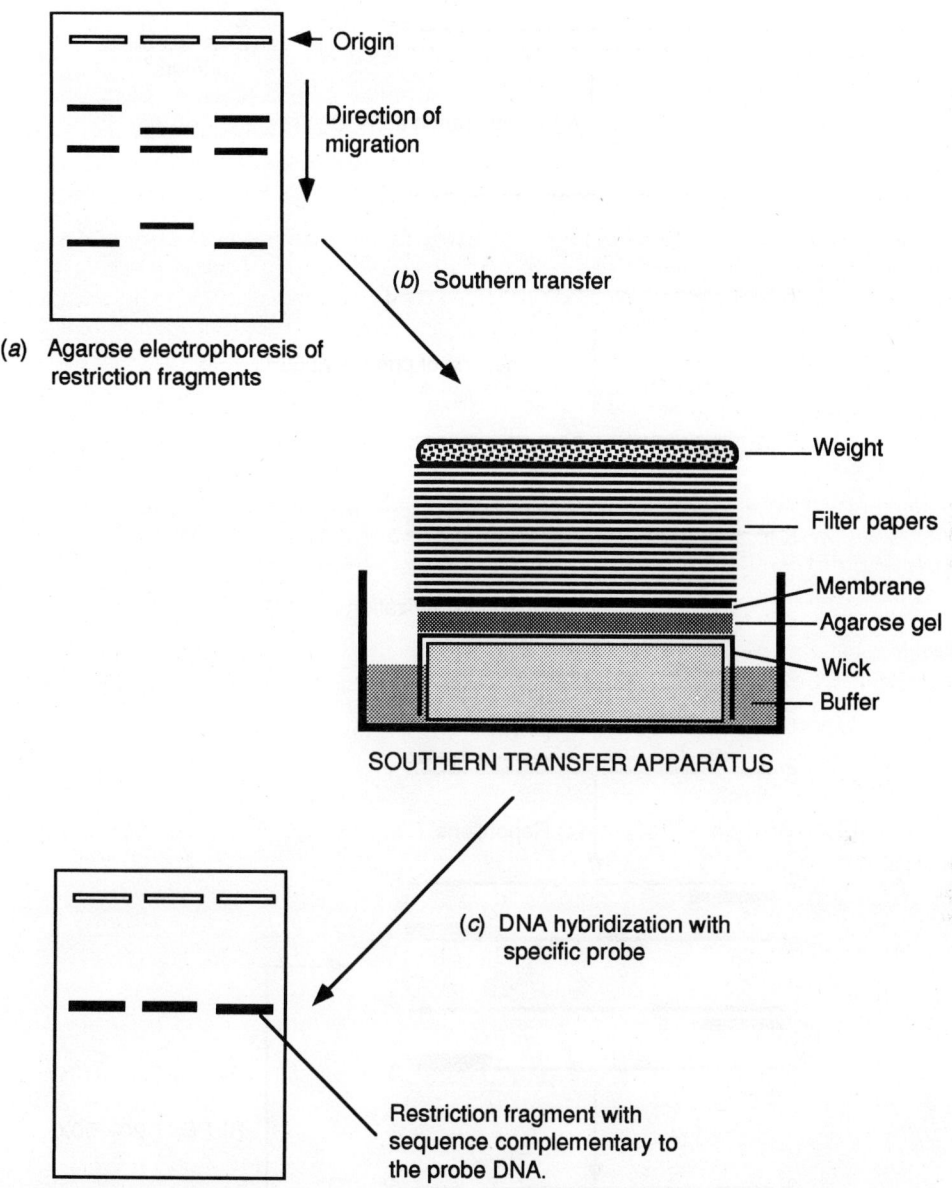

Fig. 9-3. The format of a Southern blot DNA hybridization. In Southern blot DNA hybridizations, the DNA molecule of interest is first digested with a suitable restriction enzyme. The restriction fragments are then separated (*a*) from one another by agarose gel electrophoresis. The individual fragments are transferred from the gel onto a nitrocellulose (or nylon) membrane by capillary action (*b*). Once the DNA has been transferred onto the membrane, it is hybridized (*c*) with suitable probes to detect the restriction fragment with the desired target DNA.

nealing, two primers hybridize to complementary sequences in the single strands. The primers are short (20 to 30 bases in length), synthetic stretches of single-stranded DNA. They are selected so that one primer is complementary to one end of the gene of interest on one strand, while the second primer is complementary to the opposite end on the other strand. The primers form hydrogen bonds with (anneal to) their complementary sequences, forming stable, double-stranded molecules. Annealing temperatures range between 37 and 60°C. During the third step, called **extension,** the primers are extended by a thermostable DNA polymerase at 72°C.

In a typical PCR assay, all the components required for the reaction are added at the beginning of the assay,

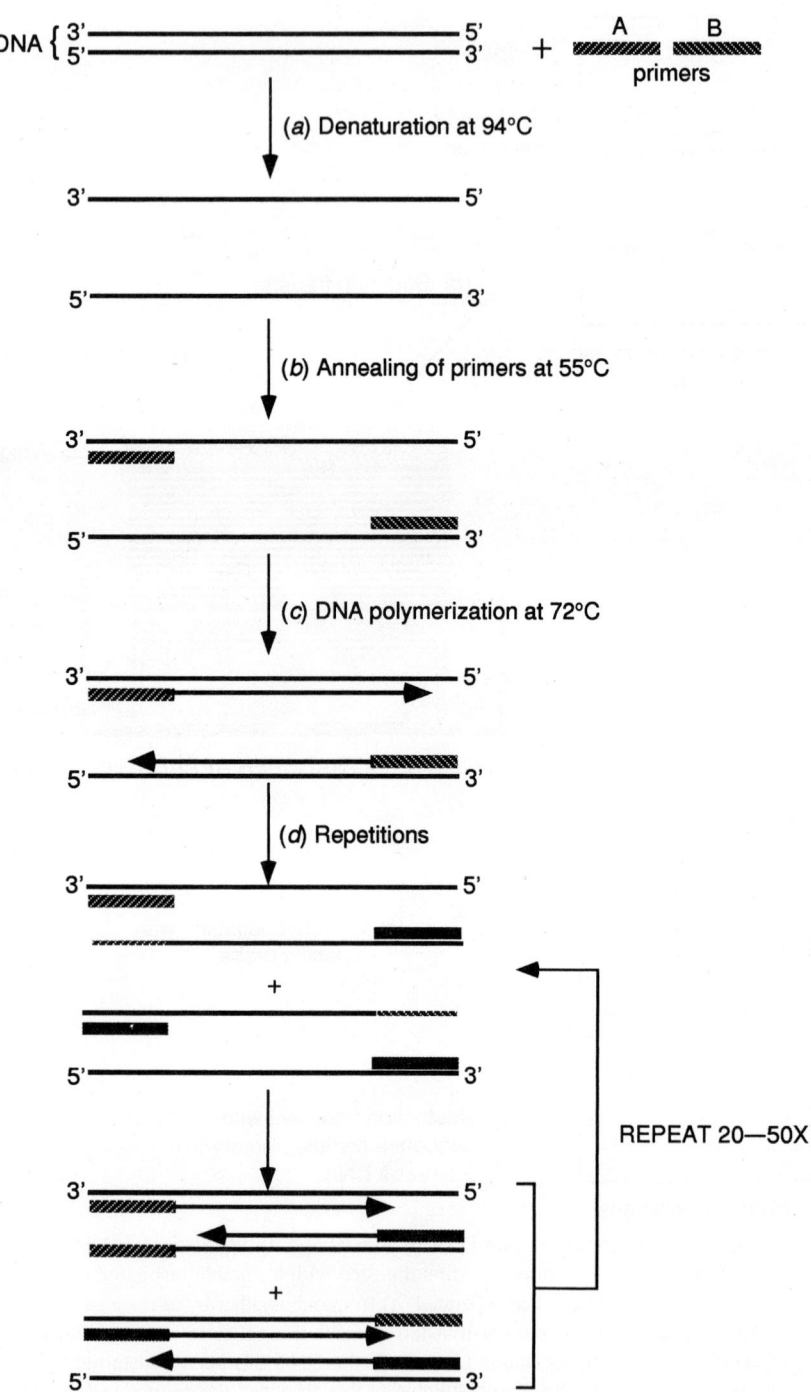

Fig. 9-4. Schematic representation of a typical PCR assay. A PCR assay involves the repeated use of temperature cycling to achieve primer-directed enzymatic amplification of desired segments of DNA. First the target DNA is denatured by heat (*a*). Upon cooling, the primers, which flank the desired segment of DNA, hybridize with complementary targets (*b*), and then DNA synthesis takes place (*c*), starting from the primer. These steps are repeated (*d*) 20–50 times to obtain large quantities of the desired DNA fragment.

and therefore, once the cycling begins the reaction need not be stopped to add more reagents. The reaction mixtures contain an excess of primers and nucleotide triphosphates, so that the polymerization reaction is not rate limiting. In a typical PCR assay the primer-to-target ratio is initially as great as $10^7 : 1$.

Each cycle is repeated 20 to 50 times in order to get an exponential synthesis of the target DNA segment flanked by the two primers. The PCR products (that is, the millions of copies of the segment of target DNA flanked by the two primers) are called **amplicons.** After each cycle, the number of templates doubles, so that if one starts with a single double-stranded DNA molecule, after 20 cycles the number of molecules synthesized by the PCR is 1×10^6, and after 30 cycles the number of molecules increases to 1×10^9. This number is obtained by applying the following formula:

$$M_f = M_i \times 2^n$$

Where M_i is the initial number of molecules (template), M_f is the final number of DNA segments produced by the PCR, and n is the number of cycles performed.

9.14. Can any DNA polymerase be used in PCR?

The DNA polymerase used in the PCR has to be thermostable in order to resist repeated 94°C heating to which the enzyme is subjected. When the concept of PCR was first devised in 1981, there were no thermotolerant DNA polymerases available. When amplification reactions were performed, DNA polymerases isolated from *E. coli* (Klenow) or from bacteriophages (Sequenase®) were used. Since these enzymes are inactivated by heat, after each denaturation cycle at 94°C, new enzyme had to be added in each cycle. Since 1985, several thermotolerant polymerases have been isolated (Table 9.1), permitting the entire reaction mixture (including the DNA polymerase) to be added at the beginning of the reaction, with no additional enzyme being needed in subsequent cycles. One such enzyme—possibly the one most widely used in PCR assays—is *Taq* polymerase. The designation *Taq* indicates that the enzyme was isolated from a thermophilic (heat-loving) bacterium called ***Thermus aquaticus.*** Some of the characteristics of this enzyme are summarized in Table 9.2. The discovery of thermotolerant enzymes made it easy to automate the PCR by means of heating blocks that heat and cool in cycles for specified periods of times.

Table 9.1. Some DNA Polymerases Used in PCR

Enzyme	Microbial Source	Thermotolerance
Klenow	*Escherichia coli*	no
Sequenase®	T7 bacteriophage	no
Taq	*Thermus aquaticus*	yes
BstE	*Bacillus stearothermophilus*	yes
Pfu	*Pyrococcus furiosus*	yes
Tth	*Thermus thermophilus*	yes
UlTma	*Teratoga maritima*	yes

Table 9.2. Characteristics of Taq Polymerase

Characteristic	Description
Origin	*Thermus aquaticus* strain *YT1*
Molecular weight	94 kd
Optimal temperature	72–80°C
Activity	150 nucleotides/sec/molecule
Stability	>50% at 95°C for 40 min
Fidelity	1.1×10^{-4} substitutions per cycle
Mg^{2+} concentration	>0.5 mM < 10 mM
Exonuclease activity	5' to 3' only
Transferase activity	5' adenosines

9.15. What factors affect the polymerase chain reaction?

While the theory of PCR is elegantly simple, the practical application of the technique is far more complex than initially surmised. Several factors influence the outcome of a polymerase chain reaction. Table 9.3 summarizes the reaction conditions for a typical PCR. Table 9.4 illustrates a typical PCR thermal profile.

Table 9.3.   Reaction Conditions for a Typical PCR Assay

Parameter	Concentration
Template	10 attomoles ($\sim 1 \times 10^6$ molecules)
pH	8.4 (10 mM Tris-HCl)
dNTPs	200 μM (each one)
Gelatin	100 g/ml (optional)
Primers	0.5 μM (each one); (0.1–1.0 μM)
Mg^{2+} concentration	>0.5 mM < 10 mM
Enzyme	1 unit
Total volume	25–100 μl

Table 9.4.   Typical PCR Thermal Profile

Program	94°C (first step)	37–60°C (second step)	72°C (third step)	Reps
Cycle 1	1 min	1 min	1 min	1×
Cycles 2–30	1 min	1 min	1 min	29×
Cycle 31*	1 min	1 min	10 min	1×

* This cycle is normally included in a PCR assay in order to allow any "unfinished" product from previous amplification to achieve its full length.

The temperature profile, with the appropriate reagent concentrations, determines the specificity of the PCR assay. Since the temperature of the reaction determines the specificity of the hybridization of primer to the template, low temperatures may allow "false priming" at sites on the DNA molecule other than the desired target site. For this reason, when the PCR assay is first carried out, a temperature-optimization experiment is performed to determine the best temperature for primer annealing and the maximum yield of the desired amplified product.

The most important chemical components of a PCR assay are the concentrations of Mg^{2+}, primers, deoxyribonucleotides, and enzyme. It has been shown that the concentration of Mg^{2+} affects the specificity of priming and primer extension by the PCR. The range of Mg^{2+} concentration that seems to yield the best results is between 1.5 and 3.0 mM. Concentrations of Mg^{2+} much above 3.0 mM generally result in the production of false products, and below 1.5 mM, no product is synthesized.

It is essential that the concentration of primers be somewhere between 0.1 and 1.0 μM. This range has been found to be suitable for most PCR assays. It is important that the primer concentration be in great excess relative to the concentration of template, since a lack of sufficient primers during the amplification process would slow down the reaction rate. Primer concentrations greater than 1 μM are excessive and represent a waste of expensive reagents. Tables 9.5 and 9.6 summarize the reagent concentrations before and after 1 million amplifications of a segment of human genomic DNA measuring 1,000 base pairs in length. Notice that even when 0.1 μM concentration of primers is used, most of it remains unused at the end of the assay.

Nucleotide triphosphate (dNTP) concentrations are by convention set at 200 μM, and the activity of the enzymes is calibrated at this concentration. As was the case with the primer concentration, the concentration of dNTPs is in great excess. Because much of it remains unused at the completion of the reaction (Tables 9.5 and 9.6), dNTPs do not become rate limiting, and therefore the polymerization reactions take place at a maximum rate.

One unit of the *Taq* polymerase in 100 μl of reaction mixture represents an enzyme concentration of 1 nM. At 1 nM concentration, even after a millionfold (10^6) amplification, there is an ample enzyme free at all times, and thus, the reaction rate is not affected by limited enzyme concentrations.

Table 9.5. Stoichiometric Analysis of PCR (Initial Conditions)

Parameter	Amount	pMoles*	Conc.	Ratio[†]
Human DNA	1 μg	5×10^{-7}	5 fM[‡]	1
Defined 1 kb fragment	0.3 pg	5×10^{-7}	5 fM	1
Each primer	65 ng	10	0.1 μM	2×10^{7}
Each dNTP	11.5 μg	2×10^{4}	200 μM	4×10^{10}
Taq polymerase	2 units	0.1	1 nM	2×10^{2}

* pMoles = picomoles = 1×10^{-12} moles.
[†] Ratio to genomic template.
[‡] 1 fM = 1 fentomole = 1×10^{-15} Molar.

Table 9.6. Stoichiometric Analysis of PCR (Conditions After 10^{6} Amplifications)

Parameter	Amount	pMoles*	Conc.	Ratio[†]
Human DNA	1 μg	5×10^{-7}	5 fM[‡]	1×10^{-6}
Defined 1 kb fragment	300 ng	5×10^{-7}	5 fM	1
Each primer	62 ng	10	0.095 μM	19
Each dNTP	11.4 μg	1.98×10^{4}	198 μM	4×10^{10}
Taq polymerase	2 units	0.1	1 nM	0.2

* pMoles = picomoles = 1×10^{-12} moles.
[†] Ratio to genomic template.
[‡] 1 fM = 1 fentomole = 1×10^{-15} Molar.

9.16. How are amplicons detected?

The end result of a PCR assay is the production of millions of copies (**amplicons**) of a segment of DNA flanked by the two primers used in the assay. One common way of detecting amplicons from a PCR assay is by agarose or polyacrylamide gel electrophoresis. In such assays, a small portion (1 to 5 μl) of the PCR reaction mixture containing the amplicons is loaded onto the wells of an agarose gel (e.g., 2% agarose) and electrophoresis is carried out. At the end of the run, the agarose gel is stained in a solution of ethidium bromide (which intercalates into the amplicons and fluoresces orange when excited with UV light) and viewed using a UV light source called a **transilluminator.** Amplicons appear as bright orange bands against the light bluish background of the unstained material. Normally, amplicons are measured for size using **molecular-weight standards.** These are solutions containing DNA molecules of known size with which to compare the molecular weight of the amplicons. Agarose gels are then photographed with high-speed film, and the size and intensity of the bands are measured. Computer-assisted programs are now available that obviate the need for photography in molecular-weight determinations of amplicons.

Amplicons can also be detected using DNA hybridization assays. In these assays, the amplicons are used as the target DNA; they are probed using oligonucleotides with complementary sequences to an internal region of the amplicon. This way the presence of the amplicon is detected by using one of the hybridization formats described above rather than by electrophoresis.

9.17. There is a method for the enzymatic amplification of selected regions of a target DNA molecule known as the ligase chain reaction. How does this method work and what is its most important application?

The **ligase chain reaction** (LCR) is a method that utilizes the ligation of oligonucleotide pairs that are complementary to adjacent sites on a DNA template (Fig. 9-5). The exponential amplification is achieved with two pairs of adjacent primers (1 and 2, 3 and 4), rather than one pair as in PCR. One pair of primers is complementary to the top strand (1 and 2), while the second pair is complementary to the bottom strand (3 and 4). In this assay, after an initial denaturation step, the two pairs of primers anneal to their corresponding complementary sequences in the target molecule under appropriate stringency conditions. Since the two upper primers and the two lower primers are contiguous with one another, a thermostable ligase can covalently bind the two contiguous primers together to produce an oligonucleotide the length of the two primers together. Because the same ligation event is concurrently taking place on both strands, a double-stranded product is produced. The multiple cycling of denaturation, primer annealing, and primer ligation can result in the production of millions of LCR products, which can be detected by electrophoresis or DNA hybridization.

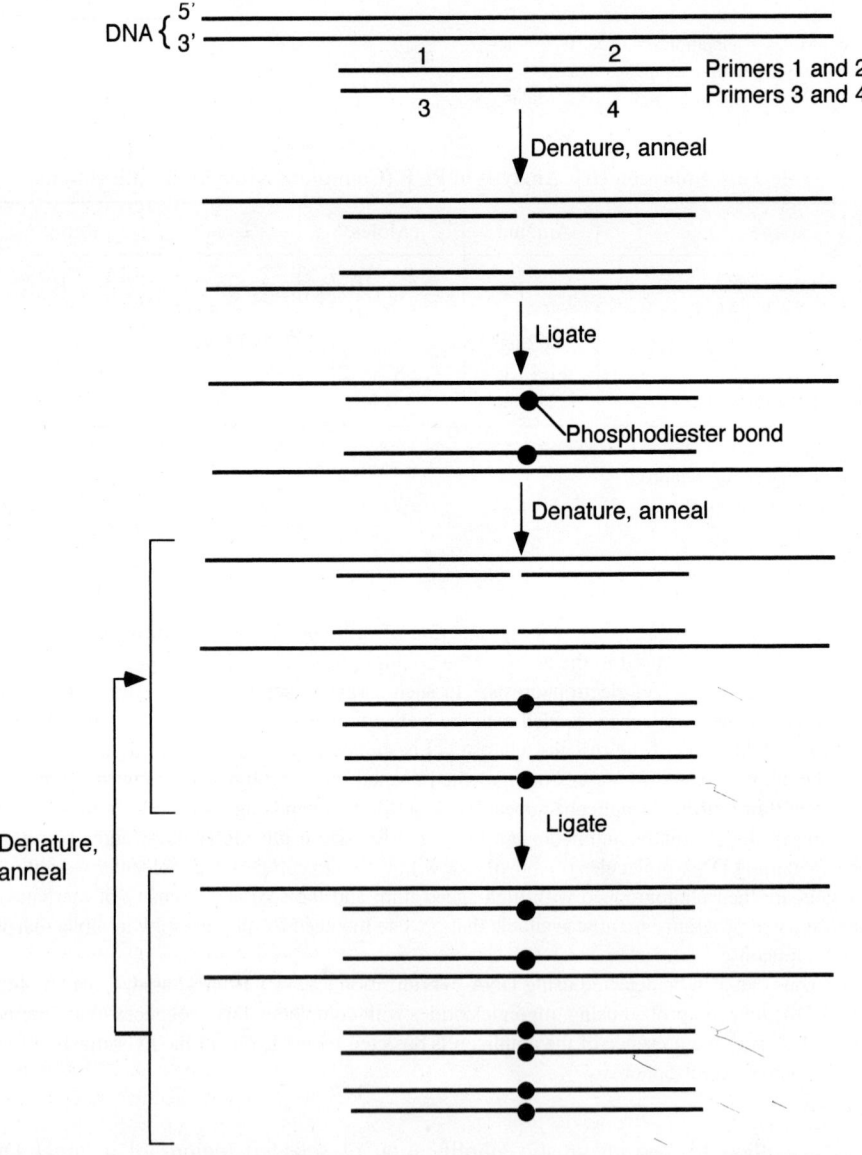

Fig. 9-5. The ligase chain reaction. The ligase chain reaction uses two primer sets (1–2, 3–4) that anneal to contiguous segments of a target DNA molecule. In this illustration primers 1 and 2 anneal to adjacent sites on one strand, and primers 3 and 4 anneal to adjacent sites on the other strand. A thermotolerant ligase forms the phosphodiester bond (black dot) between adjacent 3'-OH and 5'-PO$_4$ groups. These ligated products serve as templates for subsequent ligation reactions. If there is a mismatch at the position where the ligation should occur, the bond will not be formed and the product will not be detected.

LCR requires that the two upper and two lower primers have full complementarity along their lengths, especially at the ends where the two primers meet. If this condition is not met, the two primers cannot be joined by the ligase and no product is produced. Since the amplification process is dependent on the production of increasingly greater numbers of targets for the primers, no LCR products will be detected at the end of the assay unless the two primers are annealed end to end on the target molecule. For this reason, the LCR is particularly useful in detecting mutations at specific loci. In such applications, primer pairs are designed so that a base complementary to the mutation sought is at the 3' end of the left-hand primer (primer 1, Fig. 9-5) or at the 5' end of the right-hand primer (primer 2, Fig. 9-5). The wild-type allele will have a mismatched primer at one end, and therefore will be neither ligated nor amplified, while the mutant allele will have fully matched primers and hence will be ligated and amplified.

9.18. Can RNA targets be amplified using enzymatic amplification assays?

Yes, there are several strategies in common use to amplify RNA targets. One such strategy is known as **RT-PCR.** RT-PCR involves the use of the enzyme reverse transcriptase to make DNA copies of RNA molecules, which are then amplified using the standard PCR protocol. In this assay, RNA extracted from the cell or virus is converted into cDNA (complementary DNA) as a first step in the assay. Once this is achieved, a reaction mixture is added containing the required components for a PCR and the procedure is carried out. The cDNA target is amplified by the *Taq* polymerase, and the product is evaluated using one of the techniques previously described. A thermostable reverse transcriptase is now available (known as RTth) that serves both to synthesize cDNA from RNA and subsequently to amplify the segment of cDNA defined by the primer pair under PCR cycling conditions.

There are other RNA-based amplification systems that utilize the specificity of bacteriophage RNA polymerases and their corresponding cognate promoters. These phage RNA polymerase-based assays have a stage in which multiple copies of RNA are transcribed from the target sequence. One of these systems is called a **transcription-based amplification system (TAS).** The first step in a TAS assay generates a copy of a dsDNA template for each copy of a nucleic acid molecule present in the sample. Because the primer sequence used in the cDNA reaction contains a promoter sequence for an RNA polymerase, the resulting cDNA molecules can be transcribed. The second step involves transcribing the template sequence into multiple copies of RNA. As an example of the amplification power of this assay, Kwoh and colleagues applied the TAS assay for the detection of the human immunodeficiency virus type 1 (HIV-1) in infected cells. Only four cycles of TAS amplification resulted in a 2.5×10^6-fold increase in the copy number of the original target molecule.

9.19. Can PCR assays be used to induce mutations?

To study the effects of mutations on gene expression, researchers have developed a technique known as **site-directed mutagenesis,** which introduces point mutations at specific sites. One of the most commonly used strategies of site-directed mutagenesis takes advantage of the primer-directed amplification of DNA to introduce mutations. These mutations could be base substitutions, insertions, or deletions. To introduce a mutation at a specific site in the target gene, one of the primers is designed with a sequence complementary to the region in the target DNA, but with the desired base substitution (or gap). The mutagenic sequence within the primer must be either at the 5' end of the primer or internal to the primer, but never at the 3' end of the mutagenic primer. The 3' end region (at least 6 to 10 bases long) of the mutagenic primer must be totally complementary to the target DNA to permit full annealing of the primer to its target and allow the *Taq* polymerase to extend the primer. The PCR assay is carried out initially (first 5 to 10 cycles) under low stringency conditions, to allow the mismatch to occur. Once a few mutagenized templates are produced during the PCR, these will serve as targets and will be fully complementary to the primer. The end product of the amplification reaction will be a large number of amplicons that contain a mutation at the desired site. The mutagenized amplicons can then be cloned into plasmids or phages and introduced into cells for evaluation.

NUCLEIC ACID SEQUENCING

9.20. How is the coding sequence of a DNA molecule obtained?

Nucleic acid sequencing reveals the genetic code of a DNA molecule. It may be carried out using one of two methods, each of which results in the production of DNA fragments of various lengths, differing from each other by a single base and from which one can infer the nucleic acid sequence of the molecule. This is accomplished using **denaturing polyacrylamide gels.** Whereas agarose gels can separate DNA molecules differing in length by 30 to 50 bases, polyacrylamide gels can discriminate among DNA molecules differing in length by a single base. Denaturing gels cause the DNA molecule to become single stranded and remain that way throughout the entire process of electrophoresis. Denaturing gels contain **urea** and are run at elevated temperatures, both of which promote the separation of the double strands of the DNA molecule.

Since DNA molecules are invisible to the observer, they are usually labeled or marked in some way in order to visualize them after the electrophoresis is completed. The most common form of labeling is with radioactive isotopes, in particular ^{32}P, ^{33}P, and ^{35}S. After electrophoresis, labeled DNA molecules are visualized by placing an x-ray film next to the dried polyacrylamide gel and storing the gel in a dark place for a few hours. During this time the radioactive particles emitted from the isotope in each DNA molecule "expose" the film, and after development, a dark band is seen on the film at the position where the DNA band was located in the gel. This "picture," called an **autoradiograph,** is a mirror image of the position of the DNA bands in the gel.

9.21. There are two methods that can be used to sequence DNA molecules. These are the Maxam-Gilbert method and the Sanger method. How does the Maxam-Gilbert method work?

Allan Maxam and Walter Gilbert in 1977 developed a method for determining the sequence of DNA molecules based on cleavage of the DNA at specific sites by chemicals rather than by enzymes. The first step in this technique is to label the DNA molecule to be sequenced with a radioactive isotope such as ^{32}P or ^{35}S and then separate the two strands. Once many copies of the template DNA have been labeled, they are divided into four aliquots. Each aliquot of the labeled DNA is treated so a chemical modification occurs. One chemical treatment modifies all the guanines in the DNA, another modifies all the guanines and adenines, another modifies all the cytosines, and still another modifies all the cytosines and thymines. By controlling the chemical treatment of each aliquot, only a few bases in each DNA molecule among the population of molecules in each aliquot are modified in the same way. After modification, the DNA molecules are treated with a reagent that cleaves the DNA molecules at each of the modified bases. For example, in the aliquot in which all the cytosines were modified, after treatment with the cleaving reagent, each ^{32}P-labeled DNA fragment will have a base at the end of the molecule that corresponds to the base preceding the cleaved C*, as illustrated below:

$$5'-{}^{32}P\text{-ATGGAC*TTA}-3' \longrightarrow 5'-{}^{32}P\text{-ATGGA}-3'$$

Each of the four aliquots is then loaded separately into a different well of a denaturing polyacrylamide gel, and electrophoresis is carried out. During electrophoresis, each DNA fragment is sorted by size, from largest to smallest by increments of a single base. The larger fragments would have been cleaved closer to the 5' end of the template, and the smaller fragments closer to the 3' end. The Maxam-Gilbert method is seldom used anymore; the Sanger method is preferred.

9.22. How does the Sanger technique work for sequencing DNA molecules?

The enzymatic synthesis of DNA takes place by the sequential formation of a phosphodiester bond between the free 5' phosphate group of an incoming nucleotide and the 3' OH group of the growing chain. This process takes place throughout the length of the DNA molecule. **Dideoxynucleotides** lack a 3' OH group, and have a 3' H group instead. In the presence of a dideoxynucleotide, the synthesis of DNA stalls because the diphosphate bond cannot be formed. The chain growth terminates at that point, and the last base in the 3' end of the chain is a dideoxy terminator. This modification of Sanger's method of DNA sequencing is known as **dideoxy termination sequencing.**

In the Sanger's sequencing technique (Fig. 9-6), the template DNA is denatured and allowed to hybridize with a radioactively labeled primer. DNA polymerases require a primer to initiate the synthesis of DNA. Therefore, by selecting appropriate primers, one can determine where to begin sequencing DNA. Daughter (progeny) DNA molecules resulting from the extension of the primers will be radioactively labeled because the synthesis of each strand of DNA was initiated by a radioactively labeled primer.

After the primer hybridization, the sample is equally distributed into four different tubes, each containing a different dideoxy terminator nucleotide (ddATP, ddCTP, ddGTP, or ddTTP), along with all the other reagents re-

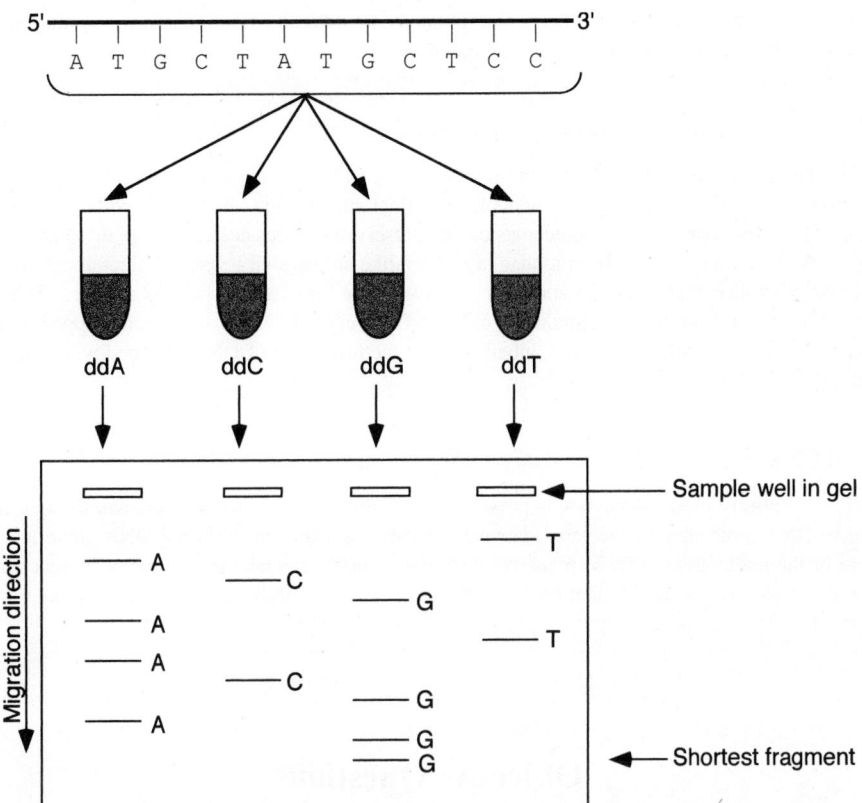

Fig. 9-6. DNA sequencing by the Sanger method. Four different reaction mixtures are used to sequence a DNA fragment by the Sanger dideoxy terminator method. Each reaction mixture contains the template DNA molecule to be sequenced, radioactively labeled primers, all four deoxynucleotides, DNA polymerase, and a different dideoxy terminator (ddA, ddC, ddG, or ddT). When one of these terminators is incorporated in the newly synthesized DNA strand, it will stop further synthesis of that strand; the result is that all the strands of various lengths in that reaction mixture end with the same base. The radioactive products are separated by electrophoresis and visualized by autoradiography. Reading from the bottom of the gel (shortest fragments terminated closest to the 5′ end) upward reveals the base sequence complementary to that of the template strand at top of figure: 3′TACGATACGAGG5′.

quired for the synthesis of DNA. These include the reaction buffer, DNA polymerase, dATP, dCTP, dGTP, and dTTP. The concentration of dideoxy nucleotides in the reaction mixture is crucial for the proper sequencing of the reaction. For example, in the tube containing ddATP, the ratio of ddATP to dATP is about 1:50. This way, not all synthesis reactions are terminated the moment the template DNA calls for an A. Instead, one of every fifty adenines will likely be a dideoxy-terminating nucleotide. This permits the elongation of the primer by the DNA polymerase to proceed until a ddNTP instead of a normal base (dNTPs) is incorporated into the growing chain. After the sequencing reaction is allowed to take place for a few minutes (usually 5 minutes) in the tube containing the ddATP, there will be a population of newly synthesized DNA molecules of various lengths, all of which end with an A. Similarly, the tubes containing the ddCTP, ddGTP, and ddTTP will have DNA molecules of various lengths ending with a C, a G, and a T, respectively.

A small aliquot of each of the tubes is loaded onto a denaturing polyacrylamide gel and electrophoresed for 2 to 6 hours to separate the various DNA molecules. The DNA molecules resulting from the synthesis reaction will be sorted according to size by electrophoresis, with the smallest molecules (those closest to the primer, which is at the 5′ end) moving further down the gel and the larger molecules (furthest away from the primer) located near the sample well at the top of the gel. After electrophoresis, an autoradiograph is performed and the band pattern is read and interpreted as in the Maxam-Gilbert method.

9.23. Can DNA molecules from extinct organisms be sequenced?

With the optimization of DNA extraction procedures, PCR, and sequencing techniques, it has become possible to retrieve relatively small DNA segments from extinct animals and plants and to determine the nucleotide sequences in these fragments. DNA molecules can be preserved in a desiccated state in dried hides, bones, and amber for millions of years. DNA from a quagga (a zebralike animal that became extinct about 100 years ago) was the first demonstration that DNA can stay in a preserved state for extended periods of time. Scientists have been able to extract DNA from human remains as old as 5,000 years old and from insects trapped in amber that have been dated 120 to 135 million years old, albeit most of the fragments of DNA obtained were only 100-1,000 base pairs in length.

9.24. How can DNA molecules resist total destruction for such long periods of time?

It is not entirely clear why DNA is preserved for millions of years. It is known, though, that most of the damage to DNA molecules is due to oxidative processes that cause nicks in the DNA chain and changes in the structure of the nucleotides. These oxidative degradative processes take place mainly in aqueous environments. DNA in desiccated states such as that bound to the hydroxyapatite in bone or in amber-forming resins is shielded from oxidative damage and can, therefore, be preserved much longer.

Objective Questions

Multiple Choice

Directions: Unless otherwise directed, choose the one best answer.

1. DNA hybridization is a technique commonly used in the detection of specific DNA sequences in a sample. This technique is based on all of the following properties of DNA *except* (*a*) double-strandedness. (*b*) denaturation-renaturation properties. (*c*) base-pairing properties. (*d*) major and minor grooves. (*e*) sequence specificity.

2. All of the following are components of a DNA hybridization assay *except* (*a*) target DNA. (*b*) probe. (*c*) DNA ligase. (*d*) reporter molecule. (*e*) nitrocellulose membrane.

3. Choose from the following list of characteristics two that are *essential* for a probe to be useful. (*a*) Complementarity to the target DNA (*b*) length of the target molecule (*c*) length of the probe (*d*) presence of a reporter group (*e*) assay format

4. Which of the following defines the fidelity and specificity of a DNA hybridization assay? (*a*) Stringency of the reaction (*b*) format of the assay (*c*) rate of polymerization (*d*) type of reporter molecule used (*e*) type of solid support used

5. The stringency of a hybridization reaction depends on all of the following *except* (*a*) NaCl concentration. (*b*) type of reporter molecule. (*c*) temperature of hybridization. (*d*) pH of the reaction. (*e*) nucleotide sequence of probe.

6. What is the principal function of reporter molecules in a DNA hybridization assay? (*a*) Enhance the stringency of the hybridization reaction (*b*) bind the target DNA to the solid support (*c*) aid in base pairing (*d*) improve annealing efficiency of the probe (*e*) aid in the detection of probe-target hybrids

7. All of the following are types of reporter molecules employed in DNA hybridization assays *except* (*a*) chemiluminescent moieties. (*b*) enzyme substrates. (*c*) ethidium bromide. (*d*) enzymes. (*e*) radioactive isotopes.

8. Which of the following DNA hybridization assay formats is most useful when one wishes to detect the presence of a pathogen's DNA in an aqueous clinical sample (e.g., cerebrospinal fluid)? (*a*) Southern blot (*b*) *in situ* hybridization (*c*) liquid-phase hybridization (*d*) dot blot (*e*) Western blot

9. Which of the following DNA hybridization assay formats is most useful when one wishes to detect the presence and location of a gene segment in a restriction digest of genomic DNA? (*a*) Southern blot (*b*) *in situ* hybridization (*c*) liquid-phase hybridization (*d*) dot blot (*e*) Western blot

10. Which of the following DNA hybridization assay formats is most useful when one wishes to detect the presence and localization of a pathogen's DNA within a cell? (*a*) Southern blot (*b*) *in situ* hybridization (*c*) in-solution hybridization (*d*) dot blot (*e*) Western blot

11. Which of the following methods is most useful for the enzymatic amplification of specific gene segments of DNA? (*a*) DNA hybridization (*b*) nucleotide sequencing (*c*) polymerase chain reaction (*d*) DNA ligation (*e*) reverse transcription

12. The polymerase chain reaction is based on the repetition of cycles of DNA replication by a thermostable DNA polymerase. These steps are DNA replication, primer annealing, and template denaturation. In what order should these steps be taken to achieve the desired results? (*a*) Replication, annealing, denaturation (*b*) replication, denaturation, annealing (*c*) annealing, denaturation, replication (*d*) annealing, replication, denaturation (*e*) denaturation, annealing, replication

13. All of the following factors affect the fidelity of the polymerase chain reaction *except* (*a*) $MgCl_2$. (*b*) annealing temperature. (*c*) pH of the reaction. (*d*) bovine serum albumin. (*e*) *Taq* polymerase.

14. The ligase chain reaction is another method used to achieve enzymatic amplification of target genes. This technique is best suited for the detection of (*a*) alleles. (*b*) pathogens. (*c*) amplicons. (*d*) RNA. (*e*) viruses.

15. The technique known as RT-PCR is best suited for the detection of (*a*) alleles. (*b*) pathogens. (*c*) amplicons. (*d*) RNA. (*e*) bacteria.

16. Site-directed mutagenesis can be achieved by which of the following procedures? (*a*) PCR (*b*) LCR (*c*) RT-PCR (*d*) RAS (*e*) TAS

17. Nucleic acid sequences are normally determined using which of the following techniques? (*a*) Agarose gel electrophoresis (*b*) capillary electrophoresis (*c*) starch gel electrophoresis (*d*) denaturing polyacrylamide gel electrophoresis (*e*) nondenaturing polyacrylamide gel electrophoresis

18. The level of resolution of sequencing gel electrophoresis is (*a*) 10 bases. (*b*) 1 base. (*c*) 100 bases. (*d*) 5 bases. (*e*) 25 bases.

19. The Maxam-Gilbert method of nucleotide sequencing is based on (*a*) PCR. (*b*) chemical modifications of DNA. (*c*) dideoxy nucleotide triphosphates. (*d*) RT-PCR. (*e*) LCR.

20. The Sanger method of nucleotide sequencing is based on (*a*) PCR. (*b*) chemical modifications of DNA. (*c*) dideoxyribonucleotide triphosphates. (*d*) RT-PCR. (*e*) LCR.

True–False

1. DNA hybridizations are techniques that rely on base complementarity in order to detect target DNA molecules.

2. Stringency is a term that defines the level of homology of one DNA molecule with another.

3. In DNA hybridization assays, target molecules must be at least twice as large as the probe.

4. Probes used in DNA hybridization assays must contain a reporter group.

5. Reporter groups are necessary because they enhance the affinity of the probe to the target DNA.

6. Oligonucleotide probes are often used when hybridization assays are required to be highly specific, although overnight hybridizations are often required.

7. $C_0t_{1/2}$ times define the number of cycles required in a PCR assay in order to optimize the yield.

8. High salt concentrations reduce the stringency of a hybridization reaction by weakening the inherent repulsive forces between the two opposite backbones of the DNA molecule.

9. Biotin reporter groups are known also as affinity-label groups.

10. Hybridization reactions are faster when the target DNA is bound to a solid support than when it is in solution.

11. Solid supports commonly used in DNA hybridization assays include nylon and nitrocellulose membranes as well as plastic microwell plates.

12. Southern blots rely on the capillary transfer of DNA from an agarose gel to a membrane.

13. The polymerase chain reaction relies on the need for DNA polymerases to initiate DNA synthesis starting from a primer.

14. Primers are important in DNA hybridization assays because they determine the length of the resulting amplified fragment.

15. PCR assays require a single primer.

16. The annealing step in a PCR assay is essentially a hybridization.

17. In PCR assays, the more cycles the larger the yield of the DNA, regardless of the reagent concentrations of the mixture.

18. Thermolabile DNA polymerases can also be used in automated PCR.

19. $MgCl_2$ and annealing temperature affect the fidelity and specificity of PCR.

20. For successful PCR reactions, primers and dNTPs should be in great excess.

21. Amplicons are detected by electrophoresis or by DNA hybridization assays.

22. The ligase chain reaction relies on a single pair of primers to flank a region to be amplified.

23. RT-PCR is based on the property of reverse transcriptase to make DNA from an RNA template.

24. Autoradiographs are "pictures" of a resulting DNA sequencing reaction and are used to infer the nucleotide sequence of a given DNA molecule.

25. Dideoxy terminators are nucleotides that lack a 2′ OH group of the deoxyribose molecule and as such cannot form phosphodiester bonds with the 5′ phosphate group of the incoming nucleotide.

Matching

Directions: Match each item in **Column A** with the one in **Column B** to which it is most closely associated. Each item in **Column B** can be used only once.

Column A	Column B
1. radioisotope	A. hybridization to target DNA
2. reporter molecule	B. agarose gel electrophoresis
3. Maxam-Gilbert method	C. Sanger method
4. dideoxy terminators	D. probe
5. denaturing gels	E. 72°C
6. stringency	F. autoradiograph
7. biotin	G. initiates DNA synthesis
8. dot blot	H. detects point mutations
9. Southern blot	I. nonradioactive label
10. in situ hybridization	J. thermotolerant enzyme
11. primer	K. DNA-sequencing electrophoresis
12. annealing step	L. RT-PCR
13. *Taq* polymerase	M. chemical modification of bases
14. extension step	N. amplicons
15. denaturation step	O. temperature and pH
16. ligase chain reaction	P. tissue sections
17. reverse transcriptase	Q. solid phase
18. PCR products	R. heat-labile DNA polymerase
19. Klenow fragment	S. 94°C
20. Only 100% base pairing allowed	T. low stringency
	U. high stringency
	V. site-directed mutagenesis

Terms

Directions: Unless otherwise specified, each correct answer is a single word.

1. Process in which two single-stranded DNA molecules of different origins come together to form a stable double-stranded molecule.

2. A single-stranded nucleic acid molecule of known sequence or origin used to detect the presence of desired target DNAs.

3. A chemical or radioactive group attached to probes.

4. Term describing the number of mismatched bases that may be allowed in a hybridization reaction while maintaining a double-stranded nucleic acid molecule.

5. The group of reporter molecules detected when its catalytic action converts substrate to product.

6. Two formats in which DNA hybridizations can be carried out. (2 words)

7. The type of hybridization assay in which DNA fragments are transferred from an agarose gel onto a membrane where hybridization takes place. (2 words)

8. The type of DNA hybridization assay in which the target DNA is placed directly on the membrane. (1 or 2 words)

9. PCR is carried out by repeating many times three temperature incubations known as steps. Name the first step.

10. The second step of PCR.

11. The third step of PCR.

12. Another name given to PCR products.

13. The thermotolerant enzyme used in the automated LCR assay.

14. The use of PCR to induce point mutations in DNA molecules. (2–3 words)

15. The gel matrix used in electrophoresis of DNA-sequencing products.

16. A method of DNA sequencing that employs base modification and cleavage strategies for sequencing. (2 words)

17. The kind of nucleotides employed in the Sanger method of sequencing. (2–3 words)

Answers to the Objective Questions

Multiple Choice

1. *d*   **2.** *c*   **3.** *a & d*   **4.** *a*   **5.** *b*   **6.** *e*   **7.** *c*   **8.** *d*   **9.** *a*   **10.** *b*   **11.** *c*
12. *e*   **13.** *d*   **14.** *a*   **15.** *d*   **16.** *a*   **17.** *d*   **18.** *b*   **19.** *b*   **20.** *c*

True–False

1. T   **2.** T   **3.** T   **4.** T   **5.** F (necessary because they permit detection of hybrids)   **6.** F (short hybridization times)   **7.** F (determines hybridization times)   **8.** T   **9.** T   **10.** F (Hybridization reactions in solution are faster.)   **11.** T   **12.** T   **13.** T   **14.** F (Primers are not involved in hybridization assays.)   **15.** F (requires two primers)   **16.** T   **17.** F (Reagent concentration is rate-limiting in PCR.)
18. F (The need to add polymerase in each cycle restricts automation.)   **19.** T   **20.** T   **21.** T
22. F (relies on two pairs of primers)   **23.** T   **24.** T   **25.** F (They lack 3′ OH group.)

Matching

1. *F*   **2.** *D*   **3.** *M*   **4.** *C*   **5.** *K*   **6.** *O*   **7.** *I*   **8.** *Q*   **9.** *B*   **10.** *P*   **11.** *G*
12. *A*   **13.** *J*   **14.** *E*   **15.** *S*   **16.** *H*   **17.** *L*   **18.** *N*   **19.** *R*   **20.** *U*

Terms

1. hybridization   **2.** probe   **3.** reporter   **4.** stringency   **5.** enzymes   **6.** solution (liquid phase), solid phase (membranes)   **7.** Southern blot   **8.** dot blot   **9.** denaturation   **10.** annealing (hybridization)
11. extension   **12.** amplicons   **13.** ligase   **14.** site-directed mutagenesis   **15.** polyacrylamide
16. Maxam-Gilbert   **17.** dideoxyribonucleotide terminators

Chapter 10

Eukaryotic Cells and Their Viruses

INTRODUCTION

Although eukaryotic cells and their viruses carry out many of the same processes extensively studied in bacteria and bacteriophages, the details of these processes differ, especially those carried out in specialized organelles. Some important processes are found almost exclusively in eukaryotes and their viruses; among these are RNA processing (exon splicing) and protein modifications (proteolytic cleavage, glycosylation, and phosphorylation). Chapter 4 discussed the regulation of transcription and RNA processing; Chapter 5, translation. This chapter first considers protein modifications and the transport of proteins to various sites and organelles. It then presents several examples of viral maturation and the interactions between viruses and cells that lead to cellular transformations and the cancerous state.

EUKARYOTIC CELLS

10.1. Eukaryotes generally modify newly formed (nascent) proteins. Where do these modifications occur?

Many protein alterations occur within the lumen of the endoplasmic reticulum (ER) and within the Golgi membranes (Golgi body). To enter these organelles, the nascent protein must have a **signal sequence** (leader sequence) at its amino terminus (Fig. 10-1a). The signal sequence 16–32 amino acids long is recognized by a complex of six proteins and a 300 nucleotide long RNA molecule (**srp RNA**) collectively called the **signal recognition particle (SRP)** that binds the nascent peptide and the ribosome, attaching them to a **SRP receptor** (SRP docking protein) on the surface of the ER. As the synthesis of a nascent polypeptide continues, the signal sequence subsequently dissociates from the SRP and becomes embedded in the membrane, where it associates with a protein in the membrane called the **signal sequence receptor (SSR)**. The signal sequence may also become associated with a **signal sequence peptidase** that cleaves it from the nascent peptide. In those cases where the protein becomes part of the membrane, the signal sequence may not be cleaved. It may function as a protein anchor.

10.2. Most proteins that will become part of the plasma membrane or be exported from the cell have sugars added. Where and how is this accomplished?

Glycosylation, which begins in the lumen of the ER, is the process of adding sugars to proteins. Short-branched polysaccharides are polymerized on an ER membrane lipid called **dolichol.** An enzyme catalyzes the transfer of polysaccharides from dolichol to specific asparagines in nascent proteins. These branching polysaccharides are referred to as **N-linked carbohydrates** because they are linked to the free amino group (NH_2—) of asparagines (Fig. 10-1b). In the Golgi membranes, the N-linked carbohydrates are modified further. *N*-acetylglucosamine, fructose, mannose, galactose and sialic acid compose the N-linked carbohydrates.

Another group of sugars, called **O-linked carbohydrates,** are added sequentially to the—OH groups of serine, threonine and sometimes hydroxyproline and hydroxylysine. The O-linked carbohydrates are built up by adding sugars one-by-one in the lumen of the ER or in the Golgi body. *N*-acetylgalactosamine, galactose, sialic acid, and fucose are commonly encountered sugars in O-linked carbohydrates.

10.3. Most of the proteins found in mitochondria and chloroplasts are encoded by nuclear genes. How do these proteins find the correct organelles, become part of the organelles, and remain in the organelles?

Proteins that are embedded in one or both of the lipid bilayers are called **integral membrane proteins.** These proteins contain one or more hydrophobic sequences called **hydrophobic domains** that "bind" the protein to the membrane when they "dissolve" into the lipid bilayers of membranes.

225

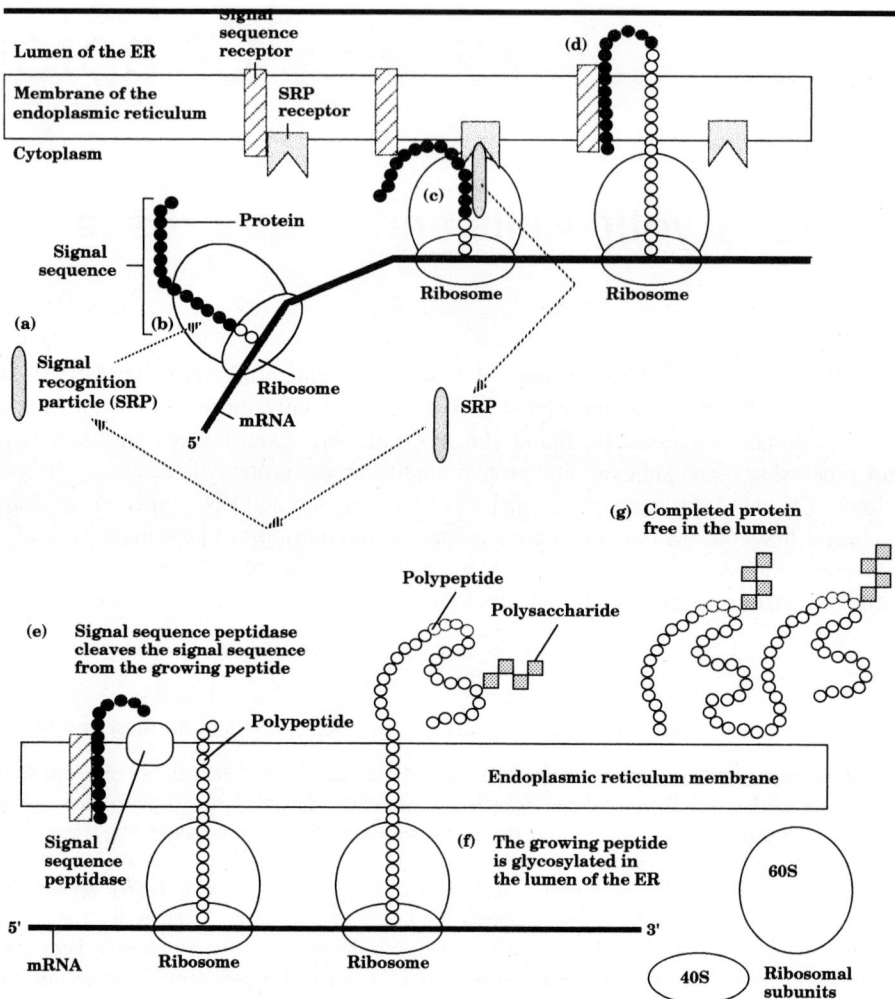

Fig. 10-1. The mechanism by which proteins pass into the lumen of the ER. (a)–(b) A signal
recognition particle (SRP) binds to the ribosome and to a signal sequence 16–32
amino acids long at the amino-terminal end of a newly forming protein. A signal
sequence is an absolute requirement if a protein is to pass into the lumen of the endoplasmic reticulum. (b)–(c) The SRP binds to a signal recognition particle receptor (SRPR) in the membrane. (c)–(d) The signal sequence binds to a signal sequence receptor in the membrane, so that it passes through the membrane. The
SRP is released from the ribosome and can be reused. (d)–(e) A signal sequence
peptidase on the lumen side of the membrane cleaves the signal sequence from the
newly forming protein. (e)–(f) The protein continues to elongate and pass through
the membrane. Generally, it becomes enzymatically glycosylated. (f)–(g) If there
are no hydrophobic regions (stop transfer regions), the protein will pass completely
through the membrane into the lumen of the ER.

A protein synthesized on the endoplasmic reticulum is able to enter the ER because of a leader region that
dissolves into the membrane or threads into a pore of some sort. As the nascent protein is synthesized, it continues to thread its way through the membrane, looping into the lumen of the ER. If the forming protein has hydrophobic domains along its length, these will remain dissolved in the membrane and the protein will loop into the
cytoplasm until it threads through the membrane again. A protein that has 6 well-spaced hydrophobic domains
might loop through a membrane 6 times. Since the hydrophobic domains cannot pass through the membrane they
are also known as **stop transfer domains.**

The amino acid sequences that retain integral membrane proteins within one-membrane systems (ER, Golgi membranes, peroxisome membrane, nuclear membrane, or plasma membrane) are just beginning to be identified. It is believed that specific amino acid sequences called **retention sequences** direct proteins to the correct membranes and keep them there. A single amino acid substitution in a retention sequence may cause the affected protein to go to the wrong membrane or be exported from the cell.

An ER retention sequence at the carboxyl terminus of a protein keeps proteins that have entered the ER attached to the ER membrane. An ER retention sequence consists of four amino acids, Lys-Asp-Glu-Leu (**KDEL sequence**). (To simplify amino acid sequences, single letters are used to represent each amino acid; see Table 5.1.) In some proteins, arginine or histidine can substitute for lysine (**RDEL** or **HDEL sequences**), and glutamic acid can replace aspartic acid (**KEEL sequence**). Removal of an ER retention sequence or the change of even a single amino acid may cause the protein to be exported from the cell. Conversely, addition of an ER retention sequence to a protein not found in the ER causes it to become part of the ER. The ER retention sequence keeps the protein in the ER by binding to an integral membrane protein referred to as the **retention sequence receptor.** The retention sequence receptor cycles the ER proteins between the ER and the *cis* Golgi membranes. *Cis* **Golgi membranes** are those that are closest to the endoplasmic reticulum.

The signal that directs specific proteins to lysosomes consists of a phosphorylated mannose in N-linked oligosaccharides. The phosphorylated mannose is recognized by a membrane protein called the mannose 6-phosphate receptor located in the *trans* Golgi membranes. *Trans* Golgi membranes are the most peripheral membranes of the Golgi body. The phosphate receptor binds phosphorylated proteins and carries the proteins to lysosomes. The acidic pH of lysosomes causes the proteins to be released from their phosphate receptors. Then, the phosphate receptors are recycled to the *trans* Golgi membranes.

10.4. How do proteins distinguish between the different compartments of an organelle such as those in a mitochondrion?

Most proteins that are routed to mitochondria and chloroplasts are synthesized in the cytoplasm on free 80S ribosomes. These proteins have two signal sequences at their amino termini, one of which binds to a signal sequence receptor protein in the outer membrane of mitochondria or chloroplasts. This same signal sequence is required for the movement of the protein through both the outer and inner membranes via a "pore" provided by a transport channel protein. To move through the outer and inner membranes, the targeted proteins must be unfolded by molecular chaperons or unfolding proteins, with energy provided by the consumption of ATP. A voltage across the inner membrane is required for transport of proteins into the matrix (in mitochondria), whereas ATP hydrolysis is required for transport of proteins into the stroma (in chloroplasts). Once in the matrix or stroma, the first signal sequence is removed by a signal sequence peptidase.

The secondary signal sequence is required for movement of the protein in the matrix or stroma into other compartments. Depending on the secondary signal, it can promote the insertion of the protein into one of the membranes. Hydrophobic sequences would anchor the protein to a membrane. In the case of mitochondria, however, the secondary signal sequence promotes the passage of the protein through the inner membrane into the intermembrane region (space between the outer and inner membranes). Another signal sequence peptidase cleaves the secondary signal sequence, so that the protein is released and remains in the intermembrane space. Protein movement into chloroplast thylakoids works in much the same way. A secondary signal sequence guides the protein from the stroma into a thylakoid.

VIRUSES THAT INFECT EUKARYOTES

10.5. Are viruses living organisms?

All viruses (**virions**) are noncellular infectious agents that proliferate only in cells. Viruses are referred to as "organisms" because some are extremely complex and highly ordered; even the simplest virus consists of hereditary material thousands of bases long covered by a protective protein coat. Yet, they are not "living" in the sense that cellular organisms are. Virions, for example, do not generate and dissipate voltages across closed membranes to concentrate nutrients, to distribute ions, or to generate usable chemical energy such as ATP. In addition, virions do not direct the synthesis from simple precursors of the many building blocks (such as amino acids, sugars, nucleotides, and lipids) needed to produce their hereditary information, coats (capsids), or membrane envelopes. All of these activities, however, are found in cellular organisms.

The Structure of Viruses

10.6. What are the physical characteristics of viruses?

Most viruses that infect eukaryotic cells consist of a nucleic acid, either DNA or RNA, covered by a protein coat. The protein coat is known as the **capsid**. A capsid is generally rigid, forming a body that is cylindrical, spherical, or icosahedral (having 20 faces and 12 vertices). Some capsids of eukaryotic viruses, however, are not easily defined since their structures and form are complex.

A single protein subunit of the capsid is referred to as a **capsomere**. A few capsids consist entirely of identical capsomeres. The capsids of most eukaryotic viruses, however, consist of a number of different proteins. The complex of nucleic acid and capsid is often designated as the **nucleocapsid**. In some cases, there is a second capsid that covers the nucleocapsid. Many of the animal viruses are surrounded by a membrane (lipid bilayer) derived from the host cell in which they proliferate. These viruses are referred to as **enveloped viruses**. Those that lack envelopes are known as **naked viruses.**

The length of nucleic acids in the simplest eukaryotic viruses averages about 3,000 nucleotides. These short nucleic acids usually specify 3 to 4 proteins: a capsomere, a nucleic acid polymerase, a helicase and a protein that promotes infection or spread of the virus from one cell to another. The most complex viruses code for as many as 200 different proteins. These extra proteins may help the virus avoid the host's immune response, infect a variety of different cell types, divert cellular metabolism toward viral proliferation, or finely tune their replication. The DNA in the complex viruses range from 100,000 to 300,000 bps.

10.7. How are eukaryotic viruses distinguished and classified?

The primary characteristics used to differentiate eukaryotic viruses are associated with their nucleic acid. First, virions are separated on the basis of whether they are **DNA viruses** or **RNA viruses**. The DNA or RNA of a virus may be single-stranded or double-stranded, depending upon the species. If the single-stranded RNA is able to function as mRNA it is referred to as **plus strand RNA** (+RNA); if it is equivalent to antisense RNA it is known as **minus strand RNA** (−RNA). Some of the genomes in plant and in animal viruses (like bromovirus and influenza virus) are fragmented, existing as separate "chromosomes" called **segments.**

Virion shape is used to differentiate among the viruses since they have a number of distinctive forms: icosahedral, cylindrical (helical), spherical, bullet, and complex. Strictly speaking, **icosahedral viruses** have a capsid with 20 triangular faces, 12 vertices, and 30 edges. The simplest icosahedral viruses have faces composed of 3 capsomeres. Such a capsid has a total of 60 capsomeres. Many viral capsids with more than 20 faces but having icosahedral symmetry are also referred to as icosahedrons.

> **Example 10.1.** Picornaviruses, the agents that are responsible for hepatitis A, poliomyelitis, the common cold virus, and foot and mouth disease are referred to as "icosahedral" viruses, yet they have 60 faces and 36 vertices. In fact, they are viruses that have icosahedral symmetry, but not a "true" icosahedral shape.

Generally, **cylindrical viruses** are referred to as **helical viruses** since their capsomeres are arranged so that they form a helix. Tobacco mosaic virus (the agent that causes mosaic disease in tobacco) is an example of a cylindrical, or helical, virus. Some of the cylindrical viruses in plants are extremely long and pliable. Because of their great length, these infectious agents are called **filamentous viruses.**

Spherical viruses (such as those in the families Togaviridae and Orthomyxoviridae) generally have a spherical membrane and an underlying protein matrix covering an icosahedral, helical, or spherical capsid or nucleocapsid.

> **Example 10.2.** Human T-cell lymphotropic viruses (HTLV-I, -II) in the family Retroviridae have a helical-like nucleocapsid, a second icosahedral capsid, and a spherical protein matrix and membrane covering the capsid. Human immunodeficiency viruses (HIV-I, -II), also in the Retroviridae, have a helical-like nucleocapsid, a second bullet-shaped capsid, a spherical matrix, and a membrane envelope.

Bullet-shaped viruses are found in the Rhabdoviridae family and include the viruses that cause rabies, vesicular stomatitis, encephalitis, and hemorrhagic fevers in animals. The capsid is classified as helical since the capsomeres are arranged so that they form a helix. One end spirals inward toward the long axis, whereas the other end spirals outward. The ends are capped, creating a bullet-shaped capsid.

Complex viruses include those found in the Poxviridae family, which cause all types of animal poxes (smallpox) and fevers (African swine fever). The largest known viruses belong to this family. To date, it is not clear whether or not poxviruses have a capsid comparable to those found in other viruses. A mass of proteins is associated with the double-stranded DNA genome, forming what at first looks like a peanut-shaped capsid. Later in viral development, the ends of the peanut-shaped capsid become enlarged. This covering and the DNA are referred to as the **viral core** or **nucleoid.** The core is covered with a number of other layers of material in addition to a membrane envelope.

The presence or absence of an enveloping membrane is another important characteristic for distinguishing viruses.

> **Example 10.3.** Some viral families that contain enveloped members are the Poxviridae (vaccinia), Herpesviridae (herpes), Paramyxoviridae (measles virus), Orthomyxoviridae (influenza virus), Rhabdoviridae (rabies virus), Retroviridae (human immunodeficiency virus), Arenaviridae (lymphocytic choriomeningitis virus), Coronaviridae (cold and diarrheal viruses), Bunyaviridae (California encephalitis virus), Flaviviridae (yellow fever virus, hepatitis C virus) and Togaviridae (equine encephalitis viruses).

The size of viruses is often helpful in distinguishing closely related viruses or viruses that have the same shape. Some of the smallest viruses with icosahedral symmetry are in the family Picornaviridae. They have diameters between 28 and 30 nanometers. Some viruses within the same family vary considerably in size; for example, those in the family Orthomyxoviridae have diameters between 90 and 120 nanometers, whereas those in the family Paramyxoviridae have diameters between 150 and 300 nanometers. Viruses in the family Poxviridae are the largest known viruses. Some of these virions can be seen with the light microscope because they generally are 300 to 450 nm by 170 to 260 nm in size.

Animal Viruses

10.8. What is involved in a viral infection?

A viral **infection** involves virion attachment to a host cell, penetration into the host cell by the virion or its hereditary information, and some sign of viral proliferation and/or cellular transformation.

10.9. How does a virion attach to a host cell?

For most animal viruses, the first step toward proliferation is virion **attachment** (or **adsorption**) to the surface of a host cell that will support its proliferation (Fig. 10-2a). Adsorption in almost all viruses that infect animal cells is mediated by specific proteins associated with the capsid or envelope, referred to as **viral binding sites** or **viral attachment proteins.** The attachment proteins interact with specific proteins or polysaccharides (usually part of glycoproteins) on the surface of the host cell. The proteins on the host cell are called **cellular receptors** or **host receptors.** The specificity of viral attachment proteins and of host receptors determine to which host cells viruses can attach (adsorb) and subsequently infect. The variety of hosts that a virus can infect and proliferate in is referred to as its **host range.** The "attraction" of a virus to certain types of host cells is known as its **cellular tropism.** This "attraction" (or ease of infection) for certain types of cells is determined primarily by the complementarity of the virus's attachment protein and the host's receptor. Since different tissues in animals often lack or have few receptors, a virus may not be able to adsorb to all tissues.

> **Example 10.4.** Variations in viral attachment proteins and host receptors explain how viruses in the same family, for example, rhinoviruses (colds), polioviruses (polio), and hepatitis viruses (cirrhosis of the liver), each absorb to specific tissues and consequently, cause very different diseases.

10.10. Once viruses have attached to the surface of a host cell, how do they enter the cell?

Virion entrance into a cell, called **penetration,** involves the movement of the virion (in some cases the membrane envelope does not enter the cell) across the host cell's plasma membrane. Penetration usually occurs through one of two processes: **endocytosis** and **membrane fusion.** Endocytosis is shown in Fig. 10-2b. In the case of endocytosis, viruses enter the cytoplasm within membrane vesicles (derived from the invagination of the plasma membrane) called **endosomes.** A lowering of the pH in endosomes is frequently the trigger that leads to capsid disintegration and nucleic acid release, a process known as **uncoating** (Fig. 10-2c). In many cases, uncoating is intimately associated with the movement of the virion's genome into the cytoplasm.

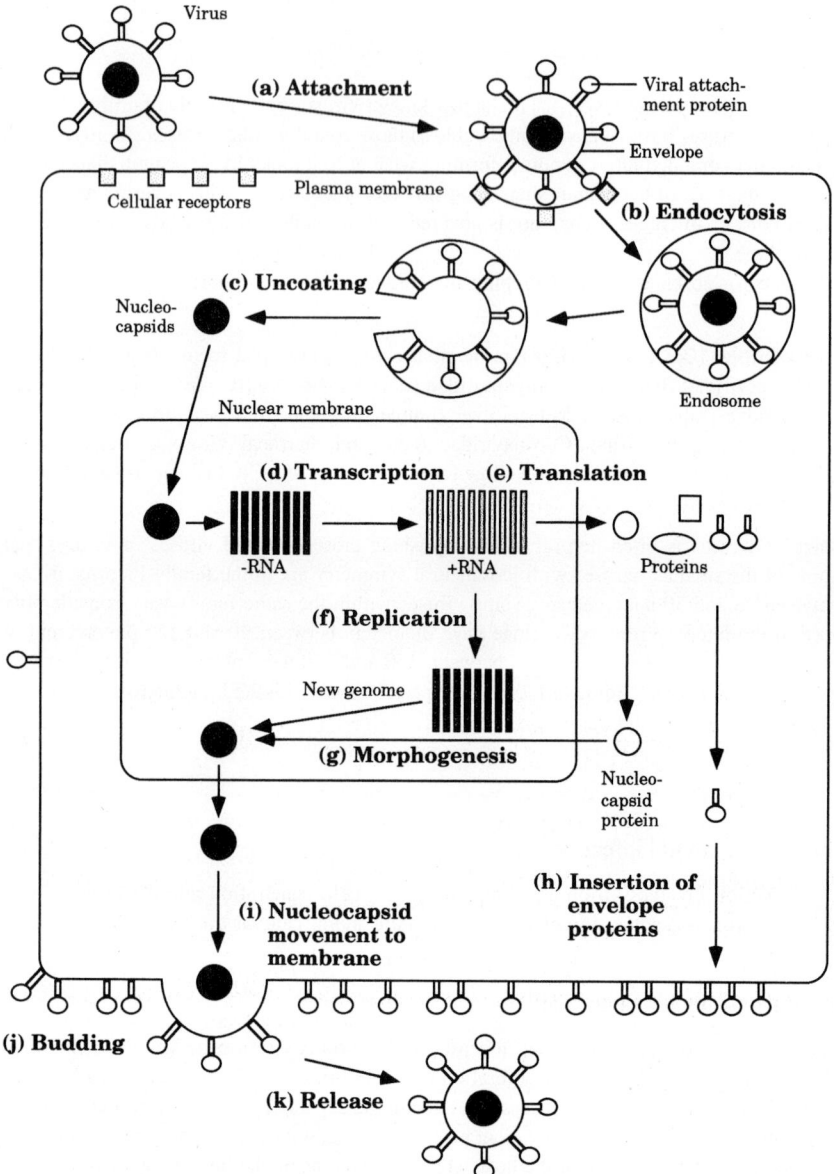

Fig. 10-2. An enveloped virus infects a eukaryote by endocytosis. (*a*) The influenza virus attaches to host receptors on the plasma membrane through its many attachment proteins. (*b*) The host's plasma membrane surrounds the virus and draws it into the cell, a process called endocytosis. (*c*) The lowering of the pH in the endosome causes the viral envelope to fuse with the endosomal membrane, releasing the nucleocapsids, a process called uncoating. (*d*) The 8 genomic RNA molecules of the virus (each containing one or two genes) are covered by capsid proteins. These nucleocapsids, represented by a large dark sphere, enter the nucleus, where the 8 minus RNA segments dissociate from the capsid proteins. The 8 minus RNA molecules are transcribed to produce 10 viral mRNAs (+RNAs). (*e*) Some of the plus RNAs enter the cytoplasm, where they serve as mRNA for various viral proteins. (*f*) The 8 full-length plus RNAs serve as templates for the synthesis of genomic minus RNAs. (*g*) Nucleocapsid proteins enter the nucleus and begin the process of morphogenesis. (*h*) Envelope proteins pass through the ER and Golgi body on their way to the plasma membrane. (*i*) The 8 nucleocapsids leave the nucleus and associate with the envelope proteins in the plasma membrane. This initiates a budding process (*j*). (*k*) A viral particle is released when the bud pinches off from the plasma membrane.

230

Almost always, viruses entering the cell through membrane fusion are enveloped. In such a virus, the envelope fuses with the plasma membrane (rather than with the membrane of the endosome) and the nucleocapsid or nucleoprotein core spills into the cytoplasm.

The genomes of most DNA viruses find their way into the nucleus. How this occurs is unknown. Exceptions include the pox viruses that replicate in the cytoplasm. The plus RNA viruses generally replicate in the cytoplasm, whereas the minus RNA viruses replicate their genomes in either the cytoplasm or the nucleus.

10.11. What happens once the viral genome reaches the cytoplasm or nucleus?

Once the viral nucleic acid reaches the appropriate compartment (cytoplasm or nucleus), it is transcribed (Fig. 10-2*d*) and the transcripts are translated (Fig. 10-2*e*). Usually, this occurs before nucleic acid replication is initiated. For the DNA viruses and a few RNA viruses, the genomes of which are in the nucleus, translation occurs in the cytoplasm on mRNAs transported through the nuclear membrane. Some of the proteins synthesized early in the infection play a role in viral nucleic acid replication.

10.12. Nucleic acid replication in viruses is extremely diverse. However, some general features of nucleic acid replication can be outlined. Generally, how do single-stranded plus RNA viruses duplicate their genomes?

Single-stranded plus RNA viruses are those having nucleic acids that can be directly translated (Fig. 10-3*a*), that is, their nucleic acids can be expressed like mRNAs. The translation of plus RNA from picornaviruses, togaviruses, and flaviviruses results in a polyprotein that is subsequently cleaved by viral and/or cellular proteases. An **RNA-dependent RNA polymerase** produced from the polyprotein catalyzes the transcription of minus RNA strands. The minus RNAs are "antisense" RNAs that serve as templates for the synthesis of genomic plus RNAs. Other proteins derived from the polyprotein function as structural components of the capsid.

Retroviruses (responsible for AIDS and leukemia) are an unusual group of plus RNA viruses in that they synthesize new plus RNA using a DNA template (Fig. 10-3*b*). Upon infection, the plus RNA genome is converted into a double-stranded DNA molecule by enzymes carried by the virus. An unusual DNA polymerase, known as **reverse transcriptase,** that can use either RNA or DNA as a template synthesizes a single-stranded DNA complementary to the plus RNA genome. A second enzyme activity associated with reverse transcriptase, designated **ribonuclease H,** digests the RNA in the RNA-DNA hybrid. Reverse transcriptase subsequently synthesizes a DNA strand complementary to the first. The resulting double-stranded viral DNA integrates into the host's DNA. This integrated DNA is known as a provirus. New plus RNA transcripts specify viral proteins. Some full-length plus RNA transcripts become part of new viral progeny forming on the plasma membrane.

10.13. How do single-stranded minus RNA viruses duplicate their nucleic acids?

Single-stranded **minus RNA** viruses are those with nucleic acids (genomes) that cannot be translated (Fig. 10-3*c*). The minus RNA strand serves as a template for the synthesis of plus RNA that functions as mRNA. Viral proteins translated from the mRNA promote the synthesis of full-length **plus RNA** strands that function as templates for the synthesis of full-length minus RNA genome strands. Usually, the virus brings into the cell an RNA-dependent RNA-polymerase for making viral mRNA. Differential messenger RNA splicing can result in multiple mRNAs, each encoding a different protein from the same region of RNA.

> **Example 10.5.** Viruses that synthesize new minus RNA genomes from plus RNA templates include the orthomyxoviruses (influenza), paramyxoviruses (measles), bunyaviruses (encephalitis, meningitis, hemorrhagic fevers), arenaviruses (meningitis, hemorrhagic fevers), and rhabdoviruses (rabies).

10.14. How do the double-stranded RNA viruses replicate their genomes?

The genomes of double-stranded RNA viruses are segmented. These viruses carry a **dsRNA-dependent RNA polymerase** which they use to transcribe their genomes (Fig. 10-4). The RNA segments and the mRNA molecules specified by them only code for one protein each. These viruses replicate their double-stranded genomes by producing many copies of plus RNA that are not translated, but serve as templates for the synthesis of complementary minus RNA strands.

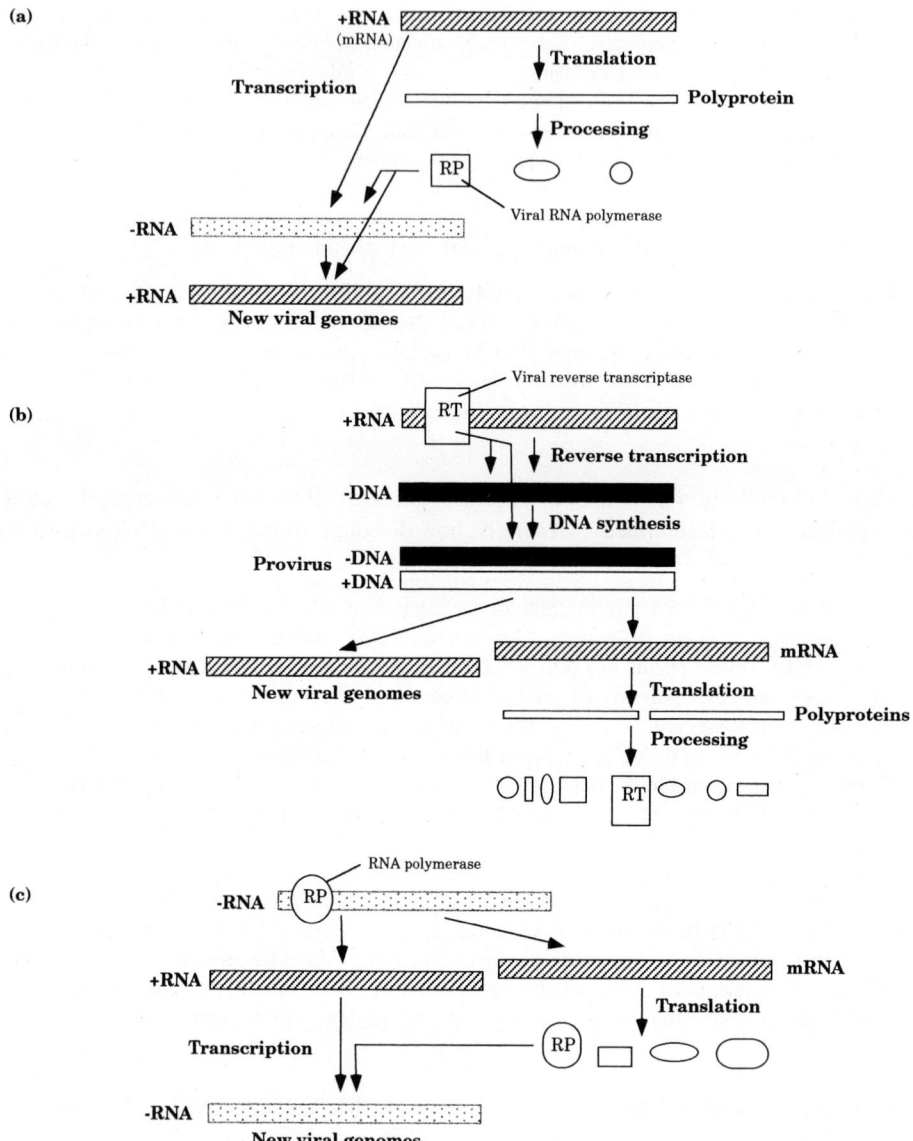

Fig. 10-3. Replication of single-stranded RNA viruses. (*a*) Virus families Picornaviridae, Togaviridae, and Flaviviridae have +ssRNA genomes. The +RNA is translated into a
polyprotein that serves as a source of an RNA polymerase. The RNA polymerase
synthesizes a −RNA (minus RNA), which serves as the template for the transcription of new viral genomes. (*b*) Retroviridae also have plus ssRNA genomes. The
+RNA is converted into ssDNA by a reverse transcriptase that is part of the virion.
The ssDNA is then made into dsDNA by the same reverse transcriptase. The
dsDNA becomes a provirus that specifies new viral genomes and +RNAs that serve
as mRNAs. (*c*) Orthomyxoviridae and Paramyxoviridae are −RNA viruses that
carry their own RNA-dependent RNA polymerases. The +RNAs that are synthesized serve as mRNA and template for the synthesis of new −RNA genomes.

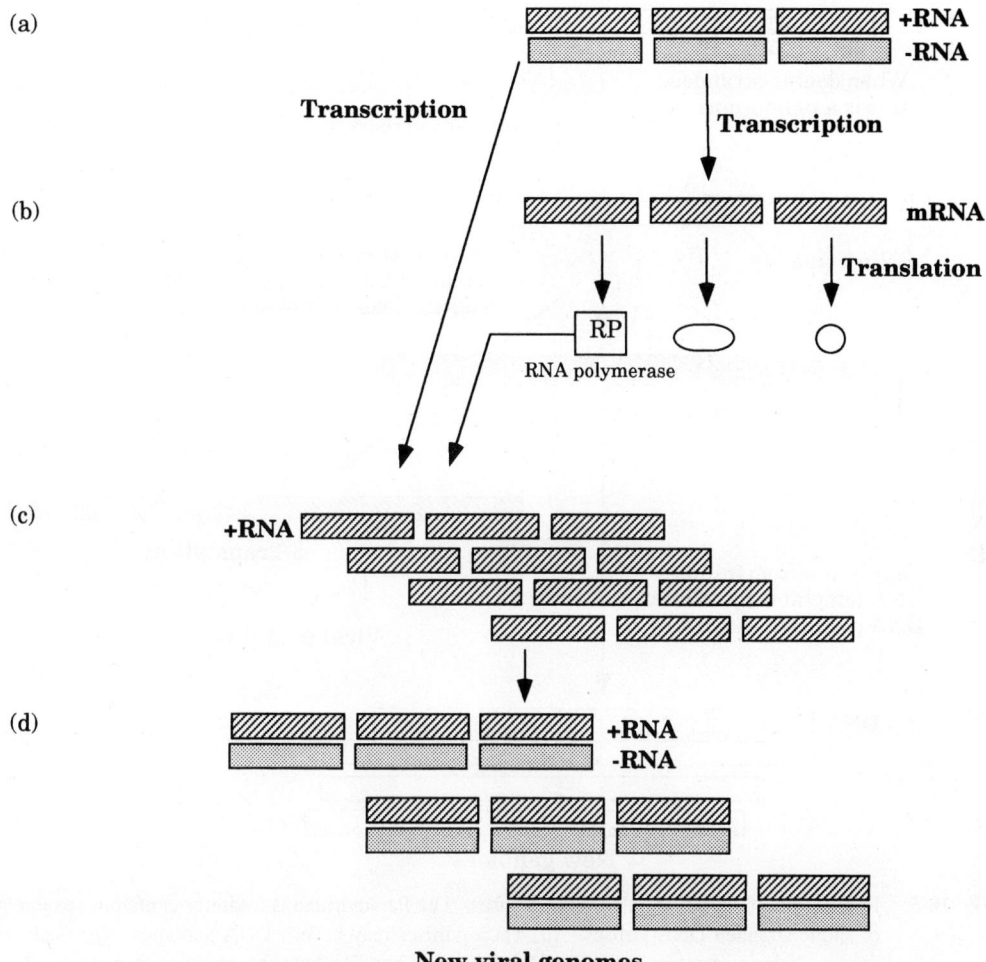

Fig. 10-4. Replication of double-stranded RNA viruses. Reoviridae, Rotaviridae, and Orbiviridae are double-stranded RNA viruses. (*a*)–(*b*) These viruses use the minus strand as a template to synthesize mRNA, and then translate it into viral proteins. (*c*)–(*d*) The viral genome is replicated by synthesizing +RNA templates, upon which negative strands are synthesized.

Example 10.6. Examples of linear, double-stranded RNA viruses include the reoviruses (no known human diseases) consisting of 10 segments, the rotaviruses (infant diarrhea and enteritis) with 11 different pieces, and the orbiviruses (Colorado tick fever) with either 10 or 12 different sections.

10.15. How do the single-stranded DNA viruses replicate their genomes?

The parvoviruses serve as an example of virions with linear, single-stranded DNA genomes. These viruses are unusual in that they may have either sense or antisense DNA (Fig. 10-5). **Sense DNA** is DNA that can serve as a template for mRNA synthesis, whereas **antisense DNA** is unable to function as such a template. Sense DNA can be transcribed immediately, but antisense DNA must serve as the template for the synthesis of the sense DNA strand.

Parvoviruses carry out synthesis of the sense strand (if needed) and replicate their DNA in the host's nucleus. Transcription by a host RNA polymerase may occur concurrently with DNA replication by a host DNA polymerase. The parvovirus genome has sequences at both ends that range in size from about 115 bases to more than 300, which can fold on themselves and form hairpin structures (Fig. 10-5*a,b*) that are palindromic. Palindromic sequences are regions of double-stranded DNA that have the same nucleotide sequence when read in opposite directions from either end of the palindrome.

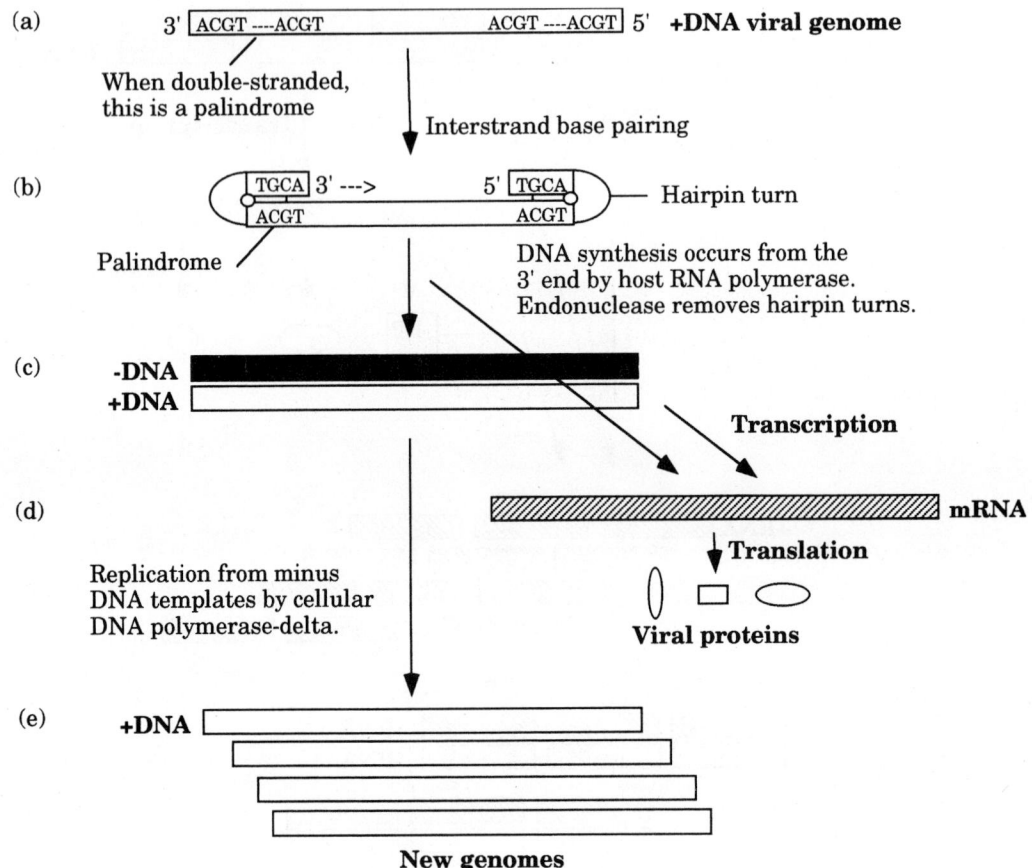

Fig. 10-5. Replication of single-stranded DNA viruses. The Parvoviridae is a family containing examples of single-stranded DNA viruses. (*a*) These viruses may have +DNA genomes. (*b*) Each end forms a hairpin structure because of the hydrogen bonding between complementary base pairs. Hypothetical hairpins with palindromes are shown. (*c*) DNA synthesis and a complex set of events lead to a double-stranded molecule. (*d*) Transcription can occur from either forms (*b*) or (*c*). (*e*) +ssDNA genomes are synthesized from the double-stranded DNA.

Synthesis of a DNA strand complementary to the viral genome begins at the 3′ end that is folded over on itself. Thus the 3′ end of the palindromic sequences serves as a primer for synthesizing a complementary strand. When ligase seals the nick, a double-stranded DNA molecule, having had its ends connected, is the result. Hairpin ends are removed by endonucleases that cut the hairpin at its end. A complicated series of DNA strand separations and DNA replications result in a double-stranded molecule that has doubled in length. DNA replication requires the host cell's DNA polymerase-δ.

10.16. How do the double-stranded DNA (dsDNA) viruses replicate their genomes?

The double-stranded DNA viruses can be divided into two catagories: those that replicate their DNA in the host's nucleus and those that replicate entirely in the host's cytoplasm.

Although there are major differences in the way dsDNA viruses replicate their genomes in their host's nucleus, some general observations can be made. The viral DNA has to be transcribed and viral proteins synthesized for DNA replication to occur (Fig. 10-6*a,b*). This initial transcription is referred to as **early transcription.** Adenoviruses have linear, double-stranded DNA genomes that specify their own DNA polymerases. Papillomaviruses have circular, double-stranded DNA genomes that encode a stimulatory protein required for the initiation of replication by a host DNA polymerase. Herpesviruses have linear, double-stranded DNA genomes that circularize during DNA replication. Their genomes encode at least 7 proteins necessary for viral DNA replication. Following DNA replication, successive rounds of transcription, referred to as **late transcriptions,** are required for the synthesis of viral structural proteins and the assembly of progeny virions.

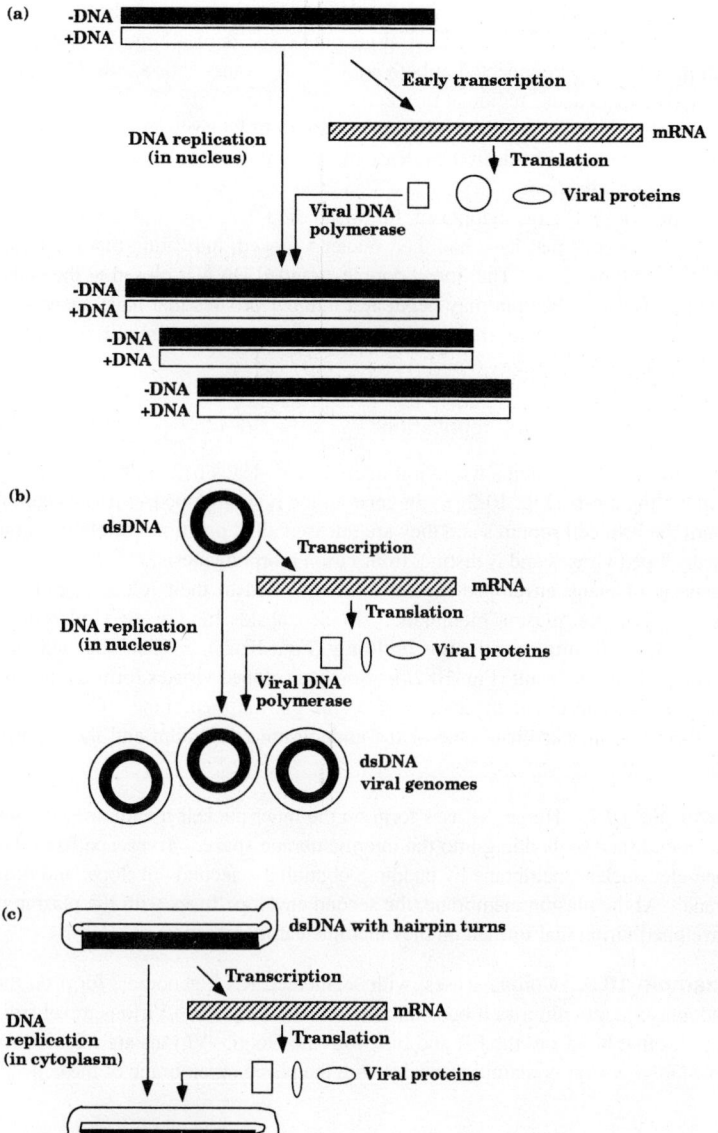

Fig. 10-6. Replication of double-stranded DNA viruses. (*a*) Aden-oviruses have linear dsDNA genomes. Early transcription and translation produce a DNA polymerase that is involved in viral DNA duplication. DNA replication initiates at the ends of the genome. (*b*) Papillomaviruses have circular dsDNA genomes within virions, whereas herpesviruses have linear dsDNA genomes that circularize after uncoating within the host. Early transcription and translation produce a DNA polymerase that is involved in DNA duplication. The papillomaviruses have a single *ori* site at which DNA synthesis is initiated. Herpes simplex viruses have three *ori* sites. Cytomegalovirus may have more than one *ori* site and replicates by a rolling circle mechanism. Epstein-Barr virus has a single *ori* site. (*c*) Poxviruses have linear dsDNA genomes with hairpin ends.

The poxviruses (including smallpox) are examples of virions that transcribe and replicate their DNA genomes in their host's cytoplasm (Fig. 10-6c). The viral genome consists of a large (86,000 bp), linear, double-stranded DNA, in which the strands are connected at both ends. Early transcription, mRNA capping, and poly-A addition occur in the virion's core, a dense region at the center of the virus defined by the **core membrane.** The early viral mRNA passes through the core membrane into the cytoplasm, where it is translated. One of the early proteins disrupts the core membrane and leads to the release of the viral genome. Another early protein serves as the DNA polymerase needed for replication.

DNA replication occurs in the cytoplasm, following the release of viral DNA from the core. Poxvirus DNA replication can occur in cells that have had their nuclei removed, indicating that poxviruses synthesize their own activators and DNA polymerases. The linear, double-stranded DNA is closed at the ends by hairpin loops. One model suggests that DNA replication may begin at a number of sites that have been nicked. Palindromes at the ends of the chromosome, which allow hairpin-loop structures to form, are required for DNA replication.

10.17. What are the final stages of viral replication?

The association of viral proteins with viral nucleic acids leads (often through a complex pathway) to **virion synthesis,** or **morphogenesis** (Fig. 10-2g). In general, the nonenveloped viruses accumulate in the cytoplasm of the host cell until the host cell ruptures and they are released all at one time. **Release** is the final step in the proliferation of nonenveloped viruses and is distinct from virion morphogenesis.

Morphogenesis of many enveloped viruses is closely tied to their release. Many enveloped viruses form on the inside surface of the plasma membrane. Nucleocapsids binding to viral proteins in the plasma membrane trigger or cause **plasma membrane budding** (Fig. 10-2j). When the bud pinches off from the cell membrane, enveloped virions result (Fig. 10-2k). Some enveloped viruses form on the inner membrane of the nucleus or on the endoplasmic reticulum instead of on the plasma membrane. Consequently, these viruses obtain their envelope from the nuclear membrane or the endoplasmic reticulum and their morphogenesis is not tied to release.

> **Example 10.7.** Herpes viruses form on the inner nuclear membrane and obtain their envelope from this membrane by budding into the intermembrane space. To escape from the cell, they pass through the outer nuclear membrane by budding, obtaining a second envelope, and diffuse to the plasma membrane. At the plasma membrane, the second envelope fuses with the plasma membrane, releasing the enveloped virion that formed on the inner nuclear membrane.

> **Example 10.8.** Coronaviruses, with segmented RNA genomes, form on the surface of the smooth endoplasmic reticulum as it buds into the lumen of the ER. Virions travel from the ER to the Golgi in vesicles that bud from the ER and fuse with the Golgi. Virions are released from the cell by fusion of post-Golgi, virion-containing vesicles with the plasma membrane of the cell.

10.18. What are retroviruses?

The retroviruses (family Retroviridae) are membraned virions with a core containing two identical chromosomes. The nucleocapsid is enclosed by a second capsid that has either helical or icosahedral symmetry. The two identical chromosomes found in these viruses are linear, single-stranded, plus-type RNA molecules. One of the characteristics that makes these viruses distinctive is the reverse transcriptase with RNaseH activity found in the core. This enzyme catalyzes the synthesis of a double-stranded DNA genome using the plus RNA from the virion as a template (Fig. 10-7). The dsDNA genome is larger than the original plus RNA template and has long terminal repeats (LTRs) bordering the viral genes. Genome expression and reproduction require that the dsDNA become a provirus within one of the host's chromosomes. Transcription of the provirus yields single-stranded plus RNAs that are translated and that serve as new viral chromosomes.

The family Retroviridae is subdivided into three subfamilies: Spumaviridae, Lentiviridae, and Oncoviridae. The **spumaviruses** cause no clinical diseases in humans, but induce characteristic vacuoles that look like foam (from the Latin *spuma*) in the cytoplasm. The **human foamy virus** (HFV) is an example of a spumavirus. The **lentiviruses** take a long time to cause disease and, because of this, are referred to as **slow viruses.** The **human immunodeficiency viruses** (HIV-I, HIV-II) that cause two different types of **acquired immunodeficiency syndrome** (AIDS) are examples of lentiviruses. The **oncoviruses,** as their name implies, cause tumors and cancers in various tissues, often because they carry an oncogene.

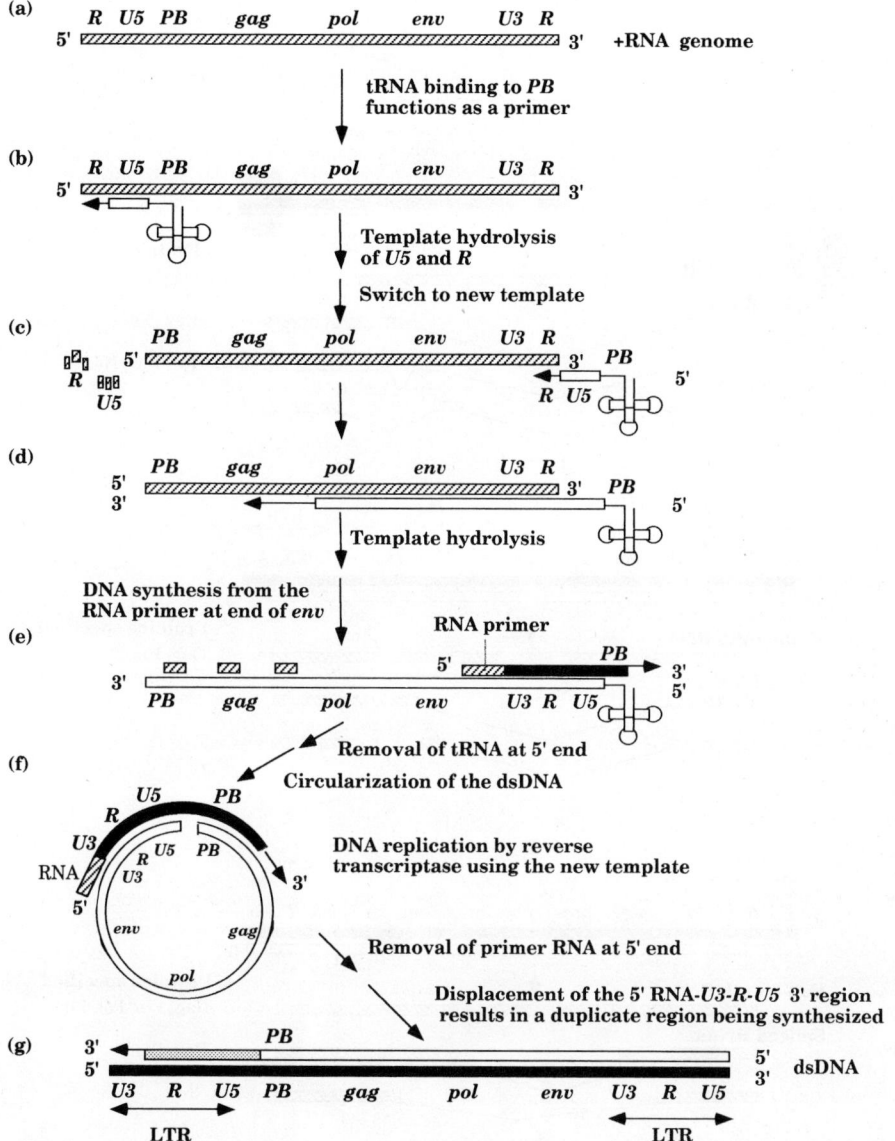

Fig. 10-7. Conversion of single-stranded RNA into double-stranded DNA. (*a–b*) A +ssRNA chromosome from a typical retrovirus is illustrated. The left end consists of regions *R, U5,* and *PB,* whereas the right end contains only regions *U3* and *R.* Only 3 (*gag, pol, env*) of the regions specifying proteins are shown for simplicity. DNA synthesis begins on the left end of the +ssRNA chromosome from a tRNA primer that binds to the *PB* region. A reverse transcriptase, using the tRNA as a primer and the left end of the ssRNA chromosome as a template, synthesizes a DNA copy of *R* and *U5.* (*b–c*) The RNaseH activity of the reverse transcriptase destroys the *R* and *U5* region of viral RNA after it has been used as a template. The newly synthesized DNA-tRNA strand switches templates, hydrogen bonding to the right-end *R* region. (*c–d*) The reverse transcriptase, using the +RNA strand as a template and the DNA-tRNA strand as a primer, completes the synthesis of the −DNA strand. (*d–e*) The RNaseH activity of the reverse transcriptase destroys the RNA template except for a small piece of RNA at the end of the *env* region, which will serve as a primer for the +DNA strand. The reverse transcriptase synthesizes a portion of the +DNA strand from a small RNA primer at the end of the *env* region. (*e–f*) The original tRNA primer is cleaved from the −DNA strand, exposing the *PB* region on the complementary piece of DNA. The newly exposed *PB* region at the right (next to the tRNA primer) hydrogen bonds to the *PB* region on the left, circularizing the molecule. Circularizing the molecule provides a new template for the reverse transcriptase, which converts the molecule to dsDNA. (*f–g*) Displacement of the 5′ RNA-U3-R-U5 region during DNA synthesis results in a duplication of the *U3-R-U5* region, so that it is found at both ends of the provirus. These regions are known as long terminal repeats (LTRs). The LTR on the left functions as a promoter region.

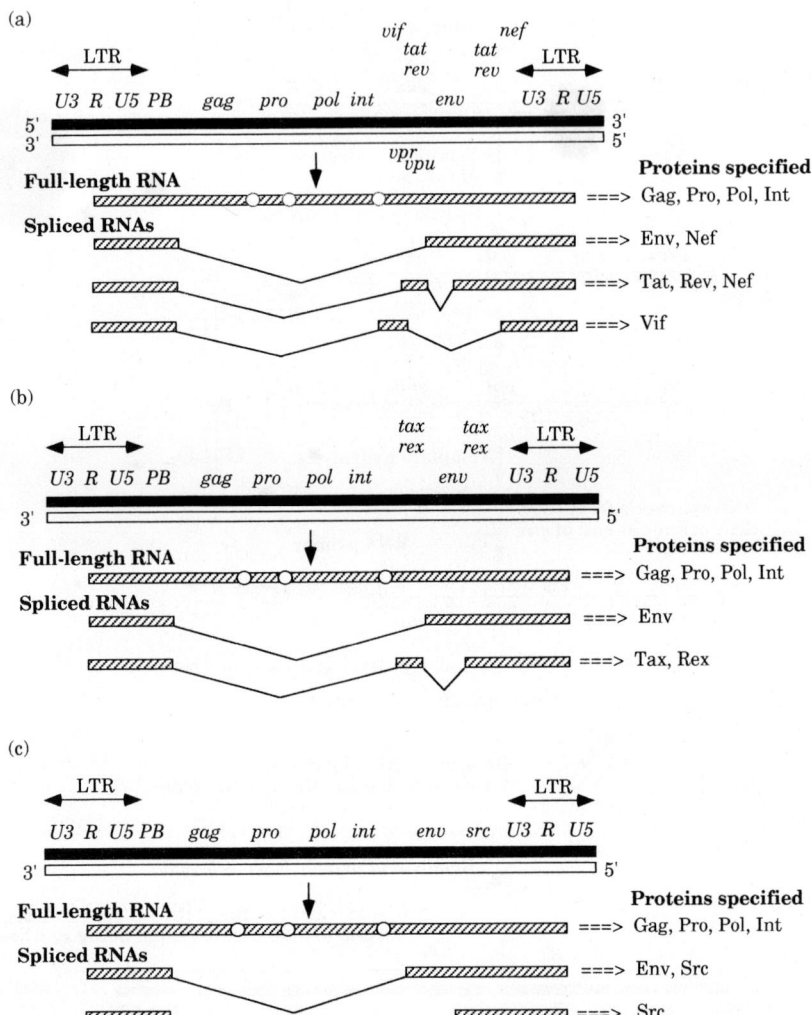

Fig. 10-8. Transcription and splicing of proviral RNA. (*a*) The provirus of the human immunodeficiency virus (HIV) is shown with its LTRs flanking a number of regions. Transcription begins in the left LTR. The full-length transcript generally specifies large amounts of the Gag polyprotein but minute amounts of the Gag-Pro and Gag-Pro-Pol-Int polyproteins because of the termination codons following *gag, pro,* and *int*. Three termination codons are symbolized by open circles on the full-length RNA. The protease activity of the polyproteins cleaves the polyprotein. A spliced transcript that includes only the *env* region specifies a polyprotein yielding the envelope proteins. The exons in a spliced transcript are symbolized by horizontal lines. The processed RNA is indicated with Vs connecting the spliced exons. Doubly-spliced transcripts are required for the synthesis of regulatory proteins such as Tat and Rev, which are specified by split exons. (*b*) The provirus of the human T-cell lymphotropic virus (HTLV) is illustrated with its LTRs flanking a number of regions. The HTLV provirus encodes many of the same proteins that the HIV provirus does: Gag, Pro, Pol, Int, and Env. The regulatory proteins Tax and Rex differ, however, from those specified by the HIV genome. HTLV does not produce proteins comparable to Vif, Vpr, Vpu, or Nef. Transcription begins in the left LTR. Full-length transcripts as well as spliced transcripts are synthesized to produce the HTLV proteins in appropriate amounts. The location of termination codons has not been determined; consequently, it is not known how similar HTLV translation and protein processing are to that of HIV. (*c*) The Rous sarcoma virus (RSV) provirus is presented with its LTRs flanking the regions all retroviruses have in common, as well as a foreign gene that was picked up originally from a host cell. In this virus, the foreign gene specifies a protein kinase that phosphorylates proteins on their tyrosines. The foreign gene is known as an oncogene since its product often transforms normal host cells and thereby promotes the development of cancers in animals. The oncogene is encoded by the spliced RNA that specifies Env and Src.

Oncogenes specify a product that transforms cells and causes some to become cancerous. One of the most studied oncoviruses is **Rous sarcoma virus** (RSV) that carries the *src* oncogene (Fig. 10-8*c*). Because of *src*, RSV readily induces chicken sarcomas (cancers of the connective tissue). The *src* gene specifies a tyrosine protein kinase that phosphorylates proteins on their tyrosine residues. These phosphorylations initiate signal transduction pathways that stimulate unsuitable transcription and translation as well as inappropriate cell proliferation (see Chapter 11). The human T-cell lymphotropic viruses such as HTLV-I and HTLV-II that have been implicated as the cause of certain leukemias are oncoviruses also.

Five regions are distinguished in all retroviral genomes (Table 10.1), however, some retroviruses (e.g., HIV) have more regions. *Gag* codes for the group-specific antigens, the nucleocapsid proteins (NC = p7−p9/p15), the second capsid proteins (CA = p24), and the matrix proteins (MA = p17−p18) that connect the capsid to the envelope. *Pro* (PRO = p11−p14) specifies a protease needed for the cleavage of the polyproteins Gag-Pro and Pol-Int. *Pol* determines the reverse transcriptase and RNaseH activities (POL = p66/p51). *Int* encodes the integration activity (INT = p32) required for viral DNA insertion into the host's DNA. *Env* specifies the polyprotein containing the envelope proteins (SU = gp120 and TM = gp41 cleaved from gp160) required for attachment to the host and for fusion with the host's plasma membrane.

Table 10.1. HIV Regions, Genes, and Proteins

Region	Gene	Protein[‡]	Function	Molecular Weight
I	*gag**	NC	Nucleocapsid (first covering)	p7−p9/p15
	gag	CA	Capsid (second covering)	p24
	gag	MA	Matrix (third covering)	p17−p18
II	*pro*	PRO	Protease	p11−p14
III	*pol*	POL	Reverse transcriptase & RNaseH (dimer)	p66/p51
IV	*int*	INT	Integrase	p32
V	*env**	SU	Attachment protein	gp120
	env	TM	Transmembrane & fusion protein	gp41
VI	*tat*	TAT	Transactivator of transcription or anti-attenuator binds to site in R of left LTR (U3-R-U5)	p14
VII	*rev*	REV	Regulator of gene expression by promoting full length & singly spliced RNA movement to cytoplasm. Binds to site in *env*.	p19
VIII	*nef*	NEF	Negative factor actually stimulates HIV transcription by freeing NF$_K$B that binds to U3. Nef stimulates the removal of CD4 from plasma membrane.	p27
IX	*vif/sor*[†]	VIF or SOR	Virion infectivity factor/short open-reading frame. Positively affects infectivity by proteolytic modification of ENV.	p23
X	*vpr*	VPR	VPR is a virion component required for growth in macrophages. Functions as a transactivator of viral genes.	p18
XI	*vpu*	VPU	VPU is a virion component that induces the degradation of CD4 on the ER. Required for normal budding and size.	p15

* Polyprotein produced by this region; p = protein; gp = glycoprotein; a slash between two molecular weights (p66/p51) indicates that a protein upon being processed gives two different sized proteins; a dash between two molecular weights (p7−p9) indicates a variation reported in the literature.

[†] Two different symbols for the same gene.

[‡] There is a trend in the scientific literature to capitalize all letters in the abbreviation of a protein.

10.19. How do the oncoviruses such as the human T-cell lymphotropic viruses (HTLV-I and HTLV-II) differ from the lentiviruses represented by human immunodeficiency viruses (HIV-I and HIV-II)?

Oncoviruses like the human T-cell lymphotropic viruses often disrupt gene expression in host cells when they insert randomly into host DNA and become proviruses. This disruption, called **insertional mutagenesis,** may explain how HTLV-I causes adult T-cell leukemia, lymphomas, and neurological disorders. No cancer has been positively associated with HTLV-II.

In addition to the *gag, pro, pol, int,* and *env* regions, the HTLVs have a number of specialized genes that finely control their development. Proteins specified by *tax* and *rex* control transcription (Fig. 10-8*b*). These viral genes also affect the expression of host genes and may contribute to cellular degeneration and promote the development of cancers. Tax is known to interact with a transcriptional activator NF_KB (nuclear factor Kappa bee) that activates host transcription. It has been shown that Tax induces to high levels the gene for an interleukin receptor. **Interleukins** are growth factors that stimulate cellular differentiation and proliferation. This inappropriate gene expression is thought to make the cell sensitive to interleukins produced by neighbors. The constitutive reproduction increases the chances of mutations to signal transduction pathways and promotes the development of cancerous cells.

Human immunodeficiency viruses such as HIV-I and HIV-II are lentiviruses. They engage in latent infections that may last 10 to 20 years before severe signs and symptoms of disease develop. HIV-I produces a more severe form of acquired immunodeficiency syndrome (AIDS) that generally kills the host 10 years after the initial infection because of its higher rate of reproduction and the cellular destruction it causes in lymph nodes. HIV-II apparently produces a milder form of AIDS that usually kills the host 15 to 20 years after the initial infection and has a lower rate of viral synthesis. In West Africa, where HIV-I and HIV-II are prevalent, HIV-I prevalence among women of reproductive age nearly doubled ($5 \rightarrow 9.2\%$) from 1988 to 1992, while HIV-II prevalence declined by 40% ($2.6 \rightarrow 1.5\%$).

HTLV and HIV differ physically. Electron micrographs show that HTLV has a capsid with icosahedral symmetry, whereas HIV has a capsid with helical symmetry. In addition to *gag, pro, pol, int,* and *env,* the genomes of human immunodeficiency viruses (HIV-I and HIV-II) contain the regulatory genes *tat, rev, vif, vpr, vpu,* and *nef* that control viral development (Fig. 10-8*a,* Table 10.1). The genomes of HTLV-I (9.032 kb) and HTLV-II (8.952 kb) are slightly smaller than those of HIV-I (9.4 kb) and HIV-II (9.2 kb). The discussion that follows will focus on HIV-I.

10.20. How does HIV adsorb to and infect cells?

The lentiviruses such as HIV have an **attachment protein** gp120 (SU) that is noncovalently bonded to a **transmembrane protein** gp41 (TM) (Table 10.1). The gp120 attaches to a specific receptor protein, CD4, found on helper T lymphocytes, monocytes, and macrophages. When gp120 binds to **CD4 receptors,** a hydrophobic region in gp41 becomes exposed and inserts itself into the cell's plasma membrane. This causes the viral membrane and the cell's plasma membrane to fuse. The fusion of the viral membrane with the plasma membrane releases the remainder of the virion (outer capsid and nucleocapsid) into the cytoplasm. Thus, the lentiviruses infect by fusing with the plasma membrane (membrane fusion) rather than by endocytosis.

10.21. What happens to HIV's RNA once it reaches the cytoplasm of an appropriate host?

Once in the cytoplasm, HIV's reverse transcriptase (associated with the two copies of the plus RNA genome) begins synthesizing a minus DNA strand from a tRNA primer (Fig. 10-7*b*). Reverse transcriptase polymerizes deoxynucleotides into DNA from a tRNA primer that binds to a region called PB near the 5′ end of the RNA genome. The **RNaseH** activity of the reverse transcriptase cuts away the template (R and U5 regions) once it has been read (Fig. 10-7*c*). This promotes the hybridization of the R region at the 3′ end of the remaining RNA and the complementary R region of the nascent DNA. The reverse transcriptase reinitiates minus DNA synthesis, copying the "new" template. This change of template is sometimes referred to as **copy-choice DNA synthesis.** This copy-choice DNA synthesis results in a minus strand DNA-tRNA "hybrid" with the following regions: 3′ PB-gag-pol-env-U3-R-U5-PB-tRNA 5′(Fig. 10-7*d*). As the reverse transcriptase elongates the DNA-tRNA strand, its RNaseH activity cuts away the RNA template (Fig. 10-7*e*).

A small portion of the RNA template, however, just 5′ of the U3 region, serves as an RNA primer for the synthesis of the second strand of DNA (Fig. 10-7*e*). The plus DNA strand synthesized from the RNA primer contains the following regions: 5′ RNA primer-U3 R U5 PB 3′. The region U3 R U5 is known as a long terminal repeat (LTR). Circularization of the duplex DNA (Fig. 10-7*f*) provides primers and templates for the synthesis of a linear, double-stranded **DNA intermediate** (Fig. 10-7g) that will be used to replicate the viral RNA genome. The DNA intermediate is synthesized in the cytoplasm within the viral capsid 4 to 8 hours after infection. Soon after the DNA intermediate is synthesized it is transported to the nucleus of the host cell. How this is accomplished is not known.

10.22. How does the DNA copy of the RNA genome become a provirus?

A number of DNA intermediates are observed in the nucleus, including linear molecules containing LTRs at both ends, as well as covalently closed circles containing either one or two LTRs (Fig. 10-7g). The linear DNA intermediate is thought to be the form that most efficiently inserts into host chromosomes. It consists of the following regions: U3-R-U5-PB-gag-pol-env-U3-R-U5. The 3′ end of each strand (of the linear DNA intermediate) is cut two bases in by the **integration protein** (Int). This same viral enzyme cuts, in a staggered manner, a target in the host's DNA that is complementary to the viral DNA and brings the staggered ends of the two DNAs together. U3-R-U5 on the left is the long terminal repeat (LTR) that contains the promoter for all HIV transcripts (Fig. 10-8a).

10.23. Where do retroviral transcripts begin and end? What is the structure of these primary transcripts and how are the transcripts modified?

Transcription is initiated from the left LTR in each of the retroviruses; more specifically from the beginning of the R region (Fig. 10-7). The **RNA polymerase-II binding site** (promoter) is located somewhere between the TATA box in U3 and the initiation of transcription in R. HIV's U3 region contains the binding site (TATA box) for the host's **TATA binding protein (TBP),** and a number of activator/enhancer and repressor binding sites: the activator SP-1 binds at GC boxes, the enhancer NF_KB binds at the NF_KB site, whereas a still uncharacterized negative regulatory factor (at one time thought to be the negative regulatory factor Nef) binds at the **negative regulatory element,** the NRE site. The NRE site is located in U3 about 300 bps from R. In fact, Nef positively stimulates HIV proliferation by promoting CD4 endocytosis by infected cells. This prevents the formation of inhibitory CD4-gp120 complexes on the surface of infected cells, as well as **syncytia** (large multinucleated cells formed by the fusion of numerous cells). Nef also stimulates signal transduction pathways in T4 cells by indirectly causing the activation of a tyrosine kinase. The signal transduction pathway activates NF_KB by freeing it from its inhibitor. NF_KB stimulates HIV's transcription. Nef promotes a 5- to 20-fold increase in productive infections. A defect in Nef results in the formation of inhibitory CD4-gp120 complexes, minimal stimulation of viral transcription by NF_KB, and a reduction in productive infections. An antiterminator protein, **Tat,** binds to a TAR site on the viral mRNA and prevents premature termination of transcription. The TAR site is found at the beginning of the R region in the left LTR. A defect in Tat allows TAR to interact with the RNA polymerase and terminate transcription. **Rev** promotes the movement of full length and singly spliced viral RNAs from the nucleus to the cytoplasm. It binds to the RNA at a site in the *env* region. A defect in Rev results in no cytoplasmic mRNA specifying Gag, Gag-Pro, Gag-Pro-Pol-Int, and Env polyproteins.

In all retroviruses, the length of the nascent plus RNA extends beyond the right LTR, but is shortened by the addition of a 3′ poly-A chain at the R-U5 boundary (Fig. 10-8). The 5′ end of the plus RNA is capped with a 7-methylguanine nucleotide. The "full length" transcript contains the following regions: m7GpppGm-R-U5-PB-*gag-pro-pol-int-env*-U3-R-poly-A. Some of the "full-length" RNA is spliced, yielding RNAs that will specify the synthesis of the envelope proteins. In the more complex retroviruses with regulatory genes, "full-length" RNA is spliced twice to specify mRNA for the following regulatory proteins: Vif, Tat, Rev, and Nef (Fig. 10-8a).

RSV does not specify regulatory proteins (such as Tat, Rev, Nef), suggesting that it lacks an attenuation site in R, as well as the mRNA translocation inhibitory site in *env* and that no inhibitory host receptor-viral attachment protein complexes form (Fig. 10.8c). Certain mutants of RSV carry a host-derived gene *(src)* for a tyrosine kinase. When overexpressed, this protein promotes inappropriate gene expression and proliferation of the host cells, often leading to malignant growths in chickens. One or more of the spliced transcripts contains the *src* oncogene that is translated.

10.24. What are some of the peculiarities of protein synthesis in cells infected with retroviruses?

In all retrovirus-infected cells, translation of the "full length" RNA leads, for the most part, to only the polyprotein Gag, since a nonsense codon just after *gag* terminates translation (Fig. 10-8). A small amount of the polyproteins Gag-Pro and Gag-Pro-Pol-Int are produced when there is **frameshift suppression** of the nonsense mutations between *gag & pro* and *pro & pol.* The amount of Pol protein produced is 1/10 to 1/20 the amount of Gag protein.

The protease activity (Pro) of the polyprotein separates Gag from Pro and Pro from Pol-Int by autocatalyses in both Gag-Pro and in Gag-Pro-Pol-Int. In HIV, the protease cleaves Gag into the nucleoprotein (NC = p7), the capsid (CA = p24) protein, and the matrix (MA = p18) proteins (Table 10.1). The protease also separates Pol (reverse transcriptase + RNaseH) from Int (integrase). The reverse transcriptase and the RNaseH activities are not split, but remain together in Pol. Almost all Env is synthesized from processed transcripts specifying only Env (Fig. 10-8).

The polyprotein specified by the *env* region enters the endoplasmic reticulum, moves through the Golgi, and to the plasma membrane. After the polyprotein reaches the plasma membrane, it is cleaved into the transmembrane protein (TM = gp41) and the attachment protein (SU = gp120) (Table 10.1).

Protein synthesis in HTLV- and RSV-infected cells is believed to be similar to that in HIV-infected cells. It is thought that nonsense codons at the end of *gag, pro,* and *int* efficiently prevent the translation of genes downstream (*env* and *src* in the case of RSV). Consequently, these genes can be translated only from spliced RNAs.

In some retroviruses like murine leukemia virus (MLV), the *gag* and *pro* regions are contiguous and in the same translational reading frame. *Gag* and *pro* are separated by an in-frame stop codon. Viruses such as HIV, HTLV, and RSV have overlapping *gag* and *pro* regions that are in different reading frames. To express the *pro, pol,* and *int* gene products, these viruses rely on a mechanism called **ribosomal frameshifting** to avoid the nonsense codon at the end of the *gag* region. Mouse mammary tumor virus (MMTV) is reported to carry out two separate frameshifts to avoid the nonsense codons at the end of *gag* and between *pro* and *pol.*

Oncogenic Viruses

10.25. What are oncogenes and how can their introduction into a normal cell by a virus promote the development of cancer?

Most **oncogenes** are mutated or inappropriately expressed cellular genes (normal cellular genes are called **proto-oncogenes** and their products are called **proto-oncoproteins**) that specify proteins in signal transduction pathways. **Oncoproteins** are oncogene products that function in signal transduction pathways to **transform** cells, allowing them to proliferate in an uncontrolled manner, generally resulting in cancerous growths (e.g., malignant tumors in solid tissues).

Viruses such as RSV that carry oncogenes derived from normal cellular genes (proto-oncogenes) are referred to as oncoviruses. Oncoviruses also include organisms such as HTLV that lack cellular oncogenes, but contain viral genes that promote cellular proliferation, thus giving more cells in which mutations in proto-oncogenes might occur. Also, these viruses transform cells when they become proviruses and disrupt tumor suppressor genes or cause the inappropriate expression of normal proto-oncogenes. Transformed cells often proliferate out of control, become potentially immortal, change shape, have new antigenic properties, and lose **contact inhibition.** Normal cells usually stop proliferating when sufficient contacts have been made with other cells. The loss of contact inhibition allows cells to wander off into other tissues and organs (**metastasis**) and spread a cancerous growth. Most oncoviruses transform cells efficiently and rapidly because of the oncogenes they carry (Fig. 10-8c). In some cases, the oncogene product is a "normal" protein, but is overproduced because it is under the control of an efficient viral promoter in the long terminal repeat (LTR). Generally, oncogene products are abnormal proteins, sometimes differing from the normal protein by a single amino acid substitution, which are not regulated correctly. Consequently, the abnormal components of signal transduction pathways (Chapter 11) stimulate the cell to express itself inappropriately. This disruptive transformation may, after several other mutations, yield cancerous cells.

Although most oncogenes are derived from proto-oncogenes, some viruses carry viral oncogenes that have no apparent relationship to cellular genes. For example, HIV's Tat protein may act as a signal peptide in conjunction with basic fibroblast growth factor (bFGF) to induce Kaposi sarcoma.

10.26. How are oncoproteins classified and how do they function in signal transduction pathways?

All oncoproteins identified to date fit into one of 8 categories: (1) peptide growth factors, (2) growth factor receptors in the plasma membrane or cytoplasm, (3) GTP-regulated proteins called G proteins, (4) membrane receptors with tyrosine kinase or with threonine-serine kinase activities, (5) cytoplasmic protein kinases with tyrosine kinase activities or with serine-threonine activities, (6) DNA-binding proteins that function as transcriptional activators or that promote DNA replication, (7) cyclins that promote the activity of protein kinases, and (8) proteins that inhibit tumor suppressor proteins.

Almost all of the oncoproteins function in various signal transduction pathways that begin with a signal (peptide or steroid hormone) and end with the activation of transcription and/or the initiation of DNA replication (Fig. 10-9).　The oncoproteins override the normal regulation of cells and continuously send signals that activate gene expression and progression through the cell cycle.　This increases the chances that mutations will occur in proto-oncogenes and in **tumor suppressor genes** (normal cellular genes whose products dampen or inhibit signal transduction pathways).　The more proto-oncogenes converted into oncogenes, the more unregulated a cell becomes.　Similarly, the more tumor suppressor genes that are damaged by mutations, the more signal transduction pathways are inappropriately expressed.

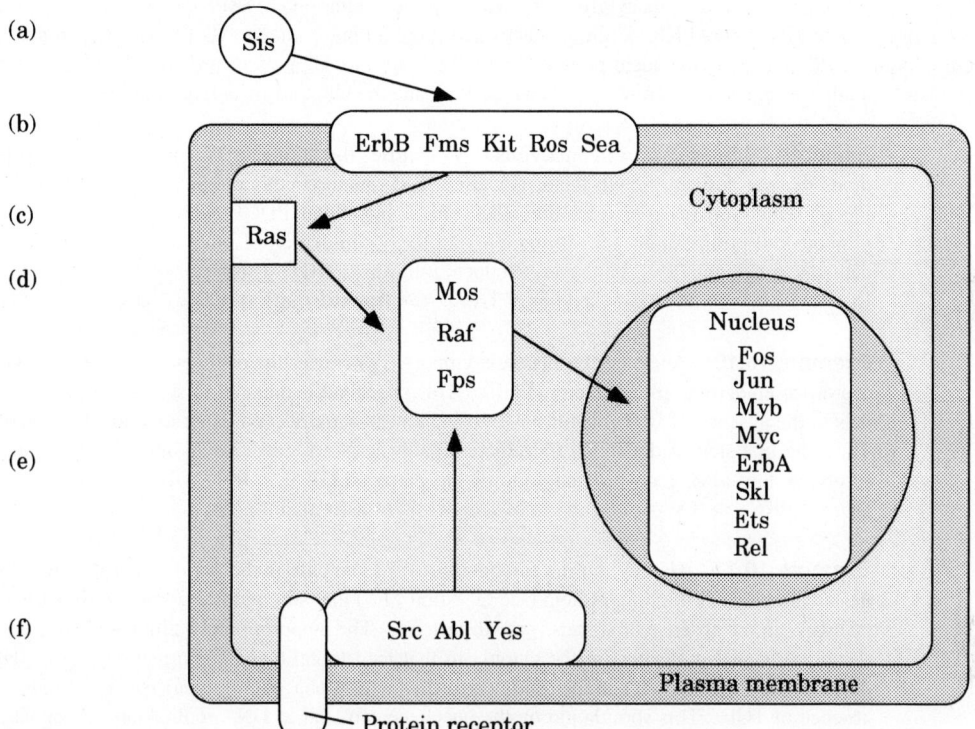

Fig. 10-9.　Location of proto-oncoproteins and abnormal oncoproteins.　(*a*) Sis is an oncoprotein that functions as an external growth factor.　The arrows indicate possible connections between the different components of signal transduction pathways.　(*b*) ErbB, Fms, Kit, Ros, and Sea are oncoproteins that function as growth factor receptors.　These receptors are located in the plasma membrane.　The outer portion of the receptor binds growth factors, whereas the cytoplasmic domain transmits the signal into the interior of the cell.　Many of the cytoplasmic domains function as tyrosine kinases.　(*c*) Ras is a monomeric G-protein that transmits a signal to cytoplasmic kinases.　(*d*) Mos, Raf, and Fps are cytoplasmic kinases that transmit a signal to proteins that function in the nucleus.　(*e*) Fos, Jun, Myb, Myc, ErbA, Skl, Ets, and Rel are proteins that function as transcriptional activators.　ErbA also functions as a steroid hormone receptor.　(*f*) Src, Abl, and Yes are tyrosine kinases that are bound to the cytoplasmic side of the plasma membrane.

Figure 10-9 illustrates where the proto-oncoproteins and oncoproteins are found in a typical cell and how some of them interact with each other and other proteins in various signal transduction pathways.　Sis represents the inappropriately expressed peptide hormone platelet-derived growth factor (PDGF) found outside the cell.　ErbB, Fms, Kit, Ros, and Sea represent mutated receptors in the membrane that happen to have constitutive tyrosine kinase activity.　Src, Abl, and Yes are membrane-bound, constitutive tyrosine kinases attached to the cytoplasmic surface of the plasma membrane.　Ras is a membrane-associated, monomeric G protein that carries signals from stimulated receptors to enzymes.　Mos, Raf, and Fps are active serine-threonine kinases free in the cytoplasm.　Most of the oncoproteins in the nucleus are transcriptional activators, the synthesis of which synthesis is induced by various signals and signal transduction pathways.

10.27. What are some oncogenes carried by oncoviruses and how do their products function in an infected cell?

Although oncoviruses generally carry only one oncogene, they may carry any oncogene. Some of the oncogene products affect the very beginning of a signal transduction pathway, whereas others affect it further along. In the signal transduction pathways discussed in the following examples, PDGF = platelet-derived growth factor, EGF = epidermal growth factor, TGF = transforming growth factor, and INF = interferon. Other abbreviations are included: R = receptor, TA = transcriptional activator, G = G-protein, PLC = phospholipase-C, PLD = phospholipase-D, DG = diacylglycerol, PKC = protein kinase-C, MAPKKK = mitogen-activated protein kinase kinase kinase, MAPKK = mitogen-activated protein kinase kinase, MAPK = mitogen-activated protein kinase, CDPK = cyclin-dependent protein kinase, TSP = tumor suppressor protein. Read the simple arrow (-->) as "stimulates," the double arrow (==>) as "yields," and the blunt arrow (--|) as "inhibits."

Example 10.9. Simian sarcoma virus (SSV) carries the oncogene *sis* that encodes the peptide hormone platelet-derived growth factor (PDGF). SSV-infected cells produce abnormally high levels of PDGF, which are released from the cell. This extracellular PDGF continually initiates signals that promote cell proliferation, transformation, and the accumulation of mutations. A signal-transduction pathway involving the *sis* gene product follows: **PDGF (Sis)** --> R --> G --> PLC ==> DG -->PKC --> TA ==> 2nd tier of TAs ==> transformation.

Example 10.10. Avian erythroblastosis virus (AEV) carries the oncogene *erbB* that encodes the receptor for epidermal growth factor (EGF). This oncogene is mutated in such a way that it sends signals in the absence of EGF. ErbB has tyrosine kinase activity even when it is not stimulated by EGF. Thus, the first step of the signal-transduction pathway is not needed. A signal-transduction pathway involving the *erbB* gene product follows the pattern EGF --> **R (ErbB)** --> G --> PLC ==> DG --> PKC --> TA ==> 2nd tier of TAs ==> transformation.

Example 10.11. Harvey murine sarcoma virus (MSV) carries the *ras* oncogene. This gene specifies an altered GTP-binding protein (see Question 11.5) that stimulates enzymes in signal-transduction pathways in the absence of signals (growth factors). The inappropriately stimulated enzymes produce secondary signal molecules in the cytoplasm that lead to cellular transformation. The signal-transduction pathway involving transforming growth factor-alpha (TGF-alpha) can be stimulated in the absence of TGF. This stimulation by the oncogene can lead to DNA replication. A signal-transduction pathway involving the *ras* gene product follows the pattern TGF --> R --> **G (Ras)** --> MAPKKK --> MAPKK --> MAPK --> CDPK --| TSP ==> DNA replication.

Example 10.12. Avian myeloblastosis virus (AMV) carries the *myb* oncogene. It is an altered transcriptional activator that does not have to be modified through a signal-transduction pathway. Thus, the Myb encoded oncoprotein directly activates transcription and possibly cell proliferation. A signal transduction pathway that is no longer required because of Myb follows the pattern INF --> R --> G --> PLD --> DG --> PKC --> **TA (Myb)** ==> transcription.

PLANT VIRUSES

10.28. What characteristics differentiate plant viruses?

The vast majority of plant viruses have a single-stranded, linear, plus RNA genome with a capsid having helical or icosahedral symmetry. Exceptions to this rule include a few double-stranded DNA viruses (caulimovirus and geminivirus), cigar-shaped viruses (alfalfa mosaic virus), and a few odd-shaped viruses (rhabdoviruses are bullet-shaped, whereas geminiviruses consist of two fused icosahedral capsids). An unusual feature of some plant viruses with segmented genomes is the segregation of the segments into separate capsids. In a few cases, the capsids vary in size. A **productive infection** (one producing virulent progeny) requires simultaneous infection of a cell by all components of the viral genome.

Example 10.13. Almovirus (*al*falfa *mo*saic *virus*) has a segmented genome divided among three different-sized particles. The largest particle is 58-nm long and contains two segments (chromosomes), #1 and #2. The middle-sized particle has a length about 2/3 that of the largest and contains one segment, #3. The smallest particle has a length about 1/3 that of the largest and contains segment #4. A fourth particle, with a length about 1/2 that of the largest, has no RNA associated with it.

The plus RNA plant viruses have small genomes coding for only 3 to 4 proteins: (1) a helicase, (2) an RNA replicase, (3) a cell-to-cell movement protein, and (4) a capsomere. The helicase is thought to be important in the unwinding and separation of the plus and minus RNA strands in the replicative form (RF). The replicase, an RNA-dependent RNA polymerase, is encoded in those viruses that are unable to use the host's enzymes. The **cell-to-cell movement protein** facilitates the spread of the viral RNA through plant tissue. The capsomere is the protein subunit of the capsid. Only one plus RNA plant virus, comovirus, specifies two capsomeres for its coat. In most cases, it has been difficult to prove that the plant viruses specify their own RNA replicase because viral infections and stress induce host-encoded, RNA-dependent RNA polymerases. The replicases often consist of two components, one determined by the viral genome and the other by the host's genome. Neither of these proteins is part of the stress-induced host replicases. Those plant viruses that do not code for replicases have to rely on the host's RNA-dependent RNA polymerases (RNA replicases) to reproduce their genome.

Example 10.14. Plant viruses that have been shown to encode part or all of their RNA replicase include tymovirus (TYMV = *t*urnip *y*ellow *mo*saic *virus*) and cucumovirus (CMV = *cucu*mber *mo*saic *virus*). The CMV replicase has been shown to consist of two viral encoded proteins (1a and 2a) and a 54K host protein.

10.29. Discuss the characteristics of tomovirus (*to*bacco *mo*saic *virus*), the cause of tobacco mosaic disease, and how it is able to infect cells.

Unlike many plant viruses that are dependent on insect vectors to infect plant cells, tomovirus (TMV) is dependent only on mechanical damage to cell walls. Damaged cell walls allow this virus to bind to the plasma membranes of host cells. It is believed that TMV enters host cells by pinocytosis or endocytosis. How its genome reaches the cytoplasm from the endosome is unknown.

The tomovirus has a helical capsid measuring 18×300 nm constructed from 4,900 identical rod-shaped capsomeres. There is a central axial channel 4 nm in diameter. The oblong capsomeres that are about 7 nm in length form the capsid "wall." The plus RNA is embedded about 5 nm in the capsid "wall" (Fig. 10-10). Remember, plus RNA is RNA that may function as mRNA. The length of the capsid is determined by the length of the viral plus RNA. Although the capsomeres can spontaneously aggregate *in vitro* into a helical capsid of various lengths, these structures are only stable between a pH of 4 to 6. RNA increases the stability of the capsid so that it is stable under normal conditions between a pH of 3 and 9. The viral RNA is arranged helically in the virion and lies in a groove found in each capsomere (Fig. 10-10).

The plus RNA genome is capped at its 5' end with 7mG, but lacks a poly-A tail on its 3' end. The 3' end of TMV-RNA resembles a cellular tRNA both in tertiary structure and in the fact that it can be modified by cellular histidyl-tRNA synthetase to acquire a covalently attached histidine at its 3' end.

Example 10.15. The 3' ends of a number of other plant viruses closely resemble tRNAs and are modified by cellular aminoacyl-tRNA synthetases. Both bromovirus and cucumovirus may have tyrosine added; tymovirus may have valine added to its 3' end by a host aminoacyl-tRNA synthetase.

10.30. How is the tomovirus plus RNA expressed?

RNA translation begins when host ribosomes initiate protein synthesis near the 5' cap and strip away capsomeres. Most of translation terminates 1/3 of the way along the genome with the production of a polypeptide labeled p126 (Fig. 10-11*a*). Occasionally, the UAG termination codon is suppressed by a cellular aminoacyl-tRNA, and a longer protein, designated p183, is produced. What these proteins do is still obscure, although there is a hint from sequence similarities to other proteins that p126 represents a helicase and p183 is a subunit for an RNA replicase. Translation may also occur from an internal start codon, yielding a small polypeptide known as p30 that somehow functions in the spread of the viral RNA through the plant. It is believed that TMV RNA moves from cell to cell through **plasmodesmata** (small cytoplasmic conduits that connect adjacent plant cells). Prelimi-

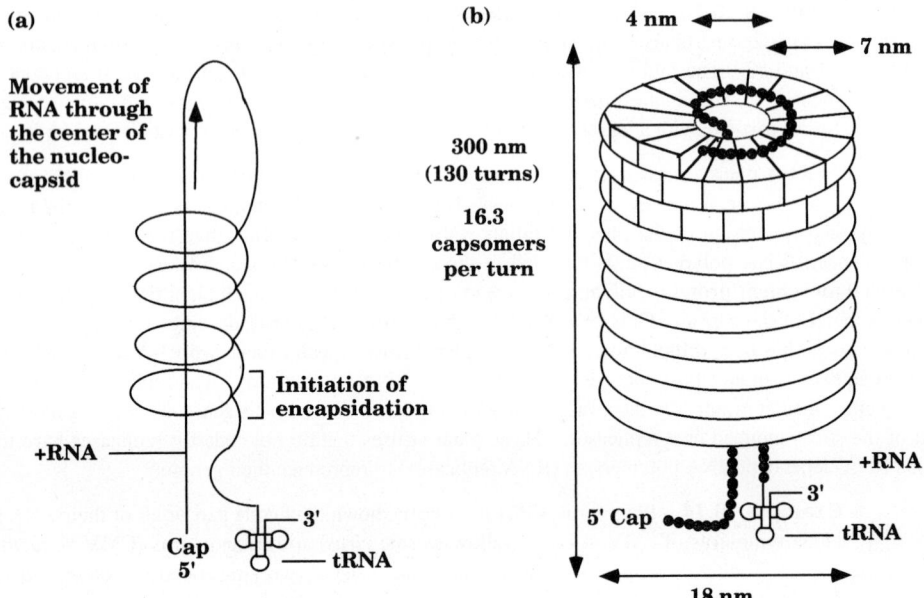

Fig. 10-10. Encapsidation of the genome of tobacco mosaic virus. (*a*) TMV has a linear +RNA genome that has a 5′ cap and a 3′ region that resembles a tRNA. Encapsidation initiates close to the 3′ end and proceeds 3′ to 5′ until the capped 5′ end is encapsidated. There is also encapsidation from the initiation site toward the 3′ end. (*b*) The capsomeres arrange themselves into a long helix, trapping the RNA between two layers of the capsomeres along the length of the virion. The virion's final dimensions are 18 nm by 300 nm. There is a hollow central volume 4 nm in diameter.

nary investigations suggest that p30 associates with the plasmodesmata, binding TMV RNA and helping its diffusion through plasmodesmata. However, the mechanism by which viral RNA moves through plasmodesmata is not known.

The RNA coding for the capsomere protein at the 3′ end of the viral genome cannot be translated while it is part of the full-length genome. A subgenomic plus RNA molecule must be made by processing the full-length plus RNA (Fig. 10-11*b*). Apparently, a 5′ cap and ribosomal binding region are required in addition to a start codon.

Viral- and host-encoded proteins make up a replicase (RNA-dependent RNA polymerase) that initiates synthesis of a viral minus strand in the cytoplasm. The double-stranded RNA is referred to as the **replicative form** (RF). The same RNA-dependent, RNA polymerase is thought to initiate the synthesis of full-length plus RNA, as well as a short RNA from the last third of the minus strand. The minus RNA template with numerous nascent plus RNAs in the process of being replicated is called the **replicative intermediate** (RI). The short plus RNAs sluffed from the RI are translated, producing the abundance of capsomeres needed for virion production.

10.31. How does the TMV virion form?

Capsomeres initially polymerize into a double-thickness disk consisting of 33 capsomeres. Plus RNA associates with the groove in the top layer of a "disk" with a helical structure at an **assembly initiation site** 800 nucleotides from the 3′ end (Fig. 10-10). New capsomeres added to the top of the disk lock the RNA into the wall of the growing capsid. New RNA toward the 5′ end associates with the newest capsomeres; the 5′ end is thus pulled through the axial channel. The capsid can grow toward the 3′ end of the RNA, but less efficiently.

Capsid synthesis proceeds until the viral genome has been completely encased in the wall of the capsid. Although capsid polymerization can occur without RNA, the disks that form are extremely unstable and depolymerize rapidly. Because of this, all TMVs are about the same length.

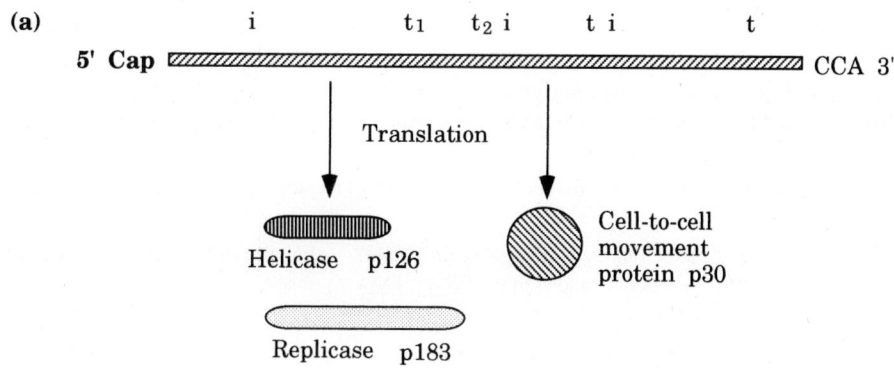

(a)

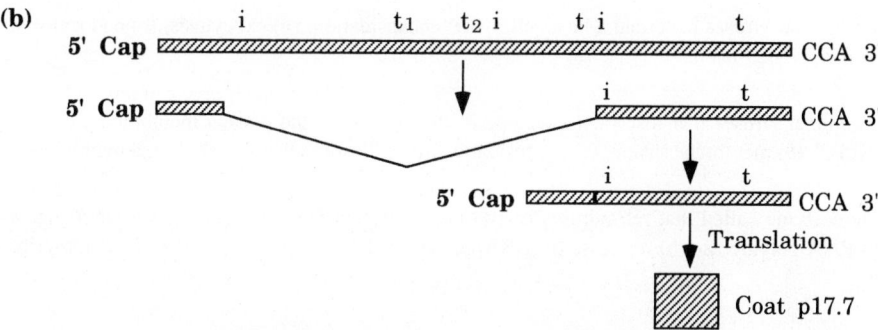

(b)

Fig. 10-11. Translation of a plus RNA genome from TMV. (*a*) A full-length +RNA genome generally encodes 3 proteins: a helicase, a replicase, and a cell-to-cell movement protein. The helicase and the replicase are specified by overlapping regions. A helicase is synthesized if there is termination at the first nonsense codon but a replicase is produced if there is suppression of the first nonsense codon. Reinitiation of protein synthesis following the replicase gene produces the cell-to-cell movement protein. (i = initiation of translation, t = termination of translation.) (*b*) A short RNA, consisting of the 3′ end of the 5′ cap and the genome, specifies the coat protein.

Objective Questions

Multiple Choice

Directions: Choose the one best answer.

1. What characteristic domain is found in eukaryotic proteins that enables them to enter the endoplasmic reticulum? (*a*) Stop transfer domain (*b*) signal sequence (*c*) signal sequence receptor (*d*) signal recognition protein (*e*) signal recognition protein receptor

2. What domain do eukaryotic proteins have that allows them to become *integral membrane proteins?* (*a*) Stop transfer domain (*b*) signal sequence peptidase (*c*) signal sequence receptor (*d*) signal recognition protein (*e*) signal recognition protein receptor

3. What enzyme removes the signal sequence domain? (*a*) Stop transfer domain (*b*) signal sequence peptidase (*c*) signal sequence receptor (*d*) signal recognition protein (*e*) signal recognition protein receptor

4. N-linked polysaccharides are attached to (*a*) serine. (*b*) glycine. (*c*) tryptophan. (*d*) cysteine. (*e*) asparagine.

5. O-linked polysaccharides are attached to (*a*) serine. (*b*) glycine. (*c*) tryptophan. (*d*) cysteine. (*e*) asparagine.

6. Which N-linked sugar is *not* added in the endoplasmic reticulum? (*a*) *N*-acetylglucosamine (*b*) mannose (*c*) glucose (*d*) sialic acid (*e*) All of these sugars are added in the ER.

7. How does HIV enter a host cell? (*a*) Endocytosis (*b*) exocytosis (*c*) membrane fusion (*d*) injects its RNA (*e*) none of the above

8. Which component in the viral envelope attaches the human immunodeficiency virus (HIV) to host cells? (*a*) gp41 (*b*) p18 (*c*) gp120 (*d*) p24 (*e*) p15

9. Most naked animal viruses penetrate a host cell by (*a*) encapsidation. (*b*) exocytosis. (*c*) endocytosis. (*d*) uncoating. (*e*) plasma membrane fusion.

10. Enveloped animal viruses that lose their envelope as they penetrate the plasma membrane are entering the cell by (*a*) endocytosis. (*b*) membrane fusion. (*c*) encapsidation. (*d*) endosome fusion. (*e*) plasma membrane melting.

11. What is the enzyme called that retroviruses carry to convert their RNA genome into a DNA genome? (*a*) RNA-dependent RNA-polymerase (*b*) replicase (*c*) DNA-dependent DNA-polymerase (*d*) reverse transcriptase (*e*) none of the above

12. What is the enzyme activity associated with reverse transcriptase that digests the RNA template after DNA has been synthesized? (*a*) Reverse discriminase (*b*) lipase-H (*c*) protease-H (*d*) RNase-H (*e*) DNase-H

13. What kind of genomes do retroviruses have? (*a*) One copy of plus ssRNA (*b*) one copy of minus ssRNA (*c*) two copies of plus ssRNA (*d*) two copies of minus ssRNA (*e*) none of the above

14. Which viral enzyme activity is in the core of retroviruses? (*a*) Reverse transcriptase (*b*) protease (*c*) RNaseH (*d*) integrase (*e*) all of the above

15. Which is *not* a retrovirus? (*a*) Oncovirus (*b*) lentivirus (*c*) spumavirus (*d*) rubivirus (*e*) The preceding are all retroviruses.

16. Transcription in HIV is initiated from (*a*) *gag.* (*b*) *pol.* (*c*) *env.* (*d*) *left LTR.* (*e*) *right LTR.*

17. Which RNA virus carries oncogenes? (*a*) Hepatitis B virus (*b*) Epstein-Barr virus (*c*) human papilloma virus (*d*) adenovirus (*e*) Rous sarcoma virus

18. The order of the products of oncogenes *myc, ras, erbB, sis* in signal-transduction pathways is (*a*) sis-erbB-ras-myc. (*b*) myc-erbB-sis-ras. (*c*) ras-sis-erbB-myc. (*d*) erbB-myc-sis-ras. (*e*) ras-sis-myc-erbB.

True–False

1. Glycosylation of proteins occurs in the endoplasmic reticulum and in the Golgi complex.

2. Carbohydrates added to proteins can be linked to the nitrogen of asparagine or through the oxygen of serine.

3. Proteins are targeted to lysosomes by methylation of a mannose on the carbohydrate chain.

4. Proteins entering the matrix of the mitochondrion do so only if there is an appropriate signal sequence and a voltage across the inner membrane.

5. Proteins entering the intermembrane space of the mitochondrion do so from the matrix.

6. A virus with a minus RNA genome must synthesize a complementary plus RNA strand before viral proteins can be produced.

7. Viruses with capsids having more than 20 faces (or even viruses that appear to be spherical) may have icosahedral symmetry.

8. Cylindrical viruses are said to have helical symmetry.

9. Generally, viral adsorption is mediated by lipid attachment sites on the virus and lipid receptors on the host cell.

10. Mutations in the attachment sites of a virus may alter the host range of the virus.

11. Single-stranded plus RNA viruses usually engage in translation before they transcribe.

12. Single-stranded minus RNA viruses usually carry an RNA-dependent RNA polymerase (replicase) to transcribe their genomes.

13. Single-stranded minus RNA viruses usually engage in transcription before they translate.

14. Retroviruses are single-stranded plus RNA viruses, and they carry a DNA polymerase that can use only DNA templates.

15. Retroviruses convert their RNA genomes into DNA that is shorter than the RNA genome.

16. Generally, double-stranded RNA viruses carry an RNA-dependent RNA polymerase known as a replicase.

17. After infection of a cell by double-stranded RNA virions (reoviruses), viral transcription precedes translation.

18. The gag region of retroviruses specifies more than one polypeptide.

19. The human T-cell lymphotropic viruses (HTLV-I, -II) are in the subfamily of oncoviruses.

20. The human T-cell lymphotropic viruses (HTLV-I, -II) have capsids with icosahedral symmetry, whereas human immunodeficiency viruses (HIV-I, -II) have bullet-shaped capsids with helical symmetry.

21. The RNA genomes of retroviruses are the same length as the DNA proviruses.

22. The RNA genomes found in HIV have a long terminal repeat on each end.

23. A number of DNA viruses (HBV, EBV, HPV, SV40, etc.) carry oncogenes, but these oncogenes are not related to genes found in normal cells.

Matching

Directions: Match the shape (*A–E* in **Column B**) with the virus (1–7 in **Column A**).

Column A	**Column B**
1. Picornaviridae (polio virus, hepatitis A virus, foot-and-mouth disease virus)	A. helical
2. Papovaviridae (papilloma virus, SV40, polyoma)	B. icosahedral or isometric
3. Rhabdoviridae (rabies virus)	C. spherical
4. Poxviridae (cow pox virus)	D. bullet
5. Retroviridae (human immunodeficiency virus, human lymphotropic virus)	E. complex
6. Togaviridae (Eastern and Western equine encephalitis viruses)	
7. Orthomyxoviridae (flu virus)	

Directions: Match the HIV region (*A–E* in **Column B**) that specifies the protein (8–13 in **Column A**).

8. Nucleoprotein	A. *pol*
9. Capsid	B. *env*
10. Matrix	C. *pro*
11. Protease	D. *int*
12. Reverse transcriptase/RNaseH	E. *gag*
13. Integrase	

Directions: Match the oncogenes (*A–E* in **Column B**) from RNA viruses with the protein (14–18 in **Column A**) that each specifies.

14. Receptor with tyrosine kinase activity	A. *sis*
15. Activator found in the nucleus	B. *erbB*
16. Growth hormone: platelet-derived growth factor	C. *ras*
17. Monomeric G-protein	D. *myc*
18. Receptor for epidermal growth factor	E. *fos*

Directions: Match the terms (*A–E* in **Column B**) to the descriptions (19–22 in **Column A**).

19. A virus's nucleic acid and protein covering	A. capsomere
20. A protein coat having 20 faces and 12 vertices enclosing a virus's nucleic acid	B. capsid
21. A membrane covering found around many animal viruses	C. nucleocapsid
	D. icosahedron
22. A subunit of a virus's protein coat	E. envelope

Terms

Directions: Unless otherwise specified, each correct answer is a single word.

1. A protein's amino terminus that allows it to enter the endoplasmic reticulum as it is being synthesized. (2 words)

2. The name of the protein that binds to the amino terminus of a nascent peptide, as well as to the 80S ribosome, attaching the complex to the endoplasmic reticulum. (3 words)

3. The name of the membrane lipid that adds polysaccharides to asparagines.

4. The organelle to which proteins are delivered when they have been labeled by a mannose 6-phosphate.

5. The term for proteins that denature nascent proteins in the cytoplasm so that they can enter organelles.

6. The term used to describe the variety of hosts that a virus can infect. (2 words)

7. The term used to describe the attraction of a virus to certain types of cells in a host. (2 words)

8. What does "gp" signify in the symbolism gp120?

9. The adjective used for a viral RNA genome that can function as a mRNA. (1 or 2 words)

10. The adjective used for viruses backing envelopes.

11. The term that describes the number of viral progeny produced per infected cell. (2 words)

12. In the case of single-stranded RNA viruses, the term used to refer to the nucleic acid template and the nascent genomic strand. (2 words)

13. DNA synthesis by retroviruses occurs initially from an RNA template. What is the enzyme called that synthesizes new DNA from the RNA template? (2 words)

14. The subfamily to which belong the viruses that carry oncogenes similar to genes found in animal cells.

15. The subfamily in which the human immunodeficiency virus (HIV) is placed.

16. The membrane protein that HIV uses to attach to host cells.

17. The region that contains the promoter for HIV transcription. (3 words)

Answers to the Objective Questions

Multiple Choice

1. *b* **2.** *a* **3.** *b* **4.** *e* **5.** *a* **6.** *d* **7.** *c* **8.** *c* **9.** *c* **10.** *b* **11.** *d*
12. *d* **13.** *c* **14.** *e* **15.** *d* **16.** *d* **17.** *e* **18.** *a*

True–False

1. T **2.** T **3.** F (by phosphorylation of mannose) **4.** T **5.** T **6.** T **7.** T **8.** T
9. F (by protein attachment sites/protein receptors) **10.** T **11.** T **12.** T **13.** T **14.** F (either
RNA or DNA as a template) **15.** F (longer than the RNA genome) **16.** T **17.** T **18.** T **19.** T
20. T **21.** F (The RNA genome is shorter [R-U5-PB-gag-pol-env-U3-R] than the provirus's [U3-R-U5-PB-gag-pol-env-U3-R-U5].) **22.** F (Only nascent DNA or proviral DNA has LTRs.) **23.** T

Matching

1. *B* **2.** *B* **3.** *D* **4.** *E* **5.** *C* **6.** *C* **7.** *C* **8.** *E* **9.** *E* **10.** *E* **11.** *C*
12. *A* **13.** *D* **14.** *B* **15.** *D* **16.** *A* **17.** *C* **18.** *B* **19.** *C* **20.** *D* **21.** *E*
22. *A*

Terms

1. signal sequence   **2.** signal recognition particle   **3.** dolichol   **4.** lysosome   **5.** chaparones
6. host range   **7.** cellular tropism   **8.** glycoprotein   **9.** plus RNA   **10.** naked   **11.** burst size
12. replicative intermediate   **13.** reverse transcriptase   **14.** Oncoviridae   **15.** Lentiviridae
16. gp120   **17.** long terminal repeat

Chapter 11

Cellular Communication

INTRODUCTION

Cells are continually receiving information from their surroundings and, in turn, responding in a way that is determined by their genes and epigenetic factors. **Epigenetic factors** are chemical or physical agents that alter the manner in which genes are expressed. For example, a sugar that activates or represses the expression of a gene could be considered an epigenetic factor. Cells are able to respond to many different chemical and physical agents in their environment. These agents will be referred to as **signals** in this chapter. Chemical signals may be as simple as an ion or as complex as a peptide or a steroid hormone. Physical signals are generally restricted to energetic blows, heat, and electromagnetic radiation (light).

Growth, proliferation, differentiation, movement, and programmed cellular death are phenomena that seem to have little in common. Each one of these activities, however, depends upon signals altering a cell's physiology, often through the activation and repression of genes. The signal pathways responding to external signals consist of proteins and a few nonprotein "second messengers," such as calcium ions (Ca^{2+}), cyclic adenosine monophosphate (cAMP), cyclic guanosine monophosphate (cGMP), diacylglycerol (DG), and inositol triphosphate (IP_3).

Signals may induce transitory or permanent changes in cells. For example, an amino acid in the environment can turn off genes involved in its synthesis. These changes, however, are transitory, since withdrawal of the amino acid reverses the expression of the genes involved. In the development of a multicellular organism from a single fertilized egg, signals permanently alter many populations of cells. This allows the development of different tissues and organs. When the genes of differentiated cells are altered by mutations, cells may grow out of control and kill the host. This is described as a cancerous condition.

This chapter will be concerned with the signals, the signal pathways, and the cellular responses found in eukaryotic cells and bacteria. In particular, the chapter will consider the mechanisms by which cells are stimulated to progress through the cell cycle stages G1 (growth), S (DNA synthesis), G2 (growth), M (mitosis), and C (cytokinesis; see Question 11.29) and those that stimulate them to differentiate and enter G_0 (no growth and maintenance) or A (programmed cell death known as apoptosis). This chapter also will consider how mutations to components of signal pathways lead to cellular transformations and malignant cancers.

GENERAL PRINCIPLES

11.1. What are some of the chemical and physical extracellular signals that affect the physiology of eukaryotic cells?

Most extracellular chemical signals fall into one of three catagories (Table 11.1): proteins and peptides; nonpeptide neurotransmitters (many derived from amino acids); and steroids and other membrane-soluble molecules such as thyroxine and retinoic acid.

Electromagnetic radiation (light) is an important nonchemical signal that affects plants and animals through many of the same signal pathways that the chemical signals utilize. For example, rod and cone cells in the retina of the eye have light receptors known as **rhodopsins.** These light receptors interact with special proteins in the cytoplasmic disc membranes of rod cells or in the plasma membranes of cone cells. These special proteins transduce the signal to deeper levels within the cell. Similarly, many plant cells have light receptors called **phytochromes.** These receptors also stimulate special proteins in the plasma membrane and these, in turn, transduce the signal. **"Transduce"** in this chapter is synonymous with "send" or "carry." Thus, signal pathways will be re-

Table 11.1. Chemical Signals

I.	**Proteins and Peptides**	
	EGF	Epidermal growth factor
	PDGF	Platelet-derived growth factor
	FGF	Fibroblast growth factor
	TNF	Tumor necrosis factor
	TGF-α	Transforming growth factor-α
	TGF-β	Transforming growth factor-β
	G-CSF	Granulocyte colony stimulating factor
	GM-CSF	Granulocyte-macrophage colony stimulating factor
	M-CSF	Monocyte colony stimulating factor
	IFN-α/β	Interferon-α/β
	IFN-γ	Interferon-δ
	IL	Interleukin (1–12)
	ERYP	Erythropoietin
	INS	Insulin
II.	**Nonpeptide Neurotransmitters**	
	SER	Serotonin (derived from tryptophan)
	MEL	Melatonin (derived from tryptophan)
	DOP	Dopamine (derived from tyrosine)
	NEPI	Norepinephrin (derived from dopamine)
	EPI	Epinephrin (derived from norepinephrin)
	ACH	Acetylcholine
	GLU	Glutamic acid
	GABA	γ-Aminobutyric acid
III.	**Steroids and Other Membrane-Soluble Molecules**	
	EST	Estrogen
	TES	Testosterone
	TYX	Thyroxine
	RA	Retinoic acid

ferred to as **signal transduction pathways.** These are pathways in cells that send or carry signals to various parts of the cell, but most often to the genes.

11.2. How do chemical signals interact with cells and signal transduction pathways?

All of the signal molecules interact with signal transduction pathways by specifically binding to protein receptors. Since the peptide and nonpeptide neurotransmitters are unable to pass through the plasma membrane, receptors for these signal molecules are found on the outer surface of cells. Because of their chemical nature, however, steroids and other membrane-soluble compounds (such as thyroxine and retinoic acid) diffuse through the plasma membrane. These signal molecules first interact with the signal transduction pathways by specifically binding to protein receptors in the cytoplasm that are referred to as cytoplasmic receptors.

11.3. How do light signals interact with cells and signal transduction pathways?

Light signals are absorbed by receptors consisting of a chromophore and a transducing protein. **Chromophores** absorb specific wavelengths of light and reflect and/or transmit others because of their numerous double-bonded carbon atoms in tandem. Since chromophores absorb specific wavelengths, they are pigmented. Their color is determined by the light that is reflected.

Example 11.1. Rhodopsin is a receptor that responds to light. It consists of the chromophore **retinal** and the protein **opsin.** Rhodopsin is found imbedded in the membranes of rod and cone cells in the retinas of animals with color vision. A group of bacteria that absorb light have a slightly different receptor called bacteriorhodopsin.

Many plant cells have light receptors called phytochromes. **Phytochromes** consist of a linear tetrapyrole that functions as the chromophore and a protein portion that is believed to anchor the chromophore to the inside surface of the plasma membrane and communicate a signal to one or more signal pathways.

11.4. How does the binding of a ligand to its receptor or the absorption of light by a receptor stimulate a signal pathway?

Many receptors imbedded in plasma membranes become active ion pores or active protein kinases when signal molecules bind to them.

> **Example 11.2.** In skeletal muscle cells, the acetylcholine receptor binds not only a neurotransmitter, but also Ca^{2+}. In the presence of acetylcholine, the receptor forms a pore that transports Ca^{2+} across the membrane. Calcium ions flow down their concentration gradient into the muscle cell, depolarizing the membrane potential. An increase in the concentration of Ca^{2+} serves as a secondary signal that stimulates contraction of the cytoskeleton and cell.

Many receptors are tyrosine kinases that autophosphorylate themselves. In this state, they interact with proteins that stimulate (or inhibit) GTP-binding proteins generally known as G proteins.

G PROTEINS

11.5. What are G proteins and what is their role in signal transduction?

G proteins are important intermediates in signal transduction pathways because they determine whether the signal will be stimulatory or inhibitory. Signals that stimulate are called **agonists,** whereas those that inhibit are known as **antagonists.** A particular chemical or physical signal may be an agonist under some conditions and an antagonist in others. Whether or not a signal inhibits or stimulates a cellular response depends not so much on the signal but upon the pathway that is affected.

> **Example 11.3.** A signal (S) binding to a receptor (R) may open or close an ion pore (P) depending on interactions between the receptor and the pore. Generally, this interaction depends upon a stimulatory (Gs) or an inhibitory (Gi) G protein. Signals passing along the pathway S --> R --> Gs --> P open ion pores; those moving along the pathway S --> R --> Gi --| P close them. An arrow (-->) signifies stimulation, a blunted arrow (--|) suggests inhibition.

11.6. What characteristics differentiate G protein families?

A major family of G proteins is **trimeric,** consisting of three subunits: α, β, and γ. The α subunit binds GDP and GTP. When a signal stimulates the receptor, the altered receptor stimulates a change in the G protein: GDP dissociates from the α subunit, and GTP takes its place. This stimulates the dissociation of the α subunit. The α subunit diffuses along the inner surface of the membrane until it contacts an enzyme or ion pore. Depending on the characteristics of the α subunit, enzymes may be inhibited or stimulated, and ion pores may be closed or opened. The inhibitory or stimulatory activity of the α subunit is blocked by the spontaneous hydrolysis of the bound GTP and its subsequent reassociation with the β and γ subunits.

A number of G proteins have been differentiated on the basis of their modification by certain bacterial toxins (Table 11.2 footnote). These bacterial toxins add ADP-ribose from nicotinaminde adenine dinucleotide (NAD^+) to the G proteins. This reaction is referred to as an ADP-ribosylation. In some cases the ADP-ribosylation activates the G protein, whereas in others, it inactivates the G protein.

A second family of G proteins consist of a single subunit. These **monomeric** (single unit) proteins are known as **Ras proteins (Ras).** Ras proteins are activated indirectly through autophosphorylation of membrane-bound tyrosine kinases and the regulatory proteins that interact with the phosphates. The relative amounts of active and inactive Ras are determined by guanine nucleotide release factors (Gnrfs) and by GTPase-activating proteins (Gaps) (Fig. 11-1). Since these proteins promote the exchange of GTP for GDP, or GDP for GTP, respectivefully, they affect Ras protein activity. Some Ras proteins are negatively regulated by **tumor suppressor gene products** such as neurofibromatosis-type-1 (NF-1) protein in humans. The product of the NF-1 gene stimulates Ras protein GTPase activity. Hydrolysis of GTP to GDP and P_i inhibits Ras protein.

> **Example 11.4.** The tumor suppressor gene, NF-1, specifies a Ras inhibitory protein and is an example of a Gap (GTPase-activating protein). A signal transduction pathway involving Ras follows: PDGF --> R --> Ras --> Raf --> Mek --> Erk --> Rsk. This pathway is modulated by NF-1: NF-1 --| Ras. An arrow (-->) signifies stimulation, a blunted arrow (--|) suggests inhibition, whereas a double arrow (==>) should be read as "yields" or "synthesizes." PDGF = platelet derived

Table 11.2. Physiological Effects Mediated by G Proteins

Signal	G Proteins*	Target Proteins	Affected Cell Type	Cellular Effect
Epinephrine	G_s (β, γ)	adenylcyclase	liver	glycogen hydrolysis
Epinephrine	G_i (β, γ)	adenylcyclase	epidermal	no glycogen hydrolysis
Glucagon	G_s (β, γ)	adenylcyclase	liver	glycogen hydrolysis
Epinephrine	G_s (β, γ)	adenylcyclase	fat	fat hydrolysis
Glucagon	G_s (β, γ)	adenylcyclase	fat	fat hydrolysis
Acetylcholine	G_i (β, γ)	potassium channel	heart muscle	low heart rate & force
Luteinizing hormone	G_s (β, γ)	adenylcyclase	ovarian follicles	synthesis of hormones estrogen & progesterone
Antidiuretic hormone	G_s (β, γ)	adenylcyclase	kidney tubules	water absorption
Enkephalins, Endorphins	G_{sip} (β, γ)	ion pores (Na, K, Ca)	brain neurons	neuron activity
Angiotensin	G_{sp} (β, γ)	phospholipase C	smooth muscle	muscle contraction & increased pressure
Odorants	G_{olf} (β, γ)	adenylcyclase	nose neurons	detection of odorants
Light	G_{st} (β, γ)	cGMP-dependent phosphodiesterase	rods & cones	detection of light
FGF‡	Ras (Gap, Gnrf)†	phospholipase-C	fibroblast	proliferation
EGF‡	Ras (Gap, Gnrf)†	adenylcyclase	epidermal	proliferation

*Some G proteins are distinguished by bacterial enzymes that chemically modify them. G_s is modified by the cholera toxin, whereas G_i, G_{sp}, and G_{sip} are modified by the pertussis toxin. G_{st} is modified by both cholera and pertussis toxins.
†Gap = GTPase-activating protein, Gnrf = guanine nucleotide release factor.
‡FGF = fibroblast growth factor, EGF = epidermal growth factor.

growth factor; R = tyrosine kinase receptor; Ras = monomeric G protein; Raf, Mek, Erk, Rsk = cascade of protein kinases.

Not all signal transduction pathways "downstream" of monomeric Ras proteins are known. Nevertheless, Ras proteins generally stimulate a cascade of protein kinases, whereas trimeric G proteins usually inhibit or stimulate enzymes such as adenylcyclase.

11.7. **What kind of proteins and enzymes do trimeric G proteins affect?**

Trimeric G proteins affect ion pores and enzymes such as adenylcyclases, guanylcyclases, and phospholipases (Table 11.2). These enzymes are important in signal pathways because they amplify weak signals by catalytically producing **second messengers** such as cyclic adenosine monophosphate (cAMP), cyclic guanosine monophosphate (cGMP), inositol triphosphate (IP_3), and diacylglycerol (DG).

Adenylcyclases and **guanylcyclases** are affected by G proteins. These enzymes synthesize cAMP and cGMP from ATP and GTP, respectively (Fig. 11-2). **Phospholipases** such as phospholipase-C (PLC) and phospholipase-A (PLA) are affected by G proteins. These enzymes catalyze the synthesis of inositol 1,4,5-trisphosphate (IP_3) and diacylglycerol (DG). Phosphatidylinositol 4,5-bisphosphate (PIP_2) is hydrolyzed to inositol 1,4,5-trisphosphate (IP_3) and diacylglycerol (DG) by the membrane-bound enzymes (Figs. 11-2c and 11-3).

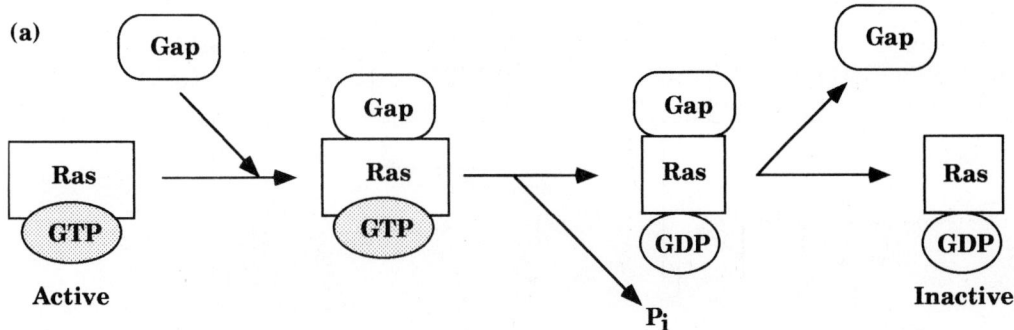

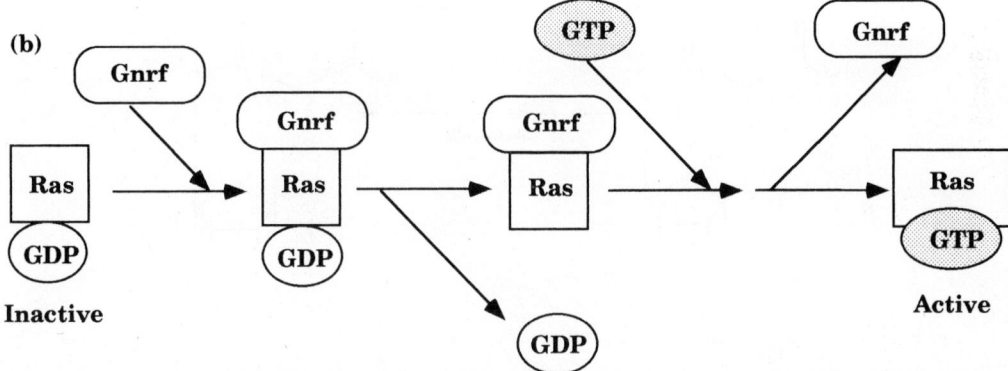

Fig. 11-1. Regulation of the *ras* gene product. The *ras* gene product is a monomeric G protein that is active
when bound by the nucleotide GTP, but inactive when bound by GDP. (*a*) Its GTPase activity is
stimulated by a GTPase-activating protein (Gap) that binds to the G protein. Hydrolysis of GTP
to GDP + P_i inactivates the G protein and inhibits signal transduction pathways that rely on the G
protein. After GTP hydrolysis, Gap dissociates from Ras. (*b*) The inactive Ras-GDP complex
can be activated by the binding of a guanine nucleotide release factor (Gnrf) that stimulates the
dissociation of GDP and the binding of GTP to Ras. Upon GTP binding to Ras, Gnrf dissociates.

PIP_2 is made from a common lipid in the plasma membrane of eukaryotes called phosphatidylinositol (PI).
Small amounts of this phospholipid are converted to phosphatidylinositol 4-phosphate (PIP) by a membrane kinase. Similarly, PIP is transformed to phosphatidylinositol 4,5-bisphosphate (PIP_2) by another kinase.

PROTEIN KINASES AND PROTEIN PHOSPHATASES

11.8. What are protein kinases and how are they categorized?

Protein kinases are so named because they phosphorylate other proteins rather than nonprotein molecules
such as sugars and lipids.

Example 11.5. The kinases that transform phosphatidylinositol to PIP_2 are *not* protein kinases.

Some protein kinases are stimulated at the beginning of signal transduction pathways by specific growth factors, whereas others are stimulated along the signal pathway by the binding of second messengers or by phosphorylation. Although some protein kinases are membrane-bound, the vast majority are free in the cytoplasm and
function in signal transduction pathways.

Protein kinases are categorized on the basis of which amino acids they phosphorylate (e.g., tyrosine kinases
and serine-threonine kinases) as well as on the basis of the factor(s) required for their activity (e.g., cAMP-depen-

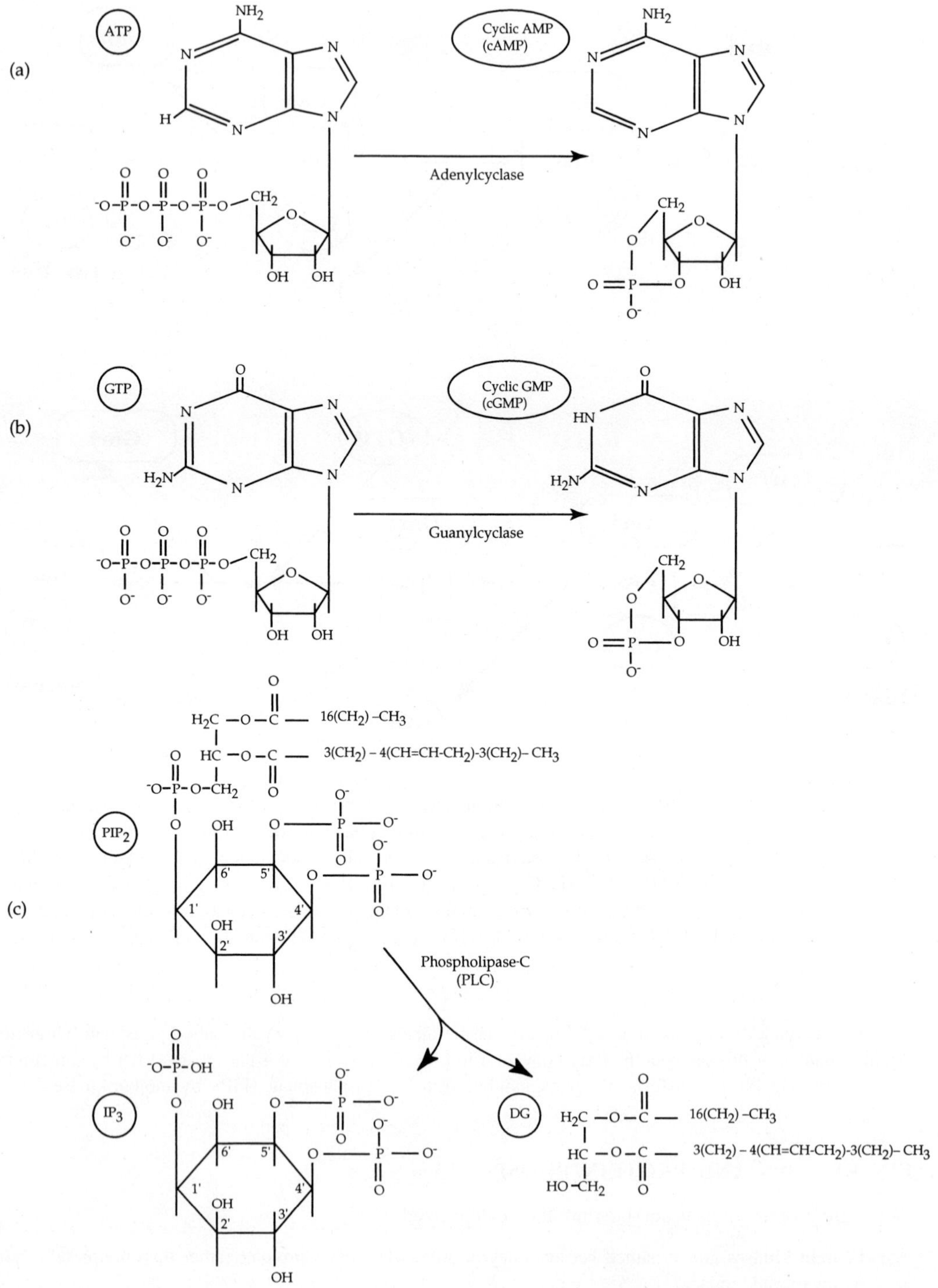

Fig. 11-2. Chemical structures of some second messengers. (*a*) Cyclic adenosine monophosphate (cAMP) is synthesized from adenosine triphosphate (ATP) by the membrane-bound enzyme adenylcyclase. (*b*) Cyclic guanosine monophosphate (cGMP) is synthesized from guanosine triphosphate (GTP) by the membrane-bound enzyme guanylcyclase. (*c*) Diacylglycerol (DG) and inositol $1',4',5'$-trisphosphate (IP_3) are synthesized from the membrane lipid phosphatidylinositol $4',5'$-bisphosphate (PIP_2) by the enzyme phospholipase-C (PLC).

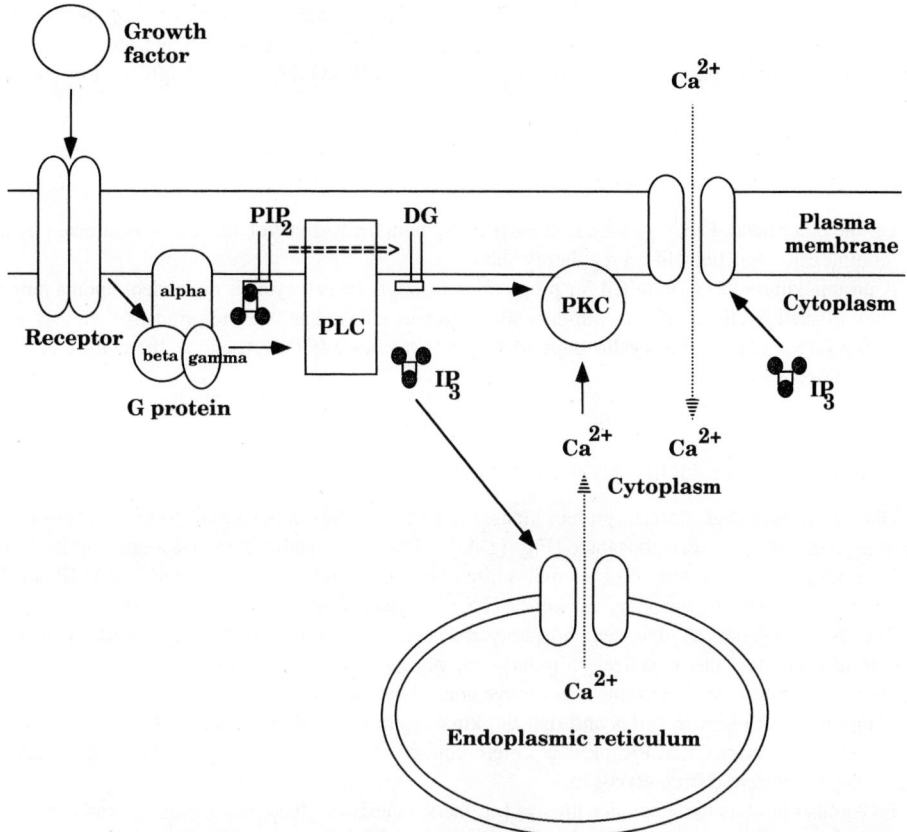

Fig. 11-3. A signal transduction pathway involving IP$_3$ and DG. A growth factor or hormone
binding to a cell-membrane receptor alters the receptor's conformation. This stimulates the dissociation of a neighboring trimeric G protein and a GDP (not shown) attached to the α subunit of the G protein. The α subunit becomes active in the signal
transduction pathway by dissociating from the β and γ subunits of the G protein and
by exchanging a molecule of GTP for the GDP (not shown). The active G protein
stimulates a membrane-bound phospholipase-C (PLC) that hydrolyzes the phosphatidylinositol 4′,5′-bisphosphate (PIP$_2$) in the membrane to diacylglycerol (DG)
and inositol 1′,4′,5′-trisphosphate (IP$_3$). IP$_3$ binding to calcium ion pores opens
these pores in the endoplasmic reticulum (ER) and the plasma membrane, allowing
calcium ions to move along their concentration gradient from the ER and from the
extracellular environment into the cytoplasm. Calcium ions and DG binding to inactive protein kinase-C (PKC) causes PKC to become active. Activated PKC phosphorylates other protein kinases in signal transduction pathways, often activating
them (not shown).

dent kinases, cGMP-dependent kinases, diacylglycerol-dependent kinases, calcium-dependent kinases, calmodulin-calcium-dependent kinases, and cyclin-dependent kinases.) Membrane tyrosine kinases (PKTyro) and serine-threonine kinases (PKST) are generally stimulated directly by a chemical signal.

Protein kinase A (PKA) and protein kinase G (PKG) are soluble serine-threonine kinases activated by the nucleotides and second messengers cAMP and cGMP, respectively.

Protein kinase C (PKC) is used to label a large family of serine-threonine protein kinases that are stimulated
directly by the second messengers diacylglycerol (DG) and/or Ca^{2+} (Fig. 11-3). In some cases, these kinases require the binding of both of these second messengers for full activity.

A protein kinase that requires the calcium ion binding protein **calmodulin** and Ca^{2+} for its activity is known
as a calmodulin-calcium-dependent protein kinase (PKCC). Ca^{2+} pores in the plasma membrane and in the endoplasmic reticulum are opened by the binding of IP$_3$ to the pores.

Most protein kinases are multimeric proteins consisting of separate catalytic and regulatory subunits. A few protein kinases, however, consist of a single polypeptide with a catalytic domain and one or more regulatory domains. The regulatory domains might bind ligands like cAMP, cGMP, and diacylglycerol or they might be sites for phosphorylation.

> **Example 11.6.** PKC, which is stimulated by diacylglycerol binding, consists of a single polypeptide with a single catalytic site.

On the other hand, PKCC, which is stimulated by both diacylglycerol and Ca^{2+}, is affected by a calmodulin-Ca^{2+} complex that functions like a regulatory subunit.

A protein kinase that is required for progression through the cell cycle is dependent upon a number of protein stimulators called **cyclins.** The cyclin-dependent protein kinase has been referred to variously as the **mitosis-promoting factor (MPF),** the **cyclin-dependent protein kinase (CDPK),** and as the **cell-division-cycle (CDC) protein kinase.**

11.9.   What is the signal transduction pathway that activates PKA?

One of the best characterized protein kinases is PKA (PKAg) in the signal pathway controlling the catabolism of glycogen to glucose 1-phosphate (Fig. 11-4a). This signal pathway includes epinephrine (EPI), its receptor (R), a stimulatory G protein (Gs), adenylcyclase (Adc), a second messenger (cAMP), cAMP-dependent protein kinase (PKAg), glycogen phosphorylase kinase (PKgpk), and glycogen phosphorylase (GP).

The phosphorylation of glycogen phosphorylase kinase (PKgpk) by PKAg converts the enzyme from an inactive form to a structure that is active. Similarly, the phosphorylation of glycogen phosphorylase (GP) by PKgpk converts GP from an inactive enzyme to an active one. This pathway illustrates the fact that signal pathways may have a number of kinases in series and that the kinases are stimulated by the binding of nucleotides and/or by phosphorylations. Notice that PKgpk is a protein kinase but GP is not, because GP phosphorylates glucose subunits as they are removed from glycogen.

PKAg also inhibits two enzymes involved in the synthesis of glycogen: glycogen synthase and phosphoprotein phosphatase (Fig. 11-4b). Glycogen synthase (GlcSyn) catalyzes the polymerization of UDP-glucose (Fig. 2-3) to glycogen. This enzyme is directly inhibited by PKAg through phosphorylation. On the other hand, phosphoprotein phosphatase (PPPP) is inhibited indirectly by PKAg through the phosphorylation of another protein that becomes an inhibitor of PPPP.

PPPP inhibits glycogen catabolism by removing phosphate groups from PKgpk and GP. Also, PPPP removes phosphate groups from glycogen synthase, activating it and stimulating glycogen synthesis.

The physiological consequences of epinephrine stimulation is the release of glucose 1-phosphate from glycogen. In the absence of epinephrine, PPPP dephosphorylates spontaneously and becomes an active enzyme to dephosphorylate PKgpk, GP, and glycogen synthase (Fig. 11-4c). This promotes the synthesis of glycogen from UDP-glucose. UDP "activates" glucose, so that it can be polymerized.

PKAg is a complex enzyme. In the absence of cAMP, it consists of two regulatory subunits and two catalytic subunits. The regulatory subunits inhibit the activity of the catalytic subunits. When cAMP binds to the two regulatory subunits, they undergo conformational changes and dissociate from the catalytic subunits. The dissociated catalytic subunits function as the active kinases.

In certain cells, acetylcholine may also stimulate glycogen catabolism, but by a slightly different signal transduction pathway (Fig. 11-4d). Notice, in this pathway PKgpk is stimulated by calcium ions rather than by phosphorylation. Different regulatory domains are involved.

11.10.   What protein kinases are central to the control of the cell cycle?

Most of the soluble protein kinases in the cytoplasm are serine-threonine kinases. One important family of cytoplasmic protein kinases is involved in stimulating the cell cycle. These protein kinases are referred to as **mitogen-activated protein kinases** (MAPKs).

Figure 11-5a shows that certain growth factors such as platelet-derived growth factor (PDGF), fibroblast growth factor (FGF), epidermal growth factor (EGF), nerve growth factor (NGF), and transforming growth factor (TGF) binding to their receptors stimulate a tyrosine kinase (PKtyro) activity on the cytoplasmic side of the membrane. In this example, the receptor autophosphorylates itself on tyrosine residues. The receptor may stimulate another tyrosine kinase (Src; attached to the plasma membrane) that is known to phosphorylate various cytoskele-

---

(a) Epinephrine stimulates glycogen catabolism.

EPI --> R --> Gs --> Adc => cAMP --> PKAg --> PKgpk --> GP --> glycogen ==> G1P

(b) Protein kinase-A inhibits glycogen synthase, but stimulates an inhibitor of PPPP.

PKAg --| GlcSyn

PKAg --> I --| PPPP

(c) Without epinephrine, phosphoprotein phosphatase stimulates glycogen synthesis.

PPPP ----| PKgpk

PPPP ----| GP

PPPP ----> GlcSyn

GlcSyn + UDP-glucose ==> Glycogen + UDP

(d) Acetylcholine stimulates glycogen synthesis

AC ---> R ---> Gsp ---> P ==> Ca^{2+}

Ca^{2+} + PKgpk ---> GP ==> G1P ==> glycogen

Fig. 11-4. Signal transduction pathways that affect glycogen synthesis and hydrolysis. EPI = epinephrine; R = receptor; Gs = trimeric G proteins; Adc = adenylcyclase; cAMP = cyclic adenosine monophosphate; PKAg = protein kinase-A, involved in glycogen breakdown. PKgpk = glycogen protein kinase; GP = glycogen phosphorylase; G1P = glucose 1-phosphate; GlcSyn = glycogen synthase; I = inhibitory protein; PPPP = phosphoprotein phosphatase; UDP-glucose = uridine diphosphate glucose (the activated precursor from which glycogen is synthesized); AC = acetylcholine; Gsp = trimeric G protein that stimulates plasma membrane calcium pore; P = calcium pore. UDP-glucose is synthesized from uridine triphosphate and glucose 1-phosphate by the enzyme UDP-glucose pyrophosphorylase. The arrows have the following meanings: --> = stimulates, ==> = yields, ---| = inhibits.

ton proteins. The phosphorylation of cytoskeleton proteins may result in cellular shape changes, detachment from other cells, or locomotion. The receptor may also stimulate Ras through the intermediary proteins Grb2 and Sos. Ras stimulates a mitogen-activated protein kinase kinase kinase (MAPKKK) to autophosphorylate itself. This enzyme, in turn, excites a mitogen-activated protein kinase kinase (MAPKK) by phosphorylation, which turns on the mitogen-activated protein kinase (MAPK) through phosphorylation. MAPK activates the protein kinase Rsk that enters the nucleus, where it inactivates the tumor suppressor protein **retinoblastoma** (Rb) by phosphorylation. Inactivation of Rb frees transcriptional factors E2f and Drtf1 required for the transcription of the transcriptional activators Fos, Jun, and Myc. E2f is also required for the synthesis of RNA primers during the initiation of DNA replication at *ori* sites.

By phosphorylation, MAPK stimulates transcriptional activators such as Jun, Fos, Myc, and Cdks (proteins necessary for progression through the different stages of the cell cycle). The **cyclin-dependent protein kinases (cdks)** along with their cyclins have also been called mitosis promoting factors (MPFs).

Certain growth factors such as epinephrine (EPI) and norepinephrine (NEPI) bind to receptors that have no

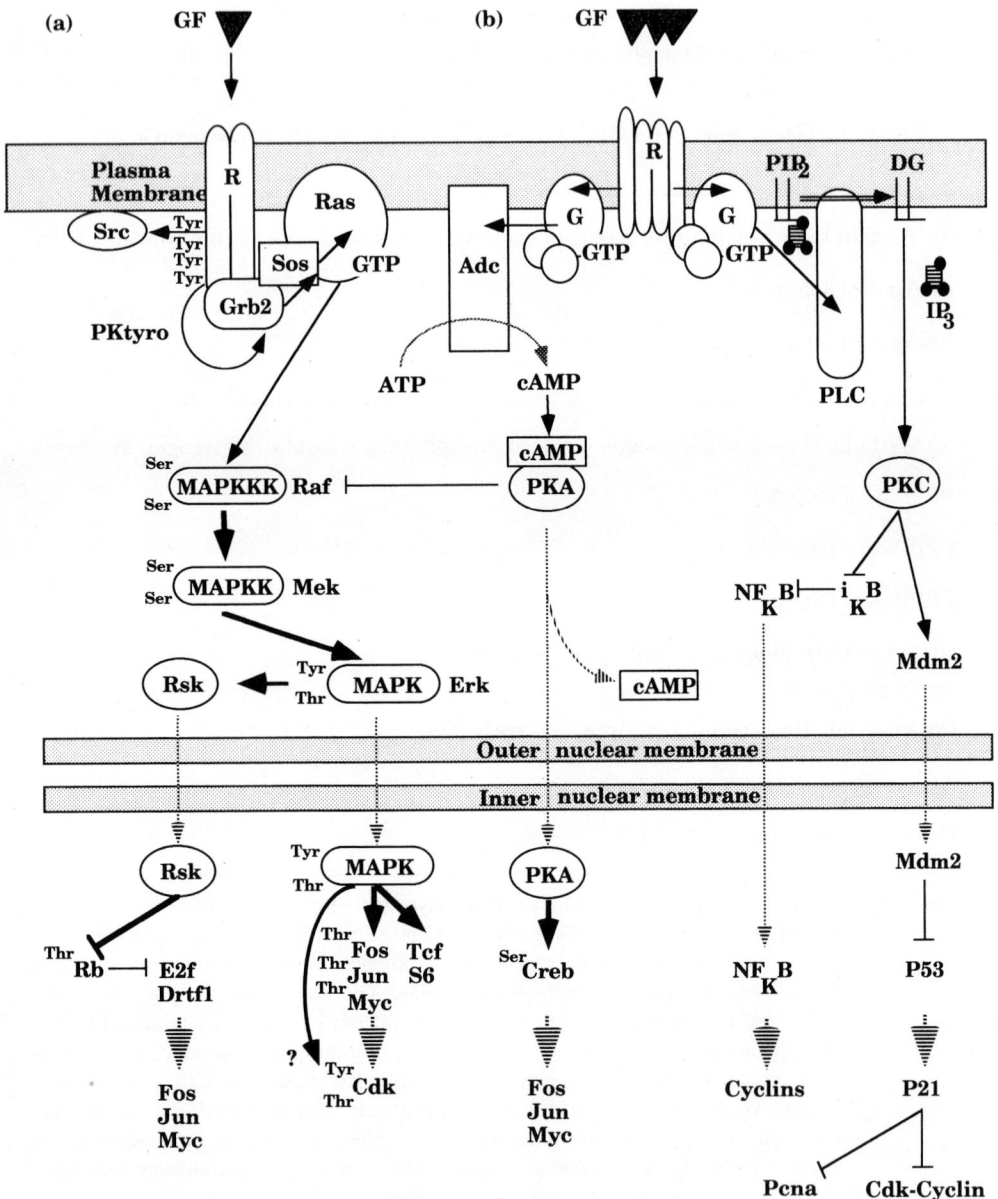

Fig. 11-5.   (Caption on page 263.)

kinase activity (Fig. 11-5*b*).   In this example, the receptor stimulates trimeric G proteins which, in turn, may stimulate an adenylcyclase or a phospholipase-C.   Adenylcyclase catalyzes the synthesis of cAMP that stimulates PKA by binding to and removing inhibitory subunits.   Activated PKA diffuses into the nucleus, where it stimulates transcriptional activators such as Creb.   Phosphoryated Creb promotes the synthesis of transcriptional activators Fos, Jun, and Myc.   Phospholipase-C catalyzes the synthesis of diacylglycerol that stimulates PKC.   Activated PKC phosphorylates an inhibitor of a transcriptional activator NF$_K$B.   Freed NF$_K$B diffuses into the nucleus, where it promotes the synthesis of cyclins.   Cyclin-dependent protein kinases (Cdks) are produced because of signals that stimulate the kinase cascade (Fig. 11-5*a*); cyclins are produced because of signals that stimulate PKC (Fig. 11-5*b*); both of them work together to promote the cell cycle.

By phosphorylation, PKC may also activate the protein Mdm2 that inhibits the tumor suppressor protein P53. P53 is a transcriptional factor that stimulates the transcription of an inhibitory protein P21, which inhibits the Cdk-cyclin complex and Pcna (proliferating cell nuclear antigen) required for the activity of DNA polymerase-δ.

Fig. 11-5. Signal transduction pathways. ➤ = stimulation caused by a physical interaction; ⇢ = movement of an effector or a protein; ➤ = stimulation caused by phosphorylation; ⇥ = synthesis of a protein. (*a*) The binding of certain growth factors (GF) to their respective receptors (R) stimulates the tyrosine kinase activities (PK-tyro) of the receptors, so that they autophosphorylate on tyrosines. The active receptor stimulates Src and/or a monomeric G protein (Ras) through two other proteins, Grb2 and Sos. Grb2 is a growth-factor receptor-bound protein, whereas Sos is the guanine nucleotide exchange factor. Ras activates the kinase cascade involving MAPKKK (a serine-threonine kinase), MAPKK (a tyrosine-threonine kinase), and MAPK (a serine-threonine kinase). MAPKKK autophosphorylates itself, whereas MAPKK, MAPK (and Rsk) are phosphorylated by other kinases. The amino acids phosphorylated on each protein are indicated. In the nucleus, MAPK stimulates by phosphorylation a number of proteins such as the kinase Rsk and the transcriptional activators Fos, Jun, and Myc. The tumor suppressor gene product retinoblastoma (Rb) is inhibited by phosphorylations on numerous threonines. This phosphorylation breaks up the complex between Rb and the transcriptional factors E2f and Drtf1, necessary for the initiation of DNA replication and the transcription of a number of transcriptional activators (Fos, Jun, Myc), transcriptional factor (Tcf), and the S6 kinase. These transcriptional activators result in the synthesis of Cdks. (*b*) The binding of certain growth factors (GF) to their respective receptors (R) stimulates trimeric G proteins. These proteins stimulate enzymes such as adenylcyclase (Adc) and phospholipase-C (PLC) to catalyze the synthesis of second messengers such as cyclic AMP (cAMP), and diacylglycerol (DG) and inositol trisphosphate (IP$_3$), respectively. cAMP stimulates PKA by binding to and removing an inhibitory regulatory subunit, whereas DG binds to and activates PKC in the cytoplasm. Some forms of PKC are stimulated by Ca^{2+} binding. Ca^{2+} is released from the endoplasmic reticulum by inositol trisphosphate (IP$_3$). PKC frees the transcriptional factor NF$_K$B from the inhibitory protein i$_K$B. In the nucleus, PKA activates the transcriptional activator Creb that promotes the synthesis of transcriptional activators, whereas NF$_K$B promotes the synthesis of cyclins. In the nucleus, a tumor suppressor protein (P53) stimulates the transcription of protein Cip1 (P21), which inhibits two proteins: proliferating cell nuclear antigen (Pcna), necessary for DNA synthesis by polymerase-δ and DNA repair by polymerases-δ or ϵ, as well as cyclin-dependent kinase when it is complexed with cyclin (Cdk-cyclin). CdK-cyclin is required for cell cycle progression. P53 is inhibited by the proto-oncogene product Mdm2.

11.11. What role do phosphatases play in signal pathways?

　　Because protein kinases are part of most signal pathways, they control almost every aspect of cellular physiology. Thus, control of a cell's physiology is affected by the phosphorylated state of its proteins. If kinases phosphorylate and this affects a response, there must be enzymes that reverse that response. **Protein phosphatases** are such enzymes; they remove phosphate groups from proteins. Thus, phosphatases are the kinases' counterparts.

　　Some phosphatases are activated by phosphorylation. Once activated, they proceed to counteract the changes the kinases have made, even to dephosphorylating and inactivating the enzymes that gave them activity. Another group of phosphatases are activated by a calmodulin-Ca^{2+} complex. Still other phosphatases have their activity controlled by inhibitory proteins.

SIGNAL TRANSDUCTION IN DISEASE AND IMMUNITY

11.12. What are bacterial endotoxins?

　　Gram-negative bacteria have an outer membrane containing unusual lipopolysaccharides that are known as **endotoxins.** These lipopolysaccharides indirectly cause inflammation and severe systemic shock (septic shock) when they induce macrophages to produce **tumor necrosis factor (TNF).** A portion of these lipopolysaccharides, designated lipid A, binds to specific receptors on macrophages and lymphocytes, stimulating them to produce and release TNF.

11.13. What is tumor necrosis factor?

　　Infections of the blood lead to the synthesis and release of at least two proteins called tumor necrosis factors (TNF-α and TNF-β). Tumor necrosis factors stimulate many cell types to kill themselves, in particular endothelial cells that line the blood vessels. This cellular death leads to circulatory failure called **septic shock.**

11.14. Which are the signal transduction pathways stimulated by bacterial endotoxins that yield TNFs?

Numerous experiments suggest that there are two signal transduction pathways leading to the synthesis of tumor necrosis factors (TNFs). One pathway involves the synthesis of diacylglycerol (DG). A second pathway by-passes PLC (phospholipase-C) and PKC (protein kinase-C). Instead, the stimulated G protein excites adenyl-cyclase (Adc), which leads to the activation of PKA (protein kinase-A). PKA frees NF_KB (transcriptional activator nuclear factor -κB) by phosphorylating i_KB (inhibitor of κB). The transcriptional factor stimulates the synthesis of mRNA for TNF. Translation of the mRNA yields TNF. These two pathways are summarized below, where R stands for receptor and G for G protein, $==>$ = yields, $-->$ = stimulates, $-|$ = inhibits.

$$\text{Lipid A} --> R --> G --> PLC ==> DG --> PKC -| i_KB ==> NF_KB ==> mRNA_{TNF} ==> TNF$$
$$\text{Lipid A} --> R --> G --> Adc ==> cAMP --> PKA -| i_KB ==> NF_KB ==> mRNA_{TNF} ==> TNF$$

11.15. What kinds of cellular changes are caused by TNF-α and TNF-β?

Two similar glycosylated proteins, designated TNF-α (**cachectin**) and TNF-β (**lymphotoxin**), belong to the same family of proteins. TNF-α in humans is 157 amino acids long, whereas TNF-β consists of 171 residues. TNF-α is synthesized principally by monocytes and macrophages. TNF-β is produced chiefly by lymphocytes. TNF-α and TNF-β can bind to the same receptors, but with different affinities. At least two TNF receptors have been identified on many different cell types.

Low levels of TNF-α released during chronic infections are responsible for the extreme weight loss associated with long-term diseases. TNF-α is the major protein in the development of septic shock following blood infections by gram-negative bacteria. Both TNF-α and TNF-β can induce programmed cell death, called **apoptosis,** in target cells (Fig. 11-6). Induced apoptosis of endothelial cells (which line the blood vessels) results in vascular collapse and septic shock. Similarly, apoptosis of tumor cells may explain how TNF-α and TNF-β kill tumors. Apoptosis is distinct from cellular **necrosis.** Necrosis is believed to be due to osmotic lysis or enzymatic attacks and is associated with increased cellular volume. Cells undergoing apoptosis shrink rather than swell, their chromatin concentrates along the nuclear membrane rather than remaining dispersed, they develop large cytoplasmic vacuoles, and they fragment into minicells rather than rupturing. Clearly, the tumor necrosis factors are misnamed, since they do not induce necrosis but apoptosis.

Apoptosis is a complex event consisting of numerous components. Chromosome degradation to nucleic acids 50–300 kilobase pairs (kbp) in length may be due to the activation of specific restriction endonucleases. The fragmentation of the nucleus into mininuclei may be caused by the activation of cysteine proteases (Ced3 in nematodes, ICE in vertebrates) that cleave the **lamin** proteins (A, B, C) that stabilize the nuclear membrane. Mutations in certain genes make cells "cell-death defective," hence the gene nomenclature, *ced.* ICE refers to a mammalian interleukin converting enzyme, a protease related to Ced3. The inhibition of superoxide dismutase, peroxide, and catalase may be the cause of plasma membrane damage that results in cell fragmentation. Mammalian Bcl2 (or Ced9 in the nematode) inhibits apoptosis: DNA fragmentation, nuclear fragmentation, cellular fragmentation, and damage caused by superoxide and hydrogen peroxide.

11.16. What signal transduction pathways are activated by TNF?

Numerous experiments suggest that the binding of TNF-α to its receptor stimulates a phospholipase-C that is specific for membrane phosphatidylcholine (PTC) (Fig. 11-7). The diacylglycerol (DG) produced stimulates another enzyme designated acidic sphingomyelinase (ASMase) that catalyzes the hydrolysis of the membrane lipid sphingomyelin (SM) into ceramide and phosphocholine (PC). Through a ceramide-dependent PK, ceramide stimulates a nuclear transcriptional activator by freeing it from an inhibitory subunit (i_KB). The transcriptional activator NF_KB (p50 + p65) induces many genes, including those for TNF-α and TNF-β. Some of the gene products may result in an increase in fat metabolism, the synthesis of inflammatory factors, or apoptosis.

According to one model, apoptosis is the result of an imbalance of growth factors and killing factors: too little epidermal growth factor activating Bcl2/Ced9 and/or too much TNF-α, activating endonucleases, proteases (ICE/Ced3), and enzyme inhibitors (of superoxide dismutase, peroxidase, and catalase) cause the cell to undergo apoptosis. Figure 11-8 illustrates a hypothetical scheme in which ceramide produced by TNF-α stimulates protein phosphatases to remove phosphates from many cytoplasmic kinases such as MAPKKK, MAPKK, and MAPK. This dephosphorylation inhibits the kinases required to stimulate transcriptional activators, which in turn are necessary for the production of proteins such as Bcl2/Ced9 that block apoptosis. A stimulatory growth factor such as **epidermal growth factor** (EGF) excites the cytoplasmic kinases. The kinases stimulate the ap-

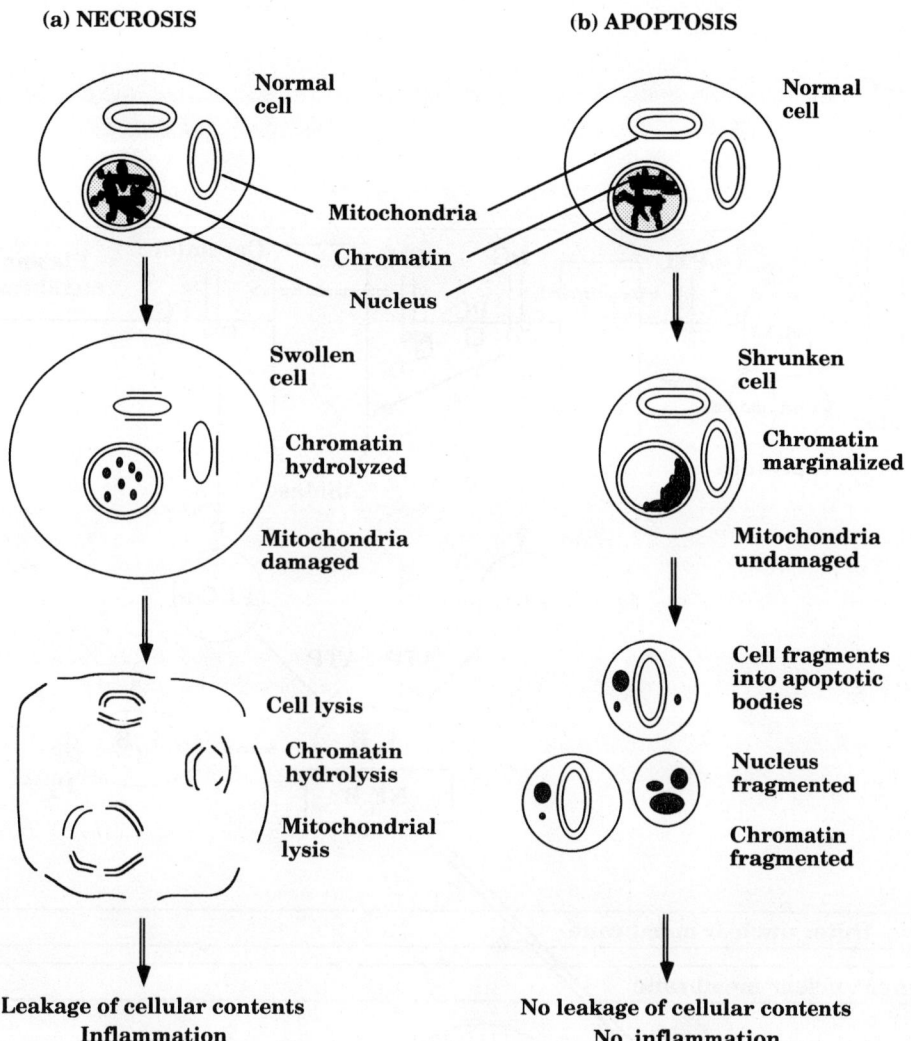

Fig. 11-6. Morphological changes in necrosis and apoptosis. (*a*) Generally, necrosis is due to unrepaired damage brought on by the loss of nutrients or by toxic compounds. Cellular necrosis is associated with an increased cellular volume, organelle fragmentation, chromatin hydrolysis, and cell lysis with the release of cellular materials. Debris is phagocytosed by macrophages, and the released cellular contents cause inflammation. (*b*) Apoptosis is due to a programmed destruction brought on by signals that fail to maintain the cell and/or signals that initiate programmed cell death. Cellular apoptosis is characterized by a decrease in volume, mitochondrial stability, chromatin condensation along the inner nuclear membrane, nuclear fragmentation, cellular fragmentation into apoptotic bodies (with intact mitochondria and without the release of cellular materials), and chromatin hydrolysis. Apoptotic bodies are phagocytosed by macrophages with minimal or no inflammation.

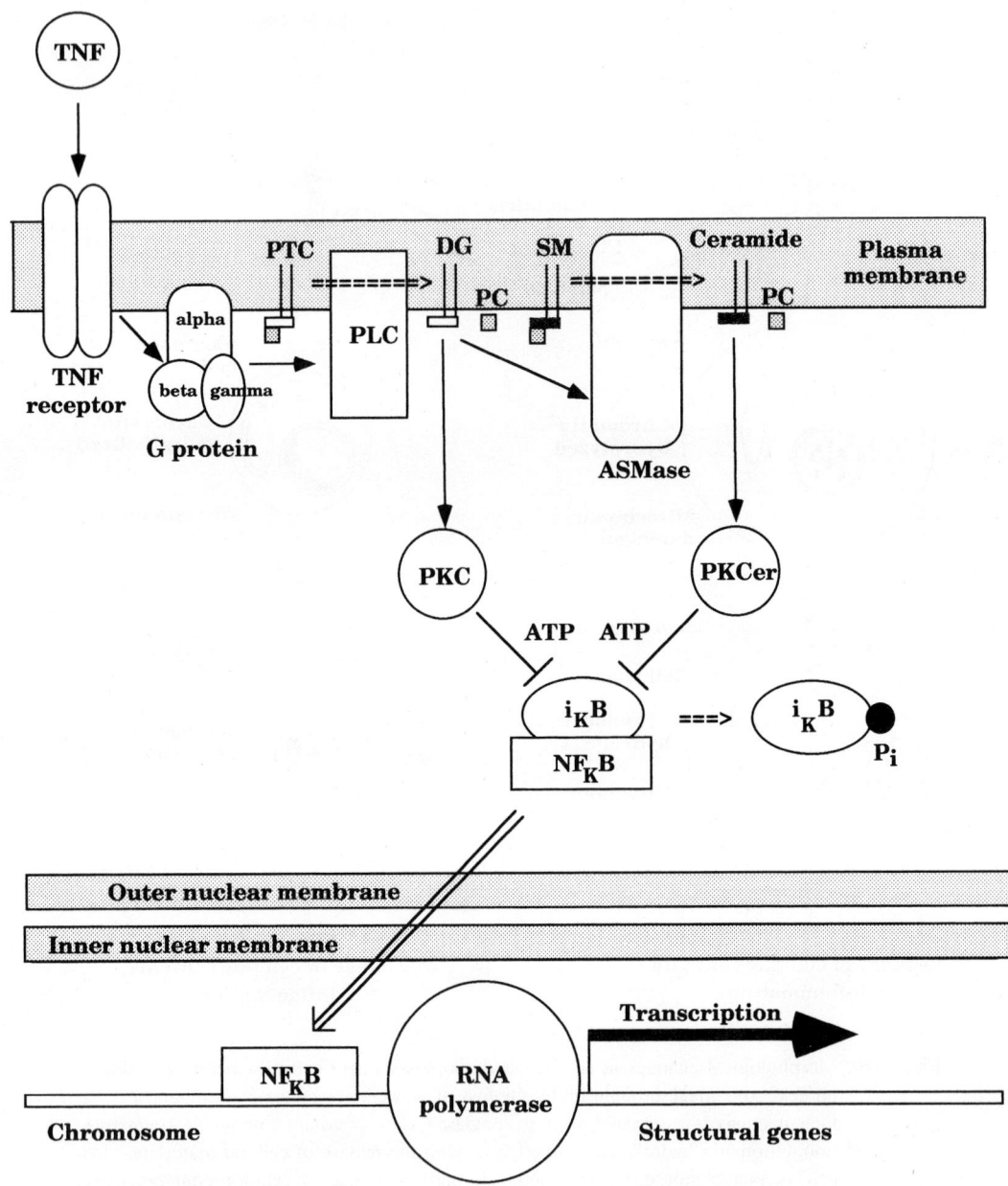

Fig. 11-7. Signal transduction pathway stimulated by tumor necrosis factor. Tumor necrosis factor (TNF) binding to its receptor activates a trimeric G protein to stimulate a plasma membrane-bound phospholipase-C (PLC). This enzyme hydrolyzes phosphatidylcholine (PTC, a membrane lipid) to diacylglycerol (DG) and phosphocholine (PC). DG, acting as a second messenger, stimulates two enzymes, membrane-bound acidic sphingomyelinase (ASMase) and soluble protein kinase-C (PKC). ASMase hydrolyzes sphingomyelin (SM, a membrane lipid) to ceramide and phosphocholine (PC). Ceramide stimulates a ceramide-dependent protein kinase (PKCer) that phosphorylates the inhibitor (i_KB) of the transcriptional factor NF$_K$B, causing the inhibitor to dissociate. PKC may also phosphorylate i_KB. NF$_K$B (consisting of subunits P50 and P65; not shown) activates transcription of various genes affecting the physiology of the cell.

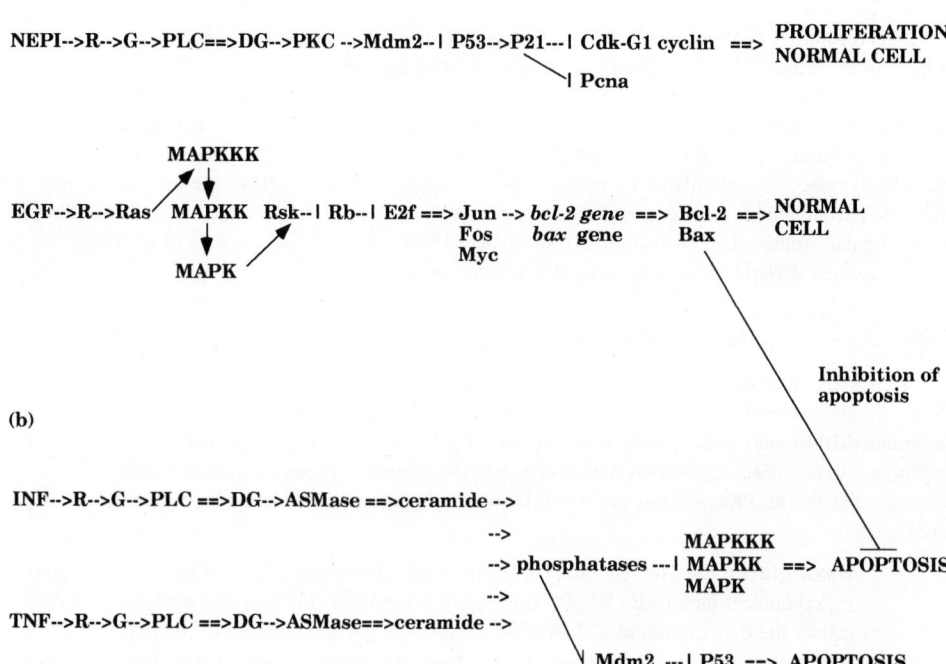

(a)

NEPI-->R-->G-->PLC==>DG-->PKC -->Mdm2--∣ P53-->P21---∣ Cdk-G1 cyclin ==> **PROLIFERATION NORMAL CELL**

⎺∣ Pcna

MAPKKK

EGF-->R-->Ras MAPKK Rsk--∣ Rb--∣ E2f ==> Jun --> *bcl-2 gene* ==> Bcl-2 ==> **NORMAL CELL**
Fos *bax* gene Bax
Myc

MAPK

Inhibition of apoptosis

(b)

INF-->R-->G-->PLC ==>DG-->ASMase ==>ceramide -->

-->

--> phosphatases ---∣ MAPKK ==> **APOPTOSIS**
MAPKKK
MAPK
-->

TNF-->R-->G-->PLC ==>DG-->ASMase==>ceramide -->

∣ Mdm2 ---∣ P53 ==> **APOPTOSIS**

Fig. 11-8. Signal transduction pathways leading to normal and apoptotic cells. Life or death is determined by the strength of signals such as those induced by norepinephrin (NEPI) and epidermal growth factor (EGF) that keep signal transduction pathways operating to inhibit apoptosis, as well as by the strength of signals (such as interferon-γ and tumor necrosis factor) that stimulate signal transduction pathways to induce programmed cell death (apoptosis). (--> = stimulates, ==> = yields, --∣ = inhibits). (*a*) Norepinephrine (NEPI) binding to its cell-surface receptor (R) initiates a signal. Mdm2 inhibits P53 so that another inhibitor, P21, is not synthesized. Consequently, the cyclin-dependent kinase (Cdk-G1 cyclin) and the proliferating cell nuclear antigen (Pcna) are not inhibited. The inhibition of P53 (required for apoptosis) and the presence of active Cdk-G1 cyclin and Pcna stimulate the proliferation of a cell and inhibit programmed cell death. Epidermal growth factor (EGF) binding to its receptor (R) initiates a different signal. Inhibition of Rb through phosphorylation frees the transcriptional factor (E2f) that promotes the synthesis of transcriptional activators (Fos, Jun, Myc). These transcriptional activators stimulate the production of Bcl-2 and Bax that are essential for the maintenance of a normal viable cell. Bcl-2 and Bax inhibit apoptosis. (*b*) Interferons (INFs) and tumor necrosis factors (TNFs) binding to their respective receptors (R) initiate signals that activate protein phosphatases. The phosphatases dephosphorylate various proteins. Dephosphorylation of the mitosis-activating protein kinases by phosphatases blocks the activation of transcriptional activators such as Fos, Jun, and Myc. This, in turn, blocks the transcription of the *bcl-2* gene. In the absence of Bcl-2 protein, apoptosis occurs. Dephosphorylation by phosphatases of the inhibitory protein (Mdm2) may result in active P53, which promotes programmed cell death.

propriate genes that specify inhibitors of apoptosis, e.g., Bax-Bcl-2 complex. Bax-Bcl-2 may limit the production or activity of nucleases that cut up DNA or the synthesis or activity of enzymes that produce toxic compounds.

11.17. When viruses infect animals, the infected cells often respond by producing various proteins that block the infection. One of the most important families of protein that inhibit viral proliferation are the interferons, which prevent viral reproduction in infected cells and prevent infection in noninfected cells. What are interferons and how many are there in humans?

Interferons (IFNs) are signal proteins, responsible for inducing the synthesis of proteins that inhibit the replication of viruses. The interferons are a type of **cytokine,** low-molecular-weight proteins involved primarily in communication between cells of the immune system. Type I interferons (interferon-α, interferon-β, and interferon-ω), encoded by genes with no introns, are synthesized and excreted by nearly all cells infected by viruses. On the other hand, type II interferons (interferon-γ) encoded by genes with introns, are synthesized almost exclusively by lymphocytes stimulated by antigens or mitogens. In humans there are at least 18 IFN-α genes, 6 of which are pseudogenes. **Pseudogenes** are closely related to active genes, but they are not preceded by a promoter or they contain mutations that make them inactive. There are 6 IFN-ω genes, five of which are pseudogenes. There is a single IFN-β gene and a single IFN-γ gene.

11.18.   How does a viral infection induce the production of interferons?

Both RNA and DNA viruses can induce interferon synthesis. Interferon induction requires double-stranded RNA. Single-stranded RNA viruses become double-stranded in the process of replicating their genomes. Double-stranded RNA may arise during transcription of a DNA virus if some of the transcripts acquire secondary structures. In any case, the dsRNA activates a dsRNA-dependent protein kinase (PKdsRNA). An important activity associated with PKdsRNA is phosphorylation of regulatory proteins that modulate the activity of transcriptional activators.

> **Example 11.7.**   The regulatory protein called inhibitor of κB (i_κB) binds a transcriptional activator called nuclear factor κB (NF$_\kappa$B), preventing it from functioning as an activator. PKdsRNA, however, causes the dissociation of i_κB from NF$_\kappa$B through phosphorylation of the inhibitor. This frees NF$_\kappa$B so that it is able to enter the nucleus and bind to enhancer sites that stimulate the expression of interferon (INF) genes.

The signal transduction pathway initiated by a viral infection involves a dsRNA-dependent protein kinase (PKdsRNA) that inactivates, through phosphorylation, the inhibitor (i_κB) of the nuclear transcriptional factor κB (NF$_\kappa$B). Free NF$_\kappa$B contributes to the activation of the INF-β gene. Transcription of INF genes requires a number of host activators, in addition to NF$_\kappa$B. In the signal transduction pathway below, the arrows indicate ==> = yields, --> = stimulates, --| = inhibits.

Virus ==> dsRNA --> PKdsRNA --| i_κB ==> NF$_\kappa$B --> INF-β gene ==> INF-β

RNA and DNA viruses induce interferon synthesis when viral antigens are presented on the surface of the infected cell. Infected cells and antigen-presenting cells (e.g., macrophages) can present viral antigens on type I and type II major histocompatibility complexes (MHC-I and MHC-II; see Chapter 13). Different T lymphocytes recognize MHC-I and MHC-II through the CD (cellular differentiation) proteins on their surface. The T lymphocyte receptors (TCRs) bind specifically to the antigens presented by the MHCs. Cytotoxic-T lymphocytes (T$_C$) with antigen-specific TCRs and CD8 receptors generally bind to infected cells with MHC-I presenting viral antigens. Helper-T lymphocytes (T$_H$), with specific TCRs and CD4 receptors, bind to macrophages with MHC-II presenting viral antigens. The binding of the MHCs and antigens to the CD and TCRs, respectively, initiates a signal in the T lymphocytes that results in the synthesis of IFN-γ. In the signal transduction pathway below, the meaning of arrows and symbols is ==> = yields, --> = stimulates, --| = inhibits. G = G protein, PLC = phospholipase-C, DG = diacylglycerol, IP$_3$ = inositol trisphosphate, PKC = protein kinase-C, TFs = transcriptional factors, IFN = interferon.

MHC-viral antigen --> TCR + CD receptors --> G --> PLC ==> DG + IP$_3$

IP$_3$ ==> Ca^{2+}

DG + Ca^{2+} --> PKC --> TFs --> IFN-γ gene ==> IFN-γ

IFN-γ secreted by T cells stimulates uninfected cells to synthesize more MHC-I molecules. This makes the immune system's attack on any cells that become infected much more efficient, since newly infected cells are capable of presenting to immune cells a high concentration of viral antigens as soon as the infection begins. This prevents viral synthesis through immediate cellular destruction due to the increased production of MHC-I molecules. In the signal transduction pathway below, GAF = interferon-γ activation factor, GAF(P) = active GAF because of phosphorylation, NF$_\kappa$B = nuclear transcriptional factor κB, IRF-1 = interferon regulatory factor, H-2Dd = heavy chain of MHC-I receptor.

$$\text{IFN-}\gamma \dashrightarrow R \dashrightarrow G \dashrightarrow ? ==> ? \dashrightarrow PK? \dashrightarrow GAF$$
$$\text{GAF(P)} + NF_KB + IRF\text{-}1 \dashrightarrow MHC\text{-I genes} ==> H\text{-2Dd}$$

11.19. A viral infection generally stimulates an RNase-L that hydrolyzes ssRNA viral genomes and viral mRNAs. How is this RNase-L stimulated?

Both RNA and DNA viral infections can induce the production of an enzyme called $2' \dashrightarrow 5'$ **poly-A synthetase** (PAS). There are at least four different types of PASs: monomers (p100), myristoylated dimers (2p69), large tetramers (4p40–46), and small tetramers (4p20–30). In the signal transduction pathways below, the arrows indicate ==> = yields, --> = stimulates, --| = inhibits, --||| = hydrolyzes.

$$\text{Virus} ==> \text{dsRNA} \dashrightarrow \text{PKdsRNA} \dashv i_KB ==> NF_KB \dashrightarrow \text{PAS gene} ==> \text{PAS}$$

The poly-A synthetases use ATP to synthesize **poly-A oligonucleotides** (PAONs), in which the nucleotides are connected through their $2'$ and $5'$ carbons.

$$\text{PAS} + \text{ATP} ==> \text{PAONs}$$

The PAONs stimulate the activity of a cellular **RNase-L**. RNase-L breaks down ssRNA and inhibits single-stranded RNA viruses by destroying the viral genome or template used to synthesize the genome. This enzyme inhibits some DNA viruses by decreasing the half-life of certain transcripts.

$$\text{PAONs} \dashrightarrow \text{RNase-L} \dashv\!\!\!\dashv\!\!\!\dashv \text{viral ssRNA}$$

Signal transduction pathways that activate RNase-L are important in the destruction of picornaviruses such as the polioviruses and the encephalomyocarditis virus (EMCV). Activation of the PAS/RNase-L pathway is not sufficient, however, to destroy ssRNA viruses such as vesicular stomatis virus (VSV) in the family Rhabdoviridae. The reason for this is unknown.

11.20. How do interferons block viral protein synthesis?

Because both IFN-α and IFN-β free NF_KB from its inhibitor, they are able to induce a number of genes that specify proteins involved in viral inhibition. INFs induce the synthesis of a protein kinase that is dependent upon dsRNA for its activation. This protein kinase (PKdsRNA, also designated PKeIF-2) is stimulated by viral infections to phosphorylate the α-subunit of eukaryotic initiation factor-2 (eIF-2), thus inhibiting it. This prevents the initiation of translation of both host and viral mRNAs.

$$\text{INF-}\alpha/\beta \dashrightarrow R \dashrightarrow G \dashrightarrow \text{PLC} ==> \text{DG} \dashrightarrow \text{PKC-}\eta \dashv i_KB ==> NF_KB$$
$$NF_KB \dashrightarrow \text{PKdsRNA gene} ==> \text{PKdsRNA}$$
$$\text{Virus} ==> \text{dsRNA} \dashrightarrow \text{PKdsRNA} \dashv \text{eIF-2} ==> \text{no cellular or viral translation}$$

Both influenza viruses in the family Orthomyxoviridae and vesicular stomatitis virus in the family Rhabdoviridae have their transcription inhibited by a protein known as Mx1. Mx1 is induced by the interferons INF-α and INF-β, once more through the freeing of NF_KB from its inhibitor.

$$NF_KB \dashrightarrow \text{Mx1 gene} ==> \text{Mx1} \dashv \text{cellular and viral transcription}$$

SIGNAL TRANSDUCTION AND THE CELL CYCLE

11.21. What signal transduction pathways are involved in the transition $G_0 \dashrightarrow G1$?

Nonproliferating cells are in the "no growth" (G_0) stage of the cell cycle. G_0 cells are expressing only those genes that are needed to maintain life and carry out any specialized function they normally have. Thus, chromosomes are mostly in the solenoid form (Chapter 3) during G_0. To begin the first growth (G1) stage of the cell cycle, cells must be signaled, usually by more than one growth factor.

> **Example 11.8.** Platelet-derived growth factor (PDGF), fibroblast growth factor (FGF), nerve growth factor (NGF), and transforming growth factor-β (TGF-β) stimulate the $G_0 \dashrightarrow G1$ transition by binding to receptors having tyrosine kinase activity.

The binding of certain growth factors to their respective receptors may stimulate the receptors' tyrosine kinase activities, resulting in autophosphorylation and in the phosphorylation of other kinases. Most of these secondary kinases are serine-threonine kinases. In some cases, activated receptors stimulate G proteins that subsequently stimulate enzymes that produce second messengers. The second messengers often stimulate serine-threonine kinases.

The serine-threonine kinases phosphorylate transcriptional activators or repressors, stimulating and inhibiting them, respectively. This results in the expression of a number of genes. In particular, the genes for G1 phase cyclins, CDPKs, RNA polymerase, DNA helicase, and DNA polymerase are expressed during G1. The accumulation of these proteins during G1 is essential for DNA replication in the S phase of mitosis.

The inactivation of transcriptional repressors is important during G1 also. An important transcription repressor is the retinoblastoma protein (Rb; p110) and a similar protein designated P107 (p107). The retinoblastoma protein, consisting of 928 amino acids, becomes highly phosphorylated near the end of G1 phase and at the beginning of S phase, ending up with 11 phosphate groups on serine and threonine residues. In its hypophosphorylated form, Rb associates with the transcriptional factors E2f and Drtf1 and removes them from circulation. Thus, Rb is not a "classical repressor" that binds to the DNA and blocks transcription by directly interferring with RNA polymerase binding.

Through its association with transcriptional factors, Rb negatively regulates the gene *c-fos* (which codes for the transcriptional activator protein Fos), the gene *c-myc* (which specifies a transcriptional activator with helicase activity that is necessary for DNA replication), the gene for **transforming growth factor-β** (TGF-β), and the gene encoding dihydrofolate reductase (Dhfr). Studies on Rb and E2f interaction suggest that hypophosphorylated Rb blocks transcription by sequestering E2f, thereby inactivating this tumor supressor. A signal transduction pathway that stimulates transcription by releasing E2f is summarized below. In the following signal transduction pathway, FGF = fibroblast growth factor and Rb = retinoblastoma protein.

FGF --> tyrosine kinase --> serine-threonine kinase --| Rb ==> E2f --> transcription

11.22. What is the importance of phosphorylation in the transition G1 --> S and what signal transduction pathways are involved?

G1 phase cyclins, synthesized during G1 in eukaryotic cells from yeasts (cyclin B) to animal cells (cyclins D, E, and A), are regulatory subunits for proteins called **cell division control kinases** (CDCKs), also known as **mitosis-promoting factors** (MPFs). In yeast, the catalytic subunit of the CDCKs are designated **Cdc** (for cell division control). Since G1 phase cyclins and a number of other cyclins synthesized in other phases of the cell cycle may stimulate Cdc activity, the catalytic subunit of the kinases is also known as a **cyclin-dependent kinase (Cdk).** The yeast G1 cyclin-Cdk complex is activated by phosphorylation and dephosphorylation of Cdk on certain threonines and tyrosines.

One model suggests that a G1 cyclin-Cdc kinase promotes the transition G1 --> S by phosphorylating H1 (histone-1) and Rb (retinoblastoma) protein. The phosphorylation of H1 causes the DNA to decondense from its solenoid form to its extended nucleosome form. The phosphorylation of the Rb protein causes the release of chelated (bound) transcriptional factors (e.g., E2f) required for binding of RNA polymerase to promoter regions. The freeing of transcriptional factors and activators from their inhibitors is followed by their binding to DNA. This promotes RNA polymerase binding to promoter sites and the synthesis of RNA primers at the origins of replication. This, in turn, leads to DNA replication. In some differentiated cells, the G1 cyclin-Cdc kinase may have to activate or inhibit other regulatory proteins before DNA replication can be initiated.

11.23. How is DNA replication limited to one round per cell cycle?

If eukaryotic cells initiated more than one round of DNA replication at the beginning of S phase, the cells would end up with multiple copies of each chromosome. This would be deleterious to the normal functioning of the cells. To prevent this, cells must have a mechanism that blocks the initiation of DNA replication after a round of DNA replication has been initiated. Research has shown that G1 cyclin-Cdk (cyclins D, E, and A in mammals) kinases called the cell division control kinases (CDCK-1) stimulate phosphatases that dephosphorylate H1 and Rb. This leads to condensation of the *ori* site and chelation of E2f. In addition, it is proposed CDCK-1 stimulates a complex of enzymes that attach the protein **ubiquitin** to target proteins, marking them for proteolysis. This results in the rapid destruction of certain G1 cyclins (yeast cyclin B; mammalian cyclins D and E). The inactivation of mammalian CDCK-1 (through the destruction of cyclins D and E) would allow Rb to become hypophosphorylated again and bind to E2f. This would stop further initiations of primer synthesis.

It has been reported that G1 cyclin (cyclin B, p36 in yeast) is required not only for the initiation of yeast DNA replication, but also for the elongation of DNA because a DNA polymerase-δ–cyclin B complex is required. Dephosphorylation of free cyclin B and/or its destruction by the ubiquitin system would help limit DNA replication to one round per cell cycle.

11.24. What protein, synthesized during G2, is required for progression through the cell cycle?

G2 is the second growth period and occurs after DNA replication (S phase). Continued signaling by growth factors, and by the new regulatory proteins synthesized during G1, turn on specific cyclin genes (cyclin A in yeasts; cyclins A and B in mammals). G2 cyclins accumulates during G2, increasing the activity of a new group of cell division control kinases (CDCK-2) that are activated by a sequence of phosphorylations and dephosphorylations comparable to those that activate the CDCK-1 kinases (Fig. 11-9). The substrate specificity of the CDCK-2 kinases differs from the specificity of the CDCK-1 kinases. CDCK-2 kinases phosphorylate proteins called lamins in the nuclear membrane, scaffold proteins in the nucleus, and a microtubule-associated protein kinase (MAP-kinase).

11.25. What stimulates the transition G2 --> M?

Phosphorylation of a chromosomal scaffold protein (believed to be topoisomerase) results in the condensation of the solenoid chromosome structure onto a scaffold. In one model for the condensed chromosome, the solenoid loops onto the scaffold 18 times, forming a disclike structure. These discs develop one next to the other until an entire chromosome is in its most condensed form, the looped-solenoid structure discussed previously (Chapter 3).

CDCK-2 --> topoisomerase ==> DNA condensation to the looped-solenoid form.

Also, CDCK-2 kinases are implicated in the phosphorylation of lamins A, B, and C. Lamins are proteins that bind the solenoid form of chromosomes to the inner nuclear membrane. Lamin B is associated with the inner nuclear membrane, whereas lamins A and C are bound to the chromosomes. Following lamin A and C phosphorylation, the chromosomes detach from the inner nuclear membrane. The phosphorylation of lamin B also stimulates the fragmentation of the nuclear membrane.

CDCK-2 --| lamins A, B, C ==> nuclear membrane fragmentation

==> chromosomes released from membrane

CDPK-2 kinases may also stimulate **caldesmon,** by phosphorylation, to chelate calcium ions. (Chelates are organic compounds that bind to inorganic ions, forming relatively stable, inert, water-soluble complexes). This leads to a low concentration in the cytoplasm of calcium ions, a signal that promotes polymerization of tubulin from the microtubule-organizing centers (MOCs) or centrosomes at each pole of the cell.

CDCK-2 --> caldesmon ==> decrease in Ca^{2+} --> tubulin polymerization

The phosphorylation of a microtubule-associated protein (MAP), called Tau stimulates polymerization of microtubules. It is believed that Tau is phosphorylated by a CDCK-2.

CDCK-2 --> Tau ==> Tubulin polymerization

The microtubules polymerize from the centrosomes toward the center of the cell. Some of the microtubules attach to the kinetochores at the centromeres on each of the chromatids of the duplicate chromosome. The growing microtubules push the chromosomes to the center of the cell, where the complementary microtubules are equally stable.

11.26. What signals promote cells to progress beyond metaphase of mitosis?

The progression from metaphase to anaphase requires the separation of the chromatids and the depolymerization of the microtubules. The molecular signals that initiate the separation of chromatids are unknown, but there are hints as to what is occurring. Separation of chromatids depends upon the rapid destruction of G2 cyclins (cyclin A in yeast; cyclins A and B in mammals) and the inactivation of CDCK-2. Presumably, the hydrolysis of G2

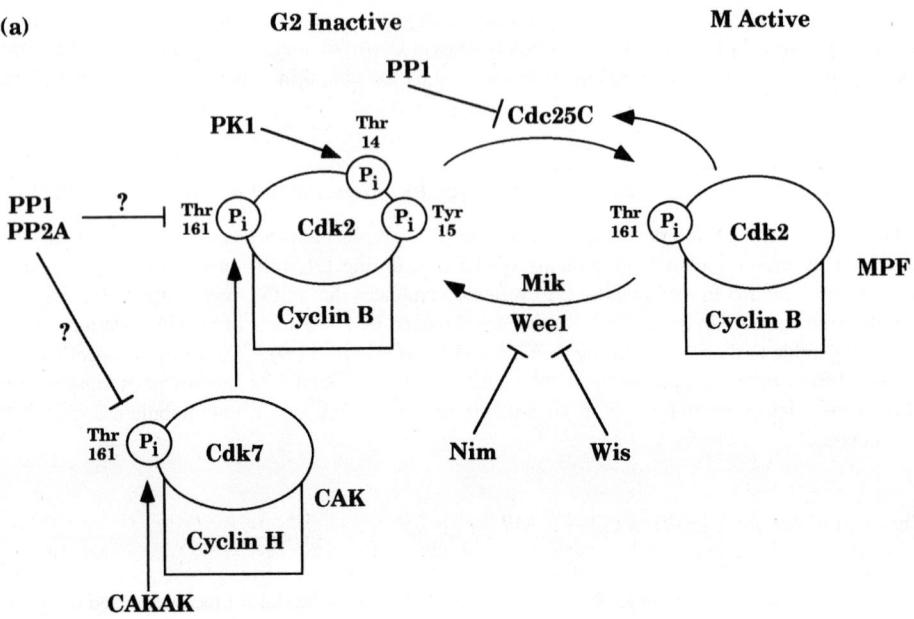

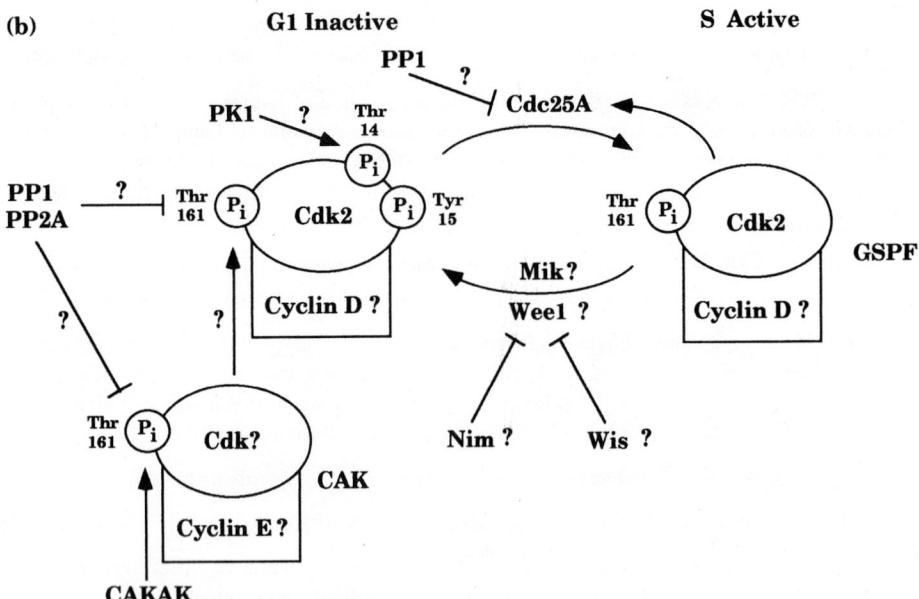

Fig. 11-9. (Caption on page 273.)

cyclins is carried out by the ubiquitin system activated by a CDCK-2. In addition, activation of phosphatases that dephosphorylate caldesmon and Tau promote the depolymerization of the microtubules.

Depolymerization of microtubules occurs at the kinetochores. One model of the kinetochore suggests that the kinetochore "walks" along a bundle of microtubules, destablizing the tubulin as it goes. Thus, the kinetochores and associated chromosomes move along a shortening microtubule bundle.

Dephosphorylation of proteins holding the chromatids together at the centromere may allow chromatid separation, whereas dephosphorylation of calcium-binding proteins and microtubule-stabilizing proteins may promote microtubule depolymerization and the movement of complementary chromatids to opposite poles.

Fig. 11-9. Activation and deactivation of cyclin-dependent kinases. (*a*) In vertebrates, when a cyclin-dependent protein kinase (Cdk2) becomes associated with cyclin B, the complex is called the mitosis promoting factor (MPF). This complex is activated by dephosphorylations in the transition from cell-cycle stages G2 to M. The phosphatase Cdc25C (cell division control) removes a phosphate from threonine 14 and another phosphate from tyrosine 15 on the Cdk2 subunit of MPF. Cdc25C is activated by the same MPF it activated, whereas Cdc25C is deactivated by an uncharacterized phosphatase PP1. The MPF may be deactivated by a phosphorylation on tyrosine 15. The phosphorylation is carried out by protein kinases related to Mik and Wee1 in yeast. Nim and Wis are proteins that inhibit Wee1. G2-inactive MPF is phosphorylated on a threonine 161 by Cdk7 complexed with cyclin H known as the *cyclin-dependent kinase activating kinase* (CAK). CAK is activated by a kinase called the CAK-activating kinase (CAKAK) upon phosphorylation of threonine 161. Protein phosphatases (PP1 and PP2A) may deactivate CAK and/or MPF by removing phosphates. An uncharacterized protein kinase (PK1) phosphorylates MPF on threonine 14, inactivating it. To progress into anaphase of mitosis from metaphase, the MPF must be inactivated. Inactivation occurs by the proteolytic breakdown of cyclin B and perhaps also by the dephosphorylation of Cdk2. (*b*) When a cyclin-dependent protein kinase (Cdk2) is associated with cyclin D, the complex is called the *G1 to S promoting factor* (GSPF). This complex is activated by dephosphorylations in the transition from G1 to S phases of the cell cycle. The phosphatase Cdc25A removes a phosphate from threonine 14 and another phosphate from tyrosine 15 on the Cdk2 subunit. Cdc25A is activated by the same GSPF it activated, whereas Cdc25A is deactivated by an uncharacterized phosphatase PP1. The GSPF is deactivated by a phosphorylation on tyrosine 15. It is postulated that the phosphorylation is carried out by protein kinases related to Mik and Wee1 in yeast. G1-inactive GSPF is phosphorylated on a threonine 161 by a Cdk complexed with cyclin E known as the *cyclin-dependent kinase activating kinase* (CAK). CAK is activated in this example also by a kinase called the CAK-activating kinase (CAKAK) by phosphorylation of threonine 161. Protein phosphatases such as PP1 and PP2A may deactivate CAK and/or GSPF by removing phosphates. An uncharacterized protein kinase (PK1) phosphorylates GSPF on its threonine 14, inactivating it. To progress into S phase from G1, the GSPF must be inactivated. Inactivation occurs by the proteolytic breakdown of cyclin D and most likely also by the dephosphorylation of Cdk2.

11.27. What signals promote the formation of new nuclei during late telophase of mitosis in animal cells?

Fragmented nuclear membranes in many cells are stored as stacks of endoplasmic reticulum. Some of the stored membranes still retain their pore complexes. The formation of nuclear membranes around the chromosomes at each pole during telophase is dependent upon the dephosphorylation of lamins A and C bound to the chromosomes and lamin B associated with the nuclear membrane vesicles. Dephosphorylated lamins A and C and the chromosomes function as "nucleating centers" (initiation centers) for the formation of a nuclear membrane. This information suggests that phosphatases, activated earlier by the cell division control kinase (CDCK-2), may be important agents for nuclear membrane synthesis.

11.28. What signals lead to cytokinesis (cell division) in animal cells?

Cytokinesis may begin in late telophase of mitosis or it may be delayed until mitosis is complete. The first sign of cytokinesis in animal cells is the invagination of the plasma membrane at the equator of the cell between the two newly formed nuclei. The invagination of the plasma membrane is powered by the tightening of an actin-myosin belt underlying the membrane.

If the actin-myosin belt functions like actin and myosin in smooth muscle, contraction of the belt is triggered by increased calcium ions at the equator and phosphorylation of a myosin light-chain kinase (MLCK). Calcium ions binding to calmodulin stimulate the movement of caldesmon so that myosin heads can associate with the actin filaments. The MLCK is phosphorylated by PKA and PKC, which are stimulated to activity by the synthesis of cAMP and IP_3, respectively. The MLCK, in turn, phosphorylates myosin light chains. This promotes myosin head interaction with actin free of caldesmon.

11.29. What signals cause the cell to enter G_0 rather than G1?

Repeated progression through the cell cycle (G1 --> S --> G2 --> M --> C --> G1 --> etc.) requires continual stimulation by growth factors. In normal cells, the signal transduction pathways promoting progression through the cell cycle can be blocked if similar cells contact each other. This phenomenon is known as contact inhibition. One model for contact inhibition proposes that, upon cellular contact, surface receptors on neighboring cells binding each other inhibit tyrosine kinases (c-Src) in the membrane. The result is the formation of **adhesion plaques** from cytoskeleton proteins just under the plasma membrane and a stable connection between the cytoskeleton and the plasma membrane. This condition blocks signal transduction pathways that stimulate a new round of the cell cycle.

It is known that tyrosine kinases (e.g., Src) on the cytoplasmic side of the membrane break the connection between the cytoskeleton and certain receptors in the membrane. This removes some of the constraints to cell division and locomotion, both of which require cell-shape changes. Adhesion plaques are regions where large numbers of connections are made between surface receptors and the cytoskeleton. Connections between surface receptors and the cytoskeleton also may contribute to adhesion plaques.

The presence of certain peptide signals and/or the lack of signal transduction may result in the activation or induction of proteins, such as P53 and Rb, that suppress the cell cycle. P53 is a transcriptional activator that induces the synthesis of P21. P21 inhibits Pcna and CDCK-1 kinases, and this, in turn, blocks the initiation of DNA replication. Similarly, Rb inhibits E2f, and this blocks the initiation of DNA replication.

In summary, contact inhibition constrains the membrane kinase, ensuring that signal transduction pathways are inactive. The lack of growth factors, inactivity of signal transduction pathways, and the accumulation of inhibitory proteins (P53, Rb) cause the cell to enter G_0. Growth factors that block contact inhibition, which stimulate the synthesis of cyclin A and activate kinases (affecting P53 and Rb activity or synthesis), cause the cell to begin the cell cycle once again.

CELLULAR TRANSFORMATION AND CANCER

11.30. How do signal transduction pathways become permanently altered and how does this lead to transformed cells?

The proteins that function in signal transduction pathways usually exist in one of two states: active and inactive. Their activity often is determined by the degree of phosphorylation and by accessory proteins such as the retinoblastoma protein.

The genes that specify signal transduction proteins are occasionally modified by mutations. Some mutations make signal transduction proteins nonfunctional, whereas others modify those domains involved in protein regulation so that the signal transduction proteins are permanently in the active state. Cells that produce nonfunctional signal transduction proteins generally die. Cells that produce signal transduction proteins that are always active behave inappropriately. They may turn on genes that should not be on, changing the cell's physiology, or they may cause the cell to repeatedly progress through successive cell cycles.

Many mutated signal transduction proteins, known as oncogene products, have been discovered in transformed and cancerous cells (Table 11.3). The cellular genes specifying the mutated signal transduction proteins are called oncogenes, whereas their normal cellular counterparts are referred to as proto-oncogenes. All oncogene products characterized to date fall into nine main catagories: (a) constitutively produced peptide hormones (Sis), (b) altered peptide hormone receptors (ErbB, Fms), (c) altered G proteins (Ras), (d) altered membrane or cytoplasmic tyrosine protein kinases (Src, Fes), (e) altered cytoplasmic serine-threonine protein kinases (Mos, Raf), (f) altered cytoplasmic hormone receptors (ErbA), (g) altered, nuclear, DNA-binding proteins and transcriptional activators (Fos, Jun, Myc), (h) altered cyclins (G1 cyclins), and (i) altered proteins that inhibit tumor-suppressor gene products or block their synthesis (Mdm).

Mutations in the proteins that negatively control signal transduction pathways (tumor-suppressor proteins such as P53 and Rb) may also lead to cellular transformation and the development of tumors and cancers (Table 11.4) because signal pathways are active that should not be. A number of proteins that negatively control signal transduction pathways have been found to be missing or abnormal in cancerous cells. When these regulatory proteins are replaced, transformed cells often regain some of their normal characteristics. The proteins that cure transformed cells are called **tumor-suppressor proteins** and include the retinoblastoma protein (Rb), the P107 protein (p107), the P53 protein (p53), the neurofibromatosis proteins (Nf1, Nf2), the Wilms tumor proteins (Wt1, Wt2), the *deleted in colon-rectal carcinoma* (Dcc) protein, the *familial adenomatous polyposis* (Fap) protein, and the *inhibin-α* (inh) protein. Figure 11-10 illustrates some of the oncogenes and tumor-suppressor genes involved in the development of colon-rectal (colorectal) carcinoma. The order of the mutations on the left in Fig. 11-10 is based on the frequency of finding a particular mutation at each stage of the cancer's development. Preliminary studies into how Rb, P107, P53, Nf, and Wt function suggest that they are each involved in negatively regulating

Table 11.3. Proto-Oncoproteins and Corresponding Oncoproteins

Type of Proto-Oncoprotein	Symbol	Oncoprotein
Hormones		
Platelet-derived growth factor	PDGF	Sis
Membrane Receptors		
Epidermal growth factor receptor	EGFR	ErbB
Macrophage colony stimulating factor receptor	MCSFR	Fms
Membrane Regulatory Proteins		
G proteins	G	H-Ras, N-Ras
		K-Ras
Membrane Kinases		
Tyrosine kinases	PKtyro	Src
Cytoplasmic Kinases		
Tyrosine kinases	PKtyro	Fes
Serine-threonine kinases	PKser-thr	Mos, Raf
Cytoplasmic Receptor		
Thyroid hormone receptor	—	ErbA
Nuclear Regulatory Proteins		
Transcriptional Activators	TA	Fos, Jun, Myc
Kinase Subunits (Activating)		
Cyclins	Cyclin	Cyclin
Inhibitors of Tumor Suppressor Proteins		
Inhibitors	Mdm	Mdm

Table 11.4. Tumor Suppressor Proteins

Symbol	Protein Name	Characteristics
Rb	Retinoblastoma protein (p110Rb)	Regulates E2f[*] Chromosomal location in humans: 13q14.1[†] Binds to promoter regions & inhibits transcription
P107	p107 protein	Regulates E2f and CDPKs[*]
P53	p53 protein	Regulates Cip (P21), P16, Ras (p21)[*], Myc[*], and DNA polymerase-α Chromosomal location in humans: 17p12−13 Binds to promoter regions & activates transcription
Wt1, Wt2	Wilms tumor proteins	Regulates transcriptional activators? Chromosomal locations in humans: 11p13 & 11p15. Binds to promoter regions
Nf1, Nf2	Neurofibromatosis proteins types 1, 2 (neurofibromin)	Regulates Ras (p21)[*] as a GTPase-activating protein (Gap)
Dcc (P190)	Deleted in colon-rectal carcinoma	Regulates cellular adhesion Chromosomal location in humans: 18q21
Fap	Familial adenomatous polyposis protein	Regulatory properties unknown Chromosomal location in humans: 5q21−22
Inh	Inhibin-α	Regulates signal transduction pathway by functioning as an inhibitory growth factor, like many members of the TGF-β family

[*]E2f = transcriptional factor, CDPK = cyclin-dependent protein kinase, Ras (p21) = monomeric G protein, Myc = DNA-binding protein believed to be a helicase, Inh = inhibins, which are family members related to transforming growth factor-β (TGF-β).

[†]The location of a gene on a human chromosome is given by the autosomal chromosome number (1 through 22), the arm (p or q), and the segment within the arm numbered from the centromere. Sometimes the segment is not known precisely and a range of segments is given.

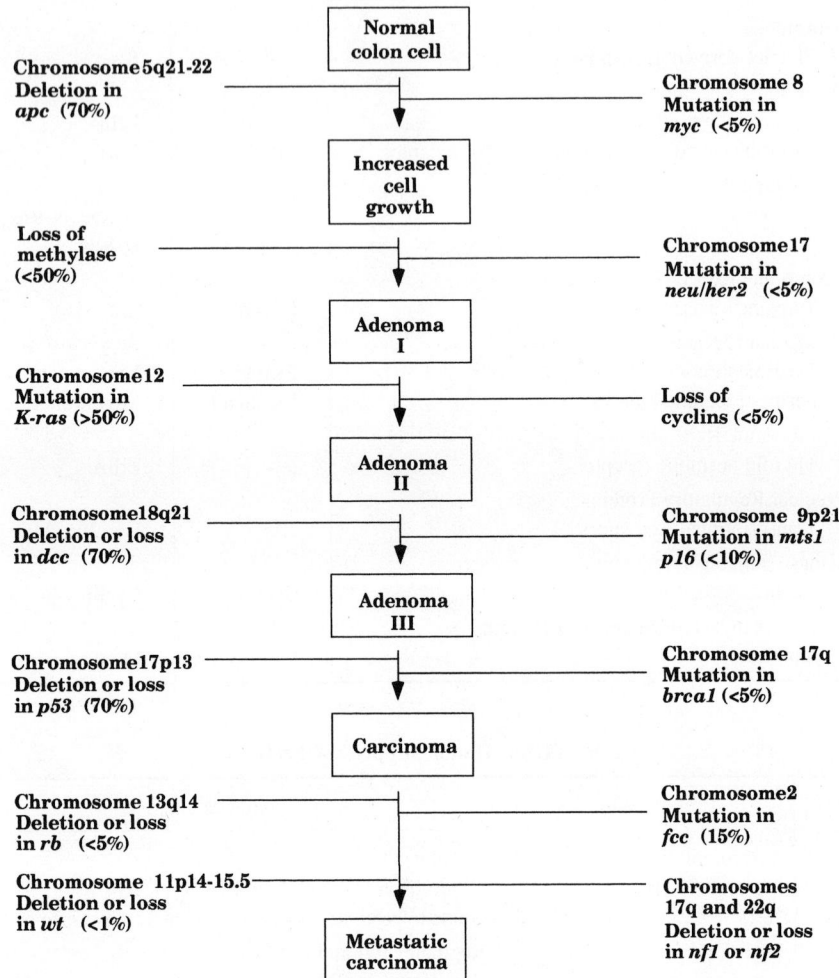

Fig. 11-10. Steps in the development of colon cancer. Normal human colon cells often begin the progression toward a cancer with a mutation in the tumor suppressor gene in a region known as the adenomatous polyposis complex (*apc*) located on the long arm (q) of chromosome 5, between segments 21 and 22. This mutation is found in more than 70% of colon cancers. Other mutations that are found in more than 70% of colon cancers include those in other tumor suppressor genes, e.g., *p53* on chromosome 17 and *dcc* (deleted in colon-rectal carcinoma) on chromosome 18. Mutations in a gene for methylase are found in about 50% of colon cancers, whereas mutations in *rb* and *wt* are infrequently found. Sometimes an entire chromosome is missing when tumor suppressor genes are not functioning. Mutations in the proto-oncogene *K-ras* are found in more than 50% of colon cancers. Mutations to a proto-oncogene (*fcc*) on chromosome 2 have been implicated in causing familial colon carcinoma. Other mutations in proto-oncogenes include those in cyclin genes (abnormal in <5% of colon cancers) on various chromosomes, *myc* (abnormal in <5% of colon cancers) on chromosome 8, and *neu/her2* (abnormal in <5% of colon cancers) on chromosome 17. An adenoma is a benign tumor of glandular epithelium in which the cells of the tumor are still arranged in a recognizable glandular structure. Adenomas I, II, and III represent the degree of disruption of the glandular structures. A carcinoma is a malignant growth of any type of epithelial cell that no longer forms recognizable tissues.

proteins: Rb and Wt inhibit transcriptional factors; P53 and P107 inhibit cyclins; P53 and Nf inhibit Ras proteins; and P107 inhibit kinases. Dcc appears to be an adhesion protein on the surface of the cell that is required for cellular stability (Table 11.4).

11.31. How do transformed cells develop into tumors and cancers?

Once a cell has acquired a mutation in a proto-oncogene or tumor-suppressor gene that stimulates the cell's proliferation, secondary mutations in other proto-oncogenes or tumor-suppressor genes occur that deregulate cells in the mutant population. These secondary mutations promote the survival and proliferation of many different types of deregulated cells (Fig. 11-10).

Spontaneous mutations in the various proto-oncogenes can convert them into oncogenes that affect many signal transduction pathways. The recombination of a proto-oncogene to an active promoter region may increase components of signal transduction pathways and disrupt their normal functioning. The maintenance of duplicate chromosomes can also increase proto-oncogenes, so that too much of their product is produced. Tumor-suppressor genes may be eliminated by spontaneous mutations or by the loss of entire chromosomes or segments of chromosomes where these genes reside.

Objective Questions

Multiple Choice

Directions: Choose the one best answer.

1. A light receptor in certain bacteria resembles that found in the eyes of animals. What is the bacterial light receptor called? (*a*) Photochrome (*b*) chlorophyll (*c*) bacteriorhodopsin (*d*) rhodopsin (*e*) none of the above

2. An enzyme that phosphorylates a protein is known as a (*a*) protein kinase. (*b*) phosphatase. (*c*) kinase. (*d*) phosphorylase. (*e*) none of the above.

3. The protein that has a cytoplasmic domain functioning as a tyrosine kinase is the receptor for (*a*) EGF (*b*) PDGF (*c*) insulin (*d*) all of the above (*e*) none of the above

4. Which subunit(s) of a G protein generally stimulates neighboring pores and enzymes? (*a*) α (*b*) β (*c*) γ (*d*) β and γ (*e*) α, β, and γ

5. What does a guanine nucleotide release factor (GNRF) do? (*a*) Inhibits Ras (*b*) stimulates Ras (*c*) inhibits trimeric G proteins (*d*) stimulates trimeric G proteins (*e*) none of the above

6. What does a GTP-activating protein (GAP) do? (*a*) Inhibits Ras (*b*) stimulates Ras (*c*) inhibits trimeric G proteins (*d*) stimulates trimeric G proteins (*e*) none of the above

7. What is an example of a GTP-activating protein (GAP)? (*a*) Retinoblastoma protein (*b*) P107 (*c*) Wilms tumor protein (*d*) neurofibromatosis protein (*e*) cyclin-A.

8. Lipid-A stimulates the synthesis of which factor? (*a*) Platelet-derived growth factor (PDGF) (*b*) epidermal growth factor (EGF) (*c*) transforming growth factor (TGF) (*d*) tumor necrosis factor (TNF) (*e*) none of these

9. Infections by RNA viruses generally result in the formation of double-stranded RNA during genome replication. Double-stranded RNA stimulates the enzymatic activity of (*a*) interferon. (*b*) double-stranded RNA-dependent protein kinases. (*c*) phospholipase-C. (*d*) sphingomyelinase. (*e*) tumor necrosis factor.

10. Interferon-γ induces the synthesis of a surface protein that presents antigens to T lymphocytes. What are the antigen-presenting molecules that stimulate cytotoxic-T cells? (a) Antibodies (b) Fc receptors (c) type I major histocompatibility complexes (MHC-I) (d) type II major histocompatibility complexes (e) none of the above

11. Which factor(s) *directly* disrupt the proliferation of viruses? (a) RNase-L (b) interferons (c) tumor necrosis factor (d) poly-A synthetases (e) interferon regulatory factors

12. The active yeast CDCK-1 (Cdk-cyclin B complex) phosphorylates histone-1 (H1). This causes the (a) DNA to decondense to an extended nucleosome form. (b) Rb protein to release E2f. (c) synthesis of transcriptional factors. (d) ubiquitin system to degrade cyclin B. (e) none of the above.

13. After the initiation of DNA replication, the *ori* sites are "closed" by (a) dephosphorylating H1 so that the DNA condenses. (b) dephosphorylating Rb so that transcriptional activators such as E2f are no longer free. (c) destroying cyclin B so that DNA polymerase-δ is unable to begin synthesis of the leading strand. (d) all of the above. (e) none of the above.

14. The active CDCK-2 (Cdk-cyclin A complex) phosphorylates lamins, topoisomerase, and microtubule-associated proteins. This causes (a) the chromosomes to detach from the nuclear membrane. (b) the nuclear membrane to fragment. (c) the solenoid DNA to condense on matrix proteins. (d) microtubules to polymerize from microtubule-organizing centers or centrosomes. (e) all of the above.

15. What happens to the Cdk-cyclin A complex at metaphase? (a) Nothing (b) only Cdk is degraded (c) only cyclin A is degraded (d) both Cdk and cyclin A are degraded (e) none of the above

16. Into which of the following categories do the tumor suppressor proteins fit? (a) Inhibitors of G proteins (b) surface receptors that inhibit signal transduction pathways (c) proteins that sequester and consequently inhibit transcriptional activators (d) all of the above (e) none of the above

True−False

1. Light receptors in the rod and cone cells of the eye are called phytochromes.

2. Peptide hormones and neurotransmitters bind to cytoplasmic receptors.

3. Serine-threonine protein kinases phosphorylate proteins on either serines or threonines.

4. The acetylcholine receptor in skeletal muscle is a sodium ion pore.

5. Some acetylcholine receptors open potassium ion pores through an intermediary known as a G protein.

6. G proteins may stimulate or inhibit proteins.

7. There are a number of protein kinases designated PKCs. Some are stimulated by diacylglycerol alone, some by just calcium ions, and still others by both DG and Ca^{2+}.

8. Phosphorylase kinase from mammalian muscle consists of four catalytic subunits, four calcium binding subunits, and eight regulatory subunits that are phosphorylated by a PKA.

9. Both pathways stimulated by lipid A result in the stimulation of a nuclear transcriptional factor known as NF_KB.

10. In the signal transduction pathway stimulated by tumor necrosis factor, diacylglycerol (DG) stimulates an enzyme that breaks down sphingomyelin.

11. The active second messenger produced from the hydrolysis of sphingomyelin is ceramide.

12. Both lipid A and tumor necrosis factor cause NF_KB to dissociate from its inhibitor and thereby activate genes requiring this factor.

13. Large amounts of tumor necrosis factor cause cells to swell with water and rupture, a process known as necrosis.

14. One model for how tumor necrosis factor works suggests that it activates protein phosphatases, which block signal transduction pathways that maintain the synthesis of an inhibitor of apoptosis.

15. The effects of tumor necrosis factor and interferon-γ are minimized by growth factors such as EGF that maintain appropriate levels of inhibitors of programmed cell death.

16. Double-stranded RNA-dependent protein kinases (PKdsRNA) free NF_KB from its inhibitor i_KB.

17. Production and presentation of type I major histocompatibility complexes (MHC-I) with viral antigens lead to the lysis of the infected cell and the destruction of a viral factory.

18. Viral double-stranded RNA stimulates the synthesis of an enzyme known as poly-A synthetase. This enzyme synthesizes poly-A oligonucleotides that stimulate RNase-L. This RNase-L degrades viral ssRNA.

19. P21, induced by the tumor suppressor protein P53, binds the cell division control kinase (CDCK) and proliferating cell nuclear antigen (Pcna), sequestering and inhibiting them so that they are unable to initiate DNA replication.

20. DNA replication is normally limited to one round per cell cycle by the dephosphorylation of histone H1 and Rb.

Matching

Directions: Match each kinase (1–4 in **Column A**) with the effector molecule (*A*–*E* in **Column B**) that stimulates its activity.

Column A	Column B
1. Protein kinase-A (PKA)	*A.* cAMP
2. Protein kinase-C (PKC)	*B.* diacylglycerol
3. Protein kinase-G (PKG)	*C.* cGMP
4. Protein kinase-CC (PKCC)	*D.* calmodulin
	E. cyclin

Directions: Match each enzyme (5–9 in **Column A**) with the product (*A*–*E* in **Column B**) that it produces.

Column A	Column B
5. Guanylcyclase	*A.* cAMP
6. Sphingomyelinase (SMase)	*B.* DG and IP_3
7. Adenylcyclase	*C.* cGMP
8. Phospholipase-C	*D.* phosphatidylinositol 4,5-bisphosphate (PIP_2)
9. Phosphatidylinositol 4-phosphate (PIP) kinase	*E.* ceramide

Directions: Match each oncogene product (10–15 in **Column A**) with its general function (*A*–*E* in **Column B**).

Column A	Column B
10. Ras	*A.* peptide hormone
11. Fos and Jun	*B.* steroid hormone receptor
12. ErbA	*C.* epidermal growth factor receptor
13. ErbB	*D.* G protein
14. Myc	*E.* transcriptional activators
15. Sis	

Terms

Directions: Unless otherwise specified, each answer is a single word.

1. Light receptor molecules in the retina cells of the eye

2. A light receptor in certain bacteria resembles that found in the eye of animals. This light receptor is called _____ .

3. The family of signal molecules to which the compound acetylcholine belongs.

4. The term used for signals that stimulate signal transduction pathways.

5. Diacylglycerol, cAMP, and calcium ions that stimulate protein kinases are known as _____ . (2 words)

6. The protein that stimulates the release of GDP from Ras. (4 words)

7. The protein that stimulates the hydrolysis of GTP attached to Ras. (3 words)

8. The protein kinase that is stimulated by cAMP. (3 words)

9. The second messenger that opens calcium ion pores in the endoplasmic reticulum and in the plasma membrane. (2 words)

10. The general term for enzymes that remove phosphate groups from proteins.

11. The portion of bacterial endotoxins that causes the synthesis and release of tumor necrosis factor from macrophages.

12. Tumor necrosis factor binding to its receptors stimulates a phospholipase-C that hydrolyzes phosphatidylcholine. What is the name of the active second messenger derived from phosphatidylcholine?

13. The term used to describe programmed cellular death.

14. Most virus infections stimulate the production of a family of proteins that block viral proliferation. These inhibitory proteins are known as _____ .

15. The enzyme that hydrolyzes single-stranded RNA.

16. The enzyme that inhibits transcriptional initiation factor eIF-2 through phosphorylation is known as _____ . (4 or 6 words)

17. The catalytic subunit of the CDPK that the cyclins stimulate. (4 words; 3 letter acronym)

18. The transcriptional factor E2f is sequestered in the nucleoplasm and inhibited by which tumor suppressor gene product? (2 words)

19. A protein that causes a cell to become transformed or cancerous.

20. Proteins the absence of which leads to cellular transformation or cancer. (2 or 3 words)

Answers to the Objective Questions

Multiple Choice

1. *c* 2. *a* 3. *d* 4. *a* 5. *b* 6. *a* 7. *d* 8. *c* 9. *b* 10. *c* 11. *a*
12. *a* 13. *d* 14. *e* 15. *c* 16. *d*

True–False

1. F (rhodopsins) 2. F (plasma membrane receptors) 3. T 4. F (calcium ion pore) 5. T 6. T
7. T 8. T 9. T 10. T 11. T 12. T 13. F (cause cells to shrink and fragment, a process
known as apoptosis) 14. T 15. T 16. T 17. T 18. T 19. T 20. T

Matching

1. *A* 2. *B* 3. *C* 4. *D* 5. *C* 6. *E* 7. *A* 8. *B* 9. *D* 10. *D* 11. *E*
12. *B* 13. *C* 14. *E* 15. *A*

Terms

1. rhodopsin 2. bacteriorhodopsin 3. neurotransmitters 4. agonists 5. second messengers
6. guanine nucleotide release factor 7. GTPase-activating protein 8. protein kinase A (PKA)
9. inositol 1,4,5-triphosphate (IP_3) 10. phosphatases 11. lipid A 12. diacylglycerol 13. apoptosis
14. interferons 15. RNase-L 16. double-stranded RNA-dependent protein kinase 17. cyclin-dependent
protein kinase (Cdk) 18. retinoblastoma protein (Rb or P110) 19. oncoprotein 20. tumor suppressor proteins

Chapter 12

Development in Multicellular Organisms

INTRODUCTION

Ontogeny is the process that leads to the development of a multicellular organism. Although the terms ontogeny and development often are used interchangeably, ontogeny is a much more specific term, referring to key steps in the development of multicellular organisms. The 19th century embryologist Ernst Haeckel coined the phrase, "ontogeny recapitulates phylogeny"; i.e., the evolutionary origin of organisms is retraced (in a modified way) in their embryological development.

> **Example 12.1.** Early human embryos develop gills and a two-chambered heart like those found in adult fish.

The mechanisms by which a single cell (zygote) develops into a multicellular organism with numerous specialized cells, tissues, and organs have been studied in detail during the 20th century. Traditionally, animal embryologists have focused on elucidating the origin and fate of three **primary germ layers:** the **ectoderm,** which develops into nerves and the outer layer of the skin; the **endoderm,** which produces the lining of the gut and associated organs; and the **mesoderm,** which is composed of the cells and tissues that lie between the ectoderm and endoderm and gives rise to connective tissues, bone cells, blood cells, and muscle cells. It was not until the late 1980s, however, that most of the very different patterns of development in animals were found to have an underlying similarity.

Ontogeny in most animals and plants is under the control of a hierarchy of genes that turn each other, as well as other, nondevelopmental genes, on and off in a sequential manner. The outcome of this gene activity is the development, in most cases, of a segmented animal or a plant with different adult tissues. Some of the genes controlling segmentation have been found in animals ranging from earthworms to mammals. One group of such genes, which determine the characteristics of each segment, are found in organisms as diverse as arthropods and mammals. DNA analysis indicates that many of these segmentation genes have been around since the nonsegmented animals (roundworms such as the nematodes) and the segmented invertebrate animals (earthworms and arthropods) diverged. The study of genes that control the development of adult tissues in plants has lagged behind similar studies in animals. Even so, we now know that master genes controlling the development of segmentlike regions exist and what their effects are. Plant development is discussed at the end of this chapter.

Upon superficial examination, fish, amphibians, reptiles, birds, and mammals may not appear to be segmented like earthworms and arthropods, but they are. Many of the same genes, or closely related genes, that direct segmentation in arthropods have been shown to control segmentation in vertebrates. The genes that direct segmentation in a vertebrate's body (head, cervix, thorax, abdomen, lumbar, tail) also are involved in the development of limbs. This discovery emphasizes the fact that limbs are segmented appendages.

A number of gene families, working in conjunction with the genes that dictate segmentation, control the development of the head and tail that are formed from terminal segments. Still other types of genes control the development of the ventral and dorsal axis of an organism. The genes that are involved in segmenting the body of an animal are referred to as **segmentation genes.** Those genes that control the anterior and posterior of an organism are known as **terminal genes,** whereas the genes that determine the characteristics of the ventral and dorsal regions are called **ventral-dorsal genes.**

To understand the importance of segmentation in the development of most multicellular animals, this chapter will begin with a discussion of the genetics and development in the fruit fly *Drosophila,* which has been extensively studied.   Not all insects develop like *Drosophila* (see Question 12.2).   The development in some vertebrates (mammals, amphibians, and birds) will be discussed in light of what has been discovered in insects.   A knowledge of the genetics underlying ontogeny has already expanded our understanding of embryology.   This bonding of genetics and classical embryology is in sharp contrast to early ideas in the field. Thomas Hunt Morgan, one of America's pioneers in classical embryology and genetics, made a distinction between the fields when he wrote in his 1926 edition *The Theory of the Gene,* ". . . genetics deals with the transmission of genes, embryology studies the development of potentialities into adult structures."   Presently, an understanding of genetics goes hand-in-hand with an understanding of embryology.

DEVELOPMENT IN AN INSECT (THE FRUIT FLY *DROSOPHILA*)

12.1.   Briefly, trace the development of *Drosophila* from oocyte to adult.

Drosophila begins its development when an unfertilized egg, or **oocyte,** in the female is fertilized by a sperm. Oocytes are produced within a long chain of **egg chambers** that constitute the ovary.   The fertilized **egg** laid by the female is only 0.5 mm long.   Within the egg, an **embryo** develops into a **larva,** hatching 24 hours after fertilization.   The larva's prime function is to consume nutrients and grow.   As it grows it molts its outer "skin" twice over a period of 4 days.   The larval stages between molts are called **instars.**   About 5 days after fertilization, the third larval stage migrates out of its food, attaches itself to a dry surface, and undergoes changes called **pupation.** The larva with its hard outer coat (or case) is referred to as the **pupa.**   The larva within the pupa undergoes a metamorphosis over a period of 4 days.   During pupation, pockets of cells known as **imaginal discs** contribute to the final development of the head parts, eyes, appendages (antennae, legs, wings, halteres), gut, exoskeleton (not shown), and genitalia (Fig. 12-1).   Notice that there are pairs of imaginal discs on each side of the midline of the larva for most of the adult structures.   One exception is an imaginal disc on the midline that develops into the genitalia.   Approximately 9 days after fertilization, an adult fly, or **imago,** emerges from the pupal case.   The adult lives for several weeks.   Its major function in perpetuating the species is to mate with a member of the opposite sex.

12.2.   What is the difference between holometabolous and hemimetabolous insects?

Flies such as *Drosophila,* moths, wasps, bees, and beetles are **holometabolous insects** that develop all their segments at about the same time.   All the segments of the final organism are present in the syncytial blastoderms (discussed in Question 12.4), while the embryo is still essentially a single cell.   When the embryos finally hatch, they become eating, growing, and molting caterpillarlike organisms (larvae).   The larvae develop hard coverings within which they (pupas) develop into adults.

Grasshoppers, locusts, and cockroaches are **hemimetabolous insects;** their embryos develop by the sequential addition of segments to the posterior end.   These insects can develop a fully formed embryo from the posterior half of the syncytial blastoderm.   Hemimetabolous insects have no pupal or caterpillarlike stages, but develop directly during embryonic formation into organisms **(nymphs)** that superficially resemble adults.   These nymphs eat, grow, and molt their exoskeleton a number of times until they become adults.

12.3.   How do *Drosophila* oocytes develop and become polarized?

The *Drosophila* monoploid (haploid) oocyte develops from a diploid **germ cell** in the fly's tubelike ovary. The germ cell divides by mitosis to become 16 daughter cells, one of which becomes the monoploid oocyte through meiosis.   The remaining 15 cells differentiate into polyploid **nurse cells** with polytene chromosomes. The origin of the diploid **follicle cells** that surround the posterior region of the oocyte is not known (Fig. 12-2). The development of the oocyte requires approximately 12 hours.

The oocyte is covered by an extracellular protein matrix called **vitelline** and a **chorionic membrane.**   Nurse cells and follicle cells that surround the oocyte are involved in establishing **polarity** (anterior–posterior axis and ventral–dorsal axis) in the oocyte and in the vitelline that surrounds the unfertilized egg.

Messenger RNA (mRNA) secreted by nurse cells is acquired by the oocyte through cytoplasmic connections located at its anterior end.   Because these mRNAs come exclusively from the female, they are referred to as **maternal mRNAs.**   They will code for **maternal proteins.**   Some maternal mRNAs are transported through the

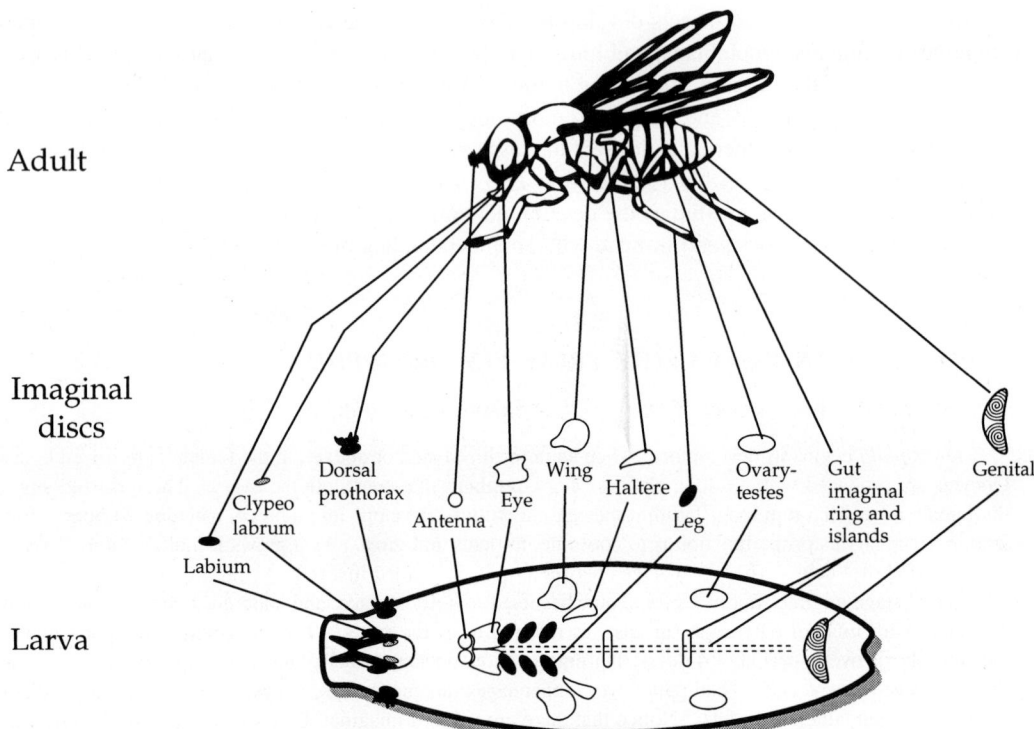

Fig. 12-1. The location of imaginal discs in a *Drosophila* larva. Imaginal discs found in insect larvae are packets of cells that develop into structures associated with the adult (imago) insect. Each imaginal disc is programmed during early embryonic development, depending upon its location, to become a specific adult structure (e.g., six imaginal discs, one for each leg). If an imaginal disc is experimentally moved to a new location in the larva, the cells will attempt to develop into the adult structure that would have developed at the old location. In some cases there is a single imaginal disc for an adult structure (e.g., genital imaginal disc).

oocyte's cytoplasm to the posterior region of the oocyte, whereas others are restricted to the anterior region (Fig. 12-2). This initiates the development of the anterior–posterior axis.

> **Example 12.2.** The mRNA for the protein bicoid is found in the anterior of the oocyte, whereas the mRNA for the protein **nanos** is attached to the posterior of the egg. The bicoid and nanos mRNAs are not translated until fertilization occurs. Some of the other maternal mRNAs are expressed, however, before fertilization and, consequently, are the first to establish an anterior–posterior axis in the oocyte.

Nurse cells also secrete other mRNAs, some that code for the proteins required for establishing the ventral–dorsal axis of the embryo after fertilization. In addition, follicle cells secrete a number of proteins into the ventral vitelline that are required for the establishment of the ventral–dorsal axis. Nurse cells and follicle cells at the poles secrete still other factors into the vitelline. These factors are required for the development of the terminal segments into a head or tail.

12.4. What important genetic events occur upon fertilization of the oocyte?

A sperm cell fertilizes the oocyte by penetrating its anterior. After the haploid nuclei (oocyte and sperm) fuse, certain genes are expressed that stimulate the new diploid nucleus to proliferate rapidly. At the same time, the repressors blocking the translation of maternal bicoid and nanos mRNAs are inactivated. The proteins bicoid and nanos are produced in the anterior and posterior of the egg (fertilized oocyte), respectively, and divide the egg into two zones. Bicoid and nanos affect transcription in the nuclei that are in their zones of influence.

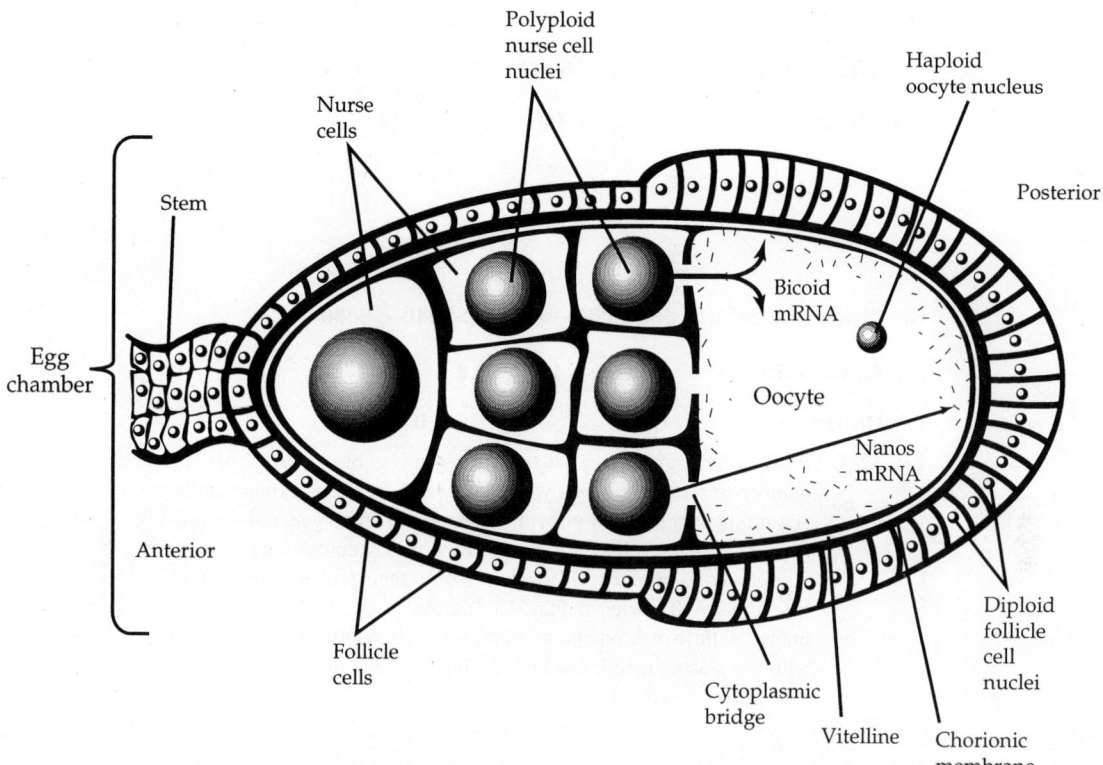

Fig. 12-2. A single egg chamber in a *Drosophila* ovary. A germline cell in this portion of a *Drosophila* ovary divided (by mitosis), producing 16 cells ($1 \rightarrow 2 \rightarrow 4 \rightarrow 8 \rightarrow 16$). Fifteen of the cells differentiated into polyploid nurse cells (only 7 are shown) with enlarged nuclei; the other one became the haploid oocyte through meiosis. It is believed that meiosis produces only one full-sized oocyte. The other products of meiosis are small and degenerate. The nurse cells and oocyte are surrounded by diploid follicle cells. The nurse cells in contact with the anterior portion of the oocyte supply it with maternal mRNAs that are important in the early development of the embryo. The mRNA diffuses through a cytoplasmic connection between the nurse cells and the oocyte. The oocyte is released from this portion of the ovary and is fertilized by a sperm penetrating the anterior region.

Bicoid activates new (previously inactive) genes and represses the expression of nanos mRNA in the anterior of the egg, whereas nanos activates still other new genes and represses the expression of bicoid mRNA in the posterior of the egg.

During a 1.5-hour period, the nuclei proliferate in the absence of any cell division. This turns the egg into a **syncytial blastoderm,** an egg containing multiple nuclei. A **syncytium** is a cell containing numerous nuclei, which in other organisms may be due to the fusion of many cells. A **blastoderm** is that portion of a fertilized egg (developing embryo) that gives rise to the various tissues of the embryo. Because the generation time for the nuclei is only 10 minutes, over 500 nuclei accumulate in 1.5 hours (Fig. 12-3*a*).

12.5. What happens to the syncytial blastoderm?

About 80 minutes after fertilization, most of the nuclei (about 256) begin to migrate to the **cortex** (outer layer of cytoplasm) of the cell, just under the plasma membrane. The nuclei continue to replicate as they migrate to the cortex. Three or four nuclei in the blastoderm's posterior end enter cells forming at the surface of the plasma membrane and become **pole cells.** Pole cells will provide the future **germ cells** (sperm and oocytes) in the adult.

Following the formation of pole cells, most of the remaining nuclei (about 380) in the cortex divide four times and then enter cells forming at the surface of the plasma membrane. These cells form a tight layer of cells,

(a) Nuclei   1-->2-->4-->8-->16-->32-->64-->128-->256-->512

Generations   1   2   3   4   5   6   7   8   9

Hours   0   0.5   1   1.5

(b) Nuclei or cells   380-->760-->1520-->3040-->6080

Generations   10   11   12   13

Hours   1.5   2.0

Fig. 12-3. Proliferation of nuclei in *Drosophila's* syncytial blastoderm. (*a*) The number of nuclei in a syncytial blastoderm increases exponentially. Assuming synchronous growth and divisions, nine generations produce 512 nuclei. Since each generation is 10 minutes long, it takes 90 minutes to produce 512 nuclei. (*b*) As their proliferation continues, about 380 nuclei of the 512 migrate toward the plasma membrane. A little over one hour later, more than 6000 nuclei associated with the plasma membrane are enveloped by the plasma membrane. This results in a 6000-cell blastoderm.

or **exoderm,** surrounding the remaining egg cytoplasm and nuclei. This structure, known as the **cellular blastoderm,** consists of approximately 6000 cells (Fig. 12-3*b*). The central volume of the egg, which contains a few nuclei, is referred to as the **yolk.**

12.6. The cellular blastoderm is a highly segmented structure. What genetic events leading to segmentation occur prior to the development of the cellular blastoderm?

Prior to cellular blastoderm formation, the syncytial blastoderm becomes segmented, undergoes terminal development, and establishes a ventral–dorsal axis. Segmentation prior to cellular blastoderm formation is controlled by a hierarchy of gene families (Table 12.1).

Table 12.1.   Genes That Control the Development of Segments in *Drosophila*

Family	Name of Gene (Abbreviation)
Maternal genes	*bicoid (bcd), nanos (nos)*
Gap genes	*hunchback (hb), kruppel (kr), knirps (kni), giant (gnt)*
Pair-rule genes	*runt (rnt), hairy (hry), even-skipped (eve), fushi tarazu (ftz)*
Terminal genes	*huckebein (hkbn), tailless (tll)*
Homeotic genes	*labrum (lab), deformed (dfd), sex combs reduced (scr), antennapedia (antp), ultrabithorax (ubx), abdominal-A (abd-A), abdominal-B (abd-B), caudal (cau)*
Segment polarity genes	*engrailed (eng), patched (ptc), wingless (wgl)*

Example 12.3.   Maternal genes are those expressed by cells of the egg chamber and involved in the very early development of the oocyte and syncytial blastoderm. **Gap genes** are so named because most mutations in these genes delete large groups of segments. **Pair-rule genes** are so named because

each pair of segments from anterior to posterior expresses one of two genes. **Segment polarity genes** are characterized by the fact that they differentiate the anterior portion of a segment from the posterior portion of a segment. **Homeotic genes** are defined as those that cause one segment to differ from a neighboring segment.

In segmentation prior to cellular blastoderm formation, maternal gene products bicoid and nanos turn on gap genes. Gap genes, in turn, regulate the expression of pair-rule genes. The embryo begins to undergo terminal development during this time. Even homeotic genes are turned on by pair-rule gene products during the end stages of syncytial blastoderm development. Segment polarity genes are activated by pair-rule genes and gap genes; however, translation of segment polarity genes is delayed until after cellular blastoderm formation.

12.7. When does the cellular blastoderm undergo gastrulation and what does this entail?

About 3.5 hours after fertilization, ectoderm in the anterior, ventral, and posterior regions of the cellular blastoderm begins to invaginate. A ringlike invagination in the anterior region called the **cephalic furrow** differentiates the head region from the rest of the body. Invaginating cells behind the head in the ventral region form the **ventral furrow** along the length of the body. The invaginating cells from the ventral ectoderm give rise to the two other primary germ layers, the endoderm and mesoderm. The newly formed endoderm defines the gut. The formation of the gut and the development of the three primary germ layers is referred to as **gastrulation.** The mesoderm provides cells that will form connective tissue such as muscle and blood vessels.

12.8. What is neurulation and when does it occur?

Neurulation in *Drosophila* is the development of the first ventral nerves, that arise from ectodermal cells and that develop after the gut-forming cells. Neurulation begins as the neurogenic ectoderm, consisting of a row of cells **(neural germ band),** subducts under the ventral ectoderm. The neurogenic cells reproduce and differentiate into various neurons and glial cells (support cells) that will form the ventral nerve cord of the insect.

12.9. What characteristics do the segmentation genes and their proteins share?

Many of the segmentation genes, or more correctly, the proteins they encode, establish segments by restricting the expression of genes within their zones of influence. In most cases, the segment-inducing proteins bind DNA and regulate transcription. These DNA-binding proteins fall into at least four groups based on the protein domains that bind to the DNA (Table 12.2; see also Chapter 3, Table 3.3).

The largest group of DNA-binding proteins have sequences of amino acids that fold into a helix-turn-helix domain (HTH domain). The HTH domain of a protein allows it to bind to DNA. The portion of the gene that codes for the HTH domain is known as the **homeobox.** The homeobox consists of 180 base pairs (bp) located near the end of the gene. Some genes with homeoboxes are referred to as **homeotic genes** (homeogenes) because these genes differentiate homologous segments in adult flies. Mutations that alter or destroy homeotic genes often result in the conversion of nonhomologous segments into nearly homologous segments in adults. The products of homeotic genes are called **homeotic proteins.** The portion of the protein that binds to DNA and is specified by the homeobox is called the **homeodomain.**

> **Example 12.4.** A combination of mutations (*abx, bx,* and *pbx*) in the *ultrabithorax* gene converts thoracic segment T3 into a segment that resembles the preceding segment, T2. The result is a fly with two pairs of wings instead of one pair of wings and one pair of halteres (Fig. 12-4*a*). Thus, this homeotic gene is important in the differentiation of homologous segments into nonhomologous segments. A single mutation *(bxd)* in the ultrabithorax gene converts a thoracic parasegment into one that resembles the preceding parasegment (parasegments are defined following Example 12.7). This results in only half a segment being like the preceding half segment (Fig. 12-4*b*).

The discovery of nonhomeotic genes, such as *bicoid,* with homeoboxes, which are necessary for the presence of more than one segment, indicates that homeotic genes cannot be defined simply as those that contain homeoboxes.

Genes that are required for the presence of one or more segments rather than the differentiation of one segment into another are not true homeotic genes.

Table 12.2.   Families of DNA Binding Protein Found in *Drosophila*

Protein Family	Example[*]	Function
Helix-turn-helix[†]	Bicoid (Bcd)	Development of head region
	Nanos (Nos)	Development of tail region
	Gooseberry (Gb)	Development of segment polarity
	Even-skipped (Eve)	Development of odd parasegments[‡]
	Fushi tarazu (Ftz)	Development of even parasegments
	Hairy (Hry)	Development of parasegments & nerves
	Engrailed (Eng)	Polarization segments
	Labrum (Lab)	Differentiation of segments
	Deformed (Dfd)	Differentiation of segments
	Sex Combs Reduced (Scr)	Differentiation of segments
	Antennapedia (Antp)	Differentiation of segments
	Ultrabithorax (Ubx)	Differentiation of segments
	Abdominal (Abd)	Differentiation of segments
	Twist (Tws)	Ventral & germ layer development
	Achaete-scute	Neurogenesis
	Daughter-less (Dl)	Neural & sex determination
Helix-zinc finger	Huckebein (Hkb)	Development of head & tail segments
(many gap genes)	Tailless (Tll)	Development of head & tail segments
	Hunchback (Hb)	Development of anterior segments
	Kruppel (Kr)	Development of middle segments
	Knirps (Kni)	Development of middle segments
	Ecdysone (Ecd) receptor	Necessary for pupation & metamorphosis
Helix & leucine zippers	Giant (Gt or Gnt)	Development of head and tail regions
β-ribbon	Runt (Rnt)?	Development of parasegments & nerves?

[*]Protein name is followed by its abbreviation in parentheses.
[†]Identical to helix-loop-helix homeodomain regions.
[‡]Parasegments are defined chemically by the presence of the pair-rule gene products Eve and Ftz. The posterior portion of a parasegment and the anterior portion of a following parasegment form the segments visible in adults. Segments are roughly defined by the pair-rule gene products Hry and Rnt.

Example 12.5.   Mutations in the gap genes *hunchback (hb)* and *kruppel (kr)* that have helix-zinc finger domains result in the loss of numerous segments, not the conversion of one segment into another like homeotic genes.

Note: in the discussions that follow, only those genes that are necessary for the "final" differentiation of segments and the development of adult structures will be referred to as homeotic genes. This is how homeotic genes are discussed in the textbooks and in the research literature.

A second group of DNA-binding proteins have one or more α-helices and amino acids bound to zinc ions. These domains allow proteins to bind DNA and are referred to as helices and zinc fingers (Table 12.2). *Hunchback (hb)* and *kruppel (kr)*, are *Drosophila* gap genes that have been reported to have zinc fingers. Most of the steroid hormone receptors have one or more helices and zinc fingers. The receptor for *Drosophila's* steroid hormone **ecdysone** is a helix–zinc-finger protein. Ecdysone stimulates pupation and metamorphosis.

Another group of DNA-binding proteins have a single bent-helix domain that binds to the major groove of the DNA. Generally, these proteins function as dimers. The dimers are held together by a second domain in each subunit that contains multiple leucines. This second domain that holds subunits of a regulatory protein together is known as the helix-leucine zipper (Table 12.2). The *Drosophila* gap gene *giant* belongs to the helix–lucine-zipper family.

A fourth group of DNA-binding proteins have a β-ribbon structure that allows them to bind DNA (Table 12.2). *Runt (rnt),* a *Drosophila* gene that functions in parasegment and nerve development, is a member of the β-ribbon group.

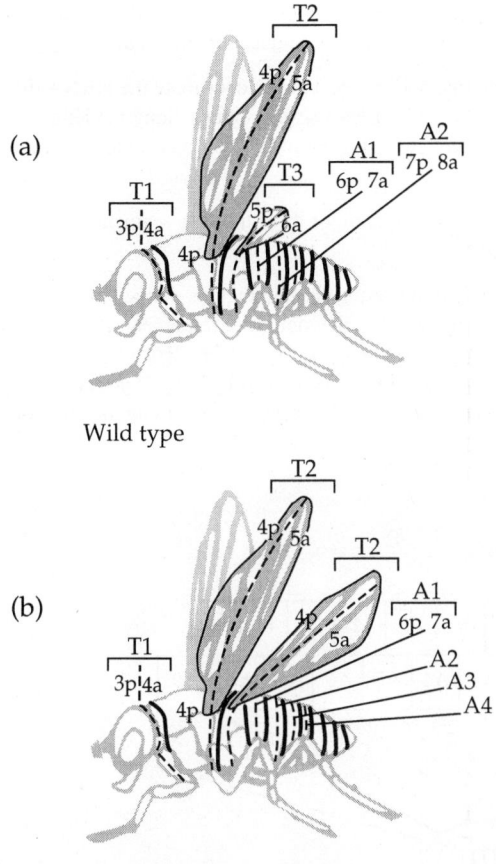

Wild type

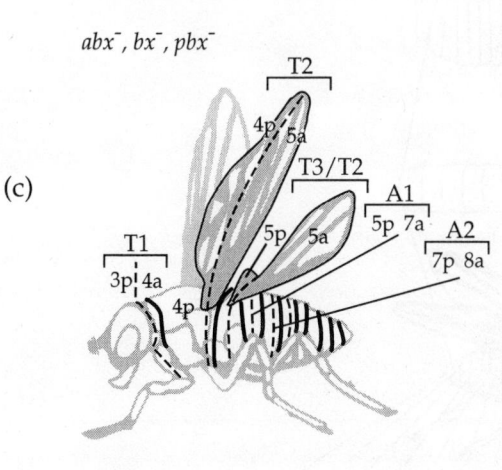

abx⁻, bx⁻, pbx⁻

bxd⁻

Fig. 12-4. Mutations in a homeotic gene affecting the development of imaginal discs. (*a*) Wild-type fly. T = thoracic segment; A = abdominal segment; p = posterior parasegment; a = anterior parasegment. (*b*) This fly has three mutations (*abx, bx, pbx*) in three controlling sites for the *ultrabithorax* (*ubx*) gene. Because of these mutations, the *ubx* gene is expressed differently in parasegments 5 and 6 so that wing development is no longer inhibited. Consequently, the imaginal discs straddling parasegments 5 and 6 develop into wings rather than halteres. (*c*) This fly has a single mutation (*bxd*) in a controlling site for the ultrabithorax (*ubx*) gene. Because of this mutation, the *ubx* gene is expressed differently in parasegment 6. In this case, the anterior portions of the halteres are still programmed to produce halteres, but the posterior portions are not. Thus a miniature wing forms from the posterior portion of the imaginal disc.

12.10. What are the different groups of segmentation genes and what effects do their mutations have on the embryo?

The development of segments is dependent on a hierarchy of genes expressed at different times and in different regions of the developing *Drosophila* embryo. The hierarchy of genes controlling segmentation is subdivided into as many as six families: maternal genes, gap genes, terminal genes, pair-rule genes, homeotic genes, and segment polarity genes (Table 12.1).

Maternal genes are those that are initially expressed from female transcripts only, generally because the mRNAs have been sequestered in the oocyte during its development. Maternal gene products divide the long axis of the fertilized egg in half. Loss of a maternal gene product results in gross abnormality in the embryo.

Example 12.6. Loss of bicoid in the anterior region results in the loss of a head and anterior segments, whereas loss of nanos causes a major shortening of the embryo and an abnormal tail.

Gap genes are generally activated after fertilization. Maternal and paternal genes contribute the mRNA that is transcribed. Gap gene products divide the syncytial blastoderm into at least seven regions along the length of the embryo. The gap genes are so named because mutations in any of them lead to the deletion of many contiguous segments and a drastic shortening of surviving embryos.

Example 12.7. Some mutations in *kruppel* result in the deletion of three thoracic segments and five abdominal segments (T1, T2, T3, A1, A2, A3, A4, A5).

Pair-rule genes are activated and repressed by gap genes, with the result that the regions defined by the gap genes are split into regions that alternately express the pair-rule genes. At this point in development, the embryo is divided along its length into at least fourteen regions called **parasegments** (PS; Fig. 12-5). Pair-rule genes are so named because they appear to work together in pairs, converting the 7 regions defined by the gap genes into 14 parasegments that alternate between expression of genes *even-skipped* and *fushi tarazu*. Some mutations in pair-rule genes result in the loss of 7 parasegments.

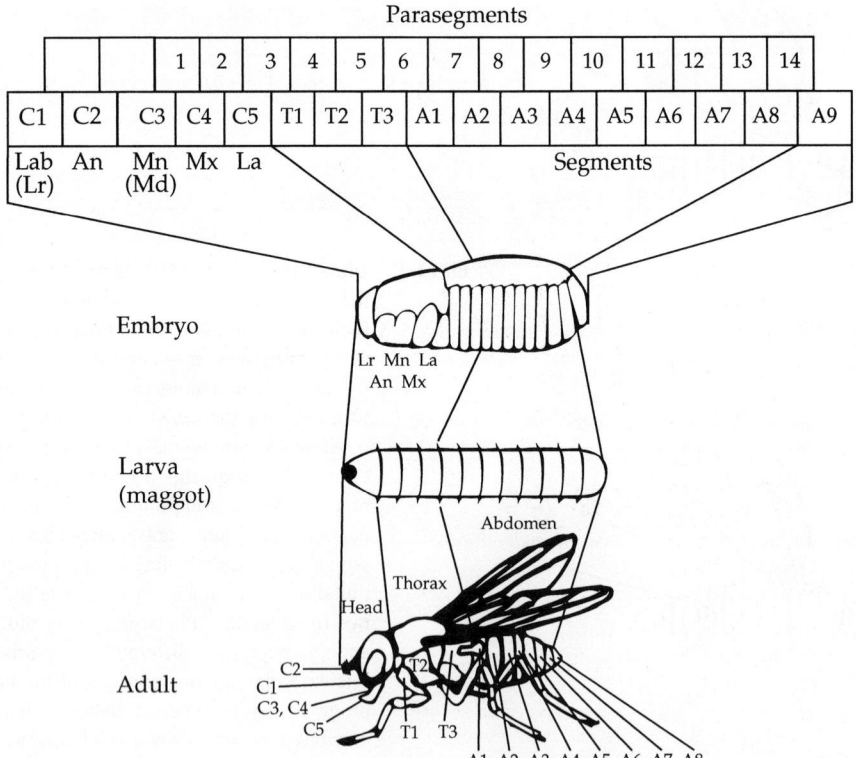

Fig. 12-5. The location of parasegments and segments in the *Drosophila* embryo, larva, and adult. There are 5 head segments (C1, C2, C3, C4, C5), 3 thoracic segments (T1, T2, T3), and 9 abdominal segments (A1–A9) in the embryo. The head segments are associated with head structures: C1 = labrum (Lab or Lr), C2 = antenna (An), C3 = mandible (Mn or Md), C4 = maxilla (Mx), and C5 = labium (La). The parasegments are numbered from 1 through 14 and are defined by the pair-rule gene products. There are more segments than parasegments because segments are defined by terminal genes as well as by segment polarity genes. For most of the segments, a posterior portion of one parasegment and the anterior portion of a subsequent parasegment constitute a segment.

Example 12.8. Mutations in the *fushi tarazu* (meaning "not enough segments" in Japanese) gene result in the loss of all even parasegments (PS2, 4, 6, 8, 10, 12, 14). The surviving embryos are half the size of normal embryos; they are missing head (PS2), thoracic (PS4), and abdominal parasegments (PS6, 8, 10, 12).

Terminal genes are activated by a protein factor in the vitelline. The same terminal gene products affect head and tail development in conjunction with maternal, gap, and pair-rule gene products.

Homeotic genes are believed to be activated by maternal, gap, and pair-rule gene products. The homeotic gene products are required for the development of imaginal discs into adult structures. They also define larval and adult **segments** (Fig. 12-5). Segments are composed of the posterior portion of a parasegment and the anterior portion of the adjoining parasegment.

Example 12.9. The maxillary (Mx) segment is composed of the posterior of parasegment 1 and the anterior of parasegment 2 (Fig. 12-5). Although homeotic genes contribute to the structures in the adult, they are initially expressed in the syncytial blastoderm.

Segment polarity genes are activated by maternal, gap, and pair-rule gene products. The segment polarity gene products further subdivide the 14 parasegments, giving them a polarity. They establish the anterior, middle, and posterior regions of each parasegment soon after the cellular blastoderm forms.

12.11. A hierarchy of genes is involved in *Drosophila* segmentation. How do these genes interact?

During oogenesis, nurse cells produce mRNA for the proteins bicoid (Bcd) and nanos (Nos). The mRNA for bicoid is sequestered within the anterior portion of the oocyte, whereas the nanos mRNA is transported through the oocyte and restricted to its posterior region. Upon fertilization, the repressors blocking transcription are eliminated and bicoid and nanos mRNAs are translated. It is believed that bicoid and nanos activate their own transcription (Bcd → Bcd and Nos → Nos), but mutually repress each other's transcriptions (Bcd ⊢⊣ Nos). Thus, bicoid is restricted to the anterior, whereas nanos is kept in the posterior of the developing syncytial blastoderm (Fig. 12-6).

The high concentration of bicoid in the anterior region of the syncytial blastoderm activates certain gap genes (*giant* and *hunchback*), whereas the high concentration of nanos in the posterior region activates other gap genes (*knirps* and *giant*). High concentrations of bicoid and nanos are believed to restrict the production of kruppel to the central region of the embryo, while medium or low concentrations promote its production. The maternal genes also function as repressors of transcription and/or translation.

Example 12.10. Nanos blocks the translation of *hunchback* transcripts (Fig. 12-6). This restricts hunchback to the anterior of the developing embryo.

The gap gene products (Gnt, Hb, Kr, Kni) also function as activators and repressors of transcription.

Example 12.11. A gradient of hunchback affects the expression of *kruppel, knirps,* and *giant*. High concentrations of the protein hunchback repress expression of the gene *kruppel* (Hb ⊣ *kr*), intermediate concentrations activate *kruppel* (Hb → *kr*), but low concentrations repress it again. Intermediate concentrations of the protein hunchback strongly repress the gene *knirps* (Hb ⊣ *kni*) but stimulate the gene *giant* (Hb → *gnt*). Kruppel and giant mutually inhibit each other's production (Kr I ⊢⊣ Gnt).

These interactions (as well as others) result in proteins Kr, Kni, and Gnt being restricted to separate bands in the anterior and posterior of the syncytial blastoderm.

It is believed that high concentrations of the gap gene products, as well as the terminal gene products tailless (Tll) and huckebein (Hkb), function as activators and repressors of transcription. It should be noted that Tll and Gnt are found at both ends of the developing organism. High concentrations of the terminal and gap gene products (Tll, Hb, Gnt, Hb, Kr, Kni, Gnt, and Tll), along the anterior-posterior axis respectively, activate the pair-rule gene *hairy (hry)*, whereas intermediate concentrations activate the pair-rule gene *runt (rnt)*. This would account for the 7 bands containing hairy that alternate with the 7 containing runt (Fig. 12-6). Hairy and runt are detected in alternating bands along the length of the embryo.

Gap gene and pair-rule gene products work together in a very complicated manner to activate and repress the second group of pair-rule genes *even-skipped (eve)* and *fushi tarazu (ftz)*. For example, Rnt activates *eve,* whereas Kr and Hry repress *eve* but activate *ftz*. In most cases, high levels of the gap gene products activate *ftz,* whereas low to intermediate levels of the gap gene products are required for *eve* activation. Parasegments are defined by

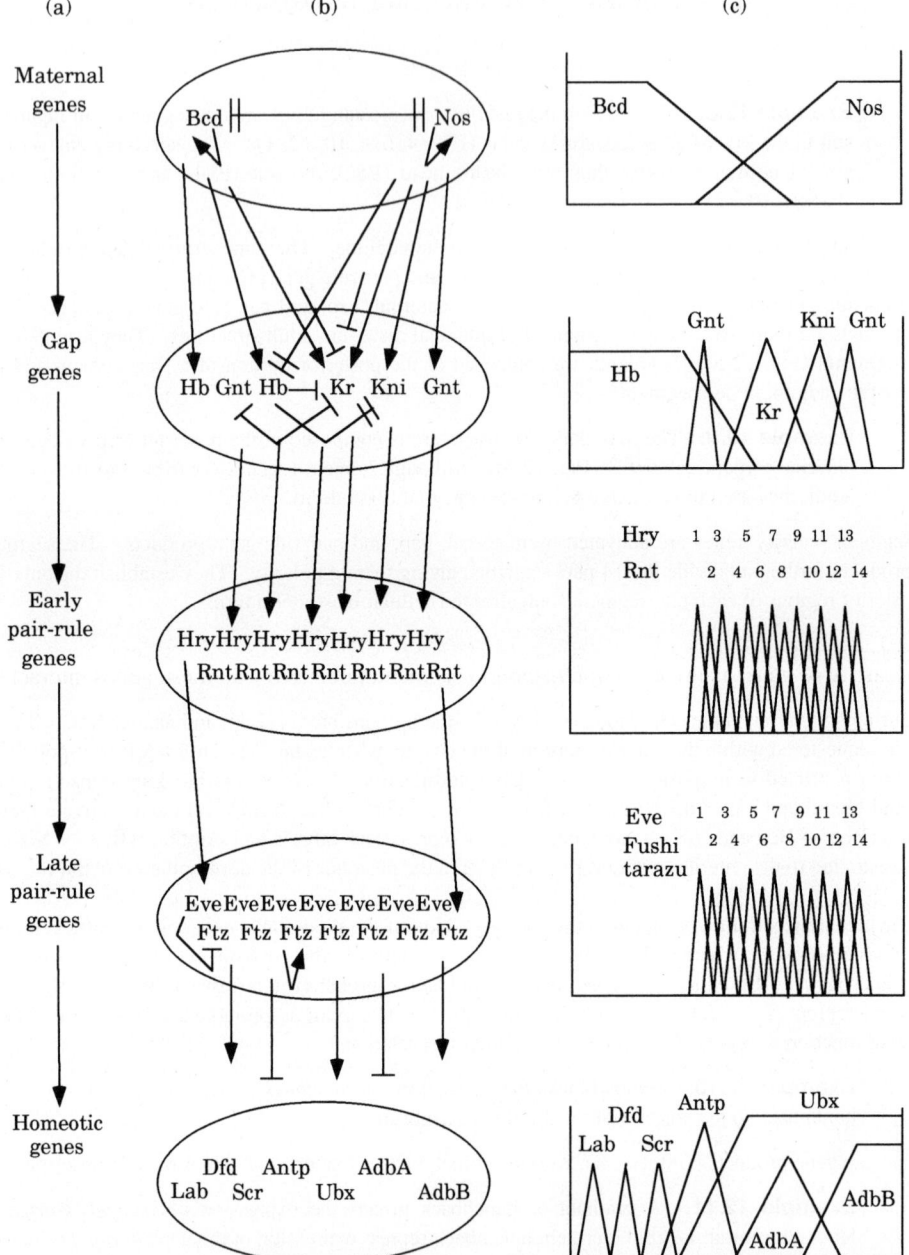

(a)   (b)   (c)

Maternal genes

Gap genes

Early pair-rule genes

Late pair-rule genes

Homeotic genes

Fig. 12-6. The hierarchy of genes that control segmentation in the *Drosophila* syncytial blastoderm and cellular blastoderm. (*a*) A hierarchy of genes: the maternal genes regulate the gap genes and the gap genes regulate the pair-rule genes, which, in turn, regulate the homeotic genes. (*b*) Arrows indicate how the various gene products regulate gene activity and restrict gene products to certain bands along the anterior–posterior axis. Gene activation or protein synthesis is indicated by the sharp arrow ($\rightarrow$), while gene inhibition is suggested by the blunt arrow ($\dashv$). A double arrow shows strong activation or repression ($\twoheadrightarrow$ or $\dashv\vert$). In some cases, low concentrations of a protein activate gene expression, whereas high concentrations inhibit. This situation is indicated by two different arrows ($\rightarrow$, $\dashv$) from Bcd to Kr. For protein symbols and names, see Table 12.2. (*c*) Protein gradients along the length of the blastoderm show how gene expression and gene products are restricted. Gradients of Bcd and Nos occupy each half of the blastoderm. Gradients of Gnt, Kr, and Kni are found in the posterior 2/3 of the blastoderm. Hry and Rnt alternate in the posterior 2/3 of the blastoderm. These 14 stripes roughly correspond to the segments that will form in the adult. Fourteen stripes of Eve and Ftz alternate in the posterior 2/3 of the blastoderm. These stripes define the parasegments.

the extent of expression of the pair-rule genes *eve* and *ftz,* whereas segments are defined roughly by expression of the pair-rule genes *hry* and *rnt.* It is believed that homeotic gene expression is initiated in the late syncytial blastoderm. Presently, there is little information regarding the controlling sites for homeotic genes. Their expression depends upon maternal, gap, and pair-rule gene products.

The pair-rule genes *even-skipped* and *fushi tarazu* stimulate a number of segment polarity genes in a very complicated manner. The result is the activation of *engrailed, patched,* and *wingless* in each parasegment from anterior to posterior. With maternal genes, gap genes, pair-rule genes, and even homeotic genes being expressed, the nuclei at the surface of the blastoderm become incorporated into cells. Clearly, the syncytial blastoderm is expressing the hierarchy of segmentation genes as it becomes a cellular blastoderm. Initially one cell expressing *engrailed,* two cells expressing *patched,* and one cell expressing *wingless* from front to back are found early in the development of each parasegment.

Although engrailed has been shown to be a DNA-binding protein, patched and wingless are thought to function as external signals (Chapter 11). Thus, complicated signal transduction pathways are required for the maintenance of polarity in the parasegments and segments.

12.12. At what point in development are the homeotic genes expressed?

Although homeotic genes are thought to regulate the development of imaginal discs into adult structures and give segments their adult appearance, most homeotic genes have been shown to be expressed in the 10-hour *Drosophila* embryo (Fig. 12-7). It is postulated that maternal, gap, and pair-rule genes are involved in the activation and repression of at least some homeotic genes much earlier, at the time of cellular blastoderm formation, 2.5 to 3.0 hours after fertilization. One reason for this belief is that protein Hb strongly represses genes *abd-A* and *abd-B.* In addition, Hb weakly represses the gene, *ultrabithorax (ubx),* that influences the development of thoracic and some abdominal segments. Also, an increase in Hb causes some abdominal segments to develop as thoracic segments.

12.13. How many homeotic genes are there, where are they expressed, and what do they do?

Homeotic genes in *Drosophila* have been mapped to chromosome 3 (Fig. 12-7), and a number of loci have been mapped within the homeotic genes: telomere—*abdB* [iab8, iab7, iab6, iab5], *abdA* [iab4, iab3, iab2], *ubx* [pbx, bxd, bx, abx], *antp, (ftz), scr, dfd, (zen), pb, lab*—centromere. As for the nonhomeotic genes *ftz (fushi tarazu)* and *zen (zerknullt), ftz* contributes to anterior–posterior axis formation, whereas *zen* is required for ventral–dorsal axis development. At one time, it was believed that the loci in *abd-B, abd-A,* and *ubx* designated different genes. For example, it was thought that the *abd-B* locus produced four gene products: infra-abdominal-8, 7, 6, and 5. Presently, there is some uncertainty about the number of proteins encoded by these complex regions. A current model suggests that there is only one gene product per region, but its concentration in each of the segments is important in determining that segment's character. The mutations that seem to define different genes affect controlling sites and the level of their expression. Figure 12-7*a* illustrates the extent of expression of each of the loci in a 10-hour *Drosophila* embryo.

Each of the homeotic genes (except for *pb*) is required for the appropriate development of one or more embryonic segments. In many cases, the deletion of a homeotic gene causes segments (or parasegments) to develop into segments that resemble more anterior segments.

> **Example 12.12.** Mutations that alter the level of *ultrabithorax* gene product (Bxd) result in the conversion of the anterior portion of abdominal segment A1 into the anterior portion of thoracic segment T3. The result is a fly with an enlarged and distorted haltere (Fig. 12-4*b*).

In contrast to mutations such as *ubx* that convert segments into more anterior segments, certain mutations in *antennapedia (antp)* convert antennae in the head into legs, appendages normally found on the more posterior thoracic segments.

12.14. What kind of genetic system guides the development of the terminal segments into a head or tail?

Although head and tail development requires many of the genes involved in segmentation, a set of genes called the terminal genes are necessary for proper head and tail development.

Terminal development is dependent on a maternal gene that codes for a signal peptide called torso-like. Torso-like is released into the vitelline matrix by nurse cells and follicle cells at each end of the oocyte. Upon fer-

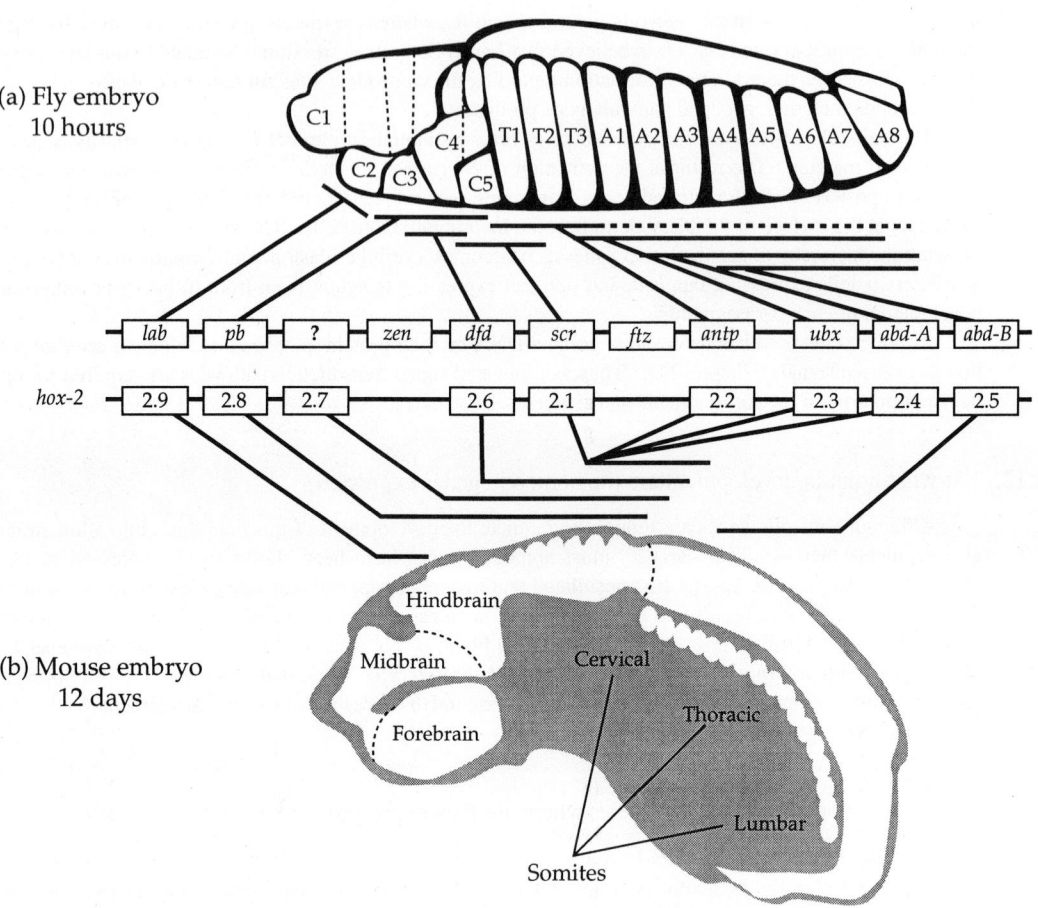

(a) Fly embryo 10 hours

(b) Mouse embryo 12 days

(c) Homeotic genes

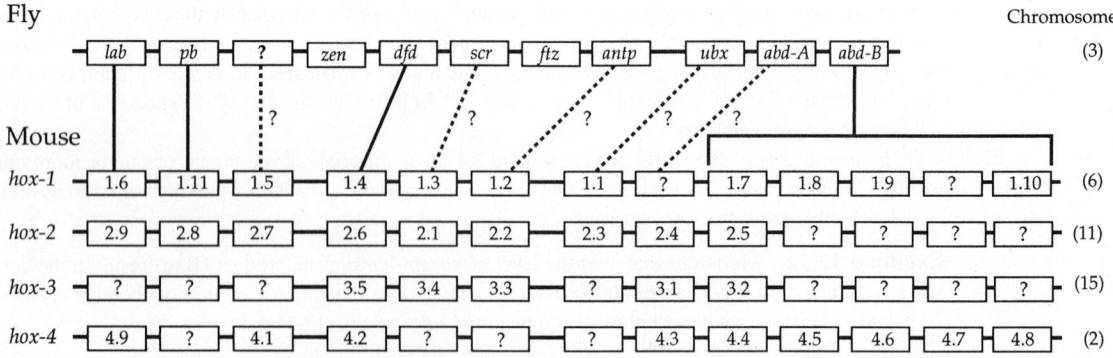

Fig. 12-7. (Caption on page 295.)

Fig. 12-7. Extent of homeotic gene expression along the anterior–posterior axis. (*a*) The extent of homeotic gene expression and the restricted presence of homeotic gene products (indicated by solid lines running the length of the embryo) in the 10-hour *Drosophila* embryo are illustrated. Dotted lines indicate the location of homeotic gene products produced at a later time or at very low concentrations. The anterior–posterior axis is from left to right. The order of the homeotic genes on chromosome 3 is the same as the order in which they are expressed in the embryo: *lab = labrum; pb = proboscis; dfd = deformed; scr = sex combs reduced; antp = antennapedia; ubx = ultrabithorax; abd-A = abdominal-A; abd-B = abdominal-B*. The genes *zerknullt* (*zen*) and *fushi tarazu* (*ftz*) shown on *Drosophila's* chromosome 3 are not homeotic genes. *Zerknullt* controls ventral–dorsal axis development, and its product is found in dorsal tissue. (*b*) The extent of homeotic gene expression and the restricted presence of homeotic gene products in the 12-day mouse embryo are shown. The mouse *hox* genes that are similar to the fly homeotic genes are compared. The *Drosophila* homeotic genes control the development of segments along the entire anterior–posterior axis. In contrast, the expression of the comparable *hox* genes are restricted to a portion of the head containing the hindbrain and a portion of the body including the cervical region. (*c*) *Drosophila's* single family of homeotic genes is compared to the mouse's 4 families of homeotic genes (*hox-1, hox-2, hox-3,* and *hox-4*). In some cases, the *hox* genes have proliferated to finely control normal development (*hox-4.8, hox-4.7, hox-4.6, hox-4.5,* and *hox-4.4*). The different families of *hox* genes reside on different chromosomes. The *hox* genes that control development of the more anterior regions are most sensitive to retinoic acid.

tilization, the egg produces torso, a membrane receptor with tyrosine kinase activity. Torso-like (signal peptide), found only at the poles, binds to torso (receptor) in the syncytial blastoderm. This sends a signal to the nuclei near the anterior and posterior that stimulates the production of the proteins tailless and huckebein. Huckebein is found only in the extreme terminal cells, whereas tailless extends further into the cellular blastoderm. These terminal gene products along with maternal and gap gene products direct the early development of the head and tail segments. This is demonstrated by mutations in segmentation and terminal genes.

> **Example 12.13.** Mutations in any one of the following genes: *bicoid* (maternal gene), *kruppel* (gap gene), and *even-skipped* (pair-rule gene) result in severely distorted head regions in embryos. The tail regions of embryos carrying any mutations in *bicoid, kruppel,* and *even-skipped* appears to be normal.

> **Example 12.14.** Mutations that eliminate *torso*, however, result in the complete elimination of the tail and some abnormalities in the head segments. In double mutants for *bicoid* and *nanos* (*bcd⁻nos⁻tor⁺*), the larvae that develop have a distorted tail-like structure at both ends.

These data suggest that the terminal genes produce a tail-like structure in the absence of segmentation genes. Segmentation genes and terminal genes working together convert the tail-like structure in the anterior region into a head, but convert the tail-like structure in the posterior region into a true tail.

12.15. How is ventral–dorsal polarity determined in *Drosophila?*

Ventral–dorsal differentiation begins during oogenesis. An extracellular signal peptide is produced by the oocyte, which remains associated with the vitelline membrane (that surrounds the oocyte) in an inactive form until it is activated and released into the vitelline matrix. Activation of the signal peptide occurs after fertilization by serine-threonine proteases that are released from the developing syncytial blastoderm. These proteases convert the inactive signal peptide to a protein called toll-ligand (Tl). Toll-ligand is restricted to the ventral vitelline matrix and consequently binds to only ventral receptors specified by toll (T). The toll receptor is part of a signal transduction pathway (Toll-ligand --> T --> G --> PLC ==> DG --> PKC), where G = trimeric GTP-binding protein, PLC = phospholipase-C, DG = diacylglycerol, and PKC = protein kinase-C. Because toll-ligand binds almost exclusively to toll in the ventral surface of the syncytial blastoderm, the highest concentration of active PKC is in the ventral region of the developing embryo.

Activated PKC separates a complex composed of the proteins cactus (Ct) and dorsal (Dl) through phosphorylation of cactus. Once freed, dorsal diffuses into ventral nuclei, where it activates certain genes and represses others. The lack of free dorsal protein in the dorsal region results in the expression of some genes not expressed in the ventral region. The consequence of this differential gene expression is the establishment of a ventral–dorsal axis.

12.16. What effect do mutations in the ventral–dorsal genes have on development?

> Mutations that eliminate dorsal protein result in highly distorted, segmented embryos that have no ventral tissue. They consist entirely of dorsal tissue, and gene products normally associated with the dorsal region are expressed throughout the embryo. Gastrulation, which normally occurs from the ventral surface, does not occur in the absence of gene products that should be in the ventral region.
>
> In contrast to mutations that inactivate dorsal, those that eliminate cactus protein result in a highly distorted embryo that has no dorsal tissue. The embryo becomes completely ventralized, and gene products associated with only ventral regions are expressed throughout.

DEVELOPMENT IN MAMMALS (HUMANS AND MICE)

12.17. What happens during the early embryonic development of mammals, particularly with regard to axis determination?

> There is no evidence that mammalian ovaries establish an anterior–posterior axis or a ventral–dorsal axis in oocytes. Similarly, there is no evidence that fertilization induces axes in the egg. As the fertilized egg moves along the fallopian tubes toward the uterus, it undergoes a number of mitotic divisions (cleavages), forming a hollow "ball" of cells called the **blastocyst.** The blastocyst becomes a polarized structure as an **inner cell mass** develops on a small portion of the inner "wall." What signals stimulate the growth of the inner cell mass is unknown. The pole containing the inner cell mass implants itself within the endometrium (uterine lining). In humans, this occurs about 6 days after fertilization. The precise way in which implantation occurs suggests that the blastocyst is highly polarized and that the inner cell mass has a ventral–dorsal axis. The dorsal surface of the inner cell mass is in contact with the blastocyst wall. Within two days, the inner cell mass becomes the **embryonic disc,** initially composed of two layers of cells. The dorsal layer of the embryonic disc has separated from the blastocyst wall, yet the embryonic disc remains attached along its circumference to the blastocyst wall. Although the embryonic disc has a ventral–dorsal axis at this point in development, it has not established an anterior–posterior axis.
>
> The **placenta** develops from cells that grow out from the blastocyst wall into the mother's endometrium. Initially, an **amniotic sac** develops over the embryonic disc and a **yolk sac** forms under the embryonic disc. The amniotic sac contains fluid that cushions the embryo; the yolk sac does not store yolk; however, blood vessels develop in the mesodermal covering of the yolk sac that transport nutritive fluids.

12.18. When is anterior–posterior polarity established in the early human embryo?

> The human embryo at 13 days after fertilization still consists of an embryonic disc with two layers of cells. The dorsal layer is the ectoderm, whereas the ventral layer is the endoderm. By day 14, some endodermal cells have elongated where the anterior (head) of the embryo will form. By day 15, a connecting stalk has developed from the yolk sac, below the posterior (tail) of the embryo, to extra-embryonic tissue. Thus, the anterior–posterior axis of the embryo has been established. Presently, little is known about the genes responsible. Biologists would like to know if genes like *Drosophila's* maternal genes (*bicoid, nanos*) and terminal genes (*torso, huckebein, tailless*) are involved in the development of vertebrates.
>
> The embryo develops its mesoderm tissue from day 15 to day 18 as dorsally located ectodermal cells migrate between the ectoderm and endoderm. The migration of ectodermal cells begins at the posterior end and progresses toward the center of the embryo. A shallow indentation in the ectoderm called the **primitive streak,** which later becomes deeper and forms the **primitive groove,** is the source of migrating cells. The primitive streak and groove develop only in the posterior half of the embryo. Ectodermal cells from the center of the primitive streak migrate forward to form the anterior mesoderm of the embryo.

12.19. When do the notochord and neural tube form?

> Beginning around day 16, the endoderm (ventral tissue) at the center of the human embryo invaginates. The invagination progresses from the center toward the anterior of the lengthening embryo. Complete invagination of the endodermal cells results in a stiff "rod" called the **notochord.**
>
> At the same time that the notochord is developing ventrally toward the embryo's anterior, the mesoderm is thickening and the ectoderm of the **neural plate** is folding upward to form the **neural folds** and a **neural groove** between the folds. The neural groove eventually extends along the entire length of the embryo. The neural folds

near the middle of the human embryo begin to fuse dorsally about day 21; this represents the beginning of the **neural tube.** The formation of the neural plate and neural tube is known as **neurulation,** and the embryo at this point is called a **neurula.**

12.20.   When does the human embryo become segmented?

The mesoderm on either side of the developing notochord and neural tube is also changing and segmenting into paired bodies called **somites.** Somites are cuboidal bodies of mesoderm that delineate parasegments. Thirty-three paired somites develop from day 20 to day 28. The neural tube, which forms at about the same time as the somites, is also divided into parasegments.

12.21.   What contributions do the somites make toward the embryo's development?

Paraxial mesoderm on either side of the notochord and neural tube gives rise to the paired somites, which in turn, provide the cells involved in the development of vertebrae, ribs, muscles connecting vertebrae, and connective tissue in the dermis of the skin. Cells in the lower middle portion of each somite (**sclerotome**) migrate toward the notochord and neural tube, where they secrete cartilage that forms the early **vertebrae.** The cartilage and bone define the segments. In the thoracic region, extensions of the vertebrae form the early ribs.

Another group of cells in the middle of each pair of somites (**myotome**) develop into cells called **myoblasts.** The myoblasts spread ventrally, where they form muscles on either side of the embryo's trunk. In addition, the myoblasts spread dorsally to produce muscle that connects the vertebrae together. These muscles define the parasegments of the embryo. Segmented muscles allow fish, amphibians, and reptiles to move their vertebral segments from side to side in a sinusoidal manner. Mammals locomote primarily by moving pairs of limbs rather than by moving the vertebral segments from side to side, and so most of the segmented muscles become fused during a later step of ontogeny. Various other cells in the paired somites (**dermatome**) also give rise to the **dermis** of the skin and connective tissue.

A number of experiments suggest that somites promote parasegment formation in the neural tube and in the **neural-crest cells,** found over the neural tube. The neural-crest cells subsequently develop into sensory nerves at the boundaries of each parasegment.

12.22.   How does the human gut develop?

The 22-day-old embryo consists of three layers of cells (ectoderm, mesoderm, and endoderm), a neural tube, a notochord, and 8 pairs of somites, but it lacks a gut. Gut development begins when a portion of the amniotic cavity above the embryo folds underneath the embryo. Folding under the anterior end results in the **head fold,** whereas folding under the posterior end yields the **tail fold.** Folding under the trunk of the embryo is known as **lateral folding.** Eventually, the amniotic cavity completely surrounds the embryo, except where the yolk sac and connecting stalk enter the embryo. The yolk sac and connecting stalk become the **umbilical cord.** Gut development and the placement of the heart in the ventral region of the human embryo are complete by day 30. The tissue that will form the heart originates in mesodermal tissue posterior to the head of the 22-day-old embryo.

12.23.   What is known about homeotic gene expression in the mammalian embryo?

Four families of homeotic genes in mammals are related to the homeotic genes in *Drosophila* (Fig. 12-7c). DNA analysis suggests that these four families of homeotic genes evolved by duplication from genes similar to those found in *Drosophila* sometime after vertebrates separated from invertebrates. The homeotic genes in mammals are called ***hox* genes,** and each family is designated by a number (1 through 4). The mouse *hox-1, hox-2, hox-3,* and *hox-4* families are found on chromosomes 6, 11, 15, and 2, respectively.

Both *Drosophila* and mouse homeotic genes have maintained their order during the millions of years they have been evolving. In addition, the homeotic gene order is related to the order of the segments where the homeotic genes are expressed. The genes at one end (such as *lab* = *hox-1.6, hox-2.9,* and *hox 4.9*) differentiate anterior segments, whereas those at the other end (such as *abd-B* = *hox-1.7, hox-2.5, hox-3.2,* and *hox-4.4*) affect posterior segments. The genes near *lab* are expressed before those near *abd-B*. In addition, the genes near *lab* show a more pronounced response to the growth factor **retinoic acid.**

Figure 12-7 illustrates how the homeotic genes in a 10-hour *Drosophila* and a 12-day mouse embryo are related and where their gene products are located. Human embryos around day 28 are comparable to mouse embryos at day 12.

12.24. How are the *hox* genes in mammals involved in the generation of limbs?

Vertebrate limbs are segmented in much the same way that the trunk of the body is. Thus, it is not surprising that homeotic genes are involved in differentiating vertebrate limb segments. Some of the same *hox* genes in vertebrates that control the segmentation of the body also determine limb development.

Developing limbs have an anterior–posterior axis, a ventral–dorsal axis, and a proximal–distal dimension (Fig. 12-8). In the very early limb bud, *hox* gene expression is segmented along the anterior–posterior axis. As the limb bud grows, *hox* gene expression become twisted so that *hox* gene products are found within diagonal stripes that cut across the limb bud.

Some of the mouse genes most closely related to *Drosophila's abd-B* gene (*hox-4.8, 4.7, 4.6, 4.5,* and *4.4*) are expressed in the segments along the lengthening limb. The expression of five *hox-4* genes defines the most distal posterior segment. Expression of only *hox-4.4* defines the most proximal and most anterior regions of the limb (Fig. 12-8*a*).

12.25. Retinoic acid is known to affect the development of limbs. What is the relationship between retinoic acid and the *hox* genes?

Retinoic acid is a ligand that binds to a cytoplasmic protein receptor. The retinoic acid receptor complex regulates transcription of some *hox* genes. This is demonstrated by placing a retinoic acid bead in the anterior portion of a mouse limb bud where only *hox-4.4* is normally expressed (Fig. 12.8*b*). Within a few hours, all the *hox* genes are expressed in an anterior segment where the retinoic acid was at its greatest concentration. Fewer *hox* genes are expressed where the retinoic acid concentration is low. Mouse limb buds treated with retinoic acid develop regions in the anterior that are nearly a mirror image of the regions in the posterior. This experiment suggests that retinoic acid or an analog may be the natural growth factor responsible for activating *hox* genes in the more distal regions of developing limbs. In addition, *hox* genes are activated by a gradient of the growth factor and this results in limb segmentation.

DEVELOPMENT IN AN AMPHIBIAN (THE TOAD *XENOPUS LAEVIS*)

12.26. What happens during the ontogeny of an amphibian like *Xenopus laevis?*

During oogenesis of Xenopus's oocytes (unfertilized eggs), a polarity known as the **animal–vegetal axis** is established (Fig. 12-9*a*). In this organism, the animal hemisphere of the oocyte, containing little or no yolk, is dark colored, whereas the vegetal hemisphere, with most of the nutritious yolk, is light colored. Penetration of the amphibian's oocyte by a sperm head occurs only in the animal hemisphere. Within 60 minutes after fertilization, the cytoskeleton detaches from the cortex and the cortex and plasma membrane rotate counterclockwise approximately 30° with respect to the cytoskeleton and cytoplasm (Fig. 12-9*b*). The **cortex** consists of a thick layer of peripheral proteins in the cytoplasm attached to the inner surface of the plasma membrane. The cortex has at least two functions: the strengthening of the plasma membrane and the binding of maternal mRNAs.

The rotation of the cortex and plasma membrane creates a region in which the animal pole cytoplasm is in contact with the vegetal cortex-membrane. This region is known as the **grey crescent,** since it often can be distinguished by color from the animal and vegetal hemispheres.

Fertilization and rotation of the cortex and plasma membrane trigger division of the zygote. Many rapid cell divisions over a period of about 7 hours result in a ball-like mass of cells called the **blastula** (Fig. 12-9*c*). In the blastula's animal hemisphere is a hollow, known as the **blastocoel.** The cells that cover the hollow constitute the ectoderm. The ectoderm will provide cells for the future **dorsal endoderm** and **dorsal mesoderm.** The vegetal pole provides cells for the future **ventral endoderm** and **ventral mesoderm.**

The grey crescent underlies the region that stimulates the process of gastrulation. The vegetal hemisphere is closely associated with the future ventral region of the embryo. **Gastrulation** forms a gut and defines the future posterior region of the embryo (Fig. 12-9*d*). Thus, gastrulation leads to the formation of the ventral–dorsal axis, as well as the anterior–posterior axis. Since the grey crescent always forms on the side opposite the place where

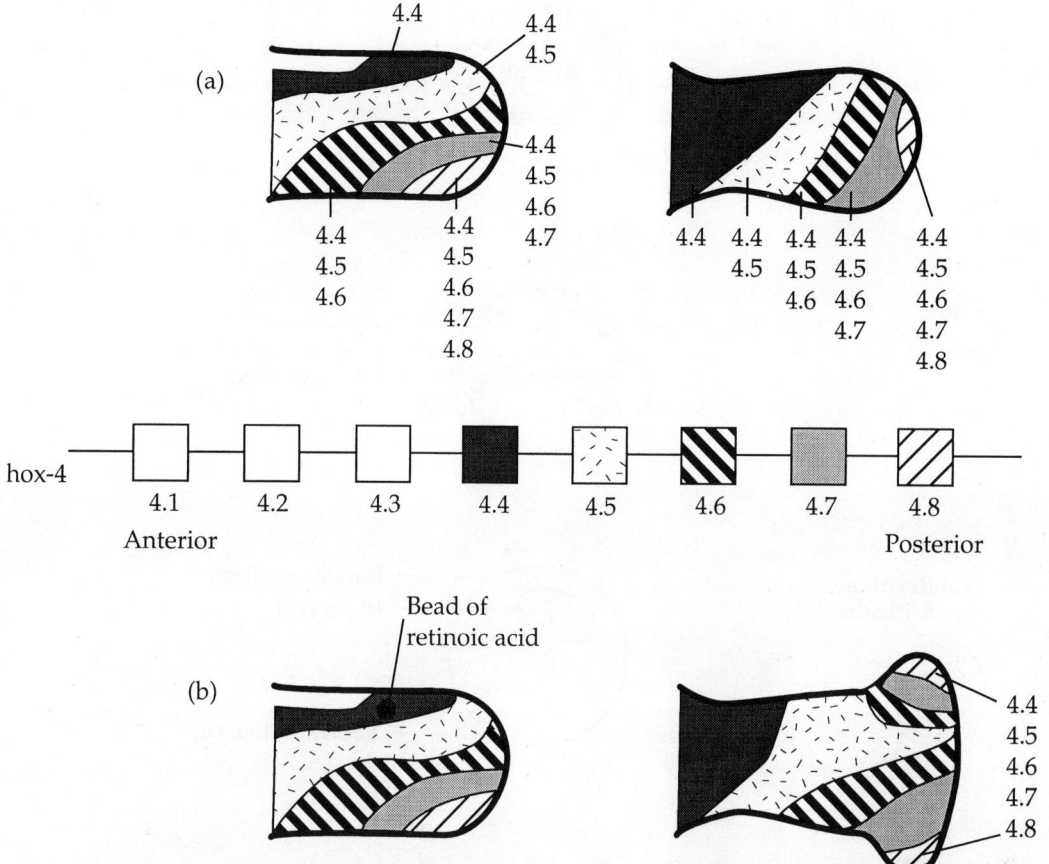

Fig. 12-8. Extent of homeotic gene expression in mouse limbs. (*a*) The extent of *hox-4* gene expression in developing mouse limbs indicates that the gene products found in the most posterior regions correspond to the genes that control the most posterior body regions. The gene product Hox-4.8 is restricted to a small, posterior region of the limb bud that is most distal from the body. The gene product Hox-4.4 is found in most of the limb bud except for the most anterior region close to the body. As the limb bud lengthens, the limb segments rotate so that they lie more perpendicular to the distal–proximal axis. More of the *hox-4* genes are expressed in the distal portions of the developing limb than in the proximal portions, presumably because the distal regions will develop digits and thus require more complex manipulations. (*b*) The *hox-4* genes, normally turned on in the posterior region of a developing limb, can be expressed inappropriately in the anterior region of a limb bud by induction with retinoic acid. A bead of retinoic acid placed in the most anterior proximal region of a growing limb bud (left figure) turns on the *hox-4.6, hox-4.7,* and *hox-4.8* genes in that region. Normally, these genes are expressed only in the posterior portion of the limb bud. As these tissues grow *hox-4.6, hox-4-7,* and *hox-4.8* are expressed inappropriately in the anterior distal portions of the limb (right figure). During the early growth of the limb, the retinoic acid bead completely dissolves and is taken up by the affected tissue. Thus, it has no effect on new tissue that forms in the anterior proximal portion of the limb. The result is the development of a deformed foot with extra digits (not shown).

the sperm enters the oocyte, the sperm entry point must be an important signal that establishes where the grey crescent forms and gastrulation occurs.

Approximately 12 hours after fertilization, the tissue above the grey crescent area, known as the dorsal ectoderm, begins to invaginate. Growth factors produced upon fertilization induce the production of secondary growth factors, which stimulate the movement and invagination of dorsal ectoderm. These invaginating cells form what is called the blastopore lip. The **blastopore lip** is the first indication of gastrulation. The lumen of the organism's primitive gut is called the **gastrocoel** or **archenteron.** Gastrulation greatly reduces the dimensions of

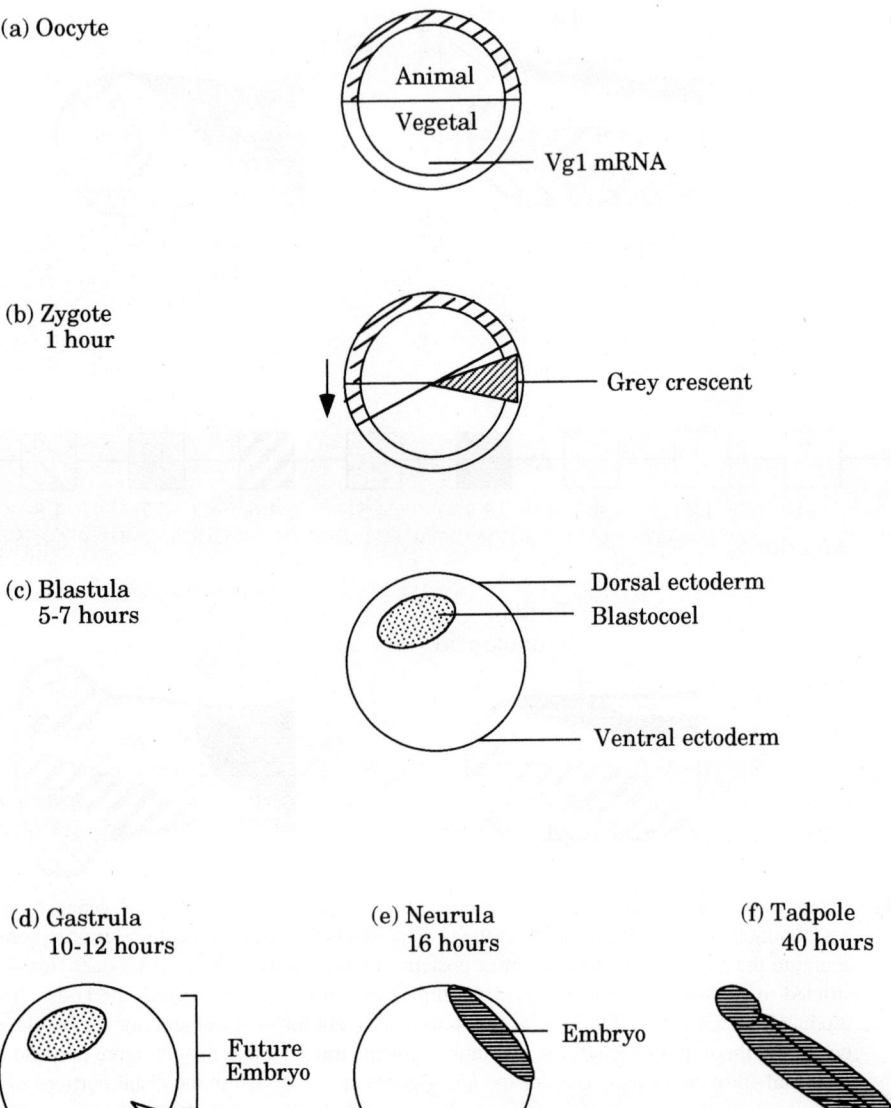

Fig. 12-9. Development of an amphibian: egg to tadpole. (*a*) The unfertilized amphibian egg, or oocyte, is polarized during its development and has an animal–vegetal axis. Vg1 mRNA—specifying one of the many transforming growth factors (TGFs)—is highly concentrated in the vegetal hemisphere, but is not translated until fertilization. (*b*) Upon fertilization, the vegetal cortex (protein cytoskeleton under the plasma membrane) and the plasma membrane rotate slightly so that a portion of the vegetal cortex and plasma membrane are in contact with the animal pole cytoplasm. This region of the cortex and membrane is known as the grey crescent. In this region, signal proteins are produced that induce the development of the anterior–posterior axis and the various tissues (ectoderm, mesoderm and endoderm). (*c*) Multiple cell divisions (cleavage divisions) result in a mass of cells called the blastula. (*d*) The invagination of dorsal ectoderm near where the grey crescent formed results in the formation of mesoderm and ectoderm, as well as a primitive gut (endoderm). This is known as gastrulation. The stage at which gastrulation occurs is known as the gastrula. (*e*) The formation of a neural tube from dorsal ectoderm is known as neurulation. The stage at which the neural tube forms is called the neurula. (*f*) The tadpole (larval feeding stage) develops on the surface of the yolk.

the blastocoel. Gastrulation in amphibians is much more than gut formation, however. It moves dorsal ectoderm inward, leading to the development of a dorsal endoderm and mesoderm. It also allows ventral ectoderm to invaginate and to develop into ventral mesoderm. There is evidence suggesting that the cells in the vegetal pole and in the region of the grey crescent produce two different growth factors that induce cells between the animal and vegetal poles to become ventral mesoderm. After 16 hours, certain cells of the dorsal mesoderm develop into a notochord, while other mesodermal tissues develop into somites and muscles.

Neurulation takes place during the development of the dorsal mesoderm and notochord. Formation of the neural tube begins when dorsal ectoderm invaginates to form a neural plate centered along the length of the embryo. Within a few hours, the neural plate cells invaginate and form the neural tube. The embryo at this stage is referred to as a neurula (Fig. 12-9e). Cells associated with the neural tube develop into **neurons** and associated **glial cells.**

A swimming, free-feeding tadpole develops about a day after neurulation is complete (Fig. 12-9f). The tail develops near the point at which gastrulation first began, whereas the body and head form from cells in the animal pole.

12.27.   What factors control tissue and axis development in the *Xenopus laevis* embryo?

The animal–vegetal polarity of the oocyte and fertilized egg (zygote) is due to the different maternal mRNAs that are sequestered in the two hemispheres. It is thought that maternal mRNA for **basic fibroblast factor** (bFGF) is attached to the cortex in the vegetal hemisphere. Upon fertilization, the inhibitor of bFGF mRNA is inactivated and the messenger is translated. In addition, experiments suggest that another maternal mRNA for a protein called noggin in the vegetal hemisphere is expressed when it contacts the animal pole cytoplasm. Noggin production occurs below the grey crescent. Thus, after fertilization and rotation of the cortex-membrane with respect to the cytoplasm, at least two growth factors are stimulating cell division and affecting cellular differentiation (Fig. 12-10). Cells in the vegetal hemisphere producing bFGF are stimulating cells at the equator to differentiate into mesoderm. The cells below the grey crescent region producing noggin are believed to define the **Nieuwkoop** (pronounced knee-OYV-kop) **center.** Noggin influences neighboring cells at the equator, where the grey crescent was, to differentiate into the **Spemann** (pronounced SPEH-mahn) **organizer.** The Spemann organizer stimulates gastrulation and is located in the dorsal lip of the blastopore. The Nieuwkoop center may also express Vg1 RNA (associated exclusively with the vegetal cortex). Vg1 RNA encodes a member of the **transforming growth factor** (TGF) family, possibly one of the many *Xenopus* signal proteins designated *Xwnt* ("X" indicates the protein is produced in *Xenopus,* "w" hints at its close relationship to *Drosophila's* protein wingless, "n" suggests that it is required for the development of nervous tissue, and "t" indicates that inappropriate expression can result in the growth of tumors).

In the 5–7 hour blastula (Fig. 12-10c), cells at the equator have differentiated into equatorial mesoderm and the Spemann organizer. Noggin produced in the Nieuwkoop center is thought to induce the synthesis of at least two proteins in the Spemann organizer belonging to the transforming growth factor (TGF) family. The proteins **Xwnt** and **activin B,** synthesized in the Spemann organizer, are thought to be responsible for promoting gastrulation and the development of an anterior–posterior axis (Fig. 12-10e). The tail develops near the Spemann organizer. In addition, these proteins contribute to the development of dorsal mesoderm.

Example 12.15. When Xwnt-1 or Xwnt-8 mRNA (coding for their respective proteins) is injected into the animal pole of *Xenopus* embryos, it stimulates a second gastrulation event and a second embryo develops. Even more amazing, when Xwnt-1 or Xwnt-8 mRNA is injected into the vegetal hemisphere, a gastrulation event occurs and a second embryo with an anterior–posterior and dorsal–ventral axes develops. Although neither of the proteins Xwnt-1 and Xwnt-8 are natural inducers produced by the Spemann organizer, a family member (so far uncharacterized) is believed to induce the activities seen when Xwnt-1 and Xwnt-8 mRNAs are injected.

A Xwnt protein is important in converting some of the dorsal ectoderm that invaginates during gastrulation into dorsal mesoderm, some of which goes on to develop into a notochord. The Xwnt-8 protein characterized in *Xenopus* is similar to the Wnt growth factor in mice and to *Drosophila's* wingless protein, one of the segment polarity gene products. Activin B mRNA injected into the animal pole or vegetal pole induces an incomplete second embryo. Thus, this second growth factor also plays a role in the induction of gastrulation and the establishment of the anterior–posterior and dorsal–ventral axes.

High concentrations of Xwnt and activin B at the entrance of the gastrocoel (lumen of the primitive gut) apparently promote tail formation in the area, whereas low concentrations of these proteins diffusing to more distal regions of the embryo (as well as ventrally produced bFGF) stimulate the production of a protein called **goosecoid** in what is to become the head. This protein resembles the proteins gooseberry and bicoid that are re-

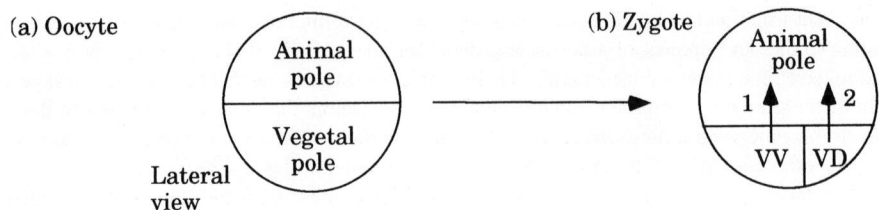

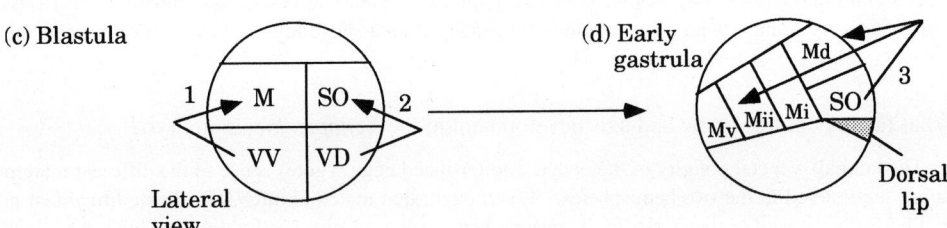

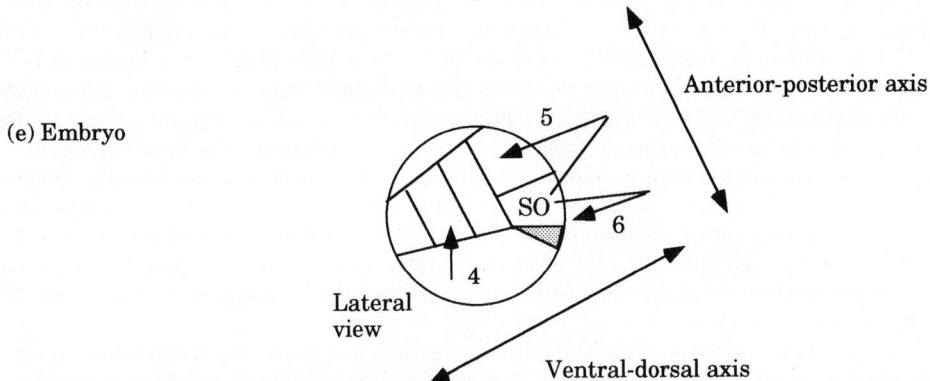

Fig. 12-10. The signals that control tissue development in amphibians. (*a*) The amphibian oocyte has an animal–vegetal axis defined by the different constituents (such as yolk) in the two hemispheres. (*b*) Fertilization of the oocyte triggers the rotation of the cortex and plasma membrane with respect to the cytoplasm, creating the grey crescent. The ventro-vegetal (VV) and the ventro-dorsal (VD) regions begin to express different signal peptides. The arrows from these regions show the diffusion of growth factors into overlying regions. The numbers next to the arrows indicate the approximate order in which the growth factors are synthesized. (*c*) The cells in the VV region produce basic fibroblast growth factor (bFGF)—first signal—while the cells in the VD region (Nieuwkoop center) produce the peptides noggin (Nog) and transforming growth factor (Xwnt)—*second signal*. Ventral mesoderm (Mv) is induced by bFGF, whereas Nog and Xwnt induce the development of the Spemann Organizer (SO) and initiate gastrulation. (*d*) The Spemann organizer in the dorsal blastopore lip produces the peptides activin B and Xwnt, which induce intermediate kinds of mesoderm (Md, Mi, and Mii)—*third signal*. (*e*) The Spemann organizer also establishes the ventral–dorsal and the anterior–posterior axes of the embryo. Mesoderm and endoderm are affected by another transforming growth factor (produced by the VV region) called bone morphogenetic protein (BMP-4)—*fourth signal*. Goosecoid—*fifth signal*—is expressed maximally in the anterior region of the embryo, and brachyury—*sixth signal*—is produced mainly in the posterior portion of the organism.

quired, respectively, for segment polarity and for head development in *Drosophila*. Since goosecoid is also necessary for head and body development in the toad, its name was derived from the first part of gooseberry and the last part of bicoid. High concentrations of activin B at the entrance of the gastrocoel seem to stimulate the production of **brachyury,** a protein required for tail development. A similar protein is found in the mouse. Mutations in the mouse gene result in improper mesoderm formation and a short tail, hence the term brachyury. There is no information as to whether this protein is related to nanos or some other tail-determining factor in *Drosophila*.

The basic fibroblast growth factor (bFGF) produced by the vegetal pole is extremely important in the development of mesodermal tissue. Together, bFGF and a gradient of activin B (produced by the Spemann organizer) produce a number of different mesodermal tissues that help define the ventral–dorsal axis (Fig. 12-10e). A high concentration of Xwnt stimulates the development of dorsal mesoderm. In the vegetal pole, where there is a high concentration of **bone morphogenetic protein-4** (BMP-4), bFGF and BMP-4 stimulate the development of ventral mesoderm. BMP-4 is in the TGF-β family of growth factors. The relationship between growth factors (numbered in the order in which they are synthesized) and development of tissues (VV, VD, Mv, SO, etc.) producing the growth factors is summarized in Fig. 12-10.

12.28. How do the primary embryonic germ layers (ectoderm, mesoderm, and endoderm) develop in amphibians?

The ectoderm, or outer layer of cells, forms during the development of the blastula. Gastrulation and the conversion of dorsal ectoderm into dorsal endoderm and dorsal mesoderm are believed to require Xwnt and activin B expression by the Spemann organizer. Dorsal or ventral ectoderm exposed to Xwnt and activin B will produce a rudimentary embryo with a head and tail, a notochord, a brain, eyes, and muscles. Clearly, the growth factors Xwnt and activin-B can induce ectoderm to undergo gastrulation and form endoderm and mesoderm. The notochord forms from dorsal mesoderm, muscles from lateral mesoderm. Nervous tissues develop from the dorsal ectoderm.

12.29. When do the amphibian homeotic genes *(xhox)* play a role in ontogeny?

One of the first homeobox genes to be transcribed in the early embryo is *xhox-3*. It is expressed in a graded fashion along the anterior–posterior axis in the "equatorial" mesoderm, maximally in the Spemann organizer and minimally in the mesoderm most distant from the Spemann organizer. Overexpression of *xhox-3* in the most distant mesoderm blocks head development. Thus, a Xhox-3 gradient from the Spemann organizer to the most distant mesoderm is important in the determination of the anterior–posterior axis.

Xhox-3 expression in mesodermal tissues is induced by bFGF, but not by platelet-derived growth factor (PDGF) found in frog embryos. Thus, Xhox-3 expression takes place no earlier than the blastula stage. In fact, significant Xhox-3 mRNA is first detected in the early gastrula. Maximal differential *xhox-3* expression occurs in late gastrula and early neurula stages, when intermediate kinds of "equatorial" mesoderm have developed. Xhox-3 mRNA is induced by low and high levels of bFGF. Lithium inhibits this induction. High levels of Xwnt seem to repress the production of Xhox-3 mRNA. The concentration of Xhox-3 mRNA continues to drop to very low levels during the tailbud and tadpole stages.

An mRNA for a homeoprotein Xlab (which resembles that from *Drosophila's* Labial, mouse's Hox-2.9 and Hox-1.6, and a chicken homeogene) is first detected in the frog early gastrula, peaks at the midgastrula stage, and is concentrated mainly in the dorsal lip of the blastopore.

Gene *mix-1,* activated soon after zygotic transcription, has a homeobox that is 53% identical to *goosecoid*. Mix-1 mRNA is not found in the dorsal lip (Spemann organizer), but is expressed mainly in the future endoderm.

DEVELOPMENT IN BIRDS (DOMESTIC CHICKEN)

12.30. What happens during the early embryonic development of birds, particularly with regard to ventral–dorsal axis determination?

During oogenesis of avian oocytes (unfertilized eggs), a polarity known as the animal–vegetal axis is established in oocytes (Fig. 12-11). The animal–vegetal axis is important in determining the ventral–dorsal axis in the developing embryo. The animal hemisphere provides the cytoplasm and subsequent cells that form the embryo, whereas the vegetal hemisphere serves as the yolk, providing nutrients to the developing organism.

Egg fertilization triggers the formation of **cleavage furrows** on the surface of the animal hemisphere in an

(a) Oocyte
(lateral
view)

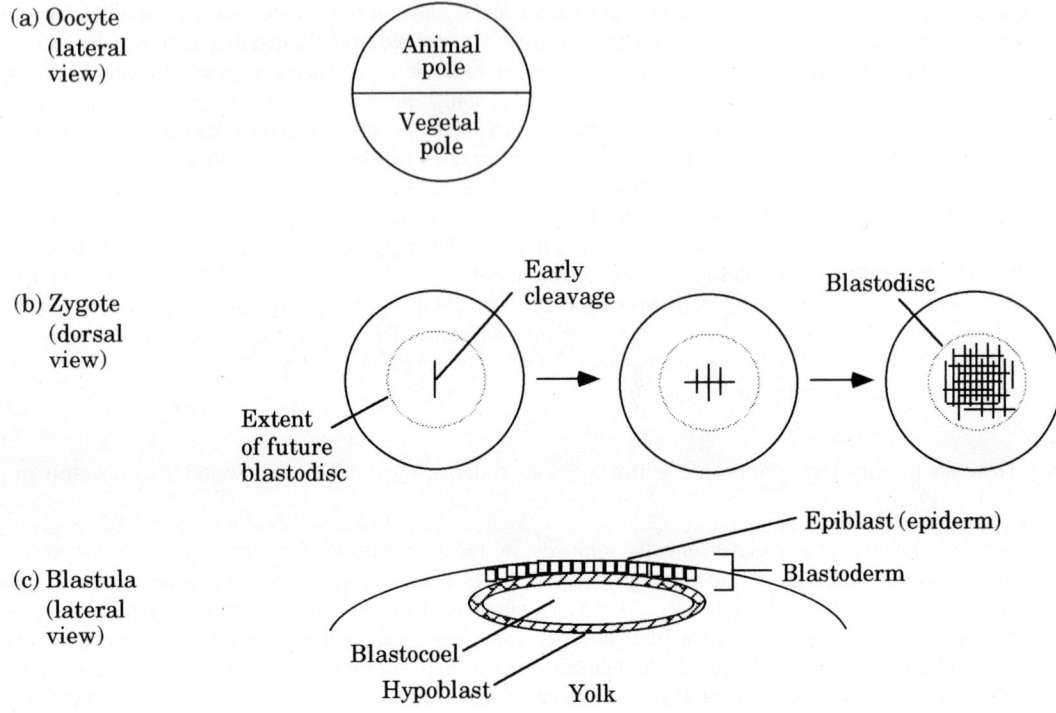

(b) Zygote
(dorsal
view)

(c) Blastula
(lateral
view)

(d) Embryo
(dorsal
view)

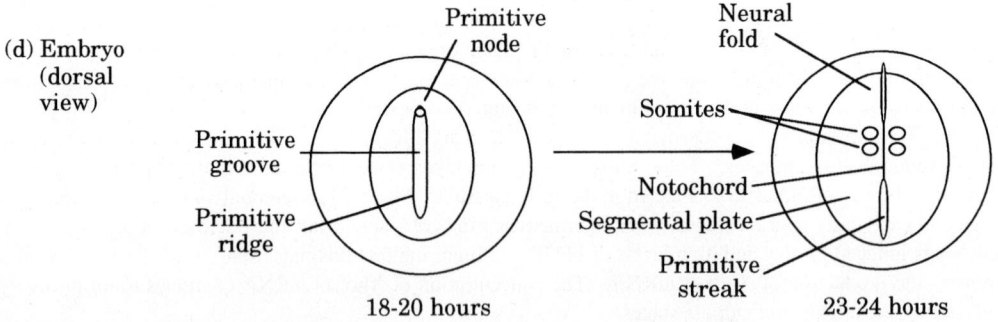

18-20 hours   23-24 hours

Fig. 12-11.   Chicken development: egg to embryo. (*a*) A chicken oocyte has an animal–vegetal axis defined
by the various constituents in the two hemispheres. The animal–vegetal axis is a precursor to the
dorsal–ventral axis. (*b*) Fertilization triggers invagination of the plasma membrane and formation
of a layer of cells in a disclike region at the animal pole. The disc of cells is known as the blastodisc. (*c*) The chicken blastula consists of a disc (two cells thick) overlying a small cavity, called
the blastocoel, in the vegetal portion of the developing chick. The bottom of the early blastocoel is
defined by the egg's plasma membrane. Just before gastrulation and mesoderm formation, a layer
of cells migrates onto the floor (the egg's plasma membrane) of the blastocoel. This layer of cells
is known as the hypoblast and is important because of the peptide signals it produces. The top
layer of cells in the blastoderm is known as the epiblast. These ectodermal cells are the source of
mesoderm. (*d*) The embryo develops a primitive streak 18–20 hours after fertilization of the
oocyte. The primitive streak is shown with its primitive node, groove, and ridge. (*e*) About
23–24 hours after fertilization, the embryo has developed a neural fold in the head region and some
of its somites on either side of a lengthening notochord. The primitive streak regresses as more
and more parasegments develop sequentially in the posterior region of the embryo.

area called the **blastodisc** (Fig. 12-11*b*). Many rapid cell divisions result in the formation of a disc of cells known as the blastoderm, with a hollow called the blastocoel beneath it at the top of the yolk (Fig. 12-11*c*). The blastoderm and blastocoel sitting on the yolk represent the early blastula. These early stages of development occur while the fertilized egg is passing down the hen's oviduct.

The early ontogeny suggests that the ventral–dorsal (VD) axis is established in the oocyte and overlaps the animal–vegetal (AnVg) axis. In a later part of this chapter, the factors involved in establishing the AnVg and VD axes are discussed.

12.31. How do the endoderm and mesoderm form in chickens?

It is believed that both the endoderm and mesoderm form from the **epiblast** (the top layer of ectodermal cells in the blastoderm; Fig. 12-11*c*). The epiblast will provide cells for the development of all embryonic tissues (endoderm, ectoderm, and mesoderm), as well as some extra-embryonic structures (amnion). Preceding the development of new tissues, however, another layer of cells forms over the yolk, but under the blastocoel. This layer of cells, called the **hypoblast,** appears to organize first at one edge of the blastodisc called the **posterior marginal zone.** The hypoblast gives rise to only extra-embryonic structures (such as the yolk sac stalk). The posterior marginal zone suggests that the embryo already has the beginnings of an anterior–posterior axis.

The endoderm forms by the migration of ectodermal cells along the anterior–posterior axis down into the blastocoel, followed by peripheral migration along the underside of the ectoderm. This results in a blastula with two layers of tissue: ectoderm and endoderm. Following the formation of the endoderm, a subsequent movement of ectodermal cells downward and into the space between the ectoderm and endoderm produces a mesoderm. Initially, the mesodermal cells are loosely connected, and because of this they are referred to as **mesenchyme.** When the mesenchymal cells are signaled to aggregate, they form various types of tissues, beginning at the anterior end: notochord, paired somites, and **lateral plate mesoderm** (mesoderm on either side of the developing notochord and somites).

12.32. When is the anterior–posterior (AP) axis defined?

It is thought that the hypoblast stimulates ectodermal cells to move down toward the hypoblast along an anterior–posterior axis. The movement of the ectodermal cells begins first near the posterior marginal zone and then spreads anteriorly. Experiments rotating the hypoblast 180° causes the anterior–posterior axis to be rotated 180° also, suggesting that the posterior marginal zone is no longer important in guiding AP axis formation. The ectodermal cells migrate downward, cutting the blastocoel in half, and then they move laterally so that they line the bottom of the blastocoel above the hypoblast. These cells represent the endoderm. The posterior marginal zone may have arisen at the site where the sperm head entered the animal hemisphere and where the first cleavage furrows formed. In any event, some signal triggers the formation of the posterior marginal zone, which, in turn, induces an anterior–posterior axis in the hypoblast. Then, the hypoblast induces an anterior–posterior axis in the embryo, about 10 hours after fertilization.

The movement of ectoderm along the anterior–posterior axis to create the endoderm and mesoderm is associated with the formation of the primitive streak, primitive groove, and primitive ridge on the surface of the developing embryo (Fig. 12-11*d*).

12.33. When does the formation of the gut occur?

The first indication of gut formation is the folding of the anterior ectoderm under the head region. This begins approximately 1.5 days after fertilization. A day later, the ectoderm and endoderm are growing under the entire embryo and the head has folded downward. Three and a half days after fertilization, the embryo is connected to the yolk by the yolk stalk.

12.34. When does neurulation occur in birds?

Approximately 18–20 hours after fertilization, a prominent notochord in the anterior 1/3 of the embryo develops from mesodermal tissue. A neural plate is visible also in the anterior 1/3, while the primitive pit, primitive groove, and primitive ridge are found in the posterior 2/3 of the embryo. The neural tube is completed two and a half days after fertilization.

A few hours later, **neural plate ectoderm** begins to form the neural tube and neural crest cells. A "wave" of neural tube formation beginning at the anterior end progresses toward the posterior end. By the second day after fertilization, more than 3/4 of the neural tube is complete. Neural crest cells associated with the neural tube develop into neurons (nerve cells) and helper glial cells.

12.35. When does the chicken become segmented?

The first segmentation genes are probably being turned on in the embryo from the posterior end (posterior marginal zone) as the hypoblast develops, 5 to 15 hours after fertilization. A hierarchy of segmentation genes turned on by the first segmentation genes may lead to the formation of a primitive streak and neural plate on the surface, to the endoderm, and to formation of mesoderm below the surface. Approximately 18 hours after fertilization, a notochord has formed. The homeotic genes have undoubtedly been turned on as neurulation proceeds and somites begin to appear 1.0 to 1.5 days after fertilization.

12.36. What are some of the signals that control embryonic development in the chicken?

Some of the same growth factors that affect early embryonic development in the frog appear to affect the chicken. High concentrations of a fibroblast growth factor (FGF) are thought to be associated with cells in the posterior marginal zone. FGF is a prime candidate for establishing the anterior–posterior axis through a gradient. Possibly, high concentrations at the posterior marginal zone induce the production of the protein activin in the ectoderm and/or hypoblast. Together, FGF and activin may turn on a gene that codes for the protein brachyury. Brachyury, in the amphibian and mouse, is associated with anterior–posterior development. In particular, it stimulates the formation of a tail and various types of mesodermal tissue.

Possibly at low concentrations of FGF, a Wnt-like protein is produced at the anterior end of the embryo and/or hypoblast. Wnt is thought to turn on the production of the homeotic gene *goosecoid.* Goosecoid is believed to stimulate anterior-posterior axis development, having its greatest effect on head tissue.

Activin, believed to be produced by the underlying hypoblast, is known to stimulate mesoderm to develop into a number of tissues. High concentrations of activin can convert mesoderm into notochord in the head region; intermediate concentrations of activin change mesoderm into somites and muscles; low concentrations of activin stimulate mesoderm to develop into loose mesoderm or mesenchyme. How other factors, working synergistically with activin, stimulate chicken tissue is unknown.

The first segmentation genes, like *goosecoid,* are turned on by the time the hypoblast has developed (15 hours after fertilization). Later segmentation gene families (such as gap, pair-rule, segment polarity) have not been characterized so far. The homeotic genes in chickens are symbolized by ***chox.*** They are expressed during neurulation and gut formation. The *chox* genes are important not only in the segmentation of the body, but also in segmentation of the appendages.

12.37. What is the relationship between *chox* gene expression and the development of cartilage and bone?

The development of chicken limb buds illustrates how the *chox* genes are involved in the formation of cartilage and bone. Most of the chicken's *chox* genes are expressed in the distal posterior region of the limb bud (Fig. 12-12*a*). In particular, the chicken *chox-4g* gene is expressed only in the most distal posterior segment. The chicken *chox-4g* gene is equivalent to the mouse *hox-4.8* gene. In more anterior segments, fewer *chox* genes are expressed. For example, in the most anterior segment of the chicken limb bud, only *chox-4d* is expressed.

Expression of all the *chox-4* genes results in the development of the cartilage and bones that will develop into the digits. Expression of all the *chox* genes except *chox-4g* results in the development of cartilage and bone for the ulna. Similarly, expression of *chox-4d, 4c, 4b,* and *4a* (not illustrated) and repression of *chox-4g, 4f,* and *4e* result in the development of cartilage and bone for the radius. The humerus develops in the region where the least number of *chox-4* genes are expressed.

The implantation of a retinoic acid bead in the anterior portion of a developing limb bud turns on all the *chox-4* genes in an anterior segment. The gradient of retinoic acid produces a gradient of gene expression, so that a region in the anterior of the limb bud mirrors a region in the posterior of the limb bud (Fig. 12-12*b*). The final result is a wing-limb that has a mirror image duplication of the three digits. Retinoic acid, or some other closely related signal molecule that uses the same receptor, functions as the natural inducer.

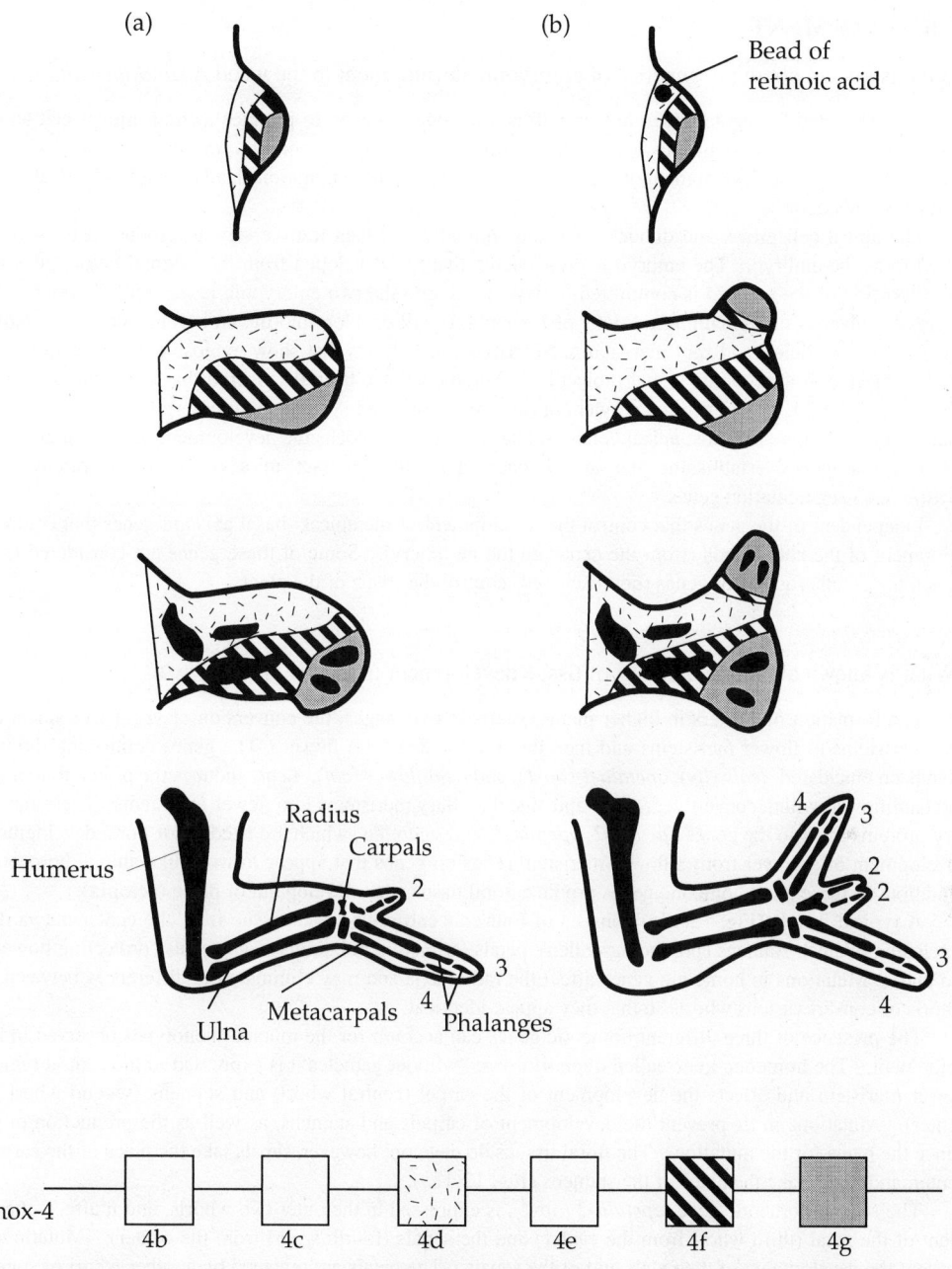

Fig. 12-12. Homeotic gene expression in developing chicken limbs. (*a*) The extent of *chox-4* gene expression in developing chicken limb buds indicates that the gene products in the most posterior regions of the developing
limbs correspond to the genes that control the most posterior body regions. The gene product Chox-4g is restricted to a small, posterior region of the limb bud that is the most distal from the body. It is found in most
of the limb bud except for the most anterior region close to the body. As the limb bud lengthens, the limb
segments rotate so that they become more perpendicular to the distal-proximal axis. More of the *chox-4*
genes are expressed in the distal portions of the developing limb than in the proximal portions, presumably
because the distal regions will develop digits and thus require more complex manipulations. (*b*) The *chox-4*
genes, normally expressed only in the posterior regions of limb buds, can be expressed inappropriately in the
anterior region of limb buds by induction with retinoic acid. A bead of retinoic acid placed in the anterior
region of a growing limb bud turns on all the *chox-4* genes that are normally expressed only in the posterior
portion of the limb bud. This results in a deformed wing with extra digits. Instead of the digits 2, 3, and 4
(anterior to posterior) the abnormal wing has digits 4, 3, 2, 2, 3, and 4 (anterior to posterior). The number of
digits has doubled, but the digits of the anterior set are backward (a mirror image of the posterior digits).

PLANT DEVELOPMENT

12.38.   What is known about the genetics of embryonic development in the weed *Arabidopsis thaliana?*

The *Arabidopsis* zygote undergoes growth and unequal division to produce a small **apical cell** and a much larger **basal cell.** This establishes the **apical–basal axis** of the future embryo. Mutations in any one of a number of genes block the development of the apical–basal axis. These mutations lead to nearly identical cells proliferating in a filament.

The apical cell grows and divides, first longitudinally and then transversely, to produce a mass of cells referred to as the embryo. The embryo is on a "stem" that has developed from the original basal cell. The most apical portion of the embryo is committed to developing into the two embryonic leaves (**cotyledons**) and a shoot meristem, whereas the remaining portions of the embryo will develop into the embryonic stem (**hypocotyl**), embryonic root (**radicle**), and root meristem. **Meristems** are tissues that show explosive cell growth and division when subjected to appropriate signal molecules. Mutations have been discovered that block the development of the hypocotyl and radicle; even the cotyledons are greatly reduced. The mutant embryo consists of cells that are unable to differentiate the most apical cells. A later-acting gene blocks the development of shoot meristems. It is believed that genes establishing the apical–basal axis may interact in a regulatory hierarchy similar to *Drosophila's* segmentation genes.

Independent of the genes that control the development of the apical–basal axis are genes that control the development of the **radial axis** (from the center to the periphery). Some of these genes are considered in the next question. Another group of genes (not discussed) control the shape of the plant.

12.39.   What is known about the genetics of tissue development in mature *Arabidopsis?*

The formation of flowers in higher plants occurs in two stages: the conversion of vegetative apical and axillary meristems to flower meristems and then the development of a flower. The genes controlling the first stage have been elucidated: *leafy* (*lfy*), *apetala-1* (*ap-1*), and *cauliflower* (*cal*). *Leafy* induces the production of apetala-1 and cauliflower, which convert vegetative apical and axillary meristems into flower meristems. These three regulatory proteins activate the genes *apetala-2, apetala-3,* and *agamous,* which are needed for floral development. The development of a flower from a flower meristem relies on genes that appear to work like animal homeotic genes. Mutations in these plant homeotic genes produce floral tissues that develop out of place (**ectopia).**

A typical flower (Fig. 12-13*a*) consists of four concentric whorls of tissue from the center outward: carpels (enclosing ovules), stamens (producing pollen), petals (attracting polinators), and sepals (protecting flower parts in the bud). Mutations in homeotic genes affect the radial axis and may eliminate the differences between adjacent whorls or convert various whorls so that they appear identical.

The presence of three different homeotic genes can account for the mutant phenotypes observed in floral development. The homeotic gene called ***agamous*** (*ag;* "without gametes") is expressed in the central region of the flower meristem and affects the development of the carpel (central whorl) and stamens (second whorl from the center). Mutations in *ag* prevent the development of carpels and stamens, as well as the production of gametes; hence the name for the mutation. The floral tissues do develop; however, sepals take the place of the carpels at the center, and petals take the place of the stamens (Fig. 12-13*b*).

The second homeotic gene, ***apetala-2*** *(ap-2),* is expressed in the outer two whorls, and it affects the development of the petal (third whorl from the center) and the sepals (fourth whorl from the center). Mutations in *ap-2* prevent the development of the petals and of the sepals. The petals are replaced by another whorl of stamens, and the sepals are replaced by carpels (Fig. 12-13*c*).

Experiments indicate that the gene products Agamous and Apetala-2 mutually restrict each other's expression to different regions of the flower meristem, much as *Drosophila's* gap genes mutually restrict each other's expression (Fig. 12-6). If one of these two homeotic genes is missing, the other's gene expression spreads throughout the flower and causes the whorls to develop along different lines. Genetic experiments and the finding of a putative DNA-binding region suggest that Apetala-2 functions as a transcriptional repressor of *agamous.* Similarly, the discovery of a DNA-binding region indicates that Agamous is a transcriptional repressor of the *apetala-2* gene.

Another homeotic gene, ***apetala-3,*** is restricted to the region where the two central whorls develop. A gene product called superman keeps apetala-3 out of the central whorl. Other uncharacterized genes must limit the expression of *apetala-3* at the flower's margin. Both *apetala-3* and *apetala-2* are necessary for the development of petals. Mutations in *apetala-3* result in the formation of sepals instead of petals. The development of stamens is dependent upon both *apetala-3* and *agamous.* Mutations in *apetala-3* result in carpels forming instead of stamens.

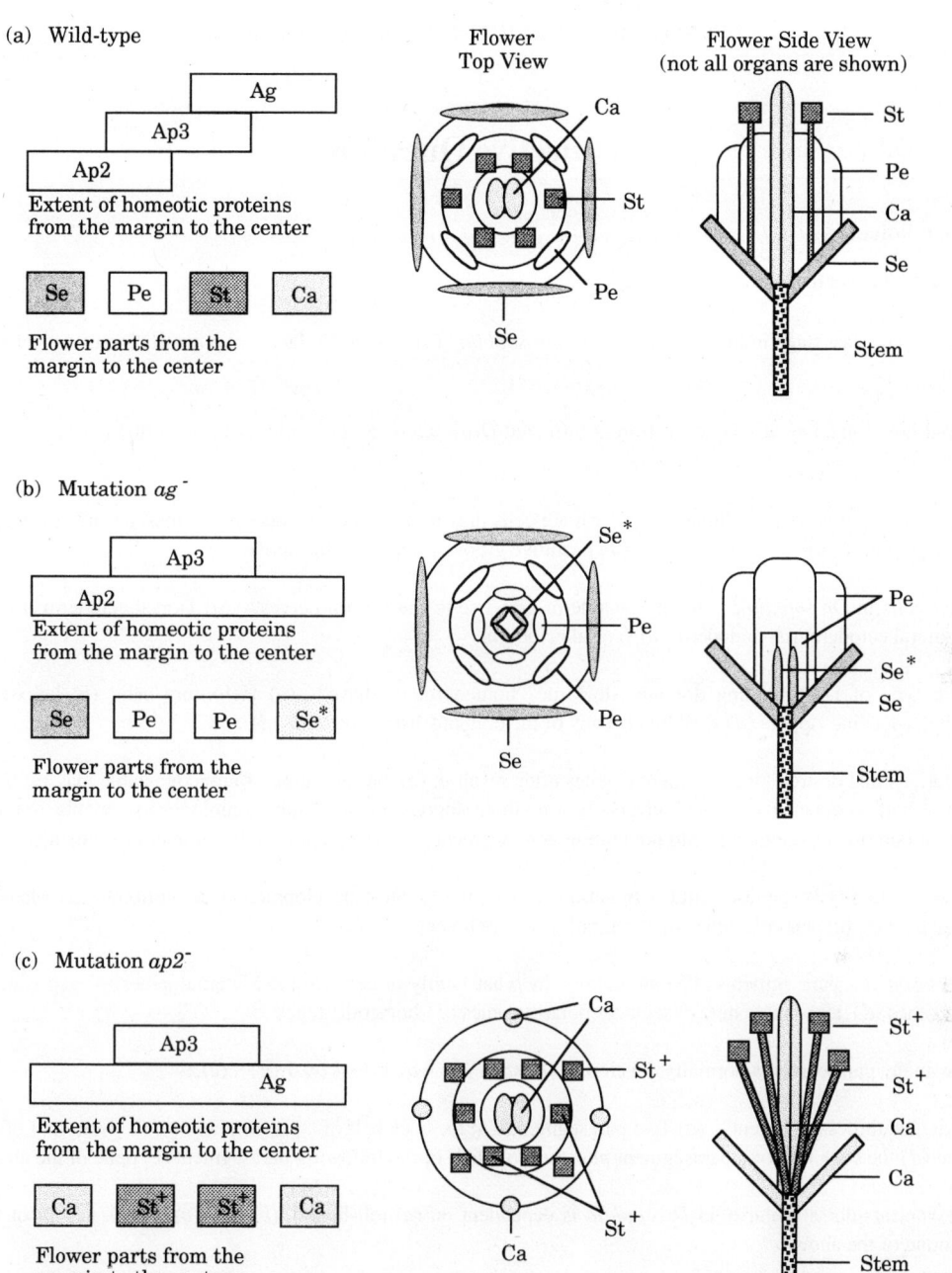

Fig. 12-13. Homeotic gene expression in developing flower parts of *Arabidopsis*. In the first column, the extent of the homeotic gene products (from the margin to the center) in floral tissues is indicated by Ap2, Ap3, and Ag. The parts that develop under the influence of the homeotic gene products are indicated by Se (sepal), Pe (petal), St (stamen), and Ca (carpel). The second and third columns show a top view and side view of the flower that forms under the influence of the homeotic gene products. The homeotic gene *agamous* (*ag*, encoding product Ag) is required for the development of carpels, whereas stamens require both *agamous* and *apetala-3* (*ap-3*, encoding product Ap3). Genes *apetala-2* (*ap-2*, encoding product Ap2) and *apetala-3* are both required for the development of petals, but only *apetala-2* is necessary for the development of sepals. (*a*) The wild-type flower from *Arabidopsis* consists of four concentric whorls. The outer whorl contains 4 sepals; the second whorl from the edge contains 4 petals. Six stamens are found in the third whorl: 4 are long in the medial positions, whereas two are short in the lateral positions. The inner whorl is occupied by a gynoecium (a collective term for all of the carpels of a flower) of two fused carpels topped by a short style (not shown) and stigmatic papillae (not shown). (*b*) In the absence of *ag*, all tissues are subjected to Ap2. Although 4 sepals should theoretically develop in place of two carpels, the identity of the organs (*) cannot be determined by their phenotype. Petals develop instead of stamens. (*c*) In the absence of *ap2*, all tissues are influenced by Ag. Stamens develop instead of petals and carpels develop instead of sepals. Some or all of the stamens fail to develop (+).

309

Objective Questions

Multiple Choice

Directions: Choose the one best answer.

1. Which is the last stage in the development of *Drosophila*? (*a*) Oocyte (*b*) larva (*c*) pupa (*d*) imago (*e*) cellular blastoderm

2. About how long does it take to go from a fertilized *Drosophila* egg to a cellular blastoderm? (*a*) 10 minutes (*b*) 2.5 hours (*c*) 12 hours (*d*) 1 day (*e*) 9 days

3. From what region of the cellular blastoderm are cells drawn to produce endoderm and mesoderm? (*a*) Cephalic furrow (*b*) ventral furrow (*c*) neural plate (*d*) primitive groove (*e*) none of the above

4. What cells in *Drosophila's* cellular blastoderm give rise to the ventral nerves? (*a*) Dorsal ectoderm (*b*) mesoderm (*c*) lateral ectoderm (*d*) endoderm (*e*) ventral ectoderm

5. What type of DNA-binding domains do "true" homeoproteins have? (*a*) Helix-turn-helix (*b*) helix-zinc finger (*c*) helix-leucine zippers (*d*) β-ribbon (*e*) any of the domains listed above (*a*–*d*)

6. Mutations that destroy "true" homeotic genes often result in (*a*) the formation of unrecognizable masses. (*b*) the deletion of half an organism. (*c*) the conversion of nonhomologous segments into homologous segments. (*d*) the conversion of homologous segments into nonhomologous segments. (*e*) an increase in the number of segments.

7. In which family of genes would a mutation lead to the greatest developmental disruption? (*a*) Maternal genes (*b*) gap genes (*c*) pair-rule genes (*d*) terminal genes (*e*) homeotic genes

8. In *Drosophila*, parasegments (PS) are defined by what family of genes? (*a*) Maternal genes (*b*) gap genes (*c*) pair-rule genes (*d*) terminal genes (*e*) segment polarity genes (*f*) homeotic genes

9. How many parasegments normally develop in *Drosophila*? (*a*) 2 (*b*) 4 (*c*) 7 (*d*) 9 (*e*) 14

10. What constitutes a segment? (*a*) Two parasegments (*b*) the back half of a parasegment (*c*) the front half of a parasegment (*d*) the back half of a parasegment plus the front half of the following parasegment (*e*) none of the above

11. The ventral–dorsal polarity in *Drosophila* is dependent on (*a*) toll-ligand. (*b*) toll. (*c*) dorsal. (*d*) all of the above. (*e*) none of the above.

12. The human embryo that consists of only two layers of cells is called (*a*) a blastocyst. (*b*) an embryonic disc. (*c*) a placenta. (*d*) an amniotic sac. (*e*) a yolk sac.

13. The human embryo develops its mesoderm from (*a*) dorsal ectoderm. (*b*) ventral ectoderm. (*c*) dorsal endoderm. (*d*) ventral endoderm. (*e*) terminal mesoderm.

14. The stiff rod that develops from mesodermal tissue in all chordates is called the (*a*) notochord. (*b*) neural tube. (*c*) spinal chord. (*d*) neural plate. (*e*) vertebral column.

15. What kind of tissue is found in somites? (*a*) Gonadal tissue (*b*) ectoderm (*c*) hypoderm (*d*) endoderm (*e*) mesoderm

16. Retinoic acid is known to affect the development of limbs. What is the relationship between retinoic acid and the *hox* genes? (*a*) It represses *hox* genes. (*b*) It activates *hox* genes. (*c*) The *hox* genes repress retinoic acid production. (*d*) The *hox* genes activate retinoic acid production. (*e*) None of the above

17. The portion of a developing limb bud where most *hox* genes are expressed is best described as (*a*) proximal anterior. (*b*) distal anterior. (*c*) proximal posterior. (*d*) distal posterior. (*e*) none of the above.

18. In amphibians, the region that "directly" stimulates gastrulation and the development of different mesodermal tissues is known as the (*a*) grey crescent. (*b*) Nieuwkoop center. (*c*) Spemann organizer. (*d*) Leeuwenhoek region. (*e*) dummkopf center.

19. In amphibians, which region is important in the development of dorsal mesoderm? (*a*) Grey crescent (*b*) Nieuwkoop center (*c*) Spemann organizer (d) Leeuwenhoek region (*e*) dummkopf center

20. In amphibians, cells that form the neural tube come from the (*a*) ectoderm. (*b*) endoderm. (*c*) mesoderm. (*d*) notochord. (*e*) somites.

21. In the absence of the plant *apetala-2* gene, what kind of tissue replaces the flower petals? (*a*) Sepals (*b*) carpels (*c*) stamens (*d*) enlarged petals (*e*) nothing

True–False

1. Fish, amphibians, reptiles, birds, and mammals are segmented in much the same way as are earthworms and arthropods.

2. During pupation in insects, pockets of cells known as imaginal discs contribute to the final development of the head parts, appendages (antennae, legs, wings, halteres), and genitalia.

3. An adult fly is called a maggot.

4. Follicle cells provide a developing *Drosophila* oocyte with mRNA.

5. The mRNA for the protein bicoid is found in the anterior of the *Drosophila* oocyte, whereas the mRNA for the protein nanos is attached to the posterior of the egg.

6. A syncytial blastoderm consists of a single cell containing multiple nuclei.

7. The formation of ectodermal tissue in *Drosophila* is referred to as gastrulation.

8. Neurulation in *Drosophila* occurs on the dorsal surface of the embryo and closely follows gastrulation.

9. Many of the genes that differentiate body segments contain regions known as a homeoboxes.

10. A homeobox codes for a homeodomain in a homeotic protein.

11. Proteins encoded by the pair-rule genes form chemical gradients that allow the *Drosophila* embryo to be divided into 14 regions called parasegments (PS).

12. It is believed that bicoid and nanos are activator/repressor proteins that activate their own transcription or translation (Bcd → Bcd & Nos → Nos), but mutually repress each other's transcription or translation (Bcd ⊢⊣ Nos).

13. Homeotic genes are not expressed in *Drosophila* until pupation.

14. Terminal development in *Drosophila* is dependent on a maternal gene that codes for a peptide hormone and a receptor protein specified by the *torso* gene in the blastoderm.

15. The ventral–dorsal differentiation in *Drosophila* depends on signal proteins and tyrosine kinase receptors such as toll.

16. Mutations that eliminate the protein Dl (dorsal) in *Drosophila* result in highly distorted segmented embryos that have no ventral tissue.

17. The mammalian blastocyst is a polarized structure because of the inner cell mass that develops.

18. The human embryo at 13 days after fertilization still consists of an embryonic disc with two layers of cells. The dorsal layer is the epiblast, whereas the ventral layer is the hypoblast.

19. The human embryo develops its mesoderm from day 15 to day 18 as ectodermal cells migrate between the ectoderm and endoderm.

20. Somites in mammals are cuboidal bodies of mesoderm that delineate parasegments.

21. Somites provide the cells involved in the development of vertebrae, ribs, muscles connecting vertebrae, and connective tissue in the dermis of the skin.

22. Human gut development begins when a portion of the amniotic cavity above the embryo folds underneath the embryo.

23. Vertebrate limbs are segmented in much the same way as is the trunk of the body.

24. Retinoic acid is known to turn on *hox* genes in the mouse.

25. Rotation of the cortex and plasma membrane after the fertilization of an amphibian's egg creates a region known as the Spemann organizer.

26. The grey crescent and the Nieuwkoop center in amphibians are believed to be the same.

27. The Spemann organizer of amphibians is in the dorsal lip of the blastopore.

28. In the chicken, the egg's animal hemisphere provides the cytoplasm and cells that form the embryo, while the vegetal hemisphere serves as the yolk, providing nutrients to the developing organism.

29. Mutations that eliminate the plant gene *agamous (ag)* block flower development totally.

30. Somites are cuboidal bodies of ectodermal tissue that develop within segments.

Matching

Directions: Match the gene family (*A–E* in **Column B**) with the name of the genes (1–6 in **Column A**).

Column A	Column B
1. *bicoid (bcd), nanos (nos)*	A. homeotic genes
2. *hunchback (hb), kruppel (kr), knirps (kni), giant (gnt)*	B. pair-rule genes
3. *huckebein (hkbn), tailless (tll)*	C. gap genes
4. *runt (rnt), hairy (hry), even-skipped (eve), fushi tarazu (ftz)*	D. segment polarity genes
5. *engrailed (eng), patched (ptc), wingless (wgl)*	E. none of the above
6. *labrum (lab), deformed (dfd), sex combs reduced (scr), antennapedia (antp), ultrabithorax (ubx), abdominal-A (abd-A), abdominal-B (abd-B), caudal (cau)*	

Directions: Match the growth factors or activators (*A–E* in **Column B**) with the region that produces them (7–9 in **Column A**).

Column A	Column B
7. Vegetal hemisphere other than the Nieuwkoop center	A. basic fibroblast growth factor (bFGF)
8. Nieuwkoop center (grey crescent)	B. noggin and transforming growth factor Xwnt
9. Spemann organizer	C. activin B and Xwnt
	D. brachyury
	E. goosecoid

Directions: Match the homeotic gene product (*A–F* in **Column B**) with its location (10–13 in **Column A**) in the *Drosophila* embryo.

	Column A		Column B
10.	Segments A5 through A8	*A.*	Antennapedia
11.	Segments T1 through T3	*B.*	Sex combs reduced
12.	Segments La (labium = C3) and T1	*C.*	Abdominal-A
13.	Segments Mn (mandible = C1), Mx (maxilla = C2), and La (labium = C3)	*D.*	Labial
		E.	none of the above

Directions: Match the chicken tissue (*A–E* in **Column B**) with the questions (14–16 in **Column A**).

	Column A		Column B
14.	What tissue overlying the yolk in the early embryo determines the anterior–posterior axis?	*A.*	yolk epithelium
		B.	epiblast
15.	What tissue produces activin?	*C.*	yolk endothelium
16.	Which tissue stimulates the development of different types of mesoderm?	*D.*	hypoblast
		E.	yolk mesoderm

Terms

Directions: Unless otherwise indicated, the answer is a single word.

1. The phrase early animal embryologists, such as Ernst Haekel, used to describe animal stages of development going through adult stages of distant ancestors. (3 words)

2. In *Drosophila,* the pockets of cells that contribute to the final development of the head parts, appendages (antennae, legs, wings, halteres), and genitalia. (2 words)

3. The term for an adult fly that emerges from the pupa.

4. The *Drosophila* oocyte is "fed" messenger RNA (mRNA) by some of the nurse cells located at the oocyte's anterior. What is the name of these mRNAs that come exclusively from the female? (1 word plus mRNAs)

5. The *Drosophila* structure that results from nuclei proliferating in the absence of any cell division. (2 words)

6. The location of the ectoderm that invaginates during neurulation in *Drosophila*.

7. That portion of a homeotic gene coding for the helix-turn-helix domain.

8. Mutations in what family of genes convert structurally different segments into nearly identical segments? (1 word plus genes)

9. Certain DNA-binding proteins use a bent-helix domain to attach to DNA. What is the name of the domain that these proteins use to attach to each other? (2 words)

10. The region, possessed by some DNA-binding proteins, which includes a metallic element. (2 words)

11. Genes that establish every other parasegment in *Drosophila* embryos. (2 words plus genes)

12. The portion of a parasegment that makes up the posterior part of segments in *Drosophila*.

13. The appendages that form on the head because of mutations in *antennapedia (antp)*.

14. The family of genes that control the development of both the head and tail. (1 word plus genes)

15. Torso's binding capability in *Drosophila*. (3 words)

16. The enzymatic activity associated with torso. (2 words)

17. The signal peptide in the ventral extracellular matrix that is responsible for inducing ventral–dorsal polarity in the *Drosophila* embryo. (2 words)

18. In mammals, the tissue that develops from the cells that grow from the blastocyst wall into the mother's endometrium.

19. The two terms used to describe the dorsal layer of cells and the ventral layer of cells in the embryonic disc of the human embryo at 13 days after fertilization. The ventral layer of cells in the 13-day-old embryonic disc.

20. The primary germ layer of cells that give rise to the mammalian embryo's mesoderm from day 15 to day 18.

21. Paired bodies (providing cells that form vertebrae, ribs, and muscle) that develop from mesoderm on either side of the notochord and neural tube in mammals.

22. In mammals, the organ that forms from the folding of the amniotic cavity under the embryo.

23. The region of chicken limb buds where most *chox* genes are expressed (2 words; hyphenated answer)

24. The polarity established during oogenesis of *Xenopus's* oocytes (unfertilized eggs). (2 words; hyphenated answer)

25. In *Xenopus*, the region created when the cortex and membrane rotate so that the animal cytoplasm is in con⸱ ⸱t with the vegetal cortex. (2 words)

26. The center of cells below the grey crescent of *Xenopus* producing noggin (and possibly Xwnt). (2 words)

27. The group of cells in the dorsal lip of the blastopore responsible for promoting gastrulation and the development of an anterior–posterior axis in *Xenopus*. (2 words)

28. Egg fertilization in the chicken triggers the formation of cleavage furrows on the surface of the animal hemisphere in an area called the blastodisc. What is the name of the disc of cells that results from the many rapid cell divisions of the blastodisc?

29. The edge of the blastodisc that appears to organize the hypoblast in the chicken. (3 words)

30. The three genes that control the development of flower parts appear to be homeotic genes because they control zones of development, and the gene products regulate gene expression by binding to _____.

Answers to the Objective Questions

Multiple Choice

1. *d*   **2.** *b*   **3.** *b*   **4.** *e*   **5.** *a*   **6.** *c*   **7.** *a*   **8.** *c*   **9.** *e*   **10.** *d*   **11.** *c*
12. *b*   **13.** *a*   **14.** *a*   **15.** *e*   **16.** *b*   **17.** *d*   **18.** *c*   **19.** *c*   **20.** *a*   **21.** *b*

True–False

1. T   **2.** T   **3.** F (imago)   **4.** F (nurse cells provide it with mRNA)   **5.** T   **6.** T
7. F (mesodermal and endodermal tissue)   **8.** F (on the ventral surface)   **9.** T   **10.** T   **11.** T

12. T   **13.** F (until late syncytial blastoderm or early cellular blastoderm)   **14.** T   **15.** T   **16.** T   **17.** T   **18.** F (dorsal layer is ectoderm, ventral layer is endoderm)   **19.** T   **20.** T   **21.** T   **22.** T   **23.** T   **24.** T   **25.** F (region known as the grey crescent)   **26.** F (Nieuwkoop center is below the grey crescent)   **27.** T   **28.** T   **29.** F (septal and petal formation occur but stamen and carpels are replaced by petals and sepals, respectively)   **30.** F (bodies of mesodermal tissue that develop within parasegments)

Matching

1. *E* (maternal genes)   **2.** *C*   **3.** *E* (terminal genes)   **4.** *B*   **5.** *D*   **6.** *A*   **7.** *A*   **8.** *B*   **9.** *C*   **10.** *E* (abdominal-B)   **11.** *A*   **12.** *B*   **13.** *E* (proboscis)   **14.** *D*   **15.** *D*   **16.** *D*

Terms

1. ontogeny recapitulates phylogeny   **2.** imaginal discs   **3.** imago   **4.** maternal mRNAs   **5.** syncytial blastoderm   **6.** ventral   **7.** homeobox   **8.** homeotic genes   **9.** leucine zipper   **10.** zinc finger   **11.** pair-rule genes   **12.** anterior   **13.** legs   **14.** terminal genes   **15.** receptor for torso-like   **16.** tyrosine kinase   **17.** toll-ligand   **18.** placenta   **19.** ectoderm/endoderm   **20.** ectoderm   **21.** somites   **22.** gut   **23.** distal–posterior   **24.** animal–vegetal   **25.** grey crescent   **26.** Nieuwkoop center   **27.** Spemann organizer   **28.** blastoderm   **29.** posterior marginal zone   **30.** DNA

Chapter 13

The Immune System

INTRODUCTION

The **immune system** protects the body from pathogenic organisms such as bacteria and viruses. It consists of many different cells and extracellular molecules that are transported throughout most of the body in **plasma** (the fluid portion of blood), lymph, and tissue fluid; some of these cells remain localized in tissues of the immune system such as the thymus gland, lymph nodes, and spleen. Cells and molecules of the immune system are capable of recognizing and destroying specific foreign substances (antigens).

A complete **antigen** (also called an **immunogen**) is any substance to which an immune response can be made (e.g., by the production of protein molecules called **antibodies** or **immunoglobulins**). An **immune response** in vertebrate animals can be defined as the acquired, transferable capacity of lymphocytes to react with specificity and memory to foreign substances. **Lymphocytes** are a class of **leukocytes** (white blood cells) that recognize specific antigens by means of cell-surface receptors which they themselves have synthesized. Other classes of leukocytes normally aid lymphocytes in their immune responses.

Unfortunately, not all immune responses are beneficial. The immune system normally responds only to foreign substances. Occasionally, however, the body mounts an immune response against its own cells or molecules, producing an **autoimmune disease. Hypersensitivity** or **allergic** reactions are usually discomforting immune responses to specific substances in the environment such as pollens, poison ivy, certain chemicals, cosmetics, dyes, foods, etc. However, some severe hypersensitivity reactions, called **anaphylactic reactions,** can be potentially lethal.

Immunology is the scientific study of antigens, antibodies, and cells of the immune systems in multicellular animals. It can be subdivided into several broadly overlapping disciplines. Historically the first scientific studies in immunology were attempts to understand the mechanisms of immunity. **Immunity** is a physiological condition, developed after contact with an immunogen, that specifically aids the body in combating pathogenic organisms of which the immunogen is a part. **Serology** is the *in vitro* study (mainly for medical diagnostic purposes) of substances in plasma and **serum** (plasma minus its clotting factors) that contribute to immunity. (Serology will not be treated as a separate discipline in this text). **Immunochemistry** is concerned with the chemical structures of and interactions between antigens, antibodies, cell receptors, cytokines, etc. **Immunobiology** includes studies of the cellular basis of immunity (**immunocytology**), transplantation rejections, immunological tolerance, autoimmune diseases, immunological aspects of cancer and **immunogenetics** (which includes the genetic mechanisms, responsible for the diversity of antigen receptors in populations of lymphocytes and for the production of cell-surface molecules that distinguish each individual as virtually unique). For pedantic purposes, however, this chapter has been divided into five sections in the following order: immunity, immunocytology, immunobiology (other than immunocytology and immunogenetics), immunochemistry, and immunogenetics.

IMMUNITY

13.1. What kinds of molecules make the best immunogens?

Large, complex, foreign biomacromolecules such as proteins, glycoproteins, lipoproteins, and carbohydrates make the best immunogens. A bacterial cell has many different antigenic molecules on its surface. Even a single foreign protein molecule contains numerous antigenic sites, called **determinants, epitopes,** or **specificities,** each of which can stimulate a specific antibody response.

13.2. There are two major branches of the immune response: cell-mediated immunity and humoral immunity. What are the characteristics of cell-mediated immunity?

Cell-mediated immunity (CMI) or **cellular immunity** is largely responsible for attacking foreign tissues (grafts) and parasitic organisms (e.g., fungi, protozoans), bacteria-infected and virus-infected cells, and abnormal or cancerous cells. Specific CMI can be experimentally transferred from one genetically identical individual to another (as between highly inbred mice or between identical human twins) by a class of lymphocytes called **T cells (T lymphocytes)** without the help of antibodies.

13.3. What are the characteristics of humoral immunity?

The **humoral** (fluid-phase) branch of the immune response consists of soluble antibodies and other proteins that are primarily responsible for protection against extracellular bacteria, toxins, and the extracellular phase of virus transmission. Antibodies can perform these functions without the presence of cells that produce antibodies. **B lymphocytes (B cells)** and their mature, antibody-secreting counterparts (**plasma cells**) produce antibodies, but antibodies alone, when transferred into another individual, cannot provide the kinds of protection (e.g., killing of virus-infected cells) afforded by cells such as killer T lymphocytes. Based on these transfer experiments, B cells and plasma cells are excluded (by default) from both the CMI and humoral branches of an immune response, even though their products (antibodies) are essential for some activities of cells contributing to CMI (e.g., specific cell killing of antibody-coated cells by other lymphocytes called NK cells). Most immune responses involve activation of both the cell-mediated and humoral branches of the immune system.

13.4. Both antigen-specific and nonspecific cells contribute to CMI. What kinds of cells are in each of these groups?

The cells involved in CMI that recognize specific antigens are called T lymphocytes (T cells). One kind of T cell can directly attack and kill virus-infected cells and cancer cells. Another kind of T cell helps B cells and other T cells to become active and/or proliferate.

Other kinds of leukocytes (including macrophages, neutrophils, natural killer cells, and eosinophils) are antigen nonspecific. Some of them, called **antigen-presenting cells (APCs)** are general phagocytic cells that can attack and digest a wide variety of foreign cells and macromolecules; they then present the processed residues on their own cell surfaces in a manner that is stimulatory to T cells and B cells.

13.5. What is the relationship between B cells, plasma cells, and their antibodies?

Antibodies are synthesized by naive B lymphocytes (before exposure to antigen) and appear as antigen receptors on their outer membrane surfaces. After contact with its cognate antigen, a B lymphocyte may be stimulated to terminally differentiate into a plasma cell that is able to secrete antibody molecules into blood or tissue fluid. In many cases, however, B cells must have the antigen presented to them in the proper format by APCs and must be stimulated by one or more chemical signals from helper T cells before the B cells can differentiate into plasma cells.

13.6. How do antibodies and antigens combine? What results from these unions?

Antibodies are multichain protein molecules produced by B lymphocytes and plasma cells. Each antibody unites specifically with its cognate antigen by forming noncovalent bonds in a lock-and-key manner analogous to the interaction of an enzyme with its substrate. The union of antibody with its cognate antigen does not directly eliminate an antigen. The destruction of antigen is usually accomplished by antibody-enhanced phagocytosis (a cellular phenomenon) or by antibody-enhanced cytolysis (by activation of soluble molecules of the complement system).

13.7. What are cytokines and what role do they play in the immune response?

Cytokines are mainly low-molecular-weight proteins (structurally unrelated to antibodies) produced and secreted by a variety of cell types (including vascular endothelium, lymphocytes, and monocytes). Cytokines regulate the intensity and duration of an immune response, but their activities are usually restricted to the local area in which they were produced. If the cytokine is produced by a lymphocyte or a monocyte/macrophage it is called a **lymphokine** or **monokine**, respectively. Note that, in contrast to antibodies, most cytokines do not directly attack the antigen inciting the immune response, but rather influence the activities of other cells of the immune system (e.g., as stimulators of mitosis or differentiation).

> **Example 13.1.** Cytokines that stimulate the proliferation of specific blood cell types are called colony-stimulating factors (CSFs). The cytokine erythropoietin, produced by liver cells, stimulates the formation of red blood cells (erythrocytes) in the bone marrow. Macrophages and fibroblast cells secrete granulocyte colony-stimulating factor (G-CSF) and macrophage colony-stimulating factor (M-CSF). Interleukin-1 (IL-1), produced by antigen-presenting cells, activates a subclass of T cells called helper T cells. Some helper T cells synthesize interleukin-2 (IL-2), which stimulates proliferation of all activated T cells. Gamma-interferon (a dimer), also produced by helper T cells, activates macrophages.

13.8. What are some of the mechanisms by which antibodies may help prevent disease?

Most foreign cells and substances to which antibodies are bound are destroyed by various mechanisms. Antibody-coated bacteria, for example, are normally engulfed by fixed (immobile) or wandering phagocytes in blood, lymph, intercellular fluid, liver, spleen, and lymph nodes. Antibodies attached to antigens may help to reduce the electrostatic repulsion between the antigenic particles and phagocytic cells and thereby enhance phagocytosis. Any substance that enhances phagocytosis is termed an **opsonin.**

Antibodies can also **agglutinate** (clump, aggregate) particulate antigens such as bacterial cells, retarding their spread through the body and making phagocytosis more efficient by causing many bacteria to be engulfed into each phagocytic vesicle.

The binding of an antibody to a bacterial toxin may block the attachment of the toxin to its cell receptor and thus **neutralizes** its harmful effects. Extracellular viruses may also be neutralized in a similar manner by antibodies. However, antibodies are usually unable to pass from the intercellular fluid into living cells. Once viruses or toxins have entered cells, antibodies cannot pass through the plasma membrane to neutralize them.

Some antibodies can activate plasma proteins collectively called **complement.** Complement activation involves a cascade of protein interactions that leads to the formation of holes or pores in the plasma membranes of target cells, resulting in their osmotic dissolution, or **lysis.** Large phagocytic cells called macrophages have receptors on their outer surfaces for a class of antibodies (IgG) and for some components of the complement system. Antigens that are coated with these antibodies or complement factors are more efficiently phagocytosed than those without the help of these molecules. Complement thus exhibits opsonic activity.

In summary, antibodies do not destroy antigens directly, but are aided in this function by phagocytic cells and complement through such processes as opsonization, agglutination, toxin or viral neutralization, and cellular lysis.

13.9. How is immunity established and maintained?

Immunization is the process of activating the immune system. This results in a physiological state, called immunity, that is better prepared to defend the body against specific potential pathogens or other foreign invaders (e.g., incompatible blood or tissue). While the capacity of the immune system to respond to a broad spectrum of antigens is inherited (and may thus vary from one individual to another), the immune response itself develops only if the genetic capacity to respond to a specific antigen is present and after contact with that antigen is made; i.e., it is an *acquired response.*

When a B cell is presented antigen in the proper format and receives a chemical signal from helper T cells it begins to proliferate, forming a clone of cells with identical antibodies as their cell receptors for that same antigen. Some of the cells in such a clone differentiate into antibody-secreting plasma cells. These cells usually die after a few weeks of intensive antibody secretion. The antibodies secreted by a plasma cell will gradually be degraded over a few weeks or months by protease enzymes. The rest of the cells in that same clone partially differentiate (by rearrangement of their immunoglobulin genes) into nonsecreting, long-lived **memory cells,** ready to respond more rapidly and in greater numbers upon the next encounter with that same antigen. This is the memory or secondary response; also known as the "booster response" or **anamnestic response** (literally, "not forgetting").

13.10. Immunity may be acquired actively or passively, naturally or artificially. What is active immunity and how can it be attained?

Lymphocytes become **sensitized** or **primed** to active status by an antigenic stimulus and respond either by secretion of antibodies and/or lymphokines. An individual that has been sensitized in this way (either by natural contact with germs or artificially by administration of a vaccine) is said to have gained **active immunity.** A **vaccine** contains dead or attenuated (weakened) microbes or antigenic components of microbes that can stimulate immunity without causing disease. In some cases, active immunity is long lasting, giving lifelong protection by a single vaccination (e.g., polio). In other instances, such as tetanus, additional doses of vaccine need to be given at regular intervals (perhaps every five years or so) to maintain effective protection.

13.11. What is passive immunity and how can it be attained?

The serum from a sensitized individual contains antibodies that can be transferred to another individual of the same or different species who then gains **passive immunity.** This form of immunity may be acquired either artificially by an injection of antibodies from another organism or naturally by passage of some antibodies from a mother to her child across the placenta. In humans and other mammals, some antibodies can also be passed to a nursling through the mother's milk. Antibodies obtained passively are normally degraded over weeks or months, leaving the individual in as much risk of becoming infected with a pathogenic organism as prior to the passive immunization.

13.12. Why is the immune response to the same antigen so variable from one individual to another and within an individual from one time to another?

The immune response is highly specific in that it is normally directed only against the foreign antigen that stimulated the response. However, the degree of protection thereby afforded to the organism may be quite variable. For reasons that are still unclear, some immune responses to vaccines (or to immunogens encountered in nature) may confer lifetime protection (e.g., vaccination for poliomyelitis), whereas other vaccinations (such as for flu viruses) may offer protection for only a year or so. This variable protection may in part be due to a relatively short lifetime of some memory cells generated by the initial antigen contact (part of the **primary immune response**). However, if the antigen is repeatedly contacted at regular intervals (e.g., yearly) each such encounter "boosts" the immune response to higher protective levels (**secondary immune responses**). If each secondary immune response is effective, the host may remain healthy and not even realize that contact with a potentially harmful microbe has occurred. (A subclinical infection is one in which no symptoms of the disease become apparent). If, on the other hand, the antigen has not been contacted for several years, the capacity of the organism to mount an effective immune response when subsequent contact is made may wane with time, perhaps because of early death of relatively short-lived memory cells.

Even a high state of immunity to a particular pathogen may not necessarily provide protection, because a sufficiently large dose of bacterial toxin or virus might overwhelm the immune defenses. The dosage of antigen can be rigorously controlled in vaccination, but may vary widely in natural infections.

The immune systems of babies are not fully mature at birth. They cannot respond to the range of immunogens of which young adults are capable. Thus, babies usually have to rely on the antibodies they received from their mothers during the first few months of their lives. Our immune systems tend to grow weaker in the advanced years of our lives. These facts may at least partly explain why our susceptibility to develop cancer is greater near the beginning and end of our lives than at other times.

Mental stress has been shown to have a deleterious effect on our immune system. Corticosteroids are potent anti-inflammatory agents that are often used medically to prolong the survival of grafted tissue by suppressing immune responses. Corticosteroids tend to be overproduced in response to stressful conditions, causing a marked depletion of circulating lymphocytes.

Some viruses, such as those causing influenza or AIDS, seem to be able to easily mutate their surface molecules (markers), so that antibodies and/or T cell receptors responsive to the original antigens of these pathogens are no longer effective. These and other possible reasons exist for the variable protection offered by the immune system to the same antigen or to different antigens in genetically different individuals of a given species, or in the same individual at various times during its life.

IMMUNOCYTOLOGY

13.13.   What are the major blood cell types and their respective functions?

Red blood cells (erythrocytes) carry oxygen from the lungs to the tissues. White blood cells (leukocytes) defend the body against pathogenic organisms. There are five major types of leukocytes: lymphocytes, monocytes, neutrophils, basophils, and eosinophils. Lymphocytes and monocytes have a globular nucleus and lack prominent granules in the cytoplasm. Hence, they are classified as mononuclear leukocytes or agranulocytes. Neutrophils, basophils, and eosinophils have a multilobed nucleus and cytoplasmic granules that stain red (eosinophils) or purple (basophils) or have small granules not easily visible in the light microscope (neutrophils). They are thus classified as polymorphonuclear leukocytes or granulocytes.

B lymphocytes produce antibodies that recognize specific antigens; both B and T lymphocytes produce lymphokines that influence the effects of one another and of other leukocytes. Blood monocytes (and their differentiated tissue counterparts called macrophages) nonspecifically phagocytose foreign particles and function as antigen-presenting cells for lymphocytes. Neutrophils are the most highly phagocytic granulocytes, but they are not antigen-presenting cells and are unable to stimulate lymphocytes. Basophils (and their differentiated tissue counterparts called mast cells) are involved in allergic reactions. The role of eosinophils is not clear, but they may aid in immunity to parasitic organisms such as worms. Eosinophils are weakly phagocytic; basophils are not phagocytic.

13.14.   There are two major functional classes of lymphocytes that cannot be distinguished by morphology in the light microscope: B lymphocytes (B cells) and T lymphocytes (T cells). Why are these cells designated B and T?

All blood cells, including lymphocytes, originate from pluripotent stem cells in the bone marrow. (The adjective "pluripotent" applies to embryonic tissue that has several possible developmental fates.) Mammalian B lymphocytes not only originate in bone marrow, but they also mature in that environment. In birds, the primary lymphoid organ where B cells mature is the *b*ursa of Fabricius, located at the junction of the hind gut and the cloaca. The designation "B" thus applies to both the bursa and bone marrow.

The other major functional class of lymphocyte is the T cell. T cells also originate from stem cells in the bone marrow, but mature and learn to recognize self from nonself in the thymus gland (hence the designation "T").

13.15.   How do B cells and T cells recognize their cognate antigens?

When released from the bone marrow of mammals or the bursa of birds, B cells are naive or virgin cells, having not yet encountered antigen, but possessing antibodies attached to the external surface of their plasma membrane. All the antibodies of a B cell have a single common specificity (monospecific); i.e., they react with only a single kind of antigenic determinant. The vast majority of B cells recognize different antigens.

T cells have protein molecules that form **T-cell receptors (TCRs)** on their plasma membranes for recognizing specific antigens. All the TCRs on a given T cell are identical, but vary widely from one T cell to another within the same individual. T cells do not make antibodies, but TCRs are distantly related to antibodies and are classified as part of the immunoglobulin supergene family. When activated by antigen contact, T cells synthesize and secrete protein growth factors called lymphokines (such as the interleukins) that affect the activities of other T cells, B cells, and antigen-presenting cells.

13.16.   At least three major T-cell subclasses develop in the thymus before they are able to participate in immune responses. What unique functions are ascribed to each of these classes?

Various lymphokines made by **helper T cells** (T_H) are stimulatory to B cells, antigen-presenting cells, and to all classes of T cells, whereas those made by **suppressor T cells** (T_S) inhibit these same cell types. Helper and suppressor T cells are thus referred to as **regulatory cells.** A third functional subclass of T cells recognizes foreign antigens on virus-infected host cells, cancer cells, and foreign grafted cells or tissues, and destroys them by producing and secreting proteins called **perforins** that form membrane channels analogous to those made of complement proteins. These cells are called **cytotoxic T cells** (T_C) or **cytotoxic T lymphocytes** (CTL) or **killer T cells.** B cells and T_C cells are the specific **effector cells** of the immune system that attach to antigens via their surface receptors. T_H and T_C cells have antigen receptors encoded by the same gene cluster, but they do not recognize the same major histocompatibility molecules (Question 13.17) on antigen-presenting cells, reflecting the different functions of the two types of cells.

13.17. What is the major histocompatibility complex and how is it involved in the function of antigen-presenting cells?

The **major histocompatibility complex (MHC)** is a chromosomal segment containing several genes the protein products of which are the main chemical markers whereby cells of the immune system recognize one another and distinguish foreign cells from self cells. When complexed with MHC molecules, processed antigen can be presented in the correct format on the surface of APCs or B cells to stimulate helper T cells to release their lymphokines (e.g., the "helper T-cell signal") that stimulates other lymphocytes (both T and B cells) to become active and to proliferate. This helper signal is essential for B-cell antibody responses to most antigens and for the generation of long-lived memory cells.

Antigen-presenting cells (APCs) include blood monocytes and their tissue counterparts (macrophages), dendritic cells in lymphoid tissue, thymus epithelial cells, and Langerhans cells in the skin. These APCs nonspecifically endocytose antigen, digest it partially, and then display this **processed antigen** on their surfaces together with self-marker glycoprotein molecules of the MHC. Antigens experimentally presented to T cells or B cells by foreign APCs (bearing MHC molecules different from those of the recipient) are not able to stimulate an immune response in the recipient. Thus each individual is **MHC restricted** to recognize antigens only in association with self-MHC molecules.

B cells can specifically bind their cognate antigen by the membrane-bound antibodies on their cell surface. After internalization and processing the antigen, B cells display the processed products on their cell surface together with protein molecules of the MHC in a form that can stimulate helper T cells to release their lymphokines (the helper T-cell signal). For B-cell activation to occur, the B cell and the helper T cell must recognize different epitopes on the same molecule of processed antigen, a phenomenon termed **associative recognition.** The B cell tends to recognize a small antigenic determinant because of its accessibility on the surface of the antigenic molecule and its hydrophilic nature. The T cell tends to recognize the more internal, hydrophobic residues of the carrier molecule that become displayed after antigen processing. Sufficient dosage of the helper T-cell signal to activate a B cell requires close cell-to-cell contact between the T cell and B cell.

13.18. Each B-cell or plasma cell clone produces a monospecific antibody, but has a short life span in culture. How then can a potentially unlimited amount of a monospecific antibody be produced experimentally?

A **myeloma** is a cancer consisting of a clone of immortalized B cells. These cells can grow indefinitely in culture. Some myeloma cells, however, are defective in their own ability to make antibodies. After B cells have been activated in a mouse immunized with a particular antigen, the spleen is removed. Spleen cells are then fused with a mutant myeloma cell line to produce immortalized hybrid cells called **hybridomas,** some of which secrete the **monoclonal antibody** of interest. The selective growth medium for supporting the hybridomas contains an inhibitor (aminopterin) that blocks the normal pathway for synthesis of nucleotides, thus forcing B cells to use an alternative (salvage) pathway for nucleotide synthesis. The chosen mutant myeloma cell line is defective in this regard and cannot survive unless it forms a hybidoma with a B cell that has an intact salvage pathway. Thus, the selective medium only allows long term growth of hybridoma cells. Each hybridoma clone is tested for the production of desired antibodies. Those clones producing the desired antibody can subsequently be subcultured indefinitely on normal growth medium.

IMMUNOBIOLOGY

13.19. How do T cells and B cells learn to distinguish self from nonself so as to avoid attacking other cells of the host organism?

An unusual genetic system controls the development of T and B cells, so that millions of different antigen receptors (antibodies on B cells; T-cell receptors on T cells) can be made. Some of these receptors can recognize self antigens. One theory suggests that, during embryological development, as immature T and B cells contact high concentrations of self antigens in primary lymphoid organs (bone marrow, thymus), the self-reactive cell clones are induced to die (resulting in **clonal deletion**) or to become immunologically unresponsive (resulting in **clonal anergy**), leaving only lymphocytes without receptors for self antigens. The host thus acquires **natural immunological tolerance** for self antigens. Another theory proposes that immunity or tolerance to self antigens may also be induced in mature lymphocytes. If a B cell or T_C cell binds antigen and also receives a signal from a T_H cell or an APC, immunity is induced. But if antigen is recognized in the absence of a T_H cell signal, self tolerance is induced (e.g., by clonal deletion).

Two types of selection of T-cell receptors are thought to occur in the thymus. Thymocytes devoid of surface receptors that recognize self-MHC molecules undergo programmed cell death **(apoptosis).** Thymocytes that bind to self-MHC molecules are thus positively selected to be MHC restricted. However, thymocytes bearing high-affinity receptors for self-MHC molecules, or self antigens complexed with self-MHC molecules, are negatively selected and die, leaving only self-restricted thymocytes that are self-tolerant.

Example 13.2. Dizygotic cattle twins that develop from two fertilized eggs are genetically nonidentical and therefore should not be tolerant of skin grafts from one another as adults. But they are. The reason for this is that their placentas and blood vessels are commonly fused, allowing exchange of blood cells at such an early age that each twin's cell markers are treated as common self antigens.

Example 13.3. Immunological tolerance can also be experimentally induced in one strain of mice by injecting newborns with immature lymphoid cells from neonates of a genetically different strain. At this stage of development, the immune system of the recipient will often learn to accept the antigens on the grafted cells as self antigens, and so will be tolerant of skin grafts from that same donor strain when it becomes an immunologically competent adult. A delay of even a few days in the experimental induction of tolerance, however, may be too late; by then the window of opportunity has closed; the animal will thereafter be intolerant of foreign cells.

Occasionally tolerance of self antigens becomes defective, causing T and/or B cells to react against their own tissue antigens, resulting in **autoimmune diseases.**

Example 13.4. In myasthenia gravis, patients make antibodies against acetylcholine receptors on their own skeletal muscles. This interaction blocks the effect of this neurotransmitter, resulting in muscular weakness and perhaps death if breathing is affected.

13.20. How does a foreign antigen stimulate a specific immune response (i.e., one directed only at that antigen) in immunologically competent individuals?

The immune system is continuously producing new populations of genetically different lymphocytes. Each T- and B-cell clone is genetically programmed to respond only to a single antigenic determinant. When foreign antigen is presented by an APC, only T and B cells that have the correct antigen receptors will recognize it. Thus, the antigen *selects* (as it were) from the multitude of lymphocyte clones, the ones that should be activated, be stimulated to differentiate, and made to proliferate. This process is known as the **clonal selection theory.**

The binding of an antigen to a lymphocyte's receptors activates the cell, causing it to proliferate into a clone of identical cells. Each member of a clone normally has identical receptors to only a single kind of epitope. Some members of such a B-cell clone will mature into antibody-secreting plasma cells, while others will become long-lived memory B cells. A similar process occurs on T-cell clones; some members become activated and release lymphokines, whereas other members remain inactive as memory cells.

13.21. Why is a specific immune response polyclonal?

Foreign proteins are the best antigens. However, large, foreign, branched polysaccharides are usually good antigens. Even small carbohydrate chains (oligosaccharides) or individual sugars (monosaccharides) can stimulate an immune response if linked to a large foreign polypeptide or protein "carrier" molecule (e.g., antigens of the ABO blood group system). A single immunogenic protein molecule usually contains multiple epitopes, each of which is able to act as a different antigenic determinant to stimulate immune responses from different lymphocyte clones. Thus, the immune response of an individual animal to a single kind of foreign protein molecule or foreign cell (with a multitude of different foreign complex antigens displayed on its surface) is usually **polyclonal;** i.e., many different lymphocyte clones respond to such a challenge, each clone recognizing a different epitope. Even a single epitope of an immunogen can stimulate a polyclonal response. Some lymphocyte clones will have receptors with a high **affinity** for the determinant, binding strongly to it by the sum of multiple noncovalent forces.

The antigen-binding regions of these receptors (T-cell receptors or membrane-bound antibodies) are called **paratopes.** Cells with high affinity paratopes are stimulated to produce larger clones than cells with lesser affinity for the same epitope.

13.22. What is affinity maturation?

Early in an immune response, when antigen concentration is relatively high, the average affinity of paratopes over all clones responding to a specific epitope is relatively low. But as time goes by and the antigen becomes scarce, the few clones with the greatest affinity for the epitope reproduce more efficiently than the others, so that most reactive cells will have higher affinities. This increase in the average affinity of receptors with time during an immune response is called **affinity maturation.** Affinity maturation can also occur over longer periods of time as new receptors come into existence and are selected by reexposure to the same antigen by recurrent or chronic infections, or by repetitive immunizations ("booster shots").

13.23. Why is it that some foreign molecules (or portions of molecules) may be unable to stimulate an immune response, yet be the target for antibodies?

Antigenic determinants that protrude from the surface of globular proteins are more accessible to lymphocyte receptors than those in the interior of the molecule. If these exposed epitopes are also capable of stimulating a strong immune response, they are said to be the **immunodominant** portions of such a molecule. Small foreign molecules (such as a single sugar or amino acid) with a molecular weight of less than about 10,000 daltons are not usually immunogenic; i.e., they cannot by themselves elicit an immune response. These kinds of molecules are called simple antigens, or **haptens.** However, haptens can artificially be made immunogenic by coupling them to a larger **carrier** molecule such as a protein. If the carrier molecule is also foreign, the immune response is likely to produce a variety of antibodies, some of which react only with epitopes of the hapten, some only with epitopes of the carrier, and possibly some that recognize parts of both molecules.

13.24. How is an immune response modulated when it is no longer needed?

Secreted antibodies bind to antigen and thereby prevent binding of antigen to membrane-bound antibody on B cells or to T-cell receptors, thus interrupting the primary signal that activates more B and T cells and continues to stimulate plasma cells to secrete antibodies. After the antigen is cleared from the body, plasma cells secreting the corresponding cognate antibody undergo apoptosis, leaving only nonsecreting memory cells awaiting the next exposure to the same antigen. Secreted antibody molecules, like many soluble proteins, have a finite life and are destroyed by proteases after a few weeks or months.

Each antigen-combining site (paratope) on an antibody molecule usually contains multiple epitopes—called **idiotopes**—which can be recognized by some membrane-bound antibodies on other B cells of the same individual. The **idiotype** of an immunoglobulin molecule is the sum of its individual idiotopes. T-cell receptors have their own idiotypes. While the concentration of antigen is the primary regulator of the intensity of an immune response, a network of interactions between idiotypes of antibodies and of T-cell receptors may also be involved. A clone of plasma cells releases thousands of molecules of monospecific antibody. Some other B cells are likely to have membrane-bound antibodies that are able to recognize the idiotypes of these soluble monospecific antibodies, thus causing these cells to release anti-idiotype antibodies. Some of these anti-idiotype antibodies may be able to react with idiotopes outside the paratope of the soluble antibodies, whereas others may be able to react with idiotopes in the paratope. The latter anti-idiotype molecules would be the internal image of the immunogen. The anti-idiotypes would be recognized by other B cells producing anti-anti-idiotypes, etc. Each level of response by B cells and T cells would be progressively weaker, so the whole system is self-damping. It is not known how some idiotype networks stimulate and others depress immune responses. It has been suggested that suppressor T cells might indirectly subdue the immune response by interacting with the idiotypes on helper T cells.

IMMUNOCHEMISTRY

13.25. What are the major structural features of a typical immunoglobulin molecule?

A typical immunoglobulin (Ig) protein molecule has a sedimentation coefficient of about 7S and a molecular weight of about 150,000 daltons. It is a heterotetramer containing two identical **heavy chains** (H chains; approximately 50,000 daltons and 440 amino acids each) and two identical **light chains** (L chains; approximately 25,000 daltons and 212 amino acids each).   The overall shape of an Ig molecule is modeled by the letter Y (Fig. 13-1). The tail end or stem of the Y consists of the carboxyl halves of two identical H chains, held together by covalent disulfide bonds as well as by numerous noncovalent interactions between side groups of amino acids.

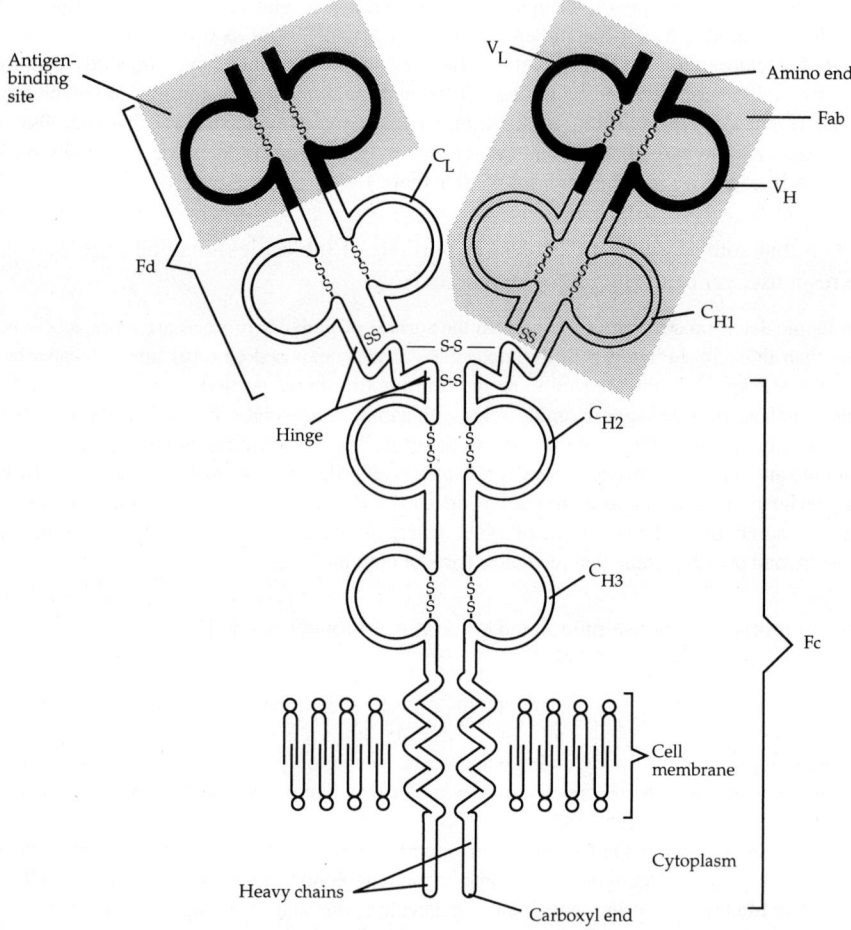

Fig. 13-1.   Diagram of a 7S immunoglobulin molecule.

13.26.   On what basis are immunoglobulins classified?   How many immunoglobulin classes are recognized?

There are five major classes of immunoglobulins, designated IgM, IgG, IgA, IgE, and IgD.   The heavy chains of Ig molecules within each class are designated γ (gamma), μ (mu), α (alpha), ϵ (epsilon), and δ (delta) for classes IgG, IgM, IgA, IgE, and IgD, respectively.   Within an Ig class, each Ig molecule has a common amino acid sequence in the **constant region** (Y stem) of its H chains (C_H).   Immunoglobulins from one species are foreign antigens when injected into another species, causing the recipient to make its own antibodies to the foreign antibodies.   Each class of immunoglobulin is thus defined by the anti-immunoglobulins with which it reacts in a serological test.

13.27.   Describe the major structural regions of H chains that do not function in binding antigen.

There are three constant domains (C_1, C_2, and C_3) in the constant region of heavy chains in classes IgG, IgA, and IgD; there are four such domains in classes IgM and IgE.   Within an Ig class, all constant domains have homologous amino acid sequences.   Each domain contains a pair of intrachain disulfide bridges.   This molecular homology among the domains suggests that their structural genes evolved from a common ancestor by duplication and have diverged by subsequent mutations at some of their amino acid positions. Between C_{H1} and C_{H2} domains is a flexible **hinge** region that allows the arms of the Y structure to form a variety of angles (Fig. 13-1); heavy chains μ and ϵ lack hinge regions and contain an extra constant domain in their central region.

13.28.   What are the structural regions of H and L chains that contribute to antigen binding?

The first 65 amino acids at the amino ends of heavy chains vary in amino acid sequence from one Ig molecule to another, if made by different B-cell clones. However, Ig molecules made by members of a fully differentiated plasma cell clone are identical in all respects; they are monoclonal antibodies (MoAbs). The variable amino end of a heavy chain is referred to as its **variable region** (V_H). Similarly, light chains have a single variable domain (V_L) at their amino ends, but only a single constant domain (C_L) at their carboxyl ends. The binding of antigen is largely restricted to three widely spaced **hypervariable regions** within both the V_H and V_L domains of an immunoglobulin molecule. Hypervariable regions are also known as **complementarity-determining regions** because the antigen-binding sites on a cognate antibody (paratopes) are of complementary structure (for forming relatively weak noncovalent or secondary bonds) to the epitopes (antibody-binding sites) on the antigen. The hypervariable regions collectively occupy 15–20% of the variable domains on H and L chains. All other regions within the V_H and V_L segments are called **framework regions.**

The V_H and C_{H1} segments of an H chain constitute the **Fd region;** it is the heavy-chain counterpart of a light chain. A light chain is connected covalently by a disulfide bond to an H chain in the C_{H1} and C_L regions. Elsewhere the H and L chains intimately associate through multiple noncovalent bonds. The variable domains of an L and an H chain associate in such a way that they form a kind of depression or pocket into which the antigen can fit in lock-and-key fashion. Each arm of the immunoglobulin molecule thus has an identical antigen-binding site. The Y-shaped 7S immunoglobulin is therefore bivalent (having two combining sites). Carbohydrates are enzymatically added to the constant domains, making immunoglobulins conjugated proteins (glycoproteins in this case), rather than simple proteins (consisting only of amino acids).

13.29.   What types of L chains are present in an immunoglobulin molecule?

There are two types of L chains classified according to the type of constant domain they have: kappa (κ) and lambda (λ). It is possible for any one of the five H-chain types to be present in the same Ig molecule with one of two types of L chains. Normally all immunoglobulins produced by a given plasma cell will have exactly the same structure, i.e., two identical H chains and two identical L chains. Thus, Ig molecules may be designated as $2\lambda2\gamma$, $2\mu2\kappa$, etc.

13.30.   What is the consequence of digesting an IgG molecule with papain?

The proteolytic enzyme papain digests Ig molecules near the junction between C_{H1} and C_{H2}, forming two **Fab** (fragment antigen-binding) molecules and one fragment that, because of its uniform amino acid sequence within a class, can be crystalized; this is designated as **Fc** (fragment crystalizable). The Fc fragment consists of the carboxyl halves of two identical H chains containing all C_H domains other than C_{H1}. These two chains are joined by disulfide bonds near the hinge region (Fig. 13-1).

13.31.   How do the kinds of antibodies produced by a B-cell clone change from the primary to the secondary response?

IgM is the antibody class produced initially by all virgin B cells. Membrane-bound IgM molecules are the antigen receptors for the primary immune response. They are the first antibodies to be secreted by plasma cells. Later, that same plasma cell may secrete an antibody of a different class (e.g., IgG). These antibodies all bear the same type of L chain and the same variable regions in both the H and L chains. Thus, the antigen-binding site (paratope) of the antibody never changes during the life of a B cell or a plasma cell. This **H-chain class switching** involves only the constant regions of H chains. The secreted form of IgM actually consists of five 7S-like molecules (Fig. 13-2) covalently joined near their carboxyl ends by disulfide bonds, together with a single polypeptide chain called **J chain.** Thus each secreted IgM molecule has ten potential paratopes or a **valence** of ten.

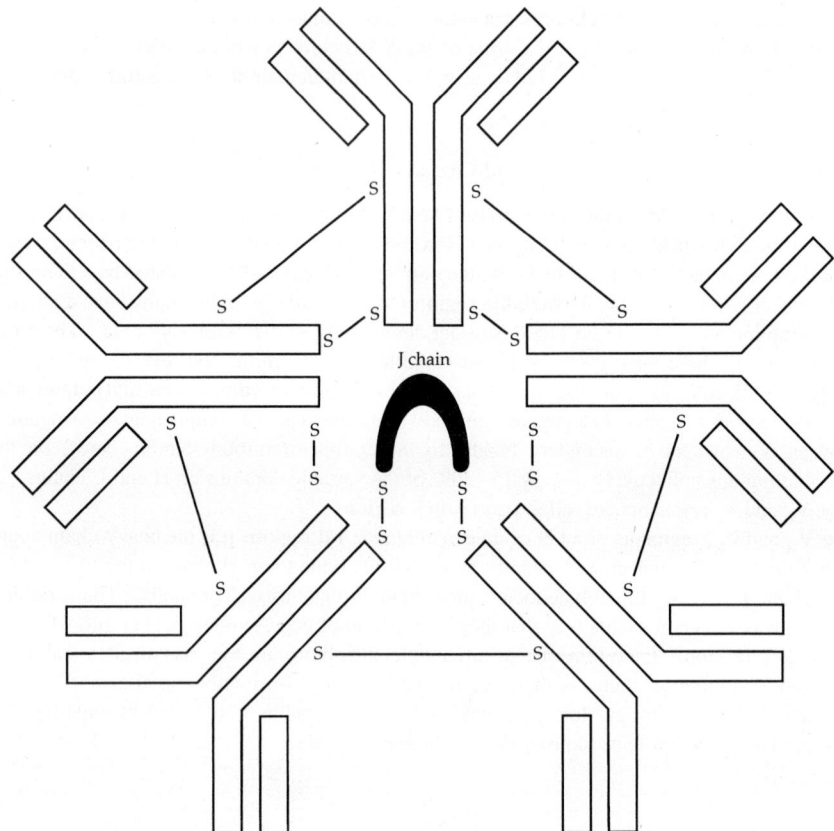

Fig. 13-2. Diagram of a pentameric 19S IgM molecule consisting of five 7S-like monomers joined by disulfide bonds (S-S) and a single J chain.

13.32. Aside from binding specific antigens, what other functions do IgG molecules perform?

Human antibodies of the IgG class predominate (approximately 75%) in secondary immune responses. Only this class of antibodies can normally pass through the placenta. Hence, any maternally derived antibodies in a baby's blood are normally of class IgG; antibodies of other classes would be products of the baby's own immune system. Both secreted IgG and IgM can activate complement, but on a molar basis IgM is much more efficient than IgG in this respect. Both macrophages and neutrophils possess cell-surface receptors for the Fc stem of IgG molecules. In contrast to macrophages, the leukocytes called neutrophils do not function as antigen-presenting cells. The stem of the IgG antibody is bound by noncovalent forces to these APC receptors, and when an antigen-IgG complex is formed, these cells will usually endocytose the complex and destroy (neutrophils) or process (macrophages) the antigen. The antigenic residues are then displayed on the macrophage surface together with MHC molecules. This complex stimulates helper T cells to provide the lymphokines needed to initiate mitotic divisions and differentiations of both specific T cells and B cells into memory cells and effector cells.

13.33. What kinds of biological responses are initiated by IgE-antigen complexes?

Blood basophils and their tissue counterparts (mast cells) have cell-surface receptors for the Fc stem of IgE molecules. When an antigen-IgE complex forms on these cells, they are stimulated to exocytose the contents of their cytoplasmic granules (degranulation) to the extracellular fluid. Histamine is stored in these granules, and its release may cause bronchial smooth muscle contraction, nasal mucus secretion, irritation of nerve endings in the skin, and other indicators of an immunological allergic reaction or hypersensitivity reaction. These granules also contain a chemotactic factor that attracts and activates other types of white blood cells called eosinophils, which attack various parasites, especially those coated with IgG antibodies.

13.34. What functions do IgA and IgD perform?

The principal class of antibodies in body secretions (milk, saliva, tears, respiratory and intestinal secretions) is IgA. This Ig class may exist as a monomeric 7S unit or as a dimer of two such units connected by a single J chain (produced by the plasma cell) and a chain called the **secretory piece** (or secretory component) that is produced by epithelial cells associated with the secretion of IgA. The J chain is the same 15 kd polypeptide used to polymerize the 7S-like monomers of IgM into pentameric form. Receptors for the Fc region of IgA molecules exist on the nonluminal surface of epithelial cells lining the intestines, bronchi, milk ducts, salivary ducts, and tear ducts. After IgA molecules have bound to these receptors by their stems, they are transported from the extracellular fluid across these epithelial cells by the same kind of process (transcytosis) that transfers IgG molecules from maternal blood to fetal blood.

IgD molecules appear early on virgin B cells, but no specific biological functions have been associated with them.

13.35. What is the complement system?

Complement is the name given to a group of nine major proteins (C1 through C9) and about an equal number of regulatory proteins. Most complement proteins are normally present in the body regardless of the state of immunization in any individual. Complement proteins are synthesized by the liver, and they circulate in blood and tissue fluids.

13.36. How is the complement system activated?

There are two pathways for activation of the complement system: the **classical pathway,** which is activated by antigen-antibody complexes involving IgG or IgM, and the **alternate pathway,** which is activated by the surface polysaccharides of infecting bacteria. Each pathway consists of a cascade of proenzymes becoming activated to enzyme status by the preceding component. The order of activation in the classical pathway is C1 → C4 → C2 → C3 → C5, 6, 7, 8, 9. Complement components are numbered in the order of their discovery rather than the order in which they are activated (C4 is the only deviant component in this regard). At each step, a proenzyme is cleaved at a specific site into two fragments labeled *a* and *b,* the latter being the largest. The larger fragment usually binds to the nearest membrane, and it must either be quickly converted to the next component of the pathway or it becomes permanently inactivated. The pathway culminates in the formation of a C56789 **membrane attack complex** that creates a pore in the bacterial outer membrane or plasma membrane, leading to osmotic lysis of the cell.

13.37. How do the classical and alternate pathways for activation of complement differ?

An antigen must be bound by more than one antibody to activate the classical pathway. Hence, there must be a cluster of antigen-antibody complexes on a cell surface to initiate this complement cascade. The alternate pathway, however, is directly activated by polysaccharides of the microbial cell wall without the help of antibodies, thereby providing a first line of defense before antibody response is possible. C3 is unstable and spontaneously splits into C3a and C3b. C3b readily binds to bacterial cell walls, to yeast cell walls, to some viral envelopes, and to another serum protein known as factor B. C3b activates factor B, which then activates another serum protein called factor D. Factor D splits C3b-bound factor B into components (Ba) and (Bb). The complex C3b (Bb) is the functional equivalent of C4b2b of the classical pathway, but it soon decomposes unless the serum protein properdin binds to and stabilizes it. From this point onward, the classical and alternate pathways are the same. A specific inhibitor protein (factor H) competes with factor B for a binding site on C3b. The polysaccharides of the microbial cell wall somehow protect C3b from inactivation by factor H, thus giving factor B the opportunity to bind to C3b and activate the rest of the pathway. The classical and alternate pathways are much more complicated both in activation and suppression than outlined here.

13.38. Does complement perform any functions other than cytolysis?

Some of the small fragments (especially C3a and C5a) cause smooth muscle contraction and can stimulate mast cells and basophils to release histamine. This chemical increases the permeability of local blood vessels, facilitating the entry of more complement, antibodies, and leukocytes into the site of infection. Phagocytic neutrophils are attracted to the site by the chemotactic activity of C5a. Macrophages have cell-surface receptors for some activated membrane-bound complement proteins on bacteria, thereby focusing their attack on such microbes. These opsonic complement components are primary mediators of inflammatory reactions.

13.39. What is the composition of the antigen receptor on T cells?

The **T-cell receptor** (TCR; Fig. 13-3) is a heterodimer (consisting of two different polypeptide chains) that exists in two forms. The most common form (over 95%) consists of a 40 kd alpha (α) chain and a 43 kd beta (β) chain; the less common form consists of a gamma γ and a delta (δ) chain. Each chain has one variable domain at its amino end, followed by one constant domain, a hydrophobic domain to anchor the TCR in the plasma membrane, and a hydrophilic domain at its carboxyl end. The V_α and V_β domains together form an antigen-binding region (paratope). The TCRs are physically associated on the cell surface with a set of polypeptide chains called the CD3 complex. The CD3 complex contains five peptides and is thought to be involved in transmission of a signal from a TCR-antigen complex to the interior of the cell. The designator CD stands for **cluster of differentiation** (cluster determinants, cluster designation) and refers to a set of surface protein antigens (called "markers") that characterize different leukocyte subpopulations according to the presence or absence of these antigens during various stages of development and/or functional activities. The CD molecules that characterize a given cell population are identified by a group ("cluster") of monoclonal antibodies. Some of the markers belong to more than one cluster of determinants. CD markers perform various functions such as the promotion of intercellular interactions and transduction of signals that lead to lymphocyte activation.

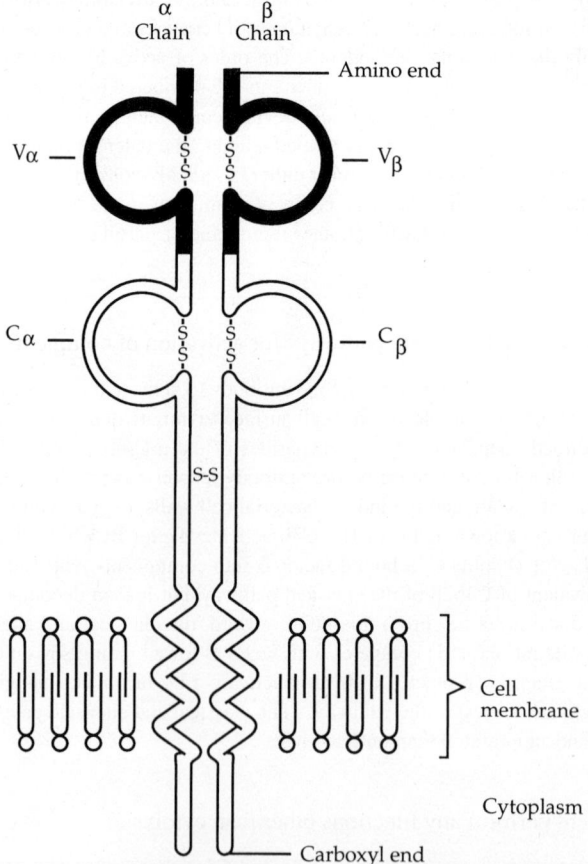

Fig. 13-3. Diagram of a T-cell receptor.

Example 13.5. Mature human T cells bear antigens 1 and 3 (CD1$^+$3$^+$). Mature helper T cells or inducer T cells also bear cluster 4 (CD4$^+$), whereas suppressor T cells and cytotoxic T cells are designated CD8$^+$. Human B cells are positive for CD19-22.

13.40. What kinds of molecules are made by the major histocompatibility complex (MHC)?

The MHC encodes three classes of molecules designated I, II, and III. Both class I and class II gene products are glycoproteins located on cell surfaces; class III genes encoded some proteins of the complement system and two cytokines called tumor necrosis factors (TNFα and TNFβ). Macrophages are the major producers of TNFα, while certain T cells produce most of TNFβ.

13.41. Describe the structure of class I MHC molecules.

Class I MHC molecules (Fig. 13-4a) are heterodimers. The α chain is a glycosylated transmembrane protein of about 345 amino acids, encoded by a MHC gene on human chromosome 6. A gene on chromosome 15 (not a member of the MHC) encodes the other, smaller (96 amino acids), nonglycosylated polypeptide, of invariate structure, called β_2-microglobulin. An α chain and a totally extracellular β_2-microglobulin chain associate noncovalently to form each class I MHC molecule. The α chain has three extracellular globular domains at the amino end, connected to a hydrophobic transmembrane domain. There is also a cytoplasmic domain at the carboxyl end. The first two domains at the amino end of the α chain contain the polymorphic (variable) residues that are recognized as foreign antigens by cytotoxic T-cell receptors during tissue transplantation rejection reactions and killing of host cells infected with intracellular parasites.

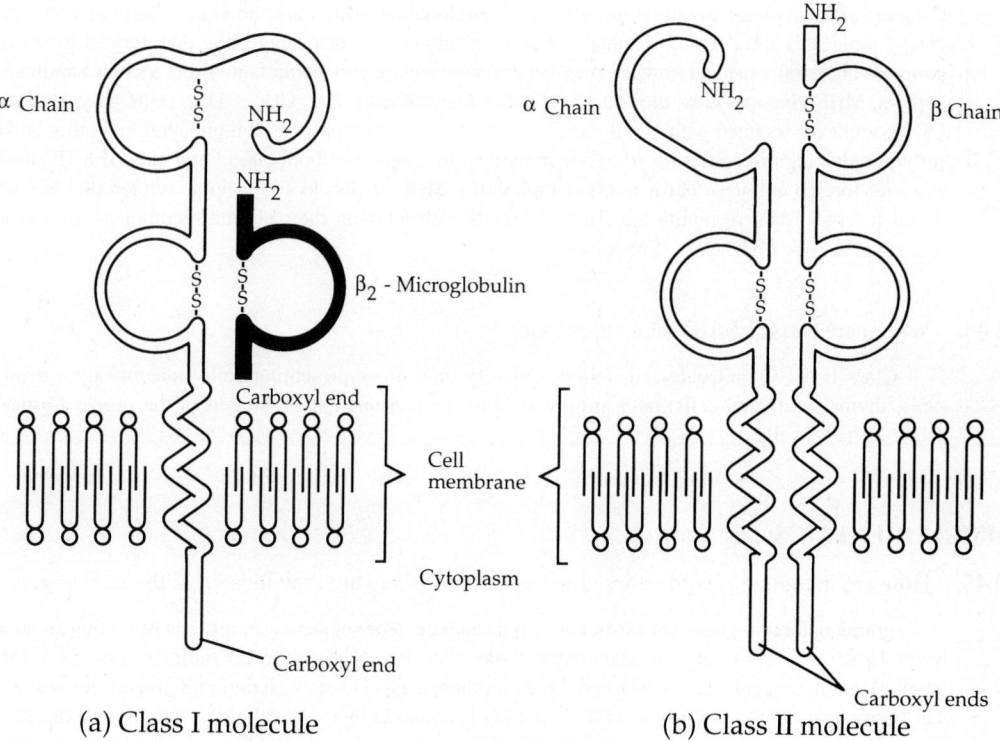

(a) Class I molecule   (b) Class II molecule

Fig. 13-4. Glycoproteins of the major histocompatibility complex (MHC) showing (a) class I and (b) class II molecules.

13.42.   Where are class I MHC molecules found?

Class I MHC molecules are found on all nucleated cells. They are not present on mature mammalian erythrocytes (red blood cells are anucleate), and that is why MHC typing and matching does not have to be performed when transfusing foreign red cells into a patient. It is also why lymphocytes, rather than erythrocytes, are typed by monoclonal antibodies in matching class I antigens in transplantation donors and recipients. Cytotoxic T cells (bearing CD8 differentiation antigens) recognize foreign antigens only when they are presented to the cell in the context of self MHC molecules of class I (dual recognition). These cells are thus said to be **MHC class I restricted.**

A few lymphocytes, called **null cells,** have neither surface immunoglobulins nor T cell receptors and thus are neither B cells nor T cells. Null cells are large granular lymphocytes (LGLs). The cytoplasmic granules of LGLs contain a pore-forming protein, cytotoxins, esterases, and proteoglycans. LGLs use these substances to kill transplanted cells, tumor cells, and virus-infected cells on first contact and therefore are also known as **natural killer (NK) cells.** NK cells are inhibited from killing cells that express self class I MHC molecules, but are uninhibited from killing any cells lacking or expressing reduced amounts of MHC class I molecules. Since the killing action of NK cells occurs in the absence of MHC class I molecules, NK cells are not MHC restricted. NK cells kill target cells by the same mechanisms employed by cytotoxic T lymphocytes (CTLs); i.e., by granule exocytosis and by induction of DNA fragmentation and apoptosis in target cells. NK cells have surface receptors for the tail or stem portions of immunoglobulins (of class IgG) and may thus recognize cells coated with these antibody molecules and kill them in a process called **antibody-dependent cell-mediated cytotoxicity** (ADCC) without the participation of MHC molecules or complement.

13.43.   What is the structure of class II MHC molecules?

Class II MHC molecules (Fig. 13-4b) are also heterodimeric glycoproteins, consisting of an α chain (different from the class I α chain) and a β chain (different from that of the T-cell receptor). The two chains are almost of equal length. Each chain has a variable domain at its amino end, followed by another extracellular domain, then by a hydrophobic transmembrane domain, and a cytoplasmic domain at its carboxyl end. The two extracellular domains on each chain are structured like immunoglobulins; so too are the extracellular domains of α chains of class I molecules and the single domain of β_2-microglobulin chains, suggesting that they all have evolved from a common ancestral gene and now are members of an extensive supergene family that includes antibodies, T-cell receptors, MHC glycoproteins, the cell-to-cell adhesion proteins (CD2, CD4, CD8), some polypeptide chains of the CD3 complex associated with T-cell receptors, as well as Fc receptors on lymphocytes and other leukocytes. The **polymorphic** regions (differing from one individual to another) of both class I and class II MHC molecules interact with foreign antigens, but it is the complex of a MHC molecule and foreign antigen that is recognized by a T-cell receptor (dual recognition). Helper T cells with CD4 on their plasma membranes are designated CD4$^+$ and are said to be MHC class II restricted.

13.44.   Where are class II MHC molecules found?

Class II MHC molecules are found primarily on antigen-presenting cells (macrophages, dendritic cells, B cells, thymus epithelial cells), and are involved in the presentation of antigen in the proper format to stimulate CD4$^+$ helper T cells.

IMMUNOGENETICS

13.45.   How are immunoglobulin genes constructed and where are they located in the human genome?

Immunoglobulin genes are assembled from separate gene segments (minigenes or modules) during B-cell development from stem cells in mammalian bone marrow. The minigene families that code for human immunoglobulin molecules are located on three chromosomes: (1) heavy-chain family on chromosome 14, (2) kappa (κ) light-chain family on chromosome 2, and (3) lambda (λ) light-chain family on chromosome 22. Each light-chain family may have up to a few hundred modules coding for the variable domains of L chains and one module for the constant portion of each light-chain type. Between the V and C modules for light chains, there is a small cluster of about 5 short J modules (J for "joining") that contribute about a dozen amino acids to the carboxyl end of the V domain. Spacer DNA exists between the V and J clusters and between the J cluster and the C gene. A similar arrangement occurs in the heavy-chain family, except that there is an additional set of about 20 short mini-genes between the V and J clusters that contribute to the "diversity" (hence the symbol D) of the variable domains of H chains, and there are at least five minigenes in the C cluster.

13.46. How is a complete immunoglobulin L chain assembled and transcribed?

During differentiation of a B cell, site-specific recombination deletes, apparently at random, a section of a κ L-chain gene between one of the V modules and one of the J modules. For example, in Fig. 13-5 the DNA se-quence between (but not including) V_3 and J_2 is excised, resulting in the placement of V_3 adjacent to J_2. Tran-scription only begins after this gene rearrangement has occurred, probably because such a change makes it possi-ble for an enhancer sequence (which may be located thousands of nucleotides distant on the same DNA molecule) to stimulate RNA polymerase binding to the promoter of the selected V module. Transcription always begins at the V gene immediately adjacent to a J gene. The primary transcript will, in our example, contain the V_3 module, all of the remaining J modules (including J_2 to the end of the J cluster), the spacer between the J cluster and the

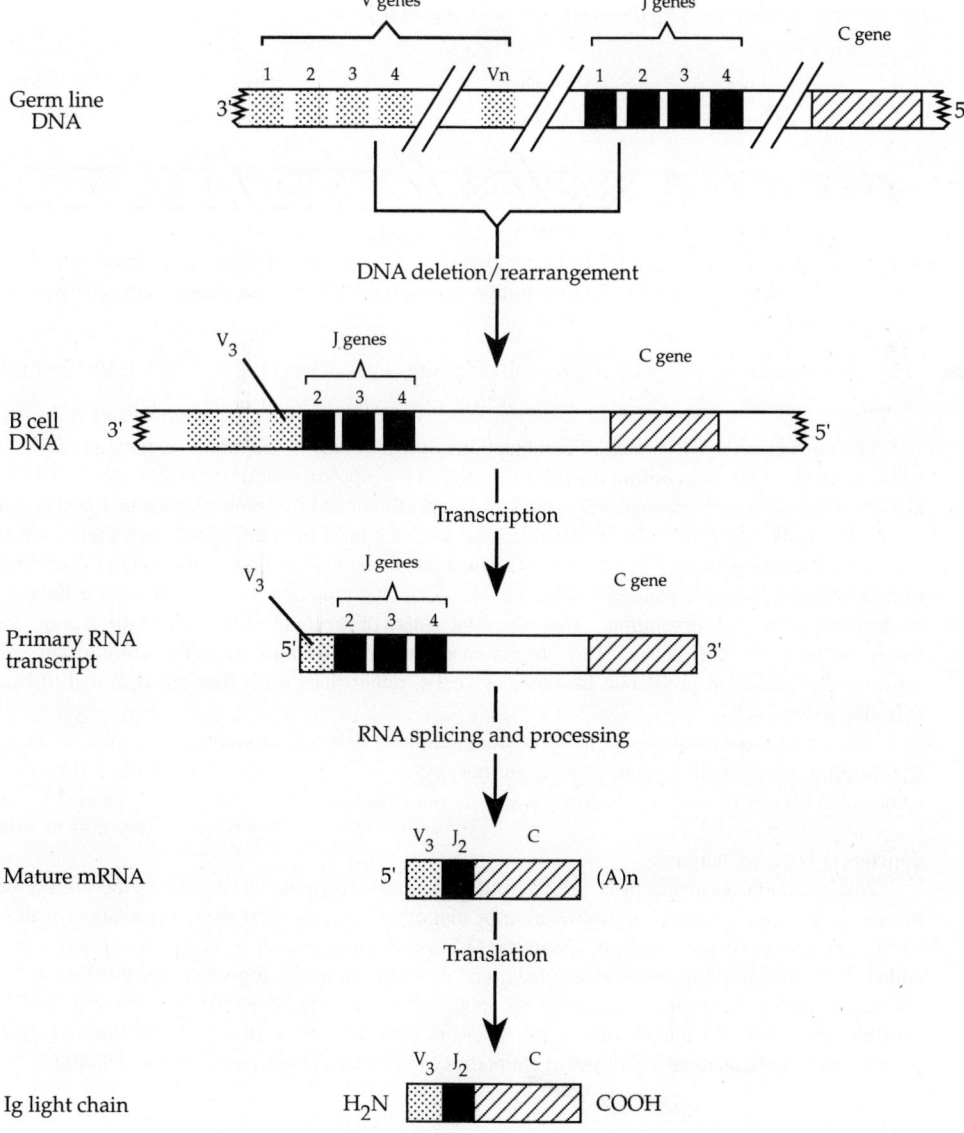

Fig. 13-5. Construction of a template encoding a κ Ig light chain, and subsequent events in Ig gene expression during development of a B lymphocyte.

single C module, and the C module itself. During nuclear processing of the transcript, mRNA splicing removes any J modules and the spacer between the rearranged V_3J_2 complex and the C module. Thus, the mature mRNA released from the nucleus contains only one each of V, J, and C modules. In the mouse, there are about 300 V modules, 4 J modules, and only one C module for a kappa light chain. The same rules apply in the synthesis of λ light chains, except there are four C modules and only two V modules.

13.47. How is a complete immunoglobulin H-chain gene assembled in the mouse?

The H-chain minigene family in the mouse has about 1000 V modules, at least 12 D (diversity) modules, 4 J modules, and a cluster of at least 5 C modules (one for each of the Ig heavy-chain types (μ, δ, γ, ϵ, α), arranged in the order given (Fig. 13-6). The D modules encode 1 to 15 amino acids in the third hypervariable segment of the V region of an H chain (V_H). Each C module contains multiple introns that must be spliced out of the primary transcript. Additionally, there are four subclasses of γ chains ($C_{\gamma 1}$, $C_{\gamma 2a}$, $C_{\gamma 2b}$, and $C_{\gamma 3}$) and three constant-region subtypes of λ light chains (λ_1, λ_2, λ_3) in all normal mice. Each of these subclasses and subtypes is coded for by separate modules, but we shall ignore them here for the sake of simplicity. Two DNA rearrangements are required to produce the template from which primary H-chain transcripts are made. The first one joins a D and a J_H module; the second one joins a V_H module to the rearranged DJ_H complex.

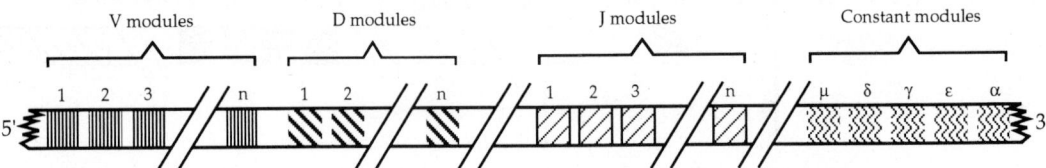

Fig. 13-6. Location of genes for Ig heavy chains in the genome. Double-slash lines indicate either the presence of more genes within a minigene cluster or spacer (noncoding) regions between clusters.

13.48. How is the great diversity of antigen-binding sites (paratopes) on different antibody molecules generated?

The number of $V_LJ_LC_L$ modular combinations of κ L chains in the mouse is roughly 300 V $\times$ 4 J $\times$ 1 C = 1,200. For H chains, the **combinatorial diversification** is approximately 1,000 V $\times$ 12 D $\times$ 4 J = 48,000. The number of possible antigen recognition sites (paratopes) obtained by combining about 1,000 L chains with about 50,000 H chains is 5 $\times$ 10^7. The mechanism that joins modules from different clusters (e.g., a V and a J module) is imprecise, causing random loss or gain of nucleotides at the joining sites, contributing to **junctional diversification** of V-region coding sequences. Also, the rate of somatic mutations in the V region is estimated to be 10^{-3} per nucleotide pair per cell generation. This is several orders of magnitude greater than the average spontaneous mutation rate for most other genes. This phenomenon, known as **somatic hypermutation,** occurs when B cells are activated by antigen to proliferate into memory cells, rather than when they are activated to become antibody-secreting plasma cells.

The above three mechanisms for generating antibody diversity (combinatorial diversification, junctional diversification, and somatic hypermutation) are not operative at the level of the germ line (i.e., they occur only in some somatic cells of the immune system and are not heritable modifications of the genome). Germ-line mutations (in gametes or their precursors), especially those that affect the hypervariable segments of either the V_H or V_L domains or both, are heritable.

Any mutation occurring in B cells during an immune response that increases the affinity of a cell-surface paratope will cause preferential proliferation of that clone, especially as the concentration of the antigen diminishes. As explained previously in Question 13.22, this phenomenon is known as affinity maturation. When added to combinatorial diversification, junctional diversification and somatic hypermutation are estimated to increase the number of potential paratopes by about a thousandfold; $(5 \times 10^7) \times (10^3) = 5 \times 10^{10}$. Fifty billion possible antibodies is a number far greater than the total number of B cells in the mouse (approximately 10^8). Clearly, no more than about 10^8 different antibodies can be made at any one time in the mouse.

13.49. B lymphocytes are diploid cells that carry genomes from their two parents. Potentially there are six minigene pools available for making antibodies: two H-chain pools (one from each parent), and four

L-chain pools (one κ and one λ from each parent). And yet the antibody made by each B lymphocyte is monospecific; i.e., each antibody contains the same heavy and light chains and thus possess the same antigen-combining sites. What is the explanation for this phenomenon?

Each human B cell makes a somewhat random choice of one chromosome 14 to code for the H chains and one chromosome (either chromosome 2 or chromosome 22) to code for either the κ or λ type of L chain. However, the κ-chain family is chosen about 60% of the time in humans (95% in mice). The expression of only a maternal or a paternal allele of a given minigene family in any cell is called **allelic exclusion**. This phenomenon is limited to the genes that encode immunoglobulins and the evolutionarily related genes that encode T-cell receptor proteins. The assembly of a V-region coding sequence begins with the H-chain family of minigenes. First, selected D modules join selected J_H modules on *both* parental chromosomes. Next, selected V_H modules join the DJ_H complex on *one* of these chromosomes. The first immunoglobulins normally made by virgin B cells contain μ heavy chains (class IgM). If the selected $V_H DJ_H$ complex allows production of complete μ chains, further rearrangements of the V_H region are prevented. Then, V_L to J_L joining can commence in one of the two κ modules. If that fails, rearrangement shifts to the other κ module. If that also fails, rearrangement will shift to first one λ module, and failing that to another λ module. Whenever a rearrangement succeeds in producing fully formed L chains, they will combine with preexisting μ chains to form class IgM molecules that prevent further rearrangements of V_L-region coding sequences. No specific biological functions have been detected for either κ or λ light chains, aside from their contribution to the antigen-recognition site (paratope) of the antibody.

13.50. Each B cell produces only monospecific antibodies during its lifetime; i.e., neither the L chains nor the V_H regions change. However, two kinds of changes can occur in the C_H region of immunoglobulins during the life of a B cell: (1) a shift from a membrane-bound form of antibody to a secreted form, and (2) a shift from μ chains (class IgM) to other H-chain classes. What structural differences characterize the membrane-bound and secreted (soluble) forms of antibodies?

All classes of immunoglobulins can be synthesized in a membrane-bound form as well as in a cell-free, soluble form. Prior to contact with antigen, a B cell manufactures the membrane-bound form that acts as its antigen receptor. After antigen contact, the B cell proliferates (when stimulated by the helper T-cell signal) into a clone, some members of which mature into plasma cells that predominately make antibodies that are liberated from the cell as soluble molecules. The H chains of membrane-bound immunoglobulins are longer than those of the secreted form and contain a hydrophobic sequence of amino acids near their carboxyl ends (antibody stem) that anchors them in the lipid bilayer of the plasma membrane with their Fab arms facing the extracellular environment. The soluble forms of antibodies do not have this hydrophobic region, and they have shorter H chains containing a hydrophilic sequence at their carboxyl ends that allows them to be released from the cell.

13.51. How is the shift from membrane-bound IgM to the soluble (secreted) form accomplished?

The most popular theory accounting for the shift from membrane-bound IgM to its secreted form involves differential processing of mRNA transcripts in the nucleus. During transcription of mRNA, the sequence AAUAAA (poly-A site) is recognized by a poly-A polymerase enzyme that cuts the molecule about 10–30 nucleotides downstream from this site and then adds about 100–200 residues of adenylic acid (a poly-A tail) to its 3' end. RNA polymerase, meanwhile, continues transcription for perhaps hundreds or even thousands of nucleotides beyond the H-chain structural gene, until it terminates by an unknown mechanism. The downstream mRNA transcript will not be capped at its 5' end and therefore will be rapidly degraded. The H-chain module for μ chains contains two poly-A sites (Fig. 13-7). Between these two sites are two exons encoding the hydrophobic region of the H chain. Before contacting antigen, a B cell selects the distal 3' poly-A site and cleaves the primary transcript a short distance downstream, then adds a poly-A tail. After mRNA splicing has removed the introns, a mRNA molecule is formed with the membrane sequence adjacent to the C_{H4} region.

After contacting antigen, however, the B cell selects (mechanism unknown) the proximal 5' poly-A site. Cleavage of the transcript at the proximal poly-A site removes the exons that code for the hydrophobic region at the end of C_{H4}. When translated into μ chains, the hydrophilic sequence at the end of C_{H4} allows secretion of the IgM molecule. The membrane-bound form of μ heavy chain is part of a 7S monomer, whereas the secreted form of IgM is a pentamer of five 7S units joined by a J chain. When synthesizing membrane μ chains, the RNA splicing process removes the region that binds to J chain from the 3' end of the C_{H4} exon. This segment is left intact when the cell secretes IgM.

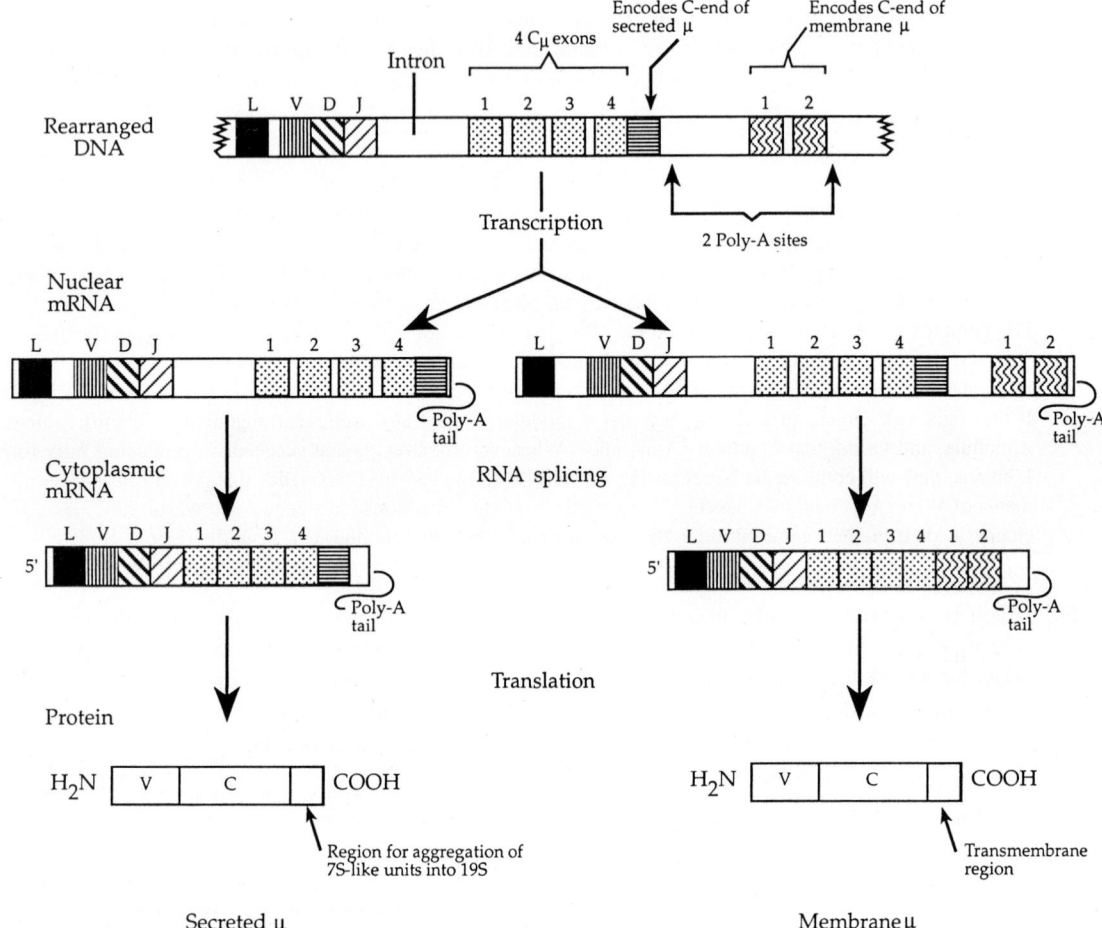

Fig. 13-7. Alternative polyadenylation scheme for production of secretory-μ vs. membrane-μ chains. L = noncoding leader sequence.

13.52. What mechanism is responsible for a B cell's shift from synthesizing only IgM to synthesizing both IgM and other Ig classes?

The first antibody made by virgin B cells is membrane-bound IgM. Somewhat later, but prior to antigen contact, most B cells also begin making membrane IgD. When stimulated by antigen, some of these cells mature into plasma cells that synthesize the pentameric secreted form of IgM that dominates the primary antibody response. Other antigen-stimulated cells may switch to making IgG, IgA, or IgE. This class switching changes only the C_H region, not the Fab region, of the immunoglobulin molecule. Thus, any variable region (VDJ) can acquire any of the biological functions associated with the C_H regions of various Ig classes.

Two mechanisms apparently control class switching. When a virgin B cell switches from making only membrane IgM to making both membrane IgM and membrane IgD, differential mRNA splicing is thought to be involved (Fig. 13-8). Once the $V_H D J_H$ complex has formed, transcription produces a large primary transcript that includes both C_μ and C_δ modules and perhaps even all of the other C_H modules as well. Production of μ chains involves splicing out the mRNA region between the $V_H D J_H$ complex and C_μ; production of δ chains involves splicing out the mRNA region between the $V_H D J_H$ complex and C_δ (C_μ in this case being treated as an intron). The same mechanism is thought to be responsible for the switch to other classes of membrane-bound Ig when virgin B cells are stimulated by antigen to mature into memory cells bearing (in addition to IgM) IgG, IgA, or IgE antigen receptors on their surfaces.

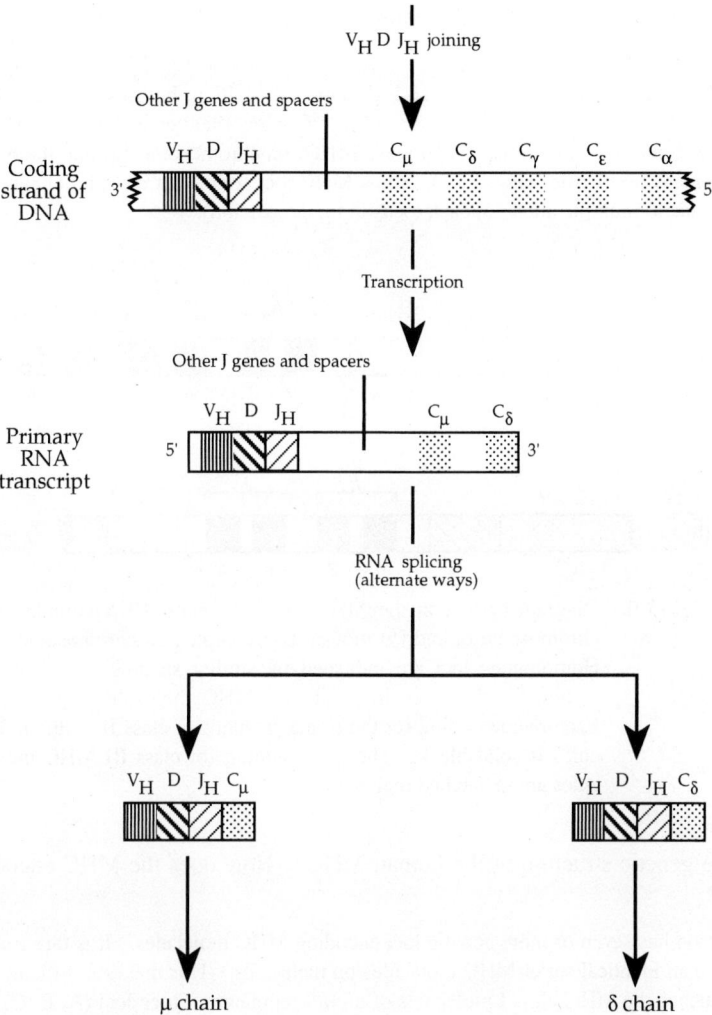

Fig. 13-8. Class switching by differential RNA splicing.

13.53. In the terminal maturation of a plasma cell, how is class switching accomplished?

A different mechanism, called **switch recombination,** is responsible for the terminal maturation of a plasma cell secreting IgG, IgA, or IgE. This process deletes all C_H minigene modules between the chosen C_H module and the $V_H DJ_H$ complex. Although the progeny of a single activated B cell may express different Ig classes, all of these antibodies contain identical Fabs. Each plasma cell synthesizes a single class of immunoglobulin; plasma cells lose their membrane-bound Ig molecules when they begin secreting their antibody.

13.54. What are the major features of the genetic system encoding T-cell receptors?

There are three families of minigenes (modules) or gene segments for $\alpha\delta$, β, and γ chains of T-cell receptors on human chromosomes 14, 7q (long arm), and 7 p (short arm), respectively, and they are on mouse chromosomes 14, 6, and 13, respectively. The δ-chain modules are interspersed among the α-chain modules. All four chain types are encoded by modules resembling V, J, and C that are assembled like those of immunoglobulin genes; only β and δ chains have additional D-like modules. Only one type of antigen-combining site exists on the T-cell receptors of any one cell because of allelic exclusion, as explained for immunoglobulin genes.

13.55. What is the name of the MHC in the mouse and in humans? Where is the MHC located in these two species?

The MHC in the mouse is called the **H-2 system;** the MHC in humans is named **HLA** (human leukocyte-associated antigens), because some of its antigens were first identified on leukocytes. These and other MHC antigens were later found on other cells. The mouse MHC (on chromosome 17) and the human MHC (on chromosome 6) have many similarities (Fig. 13-9). Most MHC loci coding for class I and class II molecules are highly polymorphic (due to multiple alleles at each locus in the population).

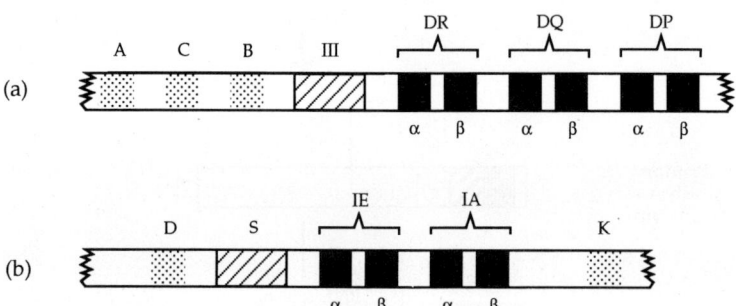

Fig. 13-9. Diagram of loci in the MHC of (*a*) humans, HLA complex on chromosome 6, and (*b*) mouse, H-2 complex on chromosome 17. Homologous loci are indicated by similar shading. Each gene coding for the α chain of class I MHC molecules is stippled. Each region coding for the α and β chains of class II MHC molecules is solid black. The genes coding for class III MHC molecules are in hatched regions.

13.56. What is the genetic structure of the human MHC? How does the MHC encode class I and class II molecules?

Each person has seven or more genetic loci encoding MHC molecules. It is rare for any two unrelated individuals to have an identical set of MHC molecules on their cells. This makes it difficult to find a tissue or organ donor that matches the MHC class I molecules of a graft recipient. Three loci (A, B, C) of the human MHC encode the α chains of class I molecules. For the class II molecules, there are two sets of loci in the mouse (IA and IE), each set coding for an α chain and one or more β chains. In humans, there are at least three such sets (DP, DQ, DR), each set also encoding an α and a β chain. A specific combination of tightly linked genes on a chromosome is referred to as a **haplotype.** Each diploid cell thus has two haplotypes for genes of the MHC. Each MHC haplotype would normally be transmitted as a unit to offspring, unless crossing over within the MHC region occurred. Since the human MHC region is less than five recombination units (centimorgans) in length, more than 95% of gametes would be expected to contain unrecombined parental haplotypes.

13.57. How many kinds of MHC molecules are on cells?

When MHC class I genes are being expressed in a given cell (as they are in all nucleated cells), each allele at every locus of the class I region is transcribed (codominant allelic interaction). The same principle operates for genes in the class II region of the MHC, but only in those cells that normally express class II antigens. Allelic exclusion does not occur in any genes of the MHC. Many alleles are present at each of the three class I loci (A, B, C) in the human population; e.g., there are over 30 alleles known at the human HLA-B locus alone. An individual who is heterozygous at all three loci would thus carry six class I MHC molecules on each cell. Each class II region (DP, DQ, DR) codes for both an α and a β chain.

If a mouse is heterozygous for a pair of alleles (1,2) at the IA locus, then four kinds of IA molecules can be produced: $\alpha 1\beta 1$, $\alpha 1\beta 2$, $\alpha 2\beta 1$, and $\alpha 2\beta 2$. If the IE locus is also heterozygous, another set of four α/β IE-chain combinations could be formed. Since there is more than one β gene at both the IA and IE loci, a mouse cell may possess between 10–20 class II molecules.

Objective Questions

Multiple Choice

Directions: Choose the one best answer.

1. Which of the following is not a member of the immunoglobulin supergene family? (*a*) Antibodies (*b*) lymphokines (*c*) MHC glycoproteins (*d*) T-cell receptors (*e*) Fc receptors on leukocytes

2. Which of the following is/are descriptive of an immune response? (*a*) Acquired (*b*) transferable (*c*) specific (*d*) more than one of the above (*e*) all of the above

3. Which of the following is normally associated with the production of hybridomas secreting a desired monoclonal antibody? (*a*) Mutant mouse myeloma cells producing antibodies (*b*) plasma cells from bone marrow of an immunized mouse (*c*) aminopterin blockage of nucleotide salvage pathway (*d*) mitogen-induced antibody diversity (*e*) none of the above

4. Which of the following is not a mechanism whereby B cells or antibodies contribute to immunity? (*a*) Presentation of antigen to T cells (*b*) agglutination (*c*) opsonization (*d*) complement activation (*e*) direct cell killing

5. One major branch of the immune response is mediated by T cells (cell-mediated immunity, CMI); the other major branch is termed (*a*) blood-borne. (*b*) soluble. (*c*) humoral. (*d*) antibiotic. (*e*) lymphokine-mediated.

6. An anamnestic response is most closely associated with (*a*) a primary immune response. (*b*) a secondary immune response. (*c*) an autoimmune response. (*d*) a cytotoxic response. (*e*) a cytokine response.

7. Macrophages are classified as (*a*) APCs. (*b*) lymphocytes. (*c*) CDs. (*d*) specific effector cells. (*e*) regulatory cells.

8. Plasma cells are (*a*) long-lived memory cells. (*b*) a subclass of T cells. (*c*) mature antibody-secreting cells. (*d*) only generated during a secondary immune response. (*e*) more than one of the above.

9. Which of the following functions is *not* primarily due to the cell-mediated branch of the immune response? (*a*) Killing of virus-infected cells (*b*) stimulation of lymphocytes to effector status (*c*) prevention of overactive immune responses (*d*) neutralization of toxic molecules via lymphokine secretions (*e*) rejection of foreign tissue grafts

10. T cells mature in the (*a*) thyroid gland. (*b*) spleen. (*c*) bone marrow. (*d*) lymph nodes. (*e*) none of the above.

11. The clonal selection theory best accounts for (*a*) the production of memory cells. (*b*) the mechanism whereby antigen and antibody unite. (*c*) the maturation of B cells into plasma cells. (*d*) the specificity of the immune response. (*e*) the phenomenon of MHC restriction.

12. Which of the following is not a formula for a 7S immunoglobulin? (*a*) $2\kappa2\lambda$ (*b*) $2\alpha2\lambda$ (*c*) $2\kappa2\gamma$ (*d*) $2\mu2\kappa$ (*e*) more than one of the above

13. Class switching pertains to (*a*) paratopes. (*b*) epitopes. (*c*) MHC molecules. (*d*) light Ig chains. (*e*) heavy Ig chains.

14. Which of the following does not contribute to the diversity of antibody structure? (*a*) Class switching (*b*) allelic exclusion (*c*) combinatorial diversification (*d*) junctional diversification (*e*) somatic hypermutation

15. The alternate pathway of complement activation (*a*) is stimulated by cell walls of microorganisms. (*b*) develops the same membrane-attack complex as the classical pathway. (*c*) produces C3b by a different route than the classical pathway. (*d*) does not require antigen-antibody interaction. (*e*) is described by all of the above.

16. The set of polypeptide chains that is thought to be involved in transmission of a signal from a TCR-antigen complex to the cell interior is designated (*a*) IL-1. (*b*) CD3. (*c*) IL-2. (*d*) CD4 and CD8. (*e*) APC.

17. How many independently assorting minigene families are potentially available for making antibodies in a B cell? (*a*) 3 (*b*) 4 (*c*) 5 (*d*) 6 (*e*) more than 6

18. The process thought to be responsible for an antibody-producing cell shifting from membrane-bound IgM to its secreted form is (*a*) choice of poly-A site. (*b*) deletion of a DNA segment. (*c*) cleavage of amino ends of H chains. (*d*) removal of exons from pre-RNA transcripts. (*e*) H-chain class switching.

19. The cells involved in allergic reactions, having surface receptors for IgE antibodies and containing histamine, are known as (*a*) neutrophils. (*b*) monocytes. (*c*) basophils. (*d*) mast cells. (*e*) more than one of the above.

20. Given the HLA types of parents shown below

Parents: (A1 B8)/(A3 B7) × (A2 B12)/(A9 B5)

which of the following offspring is least likely to be produced? *Note:* the slash line separates haplotypes (in parentheses) on homologous chromosomes. (*a*) (A1 B8)/(A9 B5) (*b*) (A1 B8)/(A3 B7) (*c*) (A3 B7)/(A9 B5) (*d*) (A2 B12)/(A1 B7) (*e*) (A9 B12)/(A3 B7)

True–False

1. Each mature B and T cell can respond to almost any foreign antigen with which it may come in contact.

2. Antibody and antigen unite by forming multiple covalent bonds.

3. All immunoglobulins possess two paratopes per molecule; i.e., they are bivalent.

4. The paratopes of immunoglobulins produced by a B cell do not change during the lifetime of the cell.

5. All classes of immunoglobulins can be synthesized in a membrane-bound form as well as in a secreted form.

6. All of the Ig genes in a plasma cell and all of the genes encoding the T-cell receptor on a T cell behave in codominant fashion; i.e., each allele is expressed in cells of heterozygous genotype.

7. Within a B cell, rearrangement of Ig genes begins with the H-chain minigene cluster on both parental chromosomes.

8. Most MHC loci coding for class I and class II molecules are highly polymorphic within each individual.

9. Allelic exclusion does not affect the genes encoding MHC molecules.

10. Human B lymphocytes are thought to mature in the bone marrow.

11. Mature B lymphocytes do not normally react with self antigens because during their development they have lost the Ig genes that could have formed self-reacting antibodies.

12. T_C cells respond to antigen associated with class I MHC molecules, whereas T_H cells respond to antigen in association with class II MHC molecules.

13. Natural killer (NK) cells are able to recognize and destroy foreign cells without the participation of MHC molecules or complement.

14. Each purified antigen (e.g., a foreign protein) is capable of stimulating only members of a single clone of B and/or T cells.

15. Immunoglobulin class switching does not involve light chains.

16. Membrane-bound IgM is a pentamer of five 7S-like subunits.

17. One gene module from each of the V, D. J, and C clusters is selected for the production of the light chains in each immunoglobulin molecule.

18. The transcription of mRNA for light chains always begins at the V gene immediately adjacent to the J gene after chromosomal rearrangement has occurred.

19. B lymphocytes bearing receptors with the greatest affinity for an antigen will be the ones that proliferate the largest clone of cells.

20. The same mechanisms are responsible for generating antibody diversity when virgin B cells are stimulated by antigen to become memory cells as when they mature into antibody-secreting plasma cells.

Matching

Directions: Match each item in **Column A** to one or more items in **Column B** with which it is necessarily associated.

Set #1

Column A

1. Complement protein
2. Functions as a dimer (2 chains)
3. Normally can exist free as well as in membrane-bound forms
4. All chains of the membrane-bound molecule are encoded by the same minigene cluster
5. Each chain contains variable and constant regions
6. Nonidentical chains are covalently bonded
7. Some chains of the membrane molecule are devoid of disulfide bonds
8. All chains of the cell-surface molecule pass through the membrane
9. Contains β_2 microglobulin
10. Cell-surface molecule contains both α and β chains
11. Found on all nucleated mammalian cells
12. Located primarily on all lymphocytes and all macrophages

Column B

A. MHC class I molecule
B. MHC class II molecule
C. MHC class III molecule
D. T-cell receptor
E. Immunoglobulin
F. None of the above

Set #2

Column A

1. May exist as molecules greater than 7S
2. Greatest percentage of serum antibodies of normal adults
3. Binds to mast cells and basophils
4. Activates complement
5. No specific functions known other than antigen-binding
6. Passes the placenta
7. First B-cell receptor
8. Main protector of mucous membranes
9. Binds to macrophages by Fc
10. May contain a J chain

Column B

A. IgA
D. IgD
E. IgE
G. IgG
M. IgM
N. None of the above

Terms

Directions: Unless otherwise specified, each correct answer is a single word.

1. A specific combination of tightly linked genes that are usually transmitted to offspring as a unit.

2. Antigenic determinants located in paratopes of immunoglobulins or T-cell receptors.

3. A cancer of B cells.

4. The commercial source of monoclonal antibodies.

5. A potentially lethal form of hypersensitivity. (1 or 2 words)

6. The mechanism responsible for the terminal maturation of a plasma cell secreting IgG that results from deletion of all C_H modules between the C_γ and VDJ modules. (2 words)

7. A protein chain, produced by epithelial cells, that is associated with IgA molecules in body fluids. (2 words)

8. The cytolytic product of T_C lymphocytes that polymerizes to form transmembrane channels in target cells.

9. A state of induced or natural immunological unresponsiveness to specific antigens.

10. Any substance, such as an antibody or C3b, that binds to an antigen and enhances its phagocytosis.

11. The process whereby the immune system is activated.

12. A polypeptide that links subunits of polymeric IgA or IgM. (1 letter and 1 word)

13. The two-letter designator for protein molecules on cell membranes used to identify human leukocyte subpopulations.

14. A term that denotes the number of antigen-combining sites on an immunoglobulin molecule.

15. A type of cellular killing in which a target cell with bound antibody is attacked by an effector cell bearing Fc receptors for IgG and is subsequently lysed without the participation of complement. (3 to 5 words; 4-letter acronym)

16. A low-molecular-weight foreign substance that is not immunogenic by itself, but can become so by being coupled to a larger carrier molecule such as a protein.

17. A large mononuclear leukocyte in tissues (analogous to monocytes in blood) that can function as an antigen-presenting cell, or as a killer cell in ADCC.

18. A granulocytic leukocyte that can participate in phagocytosis and early inflammatory reactions. It possesses Fc receptors and can also participate in ADCC, but not as an antigen-presenting cell.

19. Harmful immune reactions against self antigens. (2 words)

20. A group of low-molecular-weight proteins, secreted by a variety of cell types, that regulate the intensity and duration of an immune response.

Immunoglobulin Structure

Directions: Name the numbered parts of the immunoglobulin molecule at the right as specified by the following numbered items. (*Note:* the diagram represents half of the molecule; the mirror half is not shown.)

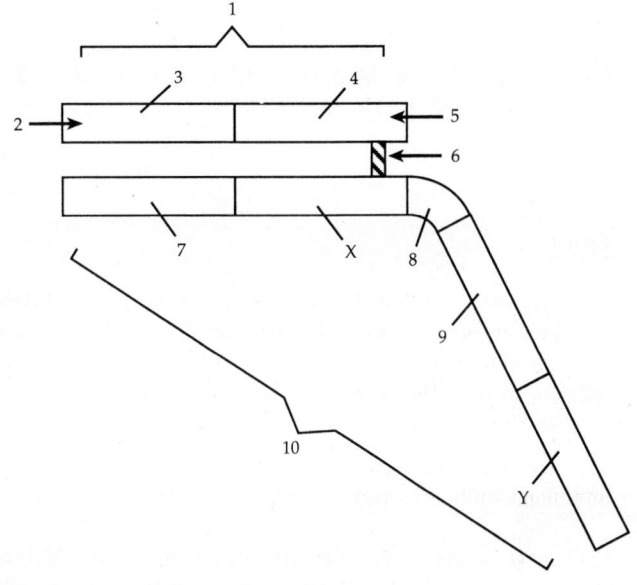

1. (3 + 4)

2. What end of the polypeptide chain is this?

3. Be very specific!

4. Be very specific!

5. What end of the polypeptide chain is this?

6. What kind of linkage is this?

7. Be very specific!

8. This is the "flexible" region.

9. Be very specific!

10. Includes 7 + X + 8 + 9 + Y

11. 7 + X

12. 1 + 11; (11 = 7 + X)

13. 2(9 + Y)

Answers to the Objective Questions

Multiple Choice

1. *b*　　2. *e*　　3. *c*　　4. *e*　　5. *c*　　6. *b*　　7. *a*　　8. *c*　　9. *d*　　10. *e (thymus gland)*
11. *d*　　12. *a*　　13. *e*　　14. *b*　　15. *e*　　16. *b*　　17. *d*　　18. *a*　　19. *e (c and d)*
20. *b*

True–False

1. F (one kind of paratope per cell)　　2. F (noncovalent bonds)　　3. F (secreted IgA and IgM have 4 and 10 paratopes, respectively)　　4. T　　5. T　　6. F (allelic exclusion allows only one allele at each locus to be expressed)　　7. T　　8. F (within a population)　　9. T　　10. T　　11. F (self-reacting cells die)　　12. T　　13. T　　14. F (polyclonal response in both B and T cells)　　15. T　　16. F (is a single 7S unit)　　17. F (no D cluster for L chains)　　18. T　　19. T　　20. F (somatic hypermutation occurs only during the development of memory cells)

Matching

Set #1: **1.** *C*   **2.** *A, B, D*   **3.** *C, E*   **4.** *B*   **5.** *D, E*   **6.** *E*   **7.** *F*   **8.** *B, D*   **9.** *A*
10. *B, D*   **11.** *A*   **12.** *B*
Set #2: **1.** *A, M*   **2.** *G*   **3.** *E*   **4.** *M, G*   **5.** *D*   **6.** *G*   **7.** *M*   **8.** *A*   **9.** *G*
10. *A, M*

Terms

1. haplotype   **2.** idiotypes   **3.** myeloma   **4.** hybridomas   **5.** anaphylaxis or anaphylactic reaction
6. switch recombination   **7.** secretory piece   **8.** perforin   **9.** tolerance   **10.** opsonin
11. immunization   **12.** J chain   **13.** CD   **14.** valence   **15.** antibody-dependent cell-mediated cytotoxicity (ADCC)   **16.** hapten   **17.** macrophage   **18.** neutrophil   **19.** autoimmune diseases
20. cytokines

Immunoglobulin Structure

1. light (L) chain   **2.** NH_2 (free amino end)   **3.** V_L (variable region of light chain)   **4.** C_L (constant region
of light chain)   **5.** COOH (free carboxyl end)   **6.** disulfide bond   **7.** V_H (variable region of heavy chain)
8. hinge region   **9.** C_{H2} (constant region 2 of heavy chain)   **10.** heavy (H) chain   **11.** Fd
12. Fab (fragment antigen-binding)   **13.** Fc (fragment crystallizable)

Chapter 14

Molecular Evolution

"The more deeply we look at the molecular level, the more similar all known life forms appear to be."*

INTRODUCTION

In the first half of the 20th century, the evidence supporting the theory that living organisms evolve was dominated by comparative studies of anatomy, physiology, and embryology. Then in 1953, Watson and Crick proposed a double helix model for the structure of DNA that set off an explosion of knowledge concerning gene structures, functions, regulations, and evolution at the molecular level. Now in addition to fossil bones, shells, petrified (mineralized) wood, and other clues to earlier life forms, we have at our disposal **molecular fossils.** These are structures in living (extant) forms (e.g., the existence of introns within transcriptional units) or processes (e.g., the utilization of a common biochemical pathway) shared by such diverse organisms that they are more likely to have been inherited from a common ancestor than independently acquired from unrelated ancestors.

Molecular studies have helped shed light on the origin of life and its subsequent evolution into a myriad of extinct and extant species. Theories regarding these early events are difficult (if not impossible) to prove conclusively with the circumstantial evidence presently available. Perhaps we shall never have the degree of "proof" we would desire regarding the emergence of life on earth. Sufficient information is already available, however, to make some very plausible scenarios. But with so much yet to be discovered, it should come as no surprise if our present models will be revised or replaced by newer ones in the continuing search for better explanations that is the heart of scientific investigation.

BEGINNINGS

14.1. What attributes should a chemical system possess to be considered "alive"?

Living cells possess, at a minimum (1) a boundary membrane separating the cell's contents from its external environment; (2) one or more DNA molecules that carry genetic information for specifying the structure of proteins involved in replication of its own DNA, in metabolism, in growth, and in cell division; (3) a transcriptional system whereby RNAs are synthesized; (4) a translational system for decoding ribonucleotide sequences into amino acid sequences, and (5) a metabolic system that provides usable forms of energy to carry out these essential activities.

The first living system(s) were undoubtedly much simpler than any cells alive today. The transition from nonliving to living was gradual, and no single event led to life in all its modern complexity. Even today biologists cannot all agree on a definition of life (or death). Though not universally accepted, the following criteria are usually included in attempts to define life, and are offered here as an introduction for further discussion. An aggregate of molecules may be considered "alive" if it (1) can use chemical energy and/or radiant (light) energy to drive energy-requiring chemical reactions; (2) can increase in mass (grow) by controlled synthesis; and (3) possesses an information coding system (genome) and a system for translating the coded information into molecules that maintain the system and allow it to reproduce one or more collections of molecules with similar properties.

---

*From *Molecular Biology of the Gene,* Fourth Edition, by Watson, et al. Copyright © 1987 by James D. Watson, Published by The Benjamin/Cummings Publishing Company.

14.2.   When did life first appear on earth?

The best estimate for the age of the earth is about 4.6 billion years b.p. (before the present). The oldest **microfossils** (visible only with the aid of light or electron microscopes) superficially resembling bacteria have been dated at about 3.5 billion years b.p. Thus, **chemical evolution** (e.g., abiotic syntheses of amino acids and their polymerization into polypeptides) during the first 1.0 to 1.5 billion years of earth history probably preceded the appearance of cellular life and its subsequent **biological evolution.**

14.3.   In what kind of environment did life originate?

According to one popular theory, the Sun, the Earth, and the other planets of our solar system are thought to have formed by the gravitational collapse of a cloud of interstellar gas and dust. The pressure of squeezing this matter together and the radioactive decay of unstable elements generated enough heat to melt the deep layers of the primitive Earth. Although the crust cooled and solidified relatively quickly, volcanic activity was much more widespread than at present. A primitive atmosphere was produced by outgassings from volcanic activity, but its most likely composition is still being debated. It was formerly thought that the primitive atmosphere was highly reducing, containing mainly hydrogen (H_2), ammonia (NH_3), methane (CH_4), carbon monoxide (CO), and large quantities of water. As the Earth cooled, most of the atmospheric water condensed to form the primitive oceans. Today, however, the major opinion favors a primitive atmosphere that was nearly neutral, containing primarily nitrogen, carbon dioxide, hydrogen sulfide (H_2S), and water. There is widespread agreement, however, that it was not an oxidizing atmosphere.

14.4.   What evidence supports the theory that the early earth's atmosphere was not oxidizing?

The oldest rocks of the earth's crust contain reduced rather than oxidized matter. Banded iron formations (sedimentary deposits of iron embedded in silicate matrix; also known as "red beds") appear for the first time in the geological record at about 2 billion years b.p. Prior to this time, iron was dissolved in sea water in its ferrous (Fe^{2+}) form. As oxygen became available, iron was converted to its ferric (Fe^{3+}) form and was precipitated as hydrous ferric oxide in rusty layers on the ocean floor. In a few hundred million years, all of the dissolved ferrous iron in the oceans had disappeared and oxygen began to accumulate in the atmosphere.

Uranites are uranium compounds (UO_2) that can only accumulate when oxygen concentrations are very low ($<1\%$). When the concentration of atmospheric oxygen exceeded 1%, the uranites are oxidized to U_3O_8 and dissolve in water. Uranites disappear from the geological record at about 2 billion years b.p., suggesting that oxygen reached concentrations greater than one percent in the atmosphere at that time.

14.5.   What caused the primitive atmosphere to become oxidizing?

Microfossils resembling modern cyanobacteria ("blue-green algae") have been found in limestone rocks dated 3.5 billion years b.p. These rocks, formed in certain intertidal regions by the layering of marine cyanobacteria and sediments, are called **stromatolites.** Stromatolites continue to form today in some parts of the world. Presumably these ancient photosynthetic bacteria produced oxygen as a by-product of splitting water, just as cyanobacteria do today. Oxygen did not immediately accumulate in the atmosphere, however, because it became bound by vast amounts of reduced iron in the oceans as quickly as it was formed. It may have required 1.5 billion years for the concentration of atmospheric oxygen to exceed one percent and lead to the disappearance of uranites from ancient sediments at about 2 billion years b.p.

Traditionally, the buildup of oxygen in the atmosphere has been attributed to the photosynthetic splitting of water by primitive cyanobacteria. But a new type of bacteria was discovered in 1993 that employs an unusual kind of photosynthesis. These bacteria acquire electrons not from water, but from iron atoms, converting iron from its ferrous (Fe^{2+}) to its ferric (Fe^{3+}) form. When ferric iron reacts with water it becomes ferric hydroxide and later becomes ferric oxide (rust). Oxygen is not produced in this process. It has been suggested that ancient bacteria with a similar type of photosynthesis could have produced the banded iron formations without releasing oxygen into the atmosphere. In this case, oxygen released into the atmosphere from the splitting of water by photosynthetic cyanobacteria might have begun less than 3.5 billion years ago.

Still another theory proposes that intense ultraviolet radiation could have knocked oxygen molecules free from water molecules. The oxygen could then have combined with iron in the oceans to form the iron oxides in sedimentary deposits. However, once oxygen began to build up in the atmosphere (whether from biological photosynthesis or from abiotic processes), high-energy solar radiation or lightning acting on oxygen (O_2) molecules created ozone (O_3). Ozone would have provided biological molecules some protection from the damaging effects of ultraviolet radiation, just as it now does for living organisms.

14.6. How did the first building blocks (monomers) for the synthesis of nucleic acids and proteins form?

There are two major scientific theories regarding how life came to be on the earth. Either life (and/or its precursors) evolved on earth from nonliving chemicals, or it evolved elsewhere in the universe and was brought to earth by comets or meteorites (**panspermia theory**). The panspermia theory does not explain the origin of life in the universe; it only shifts its origin to some place other than Earth. The belief that life was created by a supernatural force is impossible to support or refute with factual evidence and hence is outside the realm of science.

Amino acids and other precursors of modern biomacromolecules have been found inside meteorites, so chemical evolution of these small molecules might have been (and still may be) widespread in the cosmos. In his 1924 paper, the Russian biochemist Alexander I. Oparin was first to propose that these essential monomers might have been produced on earth by prebiotic chemical pathways when the earth's atmosphere was vastly different than it is today. The theory was further elaborated in his 1936 book *The Origin of Life*. Meanwhile, the British geneticist J. B. S. Haldane independently proposed many of Oparin's conjectures on the origin of life. Thus, these ideas have come to embody what is known as the Oparin-Haldane hypothesis. It was not until 1953, however, that the first successful experimental attempt was made to test the Oparin-Haldane hypothesis. Stanley Miller, at the suggestion of his mentor Harold Urey, used a reflux apparatus (Fig. 14-1) to recirculate water vapor and other gases (CH_4, NH_3, and H_2) presumed to have been present in the primitive atmosphere through a chamber where they were exposed to a continuous high voltage electrical discharge (spark) that simulated natural lightning. After a few days, the mixture was analyzed and found to contain at least ten different amino acids, some aldehydes, and hydrogen cyanide (HCN). Subsequent experiments by Miller and other researchers using different molecular mixtures and energy sources produced a variety of other building blocks of biological polymers.

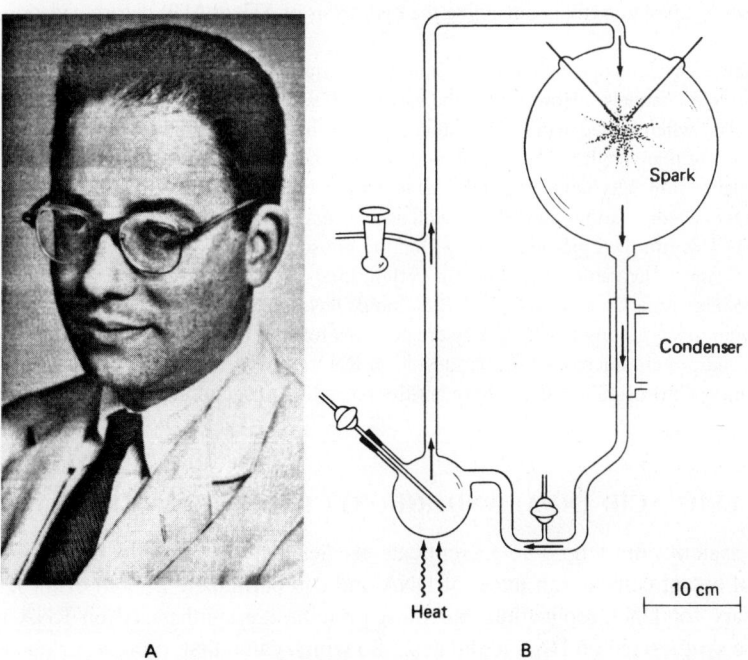

Fig. 14-1. (*a*) Stanley L. Miller and (*b*) diagram of his apparatus for synthesizing organic molecules under presumptive primitive earth conditions. (*From The Science of Evolution by W. P. Stansfield, Macmillan Publishing Company, 1977.*)

Example 14.1. Refluxing a concentrated solution of hydrogen cyanide in ammonia through a spark chamber produced the nucleic acid base adenine. Under conditions favorable to polymerization of HCN, orotic acid was produced. Sunlight is known to photochemically decarboxylate orotic acid to form the RNA base uracil. Several kinds of sugars have been produced from formaldehyde. Polymerization of formaldehyde can yield ribose, the sugar component of RNA molecules. The pathways through which abiotically synthesized monomers of macromolecules were formed are highly speculative and may indeed be incompatible. But it is likely that these or other diverse pathways might have been used in different places and/or at different times during the period of chemical evolution.

14.7. Is there any experimental evidence that biomacromolecules can spontaneously polymerize from smaller monomers?

Sidney Fox and his colleagues heated amino acids under water-free (anhydric) conditions to temperatures of 160 to 210°C and found that the amino acids polymerized into proteinlike chains which he called **proteinoids.** Unlike the linear proteins produced by modern cells, proteinoids are branched polypeptide chains. Their molecular weights range from 4,000 to 10,000 daltons, indicating that they consist of approximately 40 to 100 amino acid residues. When dissolved in water, these proteinoids exhibit several properties of biological proteins, including limited enzymatic activity and susceptibility to digestion by protease enzymes such as pepsin and trypsin.

Proteinlike peptides can also be synthesized from amino acids on clays. Clays consist of alternating layers of inorganic ions and water molecules. The highly ordered lattice structure of clays strongly attract organic molecules and promotes chemical reactions between them, especially if the clay contains an abundance of metallic ions. In the laboratory, mixtures of amino acids, water, and clay have been subjected to temperatures fluctuating between 25 and 94°C, as they might be in mud flats alternating between daytime and nighttime conditions in nature. Polypeptides have been detected in such experiments. Nucleic acid polymers have also been produced by prolonged heating of mixtures containing nucleotides and phosphates at about 60°C. Interactions would be more frequent if these macromolecules could be concentrated by the partial evaporation of water from tidal pools, inland seas, or lakes containing these aggregates.

When solutions of proteinoids are heated in water and then allowed to cool, small, spherical particles, called **microspheres,** are formed. These microspheres are about the same size and shape as spherical bacteria. Some of them are able to grow (add mass) by accretion of proteinoids and lipids and subsequently proliferate by binary fission or budding. Some of these microspheres can absorb various molecules from their aqueous environment, and some have weak catalytic activity, including the hydrolysis of ATP to ADP, decarboxylation, deamination, and amination.

For many years scientists believed that lipid cell membranes were permeable to nutrients and ions only if they contained protein channels. However, in the early 1960's, biophysicist Alec Bangham found that lipids extracted from egg yolks, when mixed with water, spontaneously become organized into double-layered bubbles approximately the size of many cells. These bubbles soon became known as **liposomes.** Subsequently it was found that liposomes made from fatty acids with tails 10 to 14 carbons long also tend to form closed vesicles, but they are leaky enough to absorb various substances such as proteins and nucleic acids from the surrounding medium. Substances trapped within the lipid bilayer or within the liposome find themselves in a hydrophobic environment that might provide more favorable conditions for certain kinds of chemical reactions. Thus, lipid bilayers may have promoted both aggregation and catalysis. Proteinoids have been observed to increase the conductance of lipid bilayer membranes, suggesting that they may create pores for ions or conduct electrons. Vesicles composed of lipid membranes and protein microspheres, but devoid of RNA or DNA molecules, are hypothesized to have existed in the early stages of the origin of life. These entities are called **progenotes.**

THE RIBONUCLEIC ACID (RNA) AND RIBONUCLEOPROTEIN (RNP) WORLDS

14.8. A "living system" must be able to replicate its genetic material and be capable of evolving. The genomes of extant cells are all made of DNA and can participate in both of these functions. Proteins are necessary for DNA replication, but most proteins are synthesized on RNA templates that themselves were synthesized on DNA templates. So which came first, proteins or nucleic acids?

Some bacteria (e.g., *Bacillus brevis*) synthesize short polypeptide antibiotic compounds (such as gramicidin S, tyrosidine, and bacitracin) without the aid of ribosomes. However, the enzymes required to carry out this unusual type of protein synthesis are much longer than the oligopeptides they produce. Because the synthesis of gramicidin S resembles fatty acid synthesis, an early theory proposed that ribosomal protein synthesis may have evolved from fatty acid synthesis. However, there must be at least one enzymatic subunit for each of the peptide bonds formed in gramicidin S. Hence the origin of these longer enzyme molecules cannot be attributed to this same type of mechanism.

A possible solution to this "chicken or egg" paradox became available in 1981 with the discovery that an intron from a rRNA transcript in the ciliate protozoan *Tetrahymena thermophila* was able to remove itself from its parent molecule without any contribution from proteins. Since that time, several other catalytic RNAs (**ribozymes**) have been found in different organisms. It has been hypothesized that RNA molecules capable of self-replication (RNA replicase) arose prebiotically by random condensation of mononucleotides into small polymers. The active sites of most modern proteins and catalytic RNAs constitute relatively small segments of the polymers to which they belong. The smaller primitive RNA replicase polymers, formed abiotically, would probably have only weak catalytic activity, and would have been subject to error-prone replication. But such a molecule might have been able to use itself or other RNA molecules as a template for polymerizing RNA nucleotides. The many errors made during replication of the early RNA replicase molecules would create a pool of genetic diversity on which natural selection could act to favor those molecules that were able to replicate faster and/or had greater accuracy. One problem with this scenario is that no replicase can copy its own active site. It is thus necessary to propose that a minimum of two RNA replicases were synthesized at nearly the same time from the "primordial soup" of precursor molecules. A primitive type of cell containing an RNA genome, called the **eugenote**, is hypothesized to have evolved from the progenote population.

14.9. How does the self-splicing intron of *Tetrahymena* remove itself from its parent rRNA molecule?

A primary transcript (pre-rRNA) of *Tetrahymena* is normally processed into 17S, 5.8S, and 26S rRNA molecules, the latter two being components of the larger 60S-like ribosomal subunit. During processing, a 413 mer (mer = nucleotide) intron in the 26S rRNA is excised and circularized by a series of **phosphoester bond transfers** or **transesterifications** (Fig. 14-2). This intron has a sequence of six contiguous nucleotides (called the **internal guide sequence**) near its 5′ end that forms base pairs with a partly complementary sequence at its 3′ end. The active site of the intron binds a *free* guanosine nucleotide (GMP, GDP, or GTP). The 3′-OH group of the G mononucleotide then initiates a nucleophilic attack on the phosphodiester bond at the splice site of the adjacent 5′ exon. The phosphate and the hydroxyl groups are exchanged in a transesterification reaction, causing a break in the pre-rRNA molecule at that point. The 5′ exon thus terminates at its 3′ end with a hydroxyl group; the 5′ end of the intron terminates with a phosphate group. The newly formed 3′ hydroxyl group on the 5′ exon now attacks the phosphodiester bond at the 3′ end of the intron. Another transesterification reaction there releases the intron (with a 3′-OH group) and the spliced RNA. Transesterification reactions do not consume any high-energy bonds. The same number of phosphodiester bonds are made as those that are broken. Thus, the intron only rearranges existing phosphodiester bonds rather than forms any new ones.

14.10. By definition, an enzyme is a molecule that can repeatedly catalyze a chemical reaction without itself being irreversibly altered in the process. The self-splicing rRNA intron of *Tetrahymena* can remove itself only once from its parent molecule. After excision, the intron no longer has the same function (a permanent change). Hence it has been claimed that the excised intron is not a true enzyme. Is this a valid argument?

After self-excision, the intron undergoes one or two rounds of autocyclization and hydrolysis, making excision of the intron irreversible. The end result is a smaller linear fragment of the intron (core). This core enzyme can polymerize poly-cytosine (poly-C) molecules from oligo-C molecules. In its role as a poly-C polymerase, the intron core behaves as a true enzyme. It does this by catalyzing the same phosphodiester bond transfers (transesterifications) it did during self-excision, except that in its new role oligo-C molecules (not part of the parent molecule from which the intron was excised) substitute for the 5′ splice site. Cleavage of RNA by either protein or RNA enzymes involves proton-exchange reactions. Transesterifications or phosphoester bond transfers also involve proton exchanges. Both hydrolysis and transesterifications involve nucleophilic attack on a specific phosphoester bond by a hydroxyl group. Hydrolysis occurs when the attacking group is water. Transesterification occurs if the attacking group is the 3′ hydroxyl group of a ribonucleotide. In essence, the core enzyme of the *Tetrahymena* intron simply moves C residues from one oligo-C molecule to another, making some oligos longer at the expense of others without creating any new phosphodiester bonds. The substrates for common RNA polymerases are monoribonucleoside triphosphates. The core intron poly-C polymerase uses oligoribonucleotide substrates. Oligonucleotides are more stable than mononucleoside triphosphates. So if oligonucleotides were relatively rich in the primitive environment, the first RNA replicases might have been capable of polymerizing oligonucleotides more readily than mononucleoside triphosphates.

14.11. The variety of peptides made by early peptide-specific protoribosomes would have been severely limited. Natural selection would be expected to favor the replacement of the internal guide sequences (functioning as mRNA sequences) of protoribosomal rRNA by separate external molecules with a vast potential for codon sequences. How might this replacement be experimentally demonstrated?

The internal guide sequence of the *Tetrahymena* intron may be a molecular fossil of the primordial RNA replicase whose primary function was (and remains) to polymerize RNA molecules. If this is true, it might be

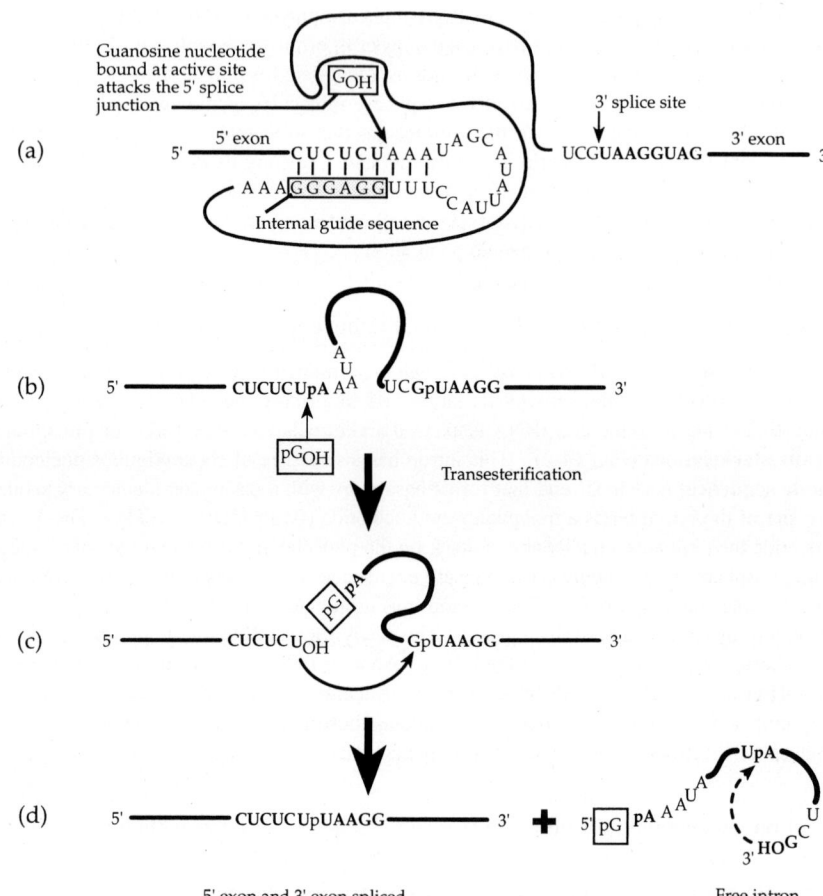

Fig. 14-2. Diagrams illustrating self-splicing activity of the *Tetrahymena* rRNA intron. (*a*) Internal guide sequence of intron pairs with 3′ end of 5′ exon. Guanosine nucleotide bound to the active site of the intron attacks the 5′ splice site. (*b*) This is a simplified diagram of (*a*). (*c*) Transesterification frees the 5′ end of the intron. A newly formed OH group at the 3′ end of the 5′ exon attacks the 3′ splice site, resulting in (*d*) spliced rRNA and free intron. The dashed arrow indicates that the 3′-OH group of the free intron can attack an internal phosphoester bond, producing a 399 mer closed circle and a 15 mer linear fragment (not shown).

possible to delete the internal guide sequence and transform the intron core into a RNA polymerase that is dependent on an external guide sequence or template. It is known the three-dimensional folding of tRNA molecules creates pockets that can bind divalent cations such as Ca^{2+}, Mg^{2+}, and Zn^{2+}. If the intron core could be redesigned to bind several such metals on its surface, these metals might be able to form ionic bridges to the negatively charged phosphates of external template backbones. This would facilitate the interaction between the catalytic domain of the intron core and the template molecule. Random oligonucleotides might serve both as primers for RNA replication and as a source of activated mononucleotides for polymerization of RNA molecules. This kind of RNA polymerase (replicase) could catalyze the synthesis of a wide variety of RNA molecules, not just the one type specified by the internal guide sequence.

14.12. The split (interrupted) genes of the more complex eukaryotes contain introns that must be removed from pre-mRNA molecules before they are released into the cytoplasm as mature mRNAs. What is the origin of this splicing mechanism?

Nuclear mRNA splicing may have evolved from self-splicing introns of either group I or group II because they all employ similar procedures (Fig. 14-3). Usually the group I introns are excised as a linear molecule that may then become circularized. Excision of group I introns is normally initiated by the 3′-hydroxyl unit of a free G nucleotide, but there is no theoretical reason why the initial attack could not be by an internal 2′-hydroxyl unit of the intron, as is the case for group II introns and for nuclear mRNA introns. In these cases, the excised intron is a lariat molecule with a branched adenine nucleotide involved in three phosphodiester bonds.

Since the splicing mechanisms for the group I nuclear rRNA intron of *Tetrahymena,* group II fungal mitochondrial introns, and nuclear mRNA introns are all very similar, these three types of introns may be descendants of a common ancestor. During this evolution, nuclear mRNA introns are hypothesized to have lost the self-splicing, intramolecular (*cis*-acting) activity of group I or group II introns. These lost functions are now performed by intermolecular *trans*-acting small nuclear ribonucleoprotein (snRNP) molecules, collectively called a spliceosome. The catalytic component of a snRNP may reside in its RNA (snRNA), rather than in its protein. The snRNAs may thus represent molecular fossils of pieces of self-splicing RNA from the RNA world. The 5′ (left) splice site of nuclear mRNA introns is recognized by the 5′ end of the snRNA component of the snRNP called U1.

The 3′ (right) splicing junction of mRNA introns is recognized by snRNPs U4, U5, and U6, but there is no region of nucleotide complementarity between the snRNAs in these snRNPs and the mRNA at this junction. This suggests that at least one of the protein components of these snRNPs is involved in the recognition of the junction. In the RNA world, the 3′ splice site might have been recognized by a complementary snRNA. However, in vertebrates, the 3′ splice site is now recognized (bound) by at least one protein of these snRNPs apart from any snRNAs. The mechanism of cleavage and ligation at the 3′ junctions of mRNA introns, however, remains unknown. Thus, during evolution of the vertebrates, at least one protein seems to have completely replaced an RNA function. It may be that today's snRNPs are still evolving and not just static remnants of the RNP world.

14.13. If RNA molecules are the presumed primordial genome/enzyme molecules of primitive living systems, what facts support this theory?

1. Ribose sugars are much easier to synthesize under simulated primordial conditions in the laboratory than are deoxyribose sugars.

2. The DNA precursors of all extant cells are produced by reduction of RNA nucleoside diphosphates by the highly conserved protein enzyme **ribonucleoside diphosphate reductase.** A **highly conserved** molecule is one that has changed little over long periods of evolutionary history as evidenced by its similar construction in a wide variety of organisms. This enzyme appears in all modern cells with few structural differences in even the most diverse organisms, suggesting that it is an ancient one that has performed the same essential task over a long evolutionary history. Living systems with RNA genomes are presumed to have evolved first. More stable DNA genomes evolved later to store genetic information. The monomers for synthesis of DNA polymers were produced in abundance from established chemical pathways only after the appearance of the ribonucleoside diphosphate reductase enzyme.

3. Ribose contains both 2′- and 3′-hydroxyl groups, whereas deoxyribose is missing a 2′-hydroxyl group. In modern tRNA molecules, both the 2′- and 3′-hydroxyl groups are involved in some unusual hydrogen bonds responsible for their cloverleaf or L-shaped structures. Single-stranded DNA would be less able than RNA molecules to form these kinds of complex three-dimensional configurations. Furthermore, the catalytic activity of some modern ribozymes is known to involve the 2′-hydroxyl group of ribose. Deoxyribose has no 2′-hydroxyl group.

4. All double-stranded DNA molecules have the same unvarying double-helical structure that would not lead us to expect them to have enzymatic properties, and no double-stranded DNA molecules have yet been discovered to have catalytic properties. However, double-stranded DNA molecules can fold back upon themselves, allowing protein-mediated interactions between segments many thousands of base pairs apart, as occurs with enhancer sequences. Single-stranded DNA molecules, however, can fold into complex tertiary shapes, as can most single-stranded RNA molecules. Superficially, they resemble proteins in this respect. By artificial selection in a cell-free molecular system (analogous to the one discussed in Question 14.21) it has been demonstrated that single-stranded DNA (ssDNA) molecules can be evolved that rapidly and specifically cut RNA molecules, as do some protein enzymes. Single-stranded DNA molecules are the genomes in some eukaryotic viruses (e.g., linear ssDNA in parvoviruses) and in some bacteriophages (e.g., circular ssDNA in ϕX-174 and M17), but they are otherwise not found in eukaryotic or prokaryotic cells.

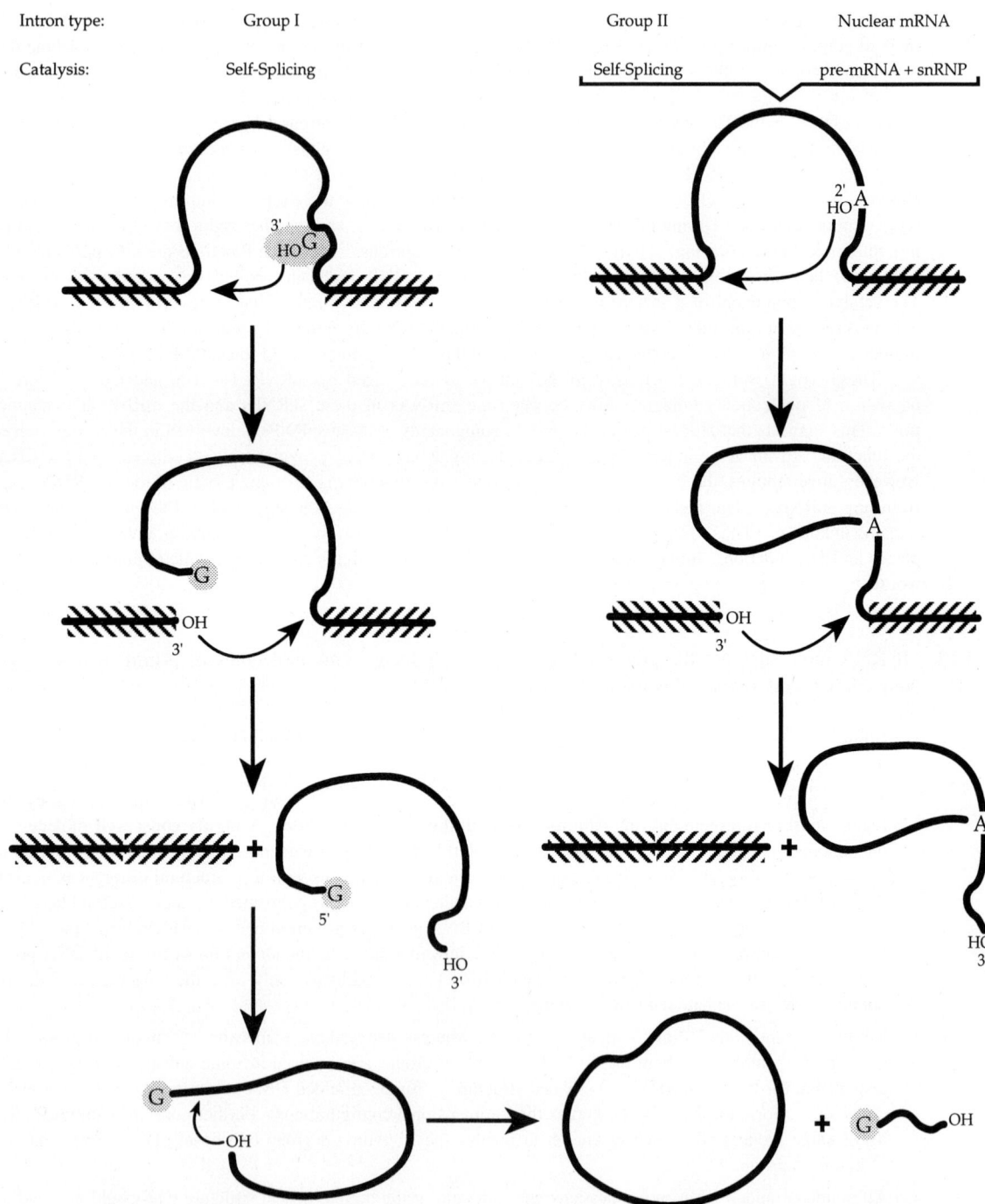

Fig. 14-3. Three types of introns and two splicing mechanisms. See Question 14.12 for explanation.

14.14. Why is it proposed that proteins eventually took over most of the catalytic functions originally performed by RNA molecules?

Modern biologically synthesized proteins can contain up to twenty different kinds of amino acids, whereas RNAs contain only four different kinds of nucleotides. The three-dimensional shapes of RNA molecules are determined by hydrogen bonding between sequences of complementary base pairs. Hence, a nucleotide sequence in one part of an RNA molecule (involved in the shape of the molecule) requires a complementary sequence (or nearly so) in another part of that same molecule. This places much greater constraints on the nucleotide sequences of functional RNA molecules (e.g., the cloverleaf and L shapes of tRNAs) than the amino acid sequences of functional proteins. It thus seems logical to assume that if an evolving system could synthesize enzymes made of proteins, there would be greater flexibility in the amino acid sequences of a functional protein than in the nucleotide sequences of a functional ribozyme. Any early ribozymes that aided in the production of useful proteins would tend to be strongly selected for retention and replication. Proteins would then gradually take over many of the functions formerly performed by RNA molecules, including most of the catalyses.

14.15. If proteins gradually took over most of the catalytic functions of early ribozymes, we might expect that during part of this transition some enzymes would be composed of both RNA and protein. Are there any such enzymes still in existence?

Perhaps the prime example of such a ribonucleoprotein is **ribonuclease P** (RNase P). This enzyme removes all *E. coli* tRNAs from their larger precursor transcripts by cleaving at the 5′ end of tRNAs. There is no common nucleotide sequence on either side of these cuts, so RNase P probably recognizes at least one of the three-dimensional features shared by all tRNA molecules. RNase P consists of a single small (377 mer) RNA molecule and one small (MW approximately 20,000) protein. Under physiological conditions, both of these components contribute to the catalytic activity of this enzyme. However, under nonphysiological conditions (e.g., in a buffer containing a high magnesium content), the RNA component alone performs catalysis as accurately and at almost the same rate as the holoenzyme. The role of the protein component is not yet known.

14.16. Two of the catalytic enzymes discussed earlier (*Tetrahymena* rRNA intron and RNase P) cleave RNA substrates. Are there any reasons for believing that the first enzymes of intermediary metabolism also contained RNA?

The catalytic activity of many protein enzymes involved in intermediary metabolism are enhanced by associated **nucleotide cofactors** that are bound (either covalently or noncovalently) to their catalytic sites. Many of these modern cofactors are structurally related to nucleotides (Fig. 14-4). Examples include nicotinamide adenine dinucleotide (NAD), flavin adenine dinucleotide (FAD), and acetyl-coenzyme A (acetyl-CoA or acyl-CoA). It is thus tempting to postulate that nucleotide cofactors are metabolic fossils of the enzymes involved in metabolic pathways of the RNA world.

14.17. The modern ribosomal mechanism of peptide bond formation is still not completely understood. For example, no single ribosomal protein has yet been shown to catalyze peptide bond formation. One explanation for this is that "peptidyl transferase" activity requires the cooperation of more than one ribosomal protein. Are there other possible explanations?

The peptidyl transferase activity of ribosomes may reside in the ribosome's rRNA component rather than in its proteins. Much of the structure and sequence of rRNA molecules is highly conserved. This finding would be expected if rRNA was not just a passive scaffold or jig, but active in peptide bond formation.

If, on the other hand, peptide bond formation is "spontaneous," the entire ribosome may act as a peptidyl transferase by aligning the activated amino acyl ~ tRNA complex and the peptidyl ~ tRNA complex on the mRNA molecule, rather than by catalyzing covalent bond formation as most anabolic enzymes do. Modern ribosomes might be molecular fossils that harbor many of the same features that were present in ribosomes of the RNP world at the time that proteins were cooperating with RNAs and/or in the process of replacing many of the catalytic functions of RNAs.

14.18. How might the genetic code have been established in the RNA world?

Early life systems that could make a variety of useful proteins would tend to have a selective advantage over those that had a more restrictive repertoire. Selection would thus promote the early protoribosomes, tRNAs, and tRNA synthetases to diversify. This process is envisioned to have produced a set of peptide-specific ribosomes, each with a different internal guide sequence serving as a mRNA sequence. A primitive genetic code would thus become established as sets of tRNA synthetases and peptide-specific protoribosomes evolved. Further refinements would tend to reduce the harmful effects of excessive mutations from transcriptional and translational

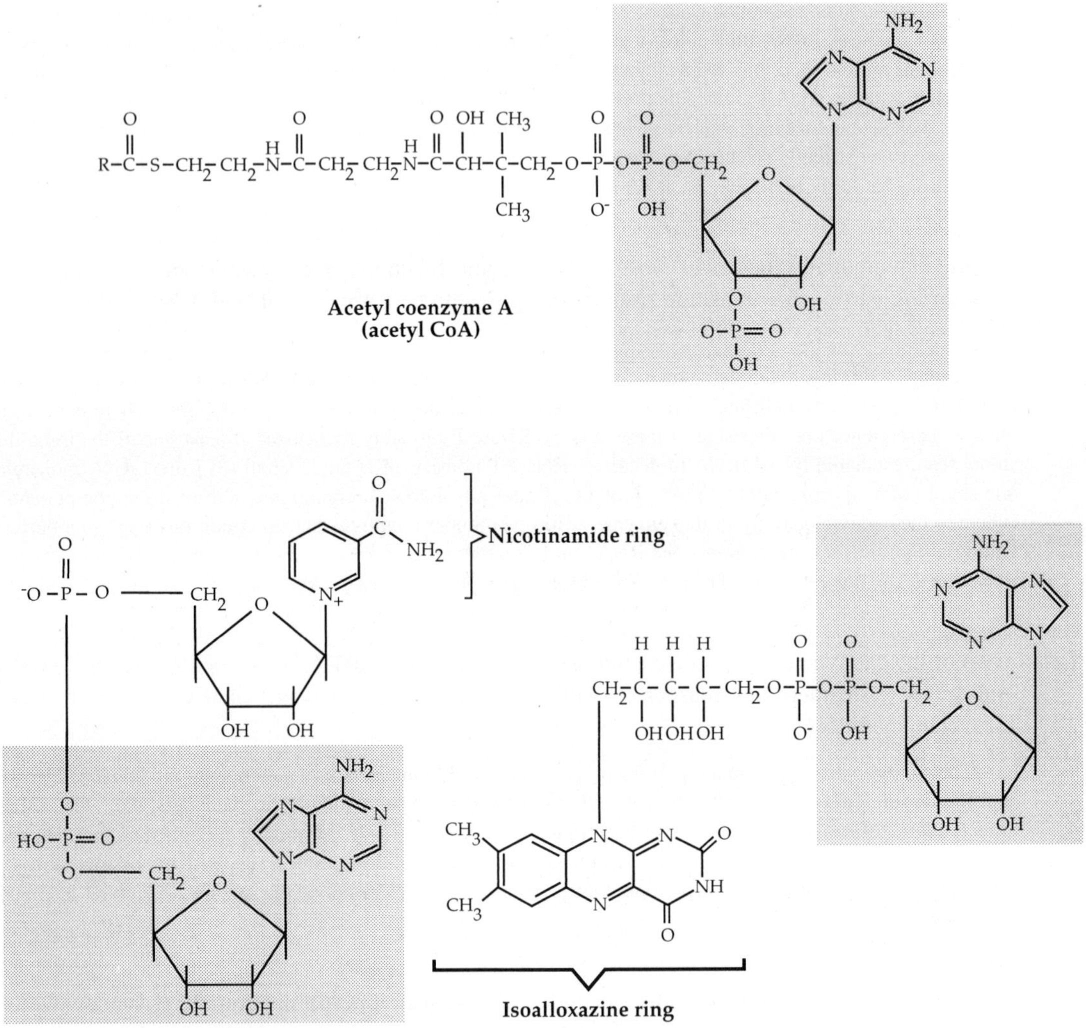

Nicotinamide adenine dinucleotide (oxidized form, NAD+)

Flavin adenine dinucleotide (oxidized form, FAD)

Fig. 14-4. Structures of three nucleotide cofactors of intermediary metabolism.

errors. In the modern genetic code, this mutation-suppression function is attributed to the degeneracy of the code (having more than one codon for most amino acids) and to wobble in the third position of the mRNA codons (usually allowing one purine to substitute for another or one pyrimidine for another).

14.19. What characteristics of amino acids may have aided in the establishment of the genetic code?

The more hydrophobic amino acids (phenylalanine, leucine, isoleucine, methionine, and valine) all have U at the center of their mRNA codons, whereas the more hydrophilic amino acids (tyrosine, histidine, glutamine, asparagine, lysine, aspartic acid, and glutamic acid) all have A at the center of their mRNA codons. Furthermore, codons with C as the central nucleotide specify more hydrophobic amino acids than those with G at the center. These facts suggest that the earliest coding might have simply specified the affinity of amino acids for water. Furthermore, in many instances, if the first two bases of a codon are established, the four possible nucleotide triplets all code for the same amino acid. It is thus tempting to speculate that early translation systems used only the first two bases in a codon to specify an amino acid. There is insufficient evidence to presently support any theories about which codons were the earliest to be established. But once a code was being used successfully, any major change in code assignments would not be tolerated. The reasons why 20 and only 20 amino acids are encoded also remain a mystery.

14.20. What inferences about the genetic code have been made from the nonenzymatic synthesis of RNA oligomers in vitro?

In the laboratory, the nonenzymatic formation of RNA oligomers occurs more readily when purines (represented by the letter R) alternate with pyrimidines (represented by the letter Y), producing the triplets RYR and YRY. Thus, primitive peptides of the early RNA world may have been partly ordered because codons with purines as the central base encode more hydrophilic amino acids, whereas codons with pyrimidines as the central base encode more hydrophobic amino acids. Artificial peptides consisting of alternating lysine and valine residues spontaneously aggregate into β-sheets like those found in silk fibroin. Those kinds of peptides have a hydrophobic and a hydrophilic face that tend to stack together into membranelike structures. Even slightly ordered peptides that contributed to adaptation of an evolving system would have allowed natural selection to operate in the perpetuation and refinement of the coding system by which they were made.

14.21. Is there any experimental evidence to support the notion that cell-free molecular systems could have evolved by natural selection?

In experimental attempts at cell-free molecular evolution, we must couple artificial selection to amplification by allowing only those molecules possessing the desired properties to reproduce. Suppose, for example, that we want to develop a ribozyme that would cleave DNA rather than RNA. The protocol we could follow would be to use reverse transcriptase to make cDNA copies of the ribozyme, and then amplify the cDNA by using the polymerase chain reaction. This would be followed by transcribing the cDNA back into RNA. By using low-fidelity versions of these procedures, we would be allowing mutations to occur and accumulate at a useful rate. Starting with a heterogeneous population of about 10^{13} ribozyme molecules, we would expose this population to a DNA substrate. Only a tiny fraction of those ribozymes would be able to cleave DNA. If we provide the right set of conditions, a ribozyme that cleaves DNA would become labeled with a portion of the cleavage product. We would amplify only the tagged molecules and allow new mutations to occur as outlined before. This cycle of events would need to be repeated ten more times. At the end of the experiment, the population of ribozyme molecules would be highly enriched with ribozymes that cut DNA very efficiently. These results have actually been attained in such an experiment.

In 1995, it was discovered that catalytic RNA molecules could be artificially selected (from a large pool of experimentally randomized RNA sequences) that rapidly ($\geq 10^5$ times accelerated) aminoacylate their $2'(3')$ termini with the amino acid phenylalanine when supplied with phenylalanyl-adenosine monophosphate. Thus, these RNA molecules catalyze the transfer of the same aminoacyl group as that of modern protein aminoacyl-tRNA synthetases. The success of these kinds of artificial selection experiments suggest that, given sufficient time and appropriate environments, natural selection could have been responsible for the production of numerous catalytic RNA molecules during the early stages in the evolution of life.

14.22. Is there anything about the life cycles of certain viruses that might be considered as molecular fossils?

Hepadnaviruses replicate their DNA genomes through an RNA intermediate, whereas retroviruses replicate their RNA genomes through a DNA intermediate. Both of these viruses may represent molecular fossils of a time when their host cells were in transition from RNA to DNA genomes.

Before the evolution of multicistronic RNA genomes, each protein or noncoding RNA (rRNA, tRNA) might have been represented by different RNA genomic fragments. Some RNA viruses have **segmented genomes** (i.e., genomes consisting of more than one kind of RNA molecule), but each segment may contain multiple genes. To function properly, each virion must contain at least one of each kind of segment. As examples, the genome of choriomeningitis virus has two segments, influenza virus has eight segments, and reovirus has ten segments. No DNA viruses are known to have segmented genomes. Perhaps these segmented RNA viral genomes may also be considered as molecular fossils of the RNP world.

Many plant viruses have single-stranded RNA genomes that are capable of self cleavage. Maybe these viruses contain RNA sequences that are molecular fossils of a time when RNA had both genomic and catalytic functions.

Example 14.2. The 324 mer genome of the Lucerne transient streak virus (LTSV) is a covalently closed single-stranded RNA molecule. It replicates by a rolling circle mechanism, producing multimeric RNA tails (containing more than one LTSV genome). Self splicing removes LTSV genomes from these multimeric molecules.

THE DNA WORLD

14.23. What is the rationale for proposing that DNA eventually replaced RNA as genomic (information storage) material?

Double-stranded DNA molecules are more stable than single-stranded RNA molecules. It would thus be advantageous for living systems to store heritable information in DNA molecules rather than in RNA molecules. The 2′-hydroxyl group of RNA can attack an adjacent phosphodiester bond, rendering RNAs much more labile than DNAs. This autocatalytic process may have been accelerated by the harsh conditions on the primitive earth. As cells became more complex, their genome sizes had to increase. If early eugenotes had segmented RNA genomes, at least one of each segment would have to be present in each daughter cell for its survival. To enhance the probability that progeny cells are provided with a full genome, natural selection would tend to favor production of polycistronic genomes. But the larger these RNA genomic segments are, the less stable they would likely become because of autocatalysis. Thus, it would be advantageous for more stable polycistronic DNA molecules to take over the genomic functions of RNA, leaving the RNAs to carry out functions that need not require long-lived molecules. The earliest anuclate cells containing DNA genomes (and all subsequent such cells) are known as prokaryotes.

14.24. What major steps were involved in the transition from RNA genomes to DNA genomes? Are any molecular fossils of this period still in existence?

At least four major processes were required to complete this transition: (1) synthesis of DNA monomers by ribonucleoside diphosphate reductase (Question 14.13); (2) reverse transcription of DNA polymers from RNA genomes; (3) replication of DNA genomes by a DNA polymerase; and (4) transcription of DNA genomes into "functional" (nongenomic) RNA molecules such as tRNA, mRNA, and rRNA. Today, reverse transcriptase (RNA-dependent DNA polymerase) is widespread in nature (plants, mammals, fruit flies, yeasts, bacteria, and in all retroviruses). Recall from Question 7.32 that the movement of retrotransposons from one chromosomal position to another only occurs with the help of reverse transcriptase. Primate DNA contains an unusually large number of repetitive sequences (such as the 500,000 copies of the ~ 300 bp *Alu* sequence in the haploid genome of humans) that require reverse transcriptase for their transposition. It is assumed that some kind of molecule with reverse-transcriptase activity was present when cellular RNA genomes were being converted to DNA genomes. Recall from Question 3.16 that a telomerase enzyme adds DNA sequences to the ends of linear chromosomes after each DNA replication. Telomerase is a ribonucleoprotein. A portion of its RNA component serves as the template for adding DNA nucleotides to the unpaired 3′ ends of copy strands left by DNA polymerase. These added telomeric DNA nucleotides are only short, repetitive sequences complementary to that of the template RNA sequence. However, since DNA is being synthesized from an RNA template, telomerase qualifies as a kind of reverse transcriptase. Are telomerases molecular fossils from the time when RNA genomes were being replaced by DNA genomes?

A very unusual RNA-DNA hybrid molecule exists in some myxobacteria and in at least one strain of *E. coli*. The two strands of nucleic acid are only joined by complementary base pairing in two localized regions, forming a branched hybrid molecule, the function of which is unknown. A gene has been isolated from myxobacteria that has 20–30% sequence homology with genes for the reverse transcriptases of mammals and *Drosophila*, as well as one encoded by a mitochondrial group II intron in yeast. Is the reverse transcriptase activity in these bacteria also a molecular fossil from the time when DNA genomes were replacing RNA genomes?

14.25. The removal of the 2′-hydroxyl groups from ribonucleoside diphosphates (rNDPs) by the early ribonucleoside diphosphate reductase enzyme would convert RNA precursors into DNA precursors. But modern DNA contains thymine instead of uracil. How is deoxythimidine triphosphate (dTTP) synthesized from the product of ribonucleoside diphosphate reductase activity? Why did DNA evolve to contain thymine rather than uracil?

Three of the dNDP products of ribonucleoside diphosphate reductase activity can be phosphorylated to triphosphate DNA precursors (dATP, dGTP, and dCTP). Deoxyuridine diphosphate (dUDP), however, must be dephosphorylated to dUMP before the enzyme **thymidylate synthase** can convert it to dTMP, from which dTTP precursors are derived. This enzyme is very highly conserved in all cells today, suggesting that the change from uracil-containing genomes to thymine-containing genomes occurred only once early in the evolution of DNA genomes.

At low frequency, cytosine (C) spontaneously becomes deaminated to uracil (U). In *E. coli* cells, these potentially harmful DNA mutations are excised by the enzyme **uracil-DNA glycosidase.** Removal of U from one strand of DNA leaves an unpaired guanine (G) on the other strand. A repair enzyme normally replaces the U with a C, thus restoring the original DNA sequence. If U had remained a component of DNA, deamination of C to U would not have been recognized as a mutation. It has been proposed that the substitution of T for U in DNA allowed the evolution of a DNA repair mechanism to correct this type of mutation and thereby increase the stability of the DNA molecule. Theorists may eventually develop more cogent explanations for the conversion of U to T in DNA. Mechanisms for repairing RNA molecules are virtually unknown in any organism (see Question 14.47 for the only known exception to this generalization). Rare transitions from C to U occurring in rRNAs or tRNAs are not expected to be very harmful, because these mutated molecules are not genomic and they become diluted by newly synthesized RNA in each cell generation.

14.26. Chirality is a fundamental characteristic of living organisms. What is chirality?

An object is said to be **chiral** if it cannot be superimposed on its mirror image by rotation and translation; for example, your right and left hands cannot be superimposed on one another, but they are mirror images of one another. The absolute configurations of amino acids and sugars are referenced to the structure of D-glyceraldehyde (Fig. 14-5). Glyceraldehyde occurs in two sterioisomeric (enantiomeric) forms (D or L) that are mirror images of one another. The D isomer rotates plane-polarized light to the right *(dextrorotary)*, whereas the L isomer rotates it to the left *(levorotary)*. Some bacteria have D-amino acids in a few specialized molecules (such as the antibiotic gramicidin S; Question 14.8) and in some amino acids of their cell walls, but all biological proteins synthesized on ribosomes consist entirely of L-amino acids, all RNA molecules contain only D-ribose, and all DNA molecules contain only D-deoxyribose.

D-glyceraldehyde   L-glyceraldehyde

Fig. 14-5. Enantiomers (mirror-image molecules) of glyceraldehyde. The middle carbon of glyceraldehyde is in the plane of the paper. The aldehyde group (CHO) and the hydroxymethyl group (CH$_2$OH) are behind the paper (denoted by dashed bond lines). The hydrogen (H) and hydroxyl (OH) groups are in front of the paper (denoted by solid bond lines).

14.27. Laboratory experiments like those of Stanley Miller that produced amino acids under presumed primitive earth conditions have resulted in an equimolar mixture of both D- and L-amino acids. Such mixtures, which may fail to rotate plane polarized light, are said to be *racemic*. So how, in the evolution of life on earth, was the choice made to construct biomacromolecules from only one sterioisomer?

The catalytic site of the first RNA polymerase enzyme must have had a "handedness," or chirality, that preferred a ribonucleotide with either the D or L form of ribose substrate, even though its own composition might have included both of these sugar isomers. The chirality preference of this enzyme would have been inherited despite this mixture of sugar isomers. The descendants of this first enzyme would continue to choose the same isomer, and natural selection would tend to perfect its ability to discriminate between the D and L forms and to favor the RNA replicase that did so in the shortest period of time with the greatest accuracy. In the transition from RNA to DNA genomes, the DNA would inherit the chirality of its RNA ancestor. In this way, the chirality of modern nucleic acids may be a "frozen accident of biological history." In the same manner, the first polypeptide polymerase made of RNA (and thus all of its descendants) by chance might have preferred L-amino acids. Furthermore, because translation evolved after RNA replication, the first tRNA molecules and first tRNA synthetase ribozymes would have contained D-ribose. If this isomer was best able to recognize the L isomer of amino acids, then that would explain its inheritance in all protein polymers by all descendant forms of life.

14.28. The split (or interrupted) genes of modern eukaryotic cells consist of coding regions (exons) and noncoding regions (introns). What potential evolutionary advantage might split genes offer? What evidence supports this theory?

Apparently exons from different genes can sometimes be recombined by natural mechanisms to code for proteins with different functions but containing related amino acid domains. Each of these exon domains may have a specific function (e.g., binding to a receptor, forming an α-helix, etc). This process, termed **exon shuffling,** is inferred to have been used extensively in the DNA world of eukaryotes. Perhaps an analogous mechanism of gene fusion was also operative when RNA was the only hereditary material (the RNA world). Exon shuffling has the potential for markedly accelerating the evolution of new proteins.

> **Example 14.3.** Several proteins having nucleotide sequence domains in common include epidermal growth factor (EGF), low-density lipoprotein receptor, blood clotting factors IX and X, complement component 9, urokinase, and tissue plasminogen activator (TPA). Complement component 9 also has another segment in common with the low-density lipoprotein receptor, the *Notch* gene product of the fruit fly, and the *lin-12* gene product of a nematode worm. Some proteins are mosaics of at least three different modules found in other proteins.

14.29. Are introns relatively new features of genomes or were they present in early forms of life?

Today introns are abundant in genomes of vertebrates, less frequent in lower eukaryotes, and absent from all common bacteria such as *E. coli*. But if introns were present in primitive genomes, most bacteria and relatively simple eukaryotes might have lost them under selection pressure to streamline their genomes for more rapid reproduction at lower energy expenditure. If, on the other hand, introns were not present in early genomes, but were inserted by recombination mechanisms into more advanced genomes, the simpler organisms may have resisted this process. Analysis of ancient, ubiquitous, highly conserved proteins may help resolve this problem.

> **Example 14.4.** The enzyme **triosephosphate isomerase** catalyzes an essential reaction in both glycolysis (the hydrolysis of glucose) and gluconeogenesis (the formation of glucose from noncarbohydrate sources such as proteins and lipids). Since the amino acid sequence of this enzyme is very similar in prokaryotes and eukaryotes, it probably was present in the progenote. In *E. coli* and budding yeast, this enzyme has no introns; in the fungus *Aspergilus* it has five introns, in humans and chickens, six, and in maize, eight. Five of the six introns in maize and vertebrates are at the same location, indicating that these introns were present before the divergence of plants and animals more than a billion years ago. The gene for triosephosphate isomerase in *Aspergilus* shares only one intron with maize and chicken; the other four fungal introns are so different that independent insertion cannot be ruled out. In all the organisms investigated so far, most of the introns are located between exons; each exon encodes a structural α/β domain consisting of an α-helix and a β-sheet. The overall structure of this enzyme suggests that it was assembled from at least ten primordial exons that were separated by introns, and the extant forms have evolved from this ancestral condition by differential intron loss. At least one intron interrupts a codon for a highly conserved (and therefore presumably functionally essential) amino acid. This fact argues against the notion that introns were inserted into the primordial gene as mutagenic events, because such random changes would most likely destroy the encoding of such an essential amino acid rather than preserve it.

14.30. Do all proteins evolve at the same rate?

Even the same protein in different species might not evolve at the same rate because of intrinsic factors (e.g., repair mechanisms) or extrinsic factors (e.g., environmental mutagens) peculiar to different populations of organisms. Complex biological problems are not likely to be solved more than once during the evolution of life. Once processes such as genome replication, mRNA transcription, protein translation, and controlled metabolism had become established, a change in any one of them that would prevent it from meshing with the rest of the organism's machinery (just as changing the teeth on a gear) would not be adaptive. Only relatively minor adjustments or refinements would subsequently be tolerated (analogous to changing the number of spokes in a gear's hub or the ornamentation outside a gear's teeth). These fundamental **housekeeping processes** are remarkably similar in all living cells; i.e., they have been highly conserved during evolution. (Read again the quote from J. D. Watson at the beginning of this chapter.) For example, triosephosphate isomerase from bacteria catalyzes the same reaction as one from humans, despite their differences in the coding sequences for this enzyme. Some relatively large changes may be possible in the gene for such an ancient enzyme without necessarily destroying the function of its protein product. If the progenote had introns in its genome, most bacteria seem to have found it advantageous to eliminate them. By genetic engineering, it is possible to create cDNA that is free of introns by reverse transcription from mRNA. This intronless cDNA can then be suitably inserted into a bacterial plasmid for amplification, transcription, and translation in a bacterial clone that does not possess the ability to splice out introns. Highly conserved proteins apparently have been able to tolerate only a few minor changes, whereas some other proteins have been able to absorb many mutations without loss of function.

14.31. How can the amino acid differences in related proteins be used to build phylogenetic trees that provide estimates of the degree of genetic divergence between various organisms?

Different proteins evolve at different rates that may be expressed as the number of point mutations per 100 amino acids per 100 million years. For example, triosephosphate isomerase has a value of 3 on such a scale, hemoglobin has 21, and fibrinopeptides have 90; nonfunctional pseudogenes are free to mutate without constraint and have a rate of about 400. Many mutations in functional proteins occur in regions outside the "active site" (e.g., outside the catalytic site of an enzyme or the binding site of a cell receptor). As long as the mutation occurs in a region that does not interfere with the normal functions of the molecule (including thermal stability, response to inhibitors or activators, etc.) it may be tolerated as a selectively **neutral mutation.** Over geological time, these neutral mutations tend to accumulate within a geological lineage. If it is assumed that such neutral mutations accumulate at a fairly constant rate for a highly conserved protein, it is possible to establish the branching pattern of a **phylogenetic tree** (also called a **cladogram,** an **evolutionary tree,** or a **geological tree**).

The **principle of parsimony** is commonly used to determine the minimum number of genetic changes required to account for the amino acid or nucleotide sequence differences between organisms sharing a common ancestor. The evolutionary distances separating organisms in a phylogenetic tree are usually expressed in units of nucleotide mutations or amino acid substitutions along each arm of the tree between branch points (Fig. 14-6).

These evolutionary distances between any two related species can sometimes be converted to chronological time units by correlation with the age at which their nearest relatives first appear in the fossil record. For deep geological time, however, the sequence differences in a highly conserved protein such as the respiratory enzyme cytochrome c or triosephosphate isomerase can be used as a **molecular (evolutionary) clock** to estimate the length of time that various organisms have been diverging from one another. The greater the number of such proteins used in constructing a phylogenetic tree, the more accurate the estimates of evolutionary divergence are likely to be.

14.32. In the construction of phylogenies, what advantages do nucleotide sequences have over amino acid sequences?

Since the development of rapid DNA sequencing techniques, phylogenetic relationships have been estimated far more from nucleotide analyses than from peptide analyses for several reasons. Nucleotide sequencing is much faster and less expensive than peptide sequencing. Even very tiny amounts of DNA in fossils over 100 million years old have been successfully multiplied a billionfold by the polymerase chain reaction to provide enough material for nucleotide sequencing. There is no comparable technique for multiplying tiny bits of polypeptides to the levels needed for amino acid sequencing.

DNA analyses can reveal samesense mutations, whereas protein analyses cannot. Furthermore, DNA analyses are not restricted to sequences coding for proteins, but can also be used for genes that are transcribed into tRNAs and rRNAs, as well as for noncoding control sequences (e.g., promotors, enhancers), introns, spacers, repetitive sequences, or any part of the genome. "Maximum likelihood" and "neighbor joining" are just two of several statistical methods that can be employed in estimating relationships between organisms. The validity

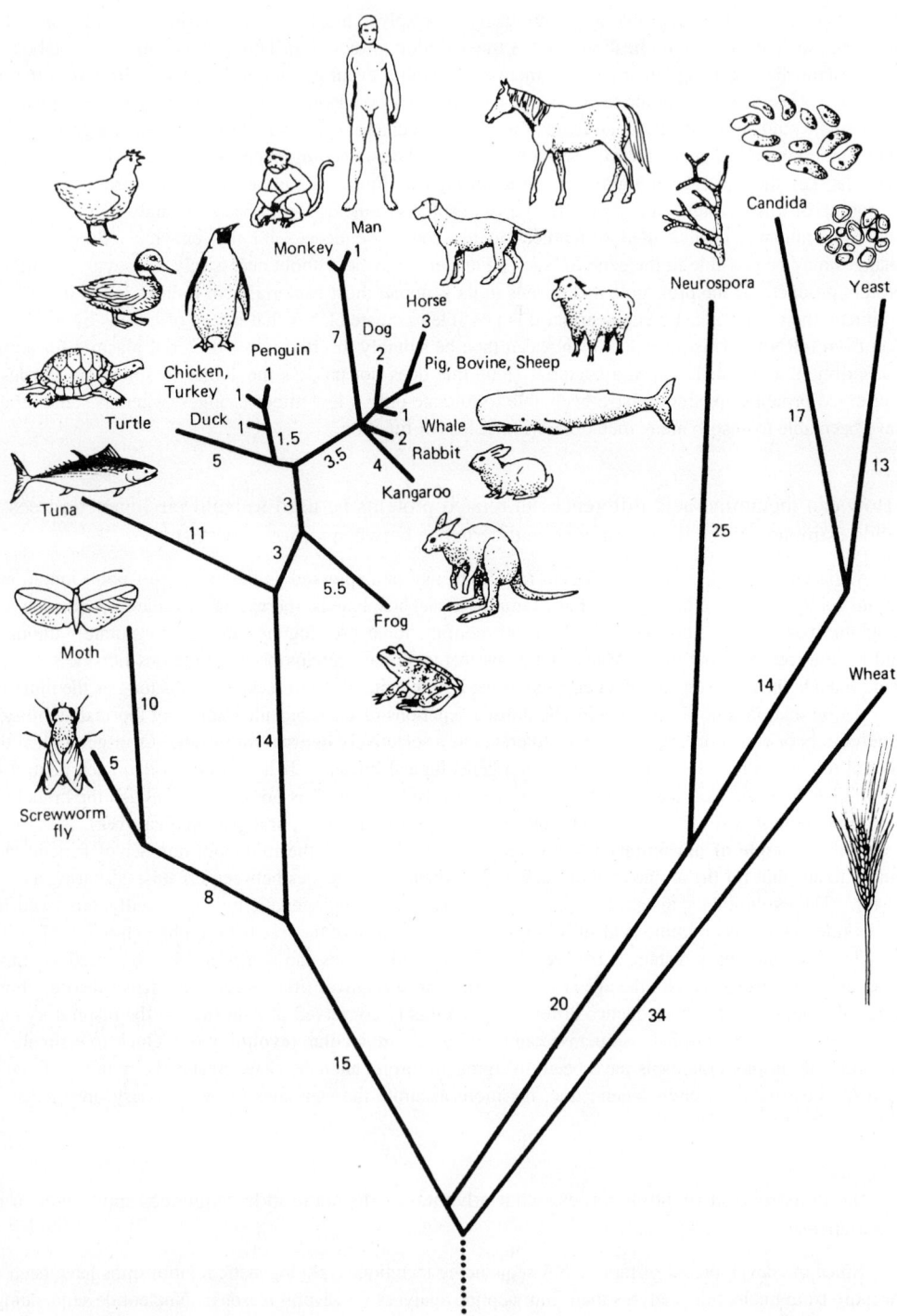

Fig. 14-6. A phylogenetic tree based on computer analysis of homologies between cytochrome *c*
molecules in various organisms. Each node (branch point) in the tree represents a common ancestral protein. The length of each branch (indicated by numbers thereon) represents the most likely number of distinctive point mutations that occurred during evolution
of these species. "Yeast" in this figure represents baker's yeast (*Saccharomyces cerevisiae*). Candida is also a yeast, but distantly related to *S. cerevisiae*. *(From The Science of
Evolution by W. D. Stansfield, Macmillan Publishing Company, 1977.)*

of a phylogenic interpretation increases with the congruence generated by multiple, independent analytical techniques.

THE EVOLUTION OF MAJOR PHYLETIC LINES

14.33.   How are bacteria classified?

The answer to this question depends on the classification system used and the taxonomic level chosen.   According to one popular system, all bacteria are grouped into the superkingdom *Prokaryotes* and into the kingdom *Prokaryotae*.   Most species of bacteria are further classified into the subkingdom *Eubacteria*.   Relatively few prokaryotes are placed in another subkingdom called *Archaebacteria*.   This latter group contains bacteria (called **archaeons**) that occupy harsh environments and/or carry out unusual kinds of metabolism.   Some of these archaeons live in hot sulfur springs, where the temperature is near the boiling point of water (90°C) and the water is strongly acidic (pH 2).   Hence they are called **thermoacidophiles** ("heat- and acid-loving").   Other archaebacteria, termed **halophiles** ("salt-loving"), prefer or require highly salty environments, such as those of the Dead Sea between Israel and Jordan or of the Great Salt Lake in the United States.   Thermoacidophiles and halophiles can metabolize both by fermentation in the absence of oxygen and by oxidative phosphorylation in the presence of oxygen.   Archaebacteria that metabolize energy anaerobically by using hydrogen gas for reducing carbon dioxide to methane are called **methanogens.**   They flourish in oxygen-free environments, such as lake bottoms or bogs, and are responsible for the liberation of flammable "marsh gas" (methane).   Other anaerobic archaebacteria metabolize energy by using hydrogen gas to reduce elemental sulfur to hydrogen sulfide (H_2S).

14.34.   At one time, prokaryotes were thought to be more closely related to a postulated progenote (the common ancestor of all cells, before there was a genome) than were eukaryotes, and all prokaryotes were thought to be more closely related to one another than to any eukaryote.   What evidence supported this belief?   What caused this idea to be overturned?

Archaebacteria today occupy the kinds of environments that were presumed to be widespread when life first evolved.   Hence, it was commonly believed that eubacteria evolved from primitive archaebacteria, and eukaryotes evolved from eubacteria.   Gradually, however, many more differences were found to separate the archaebacteria from the eubacteria (Table 14.1).   Some archaebacterial traits are shared with the eubacteria (they are both prokaryotes), whereas others are shared with eukaryotes (e.g., genes for some archaebacterial rRNAs and tRNAs contain introns; split genes are absent in eubacteria).   Based on his analysis of the nucleotide sequences in the highly conserved 16S rRNAs from many organisms, Carl Woese proposed in 1977 that archaebacteria are as different from the eubacteria as either group is from the eukaryotes.   Although this proposal was initially met with widespread skepticism, subsequent biochemical evidence supporting Woese's theory has convinced most of the scientific community that extant life consists of three separate lines of descent, so different from one another that not one of them can logically be ascribed as being the common ancestor of the other two.   Rather, all three lines are thought to have descended from the same progenote (Fig. 14-7).   The older taxonomic schemes need to be revised to reflect these three fundamental evolutionary roots.

14.35.   How did eukaryotic cells become nucleated?

Organisms with a nucleus may have evolved as long ago as 3.5 billion years b.p., but how the first nuclear membrane arose remains a mystery.   According to the **membrane proliferation hypothesis,** one or more invaginations of the plasma membrane in the progenote coalesced internally to surround the genome, became severed from the plasma membrane (Fig. 14-8), and formed a double-layered nuclear membrane.   The first hypothetical cell to possess a nucleus is referred to as the **urkaryote.**   The genome of the urkaryote is conceptualized as consisting of segmented, double-stranded DNA molecules.   The manner of infolding of the plasma membrane shown in Fig. 14-8 accounts for the fact that the nucleus of modern eukaryotic cells is enclosed within a "double membrane" consisting of two lipid bilayers.   Since the nuclear membrane has pores, the lumen of the nucleus is the topological equivalent of the cytosol.   The space between the inner and outer nuclear membranes and the lumen of the endoplasmic reticulum are continuous with one another and are the topological equivalents of the extracellular space.

A few modern eubacteria are able to form a metabolically inert **endospore** when culture conditions are unfavorable for normal growth and then germinate a vegetative cell from the endospore when favorable conditions re-

Table 14.1. Some Characteristics of Archaebacteria, Eubacteria, and Eukaryotes

Characteristics	Archaebacteria	Eubacteria	Eukaryotes
Cell length	$1-10 \ \mu m$	$1-10 \ \mu m$	$10-100 \ \mu m$
Membrane-bound organelles	None	Few (e.g., Chlorobium vesicles)	Nucleus, mitochondria, chloroplasts, lysosomes, etc.
Genome (DNA; chromosome)	Single, circular	Single, circular	Multiple, linear
Genetic recombination	Unidirectional (if present)	Unidirectional (if present) transformation, transduction, conjugation	Biparental, independent assortment, crossing over
Repetitive DNA	Little or none	Little or none	Often very high
Transcription and translation	Coupled	Coupled	Uncoupled
Cytoskeleton, endocytosis, exocytosis	Absent	Absent	Present
Asexual reproduction	Binary fission	Binary fission or budding	Mitosis
Sexual reproduction	No gametes	No gametes	Meiosis predominates in multicellular eukaryotes
Cell-wall composition	No peptidoglycans or N-acetylglucosamine acid	Peptidoglycans and muramic acid	Cellulose (plants), chitin (fungi)
Muramic acid in cell walls	No	Yes	No
Metabolism	Anaerobic (methanogens) or aerobic	Anaerobic or aerobic	Mostly aerobic, some fermenting amoeba
Cellular differentiation	None(?)	Sporulation, heterocysts, bacteroids	Extensive in multicellular forms
Introns (split genes)	Some in rRNA and tRNA	Absent	Common, especially in higher forms
Chromosome attachment for cell division	Plasma membrane	Plasma membrane	Tubulin spindle
Aliphatic chains of membrane phospholipids	Branched, ether-linked	Unbranched, ester-linked	Unbranched, ester-linked
Membrane lipid glycerophosphate	D isomer	L isomer	D isomer
RNA polymerases (relative size)	Large, multisubunit	Simpler	Large, multisubunit
EF2 sensitivity to diphtheria toxin	Yes	No	Yes

(*Table continues on page 361.*)

Table 14.1.　Some Characters of Archaebacteria, Eubacteria, and Eukaryotes (*Cont.*)

Characteristics	Archaebacteria	Eubacteria	Eukaryotes
Transfer RNAs:			
Initiating amino acid	Methionine	Formyl-methionine	Methionine
Ribothymidine in TψC loop	Absent	Present	Present
Dihydrouracil in DHU loop	Absent	Present	Present
Ribosomes:			
Subunit sizes	30S, 50S	30S, 50S	40S, 60S
Small rRNA size	16S	16S	18S
Large rRNA size	23S	23S	25–28S
CCUCC binding site at 3′ end of small rRNA	Present	Present	Absent
Sensitivity to antibiotics:			
Anisomycin	Yes	No	Yes
Chloramphenicol and kanamycin	No	Yes	No

turn. During endospore formation, the cell replicates its DNA. Cytokinesis then pinches off a smaller cell from a larger cell, both within the old cell wall. The larger cell then engulfs the smaller cell to produce an outer membrane around the inner membrane of the smaller cell. The process is more complicated than this, but this type of mechanism has been proposed as a model for the origin of the nuclear membrane in the urkaryote.

Today, however, nuclear membranes appear to dissolve and reform *de novo* during mitosis at each cell division. Perhaps the first nuclear membranes could also have developed *de novo*. During the anucleate period, an eukaryotic cell is in a prokaryotelike state, in that the chromosomes are not isolated from the cytoplasm.

More modern eukaryotes are thought to have evolved much later from the urkaryote line by acquisition of other membrane-bound organelles such as mitochondria and chloroplasts.

14.36. What function does a nuclear membrane serve in a cell that would give that cell a selective advantage over an anucleate cell?

The nuclear membrane keeps ribosomes and many other large cytoplasmic molecules confined to the cytosol. Primary mRNA transcripts from split genes must undergo several kinds of processing in the nucleus before they are released to the cytoplasm for translation into proteins. Removal of introns and splicing of exons is an important part of this nuclear processing. Without a nuclear membrane to separate ribosomes from pre-mRNA, many translated proteins would contain amino acid sequences of introns that had not yet been spliced out. The synthesis of some of these prematurely translated proteins would also be shortened if ribosomes encountered a stop codon within an intron. Normal proteins would be produced only if the introns were spliced out of the mRNA prior to being translated by ribosomes. Cells that could form a nuclear envelope to ensure that transcription and translation are uncoupled events would be at a selective advantage over anucleate cells with split genes.

14.37. Whereas the mechanism of nuclear membrane formation in urkaryotes is unclear, the origin of mitochondria in younger eukaryotes has a plausible (though not universally accepted) explanation. What is this theory?

According to the **endosymbiotic theory,** some ancient nucleated cells were capable of ingesting food particles or other cells by endocytic invaginations of their plasma membranes. Endocytosis, a hallmark process of eukaryotes, occurs only in organisms with a cytoskeletal system. These "feeder" cells had no cell wall to interfere

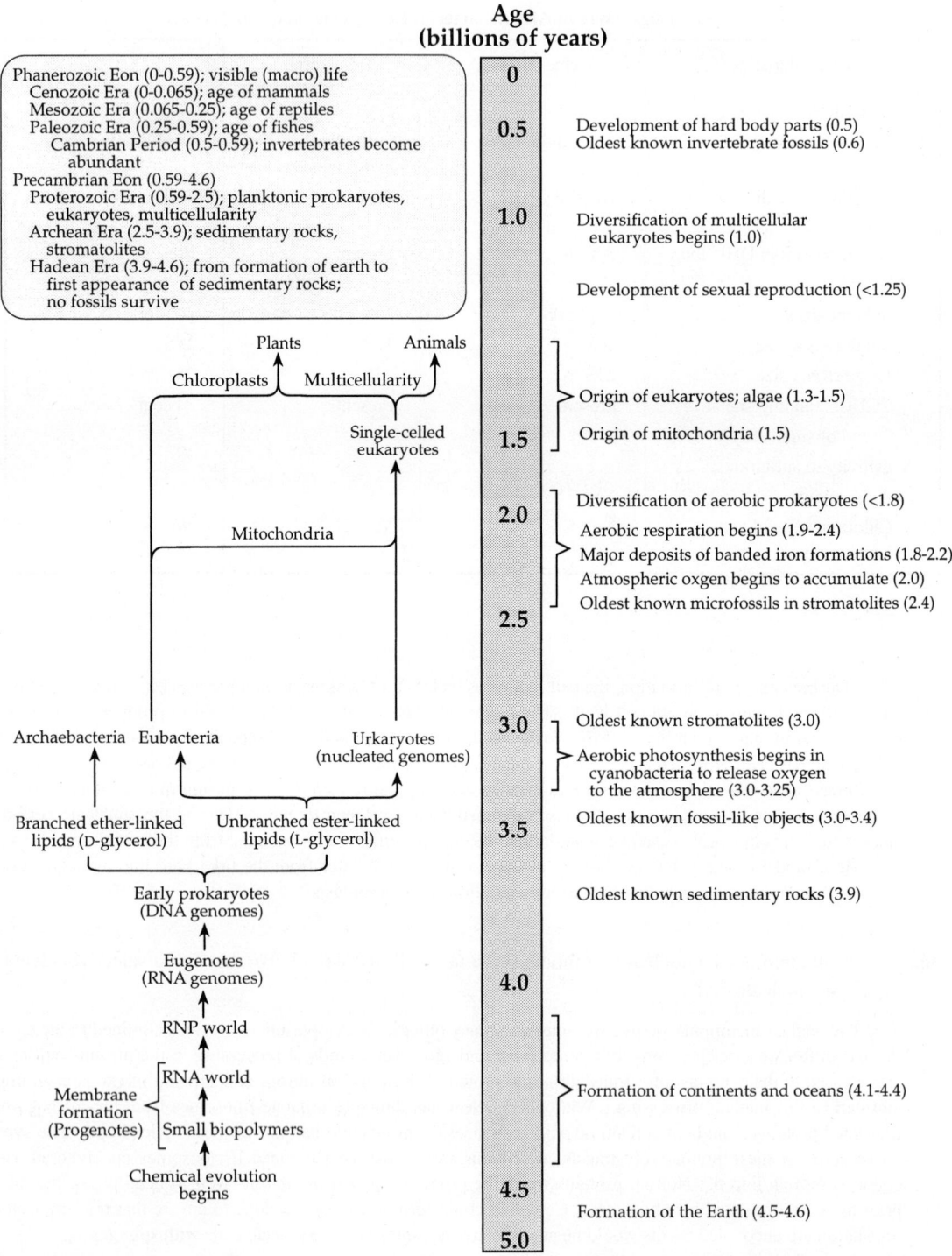

**Age
(billions of years)**

Phanerozoic Eon (0-0.59); visible (macro) life
  Cenozoic Era (0-0.065); age of mammals
  Mesozoic Era (0.065-0.25); age of reptiles
  Paleozoic Era (0.25-0.59); age of fishes
    Cambrian Period (0.5-0.59); invertebrates become
      abundant
Precambrian Eon (0.59-4.6)
  Proterozoic Era (0.59-2.5); planktonic prokaryotes,
    eukaryotes, multicellularity
  Archean Era (2.5-3.9); sedimentary rocks,
    stromatolites
  Hadean Era (3.9-4.6); from formation of earth to
    first appearance of sedimentary rocks;
    no fossils survive

0

0.5   Development of hard body parts (0.5)
      Oldest known invertebrate fossils (0.6)

1.0   Diversification of multicellular
        eukaryotes begins (1.0)

      Development of sexual reproduction (<1.25)

Plants   Animals

Chloroplasts Multicellularity

              Single-celled   Origin of eukaryotes; algae (1.3-1.5)
              eukaryotes
1.5   Origin of mitochondria (1.5)

Mitochondria

2.0   Diversification of aerobic prokaryotes (<1.8)

      Aerobic respiration begins (1.9-2.4)
      Major deposits of banded iron formations (1.8-2.2)
      Atmospheric oxgen begins to accumulate (2.0)
2.5   Oldest known microfossils in stromatolites (2.4)

Archaebacteria Eubacteria   Urkaryotes
                                  (nucleated genomes)
3.0   Oldest known stromatolites (3.0)

      Aerobic photosynthesis begins in
        cyanobacteria to release oxygen
        to the atmosphere (3.0-3.25)
Branched ether-linked   Unbranched ester-linked
lipids (D-glycerol)   lipids (L-glycerol)   Oldest known fossil-like objects (3.0-3.4)
3.5

      Early prokaryotes
      (DNA genomes)   Oldest known sedimentary rocks (3.9)

      Eugenotes
      (RNA genomes)

      RNP world
4.0

        RNA world
Membrane
formation   Formation of continents and oceans (4.1-4.4)
(Progenotes)   Small biopolymers

      Chemical evolution
      begins
4.5

      Formation of the Earth (4.5-4.6)
5.0

Fig. 14-7. A time scale of major evolutionary events in the history of life. Major geological times (billions of years) at the upper left are well established. Most of the ages for events on the right side of the scale may have large errors of estimates. The proposed sequence of events on the lower left side of the scale is hypothetical and event ages are highly speculative. The time bar in the middle of the figure, representing billions of years before the present (b.p.), has been expanded during times when a multitude of events are thought to have occurred and has been contracted when fewer events need to be recorded.

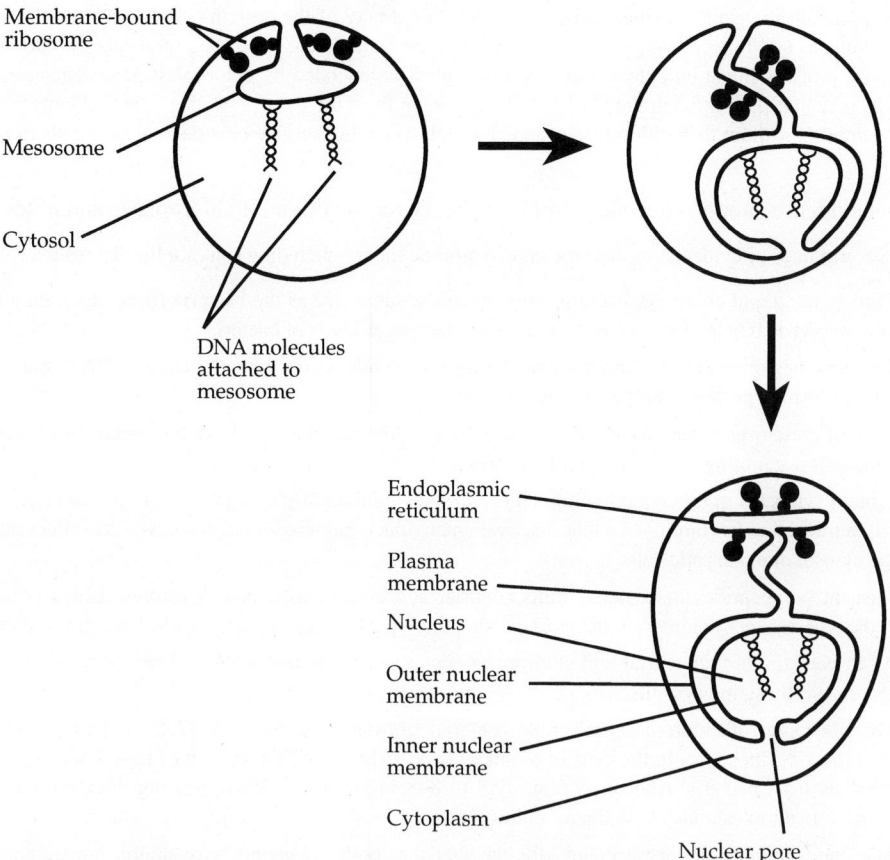

Membrane-bound ribosome

Mesosome

Cytosol

DNA molecules attached to mesosome

Endoplasmic reticulum

Plasma membrane

Nucleus

Outer nuclear membrane

Inner nuclear membrane

Cytoplasm

Nuclear pore

Fig. 14-8. Diagram of a hypothetical mechanism for the formation of a double membrane that isolates the DNA in a nucleus from the components of the cytoplasm. Each black line of a membrane represents a lipid bilayer. A portion of the endoplasmic reticulum is continuous with the outer membrane of the nuclear envelope.

with endocytosis. At least one large, fermenting, feeder cell engulfed one or more smaller respiratory bacteria, but failed to digest them. A type of purple, photosynthetic bacteria that had lost its photosynthetic ability and retained its respiratory chain is hypothesized to represent the endocytosed bacteria (endosymbiont). This endosymbiont was not digested by the feeder cell and was able to survive in an environment where nutrients were abundant and where it could hide from other predatory cells. In turn, the host feeder cell gained the energetic advantages of oxidative respiration over fermentation. These complementary advantages evolved into a **symbiotic** ("living together") relationship known as a **mutualism,** wherein neither entity can survive without the other. Part of this mutual adaptation involved the transfer of most of the genes of the bacterial endosymbiont into the nucleus of the host cell. Most negatively charged molecules, including mRNAs, tRNAs, rRNAs, and some proteins, that cannot cross the membrane of these organelles must still be encoded by the genomes of these organelles. This process is proposed to have given rise to the mitochondria of modern eukaryotic cells at least 1.5 billion years ago.

14.38. How did chloroplasts originate?

A stronger case can be made for the evolution of chloroplasts by endosymbiosis than that made for mitochondria. An aerobic, eukaryotic feeder cell (one that already had evolved mitochondria) is proposed to have engulfed one or more eubacteria (related to the cyanobacteria) that were capable of oxygenic photosynthesis. In the process of evolving into chloroplasts, the endosymbionts relinquished some of their genes to the nuclear genome, but not as many as did the endosymbionts that evolved into mitochondria. The protochloroplasts had to retain all of the genes specifying tRNAs and rRNAs for protein synthesis within the chloroplast because these negatively charged molecules cannot pass through the chloroplast membrane from the cytosol. Likewise, the mitochondria had to maintain the genes for their tRNAs and rRNAs. Many of the proteins used by chloroplasts for oxygenic photosynthesis and for replicating their own genomes are made in the cytosol and transported into the chloroplast. But those proteins made on chloroplast ribosomes must be encoded by chloroplast genes because mRNA molecules also cannot pass through the chloroplast membrane from the cytosol or vice versa. Plants cannot carry out photosynthesis without their chloroplasts, and chloroplasts cannot survive outside their host cells (a mutualism).

14.39. What evidence supports the endosymbiotic theory for the origin of chloroplasts and mitochondria?

Several lines of evidence support the endosymbiotic theory, including (but not limited to) the following:

1. Mitochondria and chloroplasts are usually about the same size as the bacteria from which they are thought to have evolved. Cyanobacteria tend to be larger than other kinds of bacteria.

2. The genomes of most mitochondria and chloroplasts reside within a single, circular DNA molecule that is devoid of histone proteins (bacterial characteristics).

3. Both of these organelles reproduce asexually by growth and division of existing organelles (never *de novo*) in a manner resembling binary fission in bacteria.

4. Protein synthesis in mitochondria and chloroplasts is inhibited by a variety of antibiotics (e.g., chloramphenicol, tetracycline, erythromycin) that inactivate many bacterial ribosomes, but have little effect on ribosomes in the cytosol of eukaryotic cells.

5. Nascent polypeptides in bacteria, mitochondria, and chloroplasts have *N*-formylmethionine at their amino ends, whereas methionine is at the amino ends of polypeptides made on cytosolic ribosomes of eukaryotes.

6. As in bacteria, mitochondrial and chloroplast genomes encode the tRNA and rRNA molecules for their own protein-synthesizing systems.

7. The ribosomes of chloroplasts resemble bacterial ribosomes in both size (70S) and rRNA sequences rather than the 80S ribosomes in the cytosol of eukaryotes. Mitochondrial ribosomes have both similarities and differences from bacterial ribosomes (e.g., 55S to 80S in various species), making their evolutionary affinities more difficult to establish than that of chloroplasts.

8. The endosymbiotic theory accounts for the fact that both organelles have double membranes. The inner membrane corresponds to the plasma membrane of the ancestral endosymbiont; the outer membrane represents the plasma membrane of the ancestral feeder host cell.

14.40. According to the endosymbiotic theory, mitochondria were acquired before chloroplasts. What is the evidence for this?

Evidence for the relative time of origins of mitochondria and chloroplasts is provided by their genomes. The modern mitochondrial genomes of various eukaryotes encode only some 10–40 proteins, whereas the chloroplast genome encodes about 75–100 proteins. Bacterial genomes encode several thousand proteins, and nuclear genomes may encode 50,000–100,000 proteins. If the eubacterial ancestors of mitochondria and chloroplasts had roughly equal genome sizes, and if the selection pressure to move genes from organelles to nucleus was equivalent, then the mitochondrial line has lost up to ten times more of its genome than that of the chloroplast line, indicating an earlier endosymbiosis of mitochondria.

14.41. What extant eubacterial groups are the closest relatives of the ancestors of mitochondria?

Introns do not exist in eubacteria, but they are present in some genes of archaebacteria, yeast mitochondria, and plant chloroplasts. Despite these facts, archaebacteria are not considered to be ancestors of either mitochondria or chloroplasts.

Some authorities suggest that mitochondria have closest affinities with the aerobic bacterium *Paracoccus denitrificans* because they share many similar membrane, respiratory, and enzymatic features. Others postulate a closer relationship between plant mitochondria and a group of anaerobic photosynthesizing purple bacteria based on 16S rRNA similarities. Many modern, purple photosynthetic bacteria can switch between respiration and photosynthesis depending on the availability of oxygen and light. It is hypothesized that a particular kind of ancient purple bacterium lost its photosynthetic ability but retained its respiratory chain. This organism became the progenitor of mitochondria through endosymbiosis.

14.42.   What are the most primitive extant eukaryotes?

Flagellated protozoans of the genus *Giardia* have rRNA sequences that are the most divergent from other eukaryotes studied so far and the most closely related to prokaryotes, thus suggesting a very ancient ancestry. Furthermore, this eukaryotic group lacks mitochondria! Perhaps this group is a descendant of the ancestral urkaryote before the endosymbiotic evolution of mitochondria. Of course, these protozoans might also have descended from a more recent ancestor that had acquired mitochondria but subsequently lost them. However, it is difficult to understand why natural selection would favor the loss of an organelle that offers such a high energy-producing advantage as does the mitochondrion. The convoluted membranes of cristae within the numerous mitochondria within a cell provides a large surface area as sites for the respiratory enzymes.

14.43.   Is there any evidence suggesting that more than one endosymbiotic event may have occurred in the evolution of mitochondria?

The mitochondrial DNA of fungi specifies RNA containing self-splicing introns, but the mitochondrial DNA of plants and multicellular animals do not. The eubacteria that are thought to have evolved into the mitochondria of plants (and perhaps also of animals) lack introns in their modern bacterial descendants. If the eubacterial progenitor of mitochondria contained introns, and the introns were lost by their free-living eubacterial relatives but were retained by the protomitochondrial endosymbiont in fungi, the discrepancy could be resolved without postulating a second endosymbiosis.

If, on the other hand, the eubacterial ancestor of mitochondria did not have self-splicing introns, perhaps after endosymbiosis began the fungal protomitochondrion acquired these introns by **horizontal transmission** (from one contemporary species of organism to another) rather than by **vertical transmission** (from parent cell to progeny cell). A precedent for this scenario lies in bacteriophage T4 containing self-splicing introns in at least three sites. Eubacteria, including *E. coli* (the host species for T4), do not contain introns. It is suggested that T4 may have acquired its introns by infecting a mitochondrion or protomitochondrion of an eukaryotic cell.

It might also be possible that endosymbiosis has occurred independently in different eukaryotic lineages. If fungal mitochondria are descendent from a different endosymbiont than that which gave rise to mitochondria in other eukaryotes, then the evolution of mitochondria would be **polyphyletic** (multiple origins) rather than **monophyletic** (having a single common ancestor). However, the mitochondrial rRNA sequences of all animals, protozoans (*Giardia* excepted), fungi, and one green alga *(Chlamydomonas reinhardi)* are closely related and form a group distinct from that of the higher plants. It might be possible that two endosymbiotic events occurred in the evolution of mitochondria in the higher plants, with the second fusion displacing the earlier endosymbiont.

14.44.   Is there any evidence suggesting that more than one endosymbiotic event may have occurred in the evolution of chloroplasts?

Cyanobacteria (cyanophytes) and prochlorophytes are two groups of prokaryotic, oxygenic photosynthesizers that have been implicated in the evolution of chloroplasts. Several modern eukaryotic species are known to engulf (endocytose) cyanophytes. Once inside the host cell, these bacteria lose their cell walls, but continue their photosynthetic activities. If these symbionts become unable to live outside the host, they are called **cyanelles.** The chloroplasts of red algae and free-living cyanophytes possess chlorophyll *a*, but not chlorophyll *b*. They both contain phycocyanin and phycoerythrin photosynthetic pigments attached to their unstacked thylakoids. They also share many homologous DNA and protein sequences. Some cyanelles retain a surface layer of peptidoglycan. All of these facts support the endosymbiotic theory for the origin of cyanelles in the red algae. Chloroplasts in most other photosynthetic eukaryotes (including green algae and green plants) more closely resemble prochlorophytes such as *Prochloron*. Both possess chlorophylls *a* and *b*, have stacked thylakoids, and are devoid of phycobiliproteins. *Prochloron* is also known to be a symbiont in ascidians (a group of primitive chordates). It is possible that red algae (Phylum Rhodophyta) and green algae (Phylum Chlorophyta) evolved autogenously into eukaryotes from different prokaryotic ancestors, and by chance developed the same nucleocytoplasmic system, but very different chloroplast systems. It seems much more likely, however, that the red and green algae evolved from a common eukaryotic ancestor (thus accounting for similar nucleocytoplasmic systems) and independently acquired different endosymbionts that evolved into different kinds of chloroplasts. This evidence favors a polyphyletic origin of chloroplasts in at least two ancestral lineages.

14.45.   How might the flagella and cilia of eukaryotes have originated?

One theory suggests that motile bacteria may have become symbiotic on the surface of ancestral eukaryotic cells. The surfaces of some modern protozoans are covered with spiral-shaped bacteria called **spirochetes** that propel the host cell by their undulations. The attachment of these bacteria to the eukaryotic cell surface seems to stimulate, in the cytoplasm, formation of structures that resemble those of basal bodies that are thought to serve as sites for microtubule elongation and as anchorage for flagella. Perhaps these modern examples of symbiotic bacteria, serving a motility function, might represent an early stage in the evolution of true flagella in eukaryotes. If so, some major unanswered questions remain. How did the bacterial flagellin system become replaced by the eukaryotic tubulin system and what, if anything, remains of the symbiont?

Pillotinas are large symbiotic spirochetes that live in the hind gut of termites. Unlike the flagella of other motile bacteria, those of pillotinas contain microtubulelike proteins. Microtubules, made of the protein tubulin, are characteristic of all eukaryotes, but are unknown in any other prokaryotes. It is not known if the common possession of microtubulelike proteins represents a true genetic homology (indicating descent from a common ancestor by vertical transmission) or if the two genes encoding microtubulelike proteins have evolved independently in spirochetes and eukaryotes. It is also possible that these genes were acquired by spirochetes from a eukaryotic host nucleus through horizontal transmission.

14.46.   The mitochondrial mRNAs for several proteins (including NADH dehydrogenase and subunits I, II, and III of cytochrome *c* oxidase) of at least three flagellated protozoans (*Leishmania, Trypanosoma,* and *Crithidia*) have been converted to cDNAs (by reverse transcriptase) and sequenced. These mRNAs have the correct coding sequences to specify these proteins, but no exactly comparable sequences in the mitochondrial DNAs have been found. What is the explanation for these facts, and how might this phenomenon relate to early biological evolution?

Apparently the mitochondrial DNA is transcribed in the normal way, but the mRNA transcripts are modified (mechanism unknown) by the addition or deletion of uridylate residues at multiple precise sites to create translatable templates for the synthesis of functional proteins. This process is termed **RNA editing.** For example, some primary transcripts contain no AUG codons to initiate translation, but RNA editing inserts a U at the precise position needed to create a start signal. The extent of this phenomenon in other organisms is not known, but it raises the question of whether it existed in the early evolution of life to make use of the larger number of imperfect RNA molecules that likely existed at that time.

Objective Questions

Multiple Choice

Directions: Choose the one best answer.

1. Which of the following is not a characteristic of all living systems? (*a*) DNA genome (*b*) growth by controlled synthesis (*c*) lipid plasma membrane (*d*) protein-synthesizing system (*e*) oxidative respiration

2. The age of the earth is estimated to be about (*a*) 10–20 billion years old. (*b*) 4–5 billion years old. (*c*) 1–2 billion years old. (*d*) 600 million years old. (*e*) 10,000 years old.

3. The first successful attempt to test the theory that precursors of biomacromolecules could be abiotically synthesized under presumed primitive earth conditions was made by (*a*) Darwin. (*b*) Huxley. (*c*) Miller. (*d*) Oparin. (*e*) Watson.

4. Which of the following may be part of the reason why thymine replaced uracil in DNA? (*a*) Cytosine spontaneously deaminates to uracil. (*b*) Thymine is easier to synthesize than uracil. (*c*) Thymine is synthesized from 5-methyl uracil. (*d*) Thymine has a stronger binding affinity for adenine than does uracil. (*e*) More than one of the above apply.

5. The first ribozyme was found in (*a*) a nuclear gene for a DNA replicating enzyme. (*b*) a mitochondrial gene for a respiratory enzyme. (*c*) a mRNA for a mitochondrial enzyme. (*d*) an intron within a pre-rRNA molecule. (*e*) an exon within a tRNA molecule.

6. Which of the following statements cannot be used to support the theory that RNA genomes preceded DNA genomes in the evolution of life on earth?   (*a*) Ribose sugars are easier to synthesize abiotically under simulated primordial earth conditions in the laboratory than are deoxyribose sugars. (*b*) All modern cells use a ribonucleoside diphosphate reductase to make DNA precursors from RNA precursors. (*c*) RNA can form more kinds of hydrogen bonds than DNA. (*d*) RNAs are more stable molecules than DNA. (*e*) Some RNA molecules have catalytic activity, whereas no double-stranded DNA molecules have been found with this property.

7. Which of the following is not characteristic of archaebacteria?   (*a*) Unbranched, ester-linked aliphatic chains of membrane phospholipids (*b*) D isomer of membrane lipid glycerophosphates (*c*) EF2 sensitivity to diphtheria toxin (*d*) methionine at the amino terminus of nascent polypeptide chains (*e*) more than one of the above

8. Which of the following is a prime example of a ribonucleoprotein enzyme, the catalytic activity of which (under non-physiological conditions), resides solely in its RNA component?   *(a)* Reverse transcriptase *(b)* cytochrome *c* *(c)* polynucleotide phosphorylase *(d)* bacterial RNA polymerase *(e)* RNase P

9. Which of the following processes offers the greatest opportunity for accelerating the evolution of new useful proteins?   (*a*) Intron excisions (*b*) transposon insertions (*c*) exon shuffling (*d*) cDNA insertions (*e*) DNA amplification

10. Which of the following statements about codons or their cognate amino acids provides a possible clue to the origin of the genetic code?   (*a*) Some amino acids offer protection against specific proteolytic enzymes. (*b*) Codons with uracil in the center tend to code for hydrophobic amino acids. (*c*) Some amino acids offer greater stability to protein structure than do others. (*d*) Wobble at the 5′ end of mRNA codons is tolerated more than wobble at the 3′ end. (*e*) Initiation and termination events of translation are almost always associated with specific amino acids.

11. In the laboratory, the kind of mRNA triplet oligomer that occurs most readily by nonenzymatic reactions is (R = purine, Y = pyrimidine) (*a*) RRR. (*b*) YRY. (*c*) YYY. (*d*) RRY. (*e*) YYR.

12. Which of the following statements pertains to nuclear mRNA introns and suggests that they may have evolved from group II self-splicing introns?   (*a*) The excision products of both types of introns may be either circular or linear. (*b*) An external cofactor is required for the excision of both types of introns. (*c*) A lariat excision product is produced by both types of introns. (*d*) Both types of introns have similar boundary sequences. (*e*) More than one of the above statements are applicable.

13. The class of molecules that excise introns from mRNA molecules at their 5′ ends in most modern eukaryotes may be described as (*a*) *cis*-acting molecules. (*b*) self-splicing introns. (*c*) snRNA molecules. (*d*) glycoprotein molecules. (*e*) more than one of the above.

14. Which of the following viruses replicates its DNA through an RNA intermediate?   (*a*) Retrovirus (*b*) influenza virus (*c*) hepadnavirus (*d*) reovirus (*e*) choriomeningitis virus

15. The most likely initial step in the transition from the RNA world to the DNA world involves (*a*) double-stranded RNA. (*b*) ribonucleoside diphosphate reductase. (*c*) DNA polymerase. (*d*) cDNA. (*e*) thymidylate synthase.

16. Which of the following pairs of sterioisomers exists in all cellular proteins and RNA molecules, respectively? (*a*) L-amino acids and D-ribose (*b*) D-amino acids and L-ribose (*c*) L-amino acids and L-ribose (*d*) D-amino acids and D-ribose (*e*) none of the above

17. Split genes exist (*a*) only in eukaryotes. (*b*) only in vertebrates. (*c*) in most eukaryotes and some archaebacteria. (*d*) in most eukaryotes, some eubacteria, some archaebacteria. (*e*) in most eukaryotes, some eubacteria.

18. The kind of mutations that can best be used as a molecular clock are (*a*) selectively neutral. (*b*) temperature-sensitive. (*c*) missense. (*d*) highly conserved. (*e*) at intron splice junctions.

19. The most likely sequence of events that produced the organelles of eukaryotes is (earliest to latest, respectively) (*a*) nucleus–mitochondria–chloroplast. (*b*) mitochondria–chloroplast–nucleus. (*c*) nucleus–chloroplast–mitochondria. (*d*) chloroplast–nucleus–mitochondria. (*e*) mitochondria–nucleus–chloroplast.

20. Which of the following statements cannot be used to support the endosymbiotic theory for the origin of mitochondria? (*a*) Mitochondrial DNA has a circular genome. (*b*) Mitochondria are about the same size as bacteria. (*c*) Mitochondrial ribosomes are 70S. (*d*) Mitochondrial genes code only for enzymes of oxidative respiration. (*e*) Mitochondrial ribosomes are inactivated by the antibiotic chloramphenicol.

21. Aerobic respiration is thought to have first appeared in some organisms of approximate age (*a*) 0.5–0.6 billion years before the present (b.p.). (*b*) 1.0–1.5 b.p. (*c*) 2.0–2.5 b.p. (*d*) 3.0–3.5 b.p. (*e*) 3.5–4.0 b.p.

22. The extant bacterial group that is most closely related to the ancestor from which mitochondria evolved is (*a*) purple photosynthetic bacteria. (*b*) archaebacteria. (*c*) cyanobacteria. (*d*) methanogens. (*e*) prochlorophytes.

23. The most primitive extant eukaryotes belong to the genus (*a*) *Plasmodium*. (*b*) *Tetrahymena*. (*c*) *Giardia*. (*d*) *Chlamydomonas*. (*e*) *Caenorhabditis*.

24. The feature(s) that distinguish(es) the most primitive extant eukaryotes is/are (*a*) circular nuclear DNA. (*b*) mitochondria absent. (*c*) bacteria type of cell wall. (*d*) highly divergent rRNA sequences from those of other eukaryotes. (*e*) more than one of the above.

25. Which of the following is a fact that argues against a symbiotic origin of eukaryotic flagella from bacterial flagella? (*a*) Bacterial flagella typically contain flagellin; eukaryotic flagella contain tubulin. (*b*) Bacterial flagella rotate; eukaryotic flagella do not rotate. (*c*) Bacterial flagella are directly energized by electrons and protons from the respiratory system of the plasma membrane; eukaryotes are not. (*d*) Eukaryotic flagella terminate in the cytoplasm at basal bodies; prokaryotes do not. (*e*) More than one of the above argue against the symbiotic theory.

True–False

1. The only place where amino acids or other precursors of biomacromolecules are known to be made is on Earth.

2. There are only two major scientific theories regarding the origin of life on Earth: creation and evolution.

3. Because protein enzymes are now required to replicate DNA, it is most likely that proteins evolved before nucleic acids.

4. The first RNA replicase constructed without a template must have been able to use itself as a template for making the first copies.

5. The self-splicing *Tetrahymena* rRNA intron is not a true enzyme because it can function only once in its own excision.

6. Clays have been demonstrated to promote linkage of amino acids into oligomeric polypeptide chains.

7. Some cell-free molecular systems are known to be capable of evolving by natural selection.

8. Like the *Tetrahymena* self-splicing intron, ribonuclease P cleaves its RNA substrate at common nucleotide sequences on either side of its target.

9. Although ribonuclease P consists of both RNA and protein, the catalytic activity of the enzyme resides entirely with the protein component.

10. No single protein in modern ribosomes has been demonstrated to catalyze peptide bond formation.

11. No circular RNA viral genomes are known to exist.

12. No DNA viruses are known to have segmented genomes.

13. The most obvious reason why DNA genomes would be expected to have replaced RNA genomes is because DNA is a more stable molecule than RNA.

14. No mechanisms for repairing RNA molecules are known.

15. Although oxygenic photosynthesis evolved long before aerobic respiration, the endosymbiotic theory has the evolution of mitochondria preceding the evolution of chloroplasts.

16. Laboratory simulations of nonbiological synthesis of amino acids under presumed primitive earth conditions produce only L sterioisomers, as found in living cells today.

17. Both the chirality of nucleic acids and amino acids in proteins are hypothesized to be a "frozen accident of biological history."

18. Loss of introns during evolution from the progenote is a more likely event than insertion of introns.

19. Different proteins of the same size are expected to evolve at about the same rate.

20. Each branch of a phylogenetic tree indicates that at least one mutation has occurred in one line that did not occur in the other line.

21. For constructing phylogenetic trees of very ancient organisms, nucleotide (or protein) sequence differences from rapidly evolving molecules are more useful than those from highly conserved molecules.

22. According to Carl Woese, eubacteria are likely to have evolved from an archaebacterial ancestor.

23. The eukaryotic nucleus evolved from the endosymbiosis of eubacteria by an anucleate archaeon.

24. Several lines of evidence support the hypothesis that mitochondria and chloroplasts evolved from eubacteria.

25. The genomes of mitochondria do not code for all the respiratory enzymes. The mRNAs for those missing enzymes are made in the nucleus and transported into the mitochondria for translation into the appropriate proteins.

26. The mitochondria of fungi contain self-splicing introns, but the mitochondria of plants and animals do not.

27. No double-stranded DNA molecule has been found to have catalytic activity.

28. Some pure RNAs are known to have self-cutting, self-splicing, and self-ligation activities.

29. Reverse transcriptase is an enzyme found only in retroviruses.

30. Chloroplasts are thought to be derived from an ancestral cyanobacterium endosymbiont because both contain chlorophylls *a* and *b*.

31. Mitochondria seem to have lost more of their genes to the nucleus during their endosymbiotic evolution than have chloroplasts.

32. All extant eukaryotic cells possess mitochondria, but not all eukaryotic cells possess chloroplasts.

Matching

Directions: Match each item in **Column A** with the one in **Column B** to which it is most closely associated. Each item in **Column B** can be used only once.

<table>
<tr><td colspan="2" align="center">**Column A**</td><td colspan="2" align="center">**Column B**</td></tr>
<tr><td>1.</td><td>uranites disappear from the geological record</td><td>A.</td><td>proteinoids</td></tr>
<tr><td>2.</td><td>panspermia theory</td><td>B.</td><td>processing of 26S rRNA of *Tetrahymena*</td></tr>
<tr><td>3.</td><td>Sidney Fox</td><td>C.</td><td>precipitated hydrous ferric oxide</td></tr>
<tr><td>4.</td><td>*Giardia*</td><td>D.</td><td>synthesis of DNA precursors</td></tr>
<tr><td>5.</td><td>red algae</td><td>E.</td><td>most primitive extant eukaryote</td></tr>
<tr><td>6.</td><td>eugenote</td><td>F.</td><td>cyanelles</td></tr>
<tr><td>7.</td><td>transesterification</td><td>G.</td><td>U4</td></tr>
<tr><td>8.</td><td>snRNA</td><td>H.</td><td>atmospheric oxygen content greater than 1%</td></tr>
<tr><td>9.</td><td>ribonucleoside diphosphate reductase</td><td>I.</td><td>earliest cell containing a genome</td></tr>
<tr><td>10.</td><td>red beds</td><td>J.</td><td>meteorites</td></tr>
</table>

Terms

Directions: Unless otherwise specified, each answer is a single word.

1. The enzyme used by all modern cells to convert RNA nucleoside diphosphates into DNA precursors. (2 or 3 words)

2. Proteinlike molecules produced in the laboratory by heating amino acids under anhydric conditions.

3. An extant ribonucleoprotein enzyme that cleaves all *E. coli* tRNA molecules from their larger precursor transcripts. (single word plus one letter)

4. The collective name for molecules (such as NAD, FAD, and acetyl-CoA) that enhance the activity of modern metabolic enzymes. (2 words)

5. Recombination of protein coding domains from different genes into new genes, the protein products of which share common or related amino acid sequences with other proteins (sometimes completely unrelated in function). (2 words)

6. The collection of molecules that process pre-mRNAs. (1 word; or 3 or 4 words; or 5-letter acronym)

7. An adjective describing a genome that consists of two or more different RNA or DNA molecules.

8. The enzyme that converts deoxyuridine monophosphate (dUMP) to deoxythymidine monophosphate (dTMP). (2 words)

9. A term describing ancient, presumably essential proteins that have undergone little change in living descendants since their first appearance in a common ancestor. (2 words)

10. The existence of only one stereoisomeric form (levorotary) of amino acid in biological proteins.

11. A term descriptive of mixtures of D and L sterioisomers, such as those produced in laboratory attempts to simulate primitive Earth conditions.

12. The single ancestral life form, devoid of a genome, from which all cells are presumed to have evolved.

13. The concept that the rate at which selectively neutral mutational changes accumulate in DNA or protein molecules is constant over time. (2 words)

14. A kingdom-level taxonomic group of prokaryotes that includes thermoacidophiles and methanogens.

15. The earliest eukaryote, devoid of mitochondria or chloroplasts.

16. A process involving horizontal genetic transmission postulated for the origin of mitochondria and chloroplasts.

17. An adjective applied to a phylogenetic lineage that traces back to a single common ancestor.

18. Matlike layers of cyanobacteria and sediments thought to form some of the oldest fossils (3.0–3.5 billion years old).

19. The addition or deletion of uridylate residues at precise sites in mitochondrial pre-mRNAs to create translatable templates for the synthesis of functional proteins. (2 words)

Answers to the Objective Questions

Multiple Choice

1. *e* **2.** *b* **3.** *c* **4.** *a* **5.** *d* **6.** *d* **7.** *a* **8.** *e* **9.** *c* **10.** *b* **11.** *b*
12. *e* (*c* and *d*) **13.** *c* **14.** *c* **15.** *b* **16.** *a* **17.** *c* **18.** *a* **19.** *a* **20.** *d*
21. *c* **22.** *a* **23.** *c* **24.** *e* (*b* and *d*) **25.** *e* (*a, b, c, d*)

True–False

1. F (Amino acids have been found in meteorites.)   **2.** F (Creation involves supernatural force outside the scope of science.)   **3.** F (The earliest enzymes are thought to be ribozymes.)   **4.** F (A ribozyme cannot copy its own active site.)   **5.** F (The intron core can act as a poly-C polymerase.)   **6.** T   **7.** F (by artificial selection)
8. F (There are no common nucleotide sequences on either side of these cuts.)   **9.** F (The enzyme activity is associated entirely with the RNA component under nonphysiological conditions.)   **10.** T   **11.** T   **12.** T
13. T   **14.** F (see RNA editing)   **15.** T   **16.** F (They produce a mixture of L and D forms.)   **17.** T
18. T   **19.** F (Some proteins have many nonessential sites at which one or more different amino acids will not destroy the biological activity of the protein; other proteins can tolerate few such changes.)   **20.** T   **21.** F (vice versa)   **22.** F (Both evolved independently from a common ancestor.)   **23.** F (origin unknown)   **24.** T
25. F (Negatively charged mRNAs cannot pass the inner mitochondrial membrane.)   **26.** T   **27.** T   **28.** T
29. F (It is found in some yeasts, plants, and animals.)   **30.** F (Cyanobacteria have no chlorophyll *b*.)   **31.** T
32. F (*Giardia* has no mitochondria.)

Matching

1. *H* **2.** *J* **3.** *A* **4.** *E* **5.** *F* **6.** *I* **7.** *B* **8.** *G* **9.** *D* **10.** *C*

Terms

1. ribonucleoside diphosphate reductase   **2.** protenoids   **3.** ribonuclease P (RNase P)   **4.** nucleotide cofactors   **5.** exon shuffling   **6.** spliceosome; small nuclear ribonucleoproteins; snRNPs   **7.** segmented
8. thymidylate synthase   **9.** highly conserved   **10.** chirality   **11.** racemic   **12.** progenote
13. molecular (evolutionary) clock   **14.** Archaebacteria   **15.** urkaryote   **16.** endosymbiosis
17. monophyletic   **18.** stromatolites   **19.** RNA editing

Index

Salmonella sp., conjugation in, 170
Samesense mutations, 120, 144–145
Sample wells, 189
Sampling error, 149
Sanger method, 218–219
Saturated, as term, 39
Sausage-shaped mitochondria, 11
Scaffold, 61–62
Schizosaccharomyces pombe, centromeric regions from, 63
Sclerotome, 297
SDS (see Sodium dodecyl sulfate)
Secondary immune response, 319
Second messengers, 256, 258–259
Secretory piece, 327
Seed parent, plant, 25
Segment polarity genes, 287
Segmentation genes, 282, 287–293, 295, 297, 306
Segmented genomes, 353
Segments, 228
Selectin, 32
Self-assembly, 161
Selfish DNA, 75
Self-ligation, 196
Self-splicing, 350
Sense codons, 121
Sense DNA, 233
Sense RNA, 87
Sense strand, 87
Sensitivity, 210
Sepals, 309
Septic shock, 263
SER (see Smooth endoplasmic reticulum)
Serine-threonine kinases, 257–258
Serology, 316
Serratia marcescens, nonselectable mutations in, 151
Sex chromosomes, 24, 27
Sexduction, 175
Sex pili, 7, 170
Sexual reproduction, 20, 23–24
Sheath, polysaccharide, 9
Shine-Dalgarno sequence, 128, 130, 136
Shuttle vectors, 200
Sickle cell anemia, missense mutations in, 144
Sigma factors, 90–91
Sigmoidal curve, 163
Signal peptidase, 135
Signal recognition particle (SRP), 225–226
Signal recognition particle RNA (srpRNA), 107–108, 225
Signal sequence peptidase, 225
Signal sequence receptor (SSR), 225–226
Signal sequence (SS), 78, 135, 225
Signal transduction, 242–243, 253–277
Silent mutations, 120
Simian sarcoma virus (SSV), 244
Simple operons, 88
Single base substitution, 144
Single-carbon compounds, 13
Single-stranded DNA (ssDNA), 72, 74, 79, 167
Single-stranded DNA (ssDNA) viruses, 233–234
Single-stranded minus RNA viruses, 231
Singlet microtubules, 8
Site-independent recombination, 78–79
Site-specific recombination, 77–78, 163–165, 170, 175
Skin cells, 154
Sliding filament theory, 23
Slippage, 74
Slow viruses, 236
Small nuclear ribonucleoprotein (snRNP), 108, 111

Small nuclear RNA (snRNA), 106–107, 114
Smallpox, 228
Smooth endoplasmic reticulum (SER), 2, 5
Sneak synthesis, 94–96, 105
snRNA (see Small nuclear RNA)
snRNP (see Small nuclear ribonucleoprotein)
Sodium chloride, 208
Sodium dodecyl sulfate (SDS), 187
Sodium hydroxide, 187
Solenoid, 61–62, 67, 269
Solid-phase hybridization, 209–210
Solute, 5
Solvents, 5, 188
Somatic cells, 23, 154
Somatic hypermutation, 332
Somites, 297
sor gene, 239
SOS regulon, 89, 99, 147
Southern blot hybridization, 193–194, 210–211
Sp1 protein, 108
Spacer DNA, 110–111
Specialized (restricted) transduction, 170
Specificities, 316
Spemann organizer, 301–302
Spermatozoa (sperm cells), 20, 27, 154
Sperm cells (see Spermatozoa)
Spermidine, 60
Spermine, 60
Sperm tails, 7, 9
Sphaeroplasts, 200
S phase, cell cycle, 20
Spi-1 protein, 108
Spindle apparatus, 21, 23, 63
Spirochetes, 366
Spliceosome, 110, 113–114
Splicing, 111–113, 194, 238, 349–350
Split genes, 356
Spontaneous mutations, 150
Spumaviruses, 236
Src oncogene, 239, 241
SRP (see Signal recognition particle)
srpRNA (see Signal recognition particle RNA)
SS (see Signal sequence)
ssDNA (see Single-stranded DNA)
SSR (see Signal sequence receptor)
ssRNA (see Single-stranded RNA)
SSV (see Simian sarcoma virus)
Stacked bases, 190
Staggered cleavages, 166
Staggering, 166
Stamens, 309
Staphylococcus aureus, 35, 176
Starch granules, 3–4, 32–33
Start codon, 120, 146, 196
Stems, 55–56
Steroid hormones, 76, 106, 254
Sterols, 39–40
Sticky ends (see Cohesive ends)
Stock culture, 149
Stop codon, 121
Stop transfer domain, 226
Streptavidin, 208–209
Streptococcus pneumoniae, 37–38, 167
Streptomycin, 133, 151, 171, 174
Strict anaerobes (see Obligate anaerobes)
Stringency, 208
Stringent response, 134
Stroma, 11–12, 227
Stromatolites, 344
Structural genes, 87–100
S-type pneumococci, 37–38
Submetacentric chromosomes, 65
Submetacentric homologues, 25
Substrate-level phosphorylation, 15–16

Succinic acid, 15
Sucrose (cane sugar), 31
Sugars, 13, 93
Sulfate, 14
Supercoiling, 21, 60, 68, 73
Superhelical turns, 69
Super-repressors, 100
Suppressor mutations, 146
Suppressor T cells, 320
Switching, 325
Switch recombination, 335
Symbiosis, 363
Synapse, 80
Synaptonemal complex, 24, 81
Syncytia, 241
Syncytial blastoderm, 285–286
Syngamy, 20
Synonymous codons, 120

Tadpole, 300–301
Tail filament, 9
Tail fold, 297
Tails, nucleolar, 66
Taq polymerase, 213–215, 217
Target DNA, 207–208
TAS (see Transcription-based amplification system)
TATA binding protein, 241
TATA box, 107–108
tat gene, 238–240
Tau protein, 272
Tautomeric shifts, 141–143
tax gene, 240
Taxonomy, 359
T-cell receptors, 320, 328–329, 335
Teichoic acid, 36
Telocentric chromosomes, 65
Telomerase, 73–74
Telomere, 64, 73–74
Telophase, 21, 23, 273
Temperate phages, 162
Temperature cycling, 212
Temperature-sensitive proteins, 47
Template, 70, 74, 76, 87, 198
Teratoga maritima, PCR and, 213
Terminal electron acceptors, 14–15
Terminal genes, 282
Terminal transferase, 208
Termination, in eukaryotes, 136
Termination codons, 121, 144
Terminators, 88–91, 93, 196–197
Tertiary structure, 46
Tetrad, 24
Tetrahymena sp., DNA replication in, 73, 75
Tetrahymena thermophila, molecular evolution and, 346–349, 351
Tetraploidy, 154
Tetrapod vertebrates, 12
TFIID transcriptional factor, 108
TGF (see Transforming growth factor)
TGF-β (see Transforming growth factor β)
Thermotolerance, 213, 216
Thermus aquaticus, PCR and, 213
Thermus thermophilus, PCR and, 213
Thickening, chromosomal, 81
13-mers, 68–69
Thylakoid compartment, 11
Thylakoids, 9, 11, 227
Thymidylate synthase, 355
Thymine, 50, 74
Time scale, of major evolutionary events in history of life, 362
Tissue-specificity factor, 108
T lymphocytes, 317, 320–321
TMV (see Tobacco mosaic virus)
TNF (see Tumor necrosis factor)